# FUNDAMENTALS OF SEMICONDUCTOR PROCESSING TECHNOLOGY

# FUNDAMENTALS OF SEMICONDUCTOR PROCESSING TECHNOLOGY

by

**Badih El-Kareh**

IBM Corporation

Graphics and Layout: Richard J. Bombard

**KLUWER ACADEMIC PUBLISHERS**
**Boston / Dordrecht / London**

**Distributors for North America:**
Kluwer Academic Publishers
101 Philip Drive
Assinippi Park
Norwell, Massachusetts 02061 USA

**Distributors for all other countries:**
Kluwer Academic Publishers Group
Distribution Centre
Post Office Box 322
3300 AH Dordrecht, THE NETHERLANDS

**Library of Congress Cataloging-in-Publication Data**

CIP

*Printed on acid-free paper.*

Printed in the United States of America

# Contents

# Preface

The drive toward new semiconductor technologies is intricately related to market demands for cheaper, smaller, faster, and more reliable circuits with lower power consumption. The development of new processing tools and technologies is aimed at optimizing one or more of these requirements. This goal can, however, only be achieved by a concerted effort between scientists, engineers, technicians, and operators in research, development, and manufacturing. It is therefore important that experts in specific disciplines, such as device and circuit design, understand the principle, capabilities, and limitations of tools and processing technologies. It is also important that those working on specific unit processes, such as lithography or hot processes, be familiar with other unit processes used to manufacture the product.

Several excellent books have been published on the subject of process technologies. These texts, however, cover subjects in too much detail, or do not cover topics important to modern technologies. This book is written with the need for a "bridge" between different disciplines in mind. It is intended to present to engineers and scientists those parts of modern processing technologies that are of greatest importance to the design and manufacture of semiconductor circuits. The material is presented with sufficient detail to understand and analyze interactions between processing and other semiconductor disciplines, such as design of devices and circuits, their electrical parameters, reliability, and yield.

The book was developed from notes prepared for courses taught at IBM and the University of Vermont. It serves as a base on which to build an understanding of the manufacture of semiconductor products. It is written in a form to satisfy the needs of engineers and scientists in semiconductor research, development and manufacturing, and to be conveniently used for a one-semester graduate-level course in a semiconductor engineering or material science curriculum. The book consists of eight chapters on unit processes that are arranged in a conventional sequence that reflects typical integrated process technologies. It begins with the preparation of semiconductor crystals and continues with thermal oxidation, thin-film deposition, lithography, etching, ion implantation, diffusion, and contact and interconnect technology. The last chapter is co-authored by J. G. Ryan, manager of Thin Film and Chemical-Mechanical Polishing Development at IBM.

One of the challenges faced when writing a book on semiconductor technologies is finding satisfactory explanations to observations reported in the literature. Semiconductor processes are based more heavily on empirical data than on prediction by simulation of physical or chemical phenomena. This by itself, requires a thorough review and comparison of published data and observations. It is easier to predict electrical parameters from given horizontal and vertical device geometries than to define processing conditions that result in such geometries. The reader should therefore not be surprised·to find sections describing "how" a process is performed without the accompanying "why".

The author is indebted to all those who helped shaping the book in its present form. He is very grateful to his friends and colleagues, Albert Puttlitz and Ashwin Ghatalia, for relentlessly checking for flaws in the manuscript and making valuable suggestions. He also thanks Robert Simonton (Eaton Corporation) and Larry Larson (SEMATECH) for spending a great deal of time with him on ion implantation equipment. The author gratefully acknowledges all his IBM colleagues for their invaluable inputs and discussions; Don Chesebro, Burn Lin, Tim Brunning, Mike Hibbs for their inputs on advanced lithography, Geoff Akiki for providing information on advanced mask preparation, Hans Pfeiffer on electron-beam lithography, Randy Mann and Robert Geffken on silicides and metallization.

The author and R. J. Bombard are indebted to IBM and SEMATECH for their tremendous support.

The book was prepared using computer facilities at IBM. B. El-Kareh, R. J. Bombard, and J. G. Ryan (Chap. 8), however, take full responsibility for its contents.

# FUNDAMENTALS OF SEMICONDUCTOR PROCESSING TECHNOLOGY

# Chapter 1

# Semiconductor Crystals

## 1.0 Introduction

The first step in the manufacture of modern integrated circuits is the preparation of a single crystal of semiconductor material. In crystalline solids, the elements are stacked in a periodic pattern as illustrated in Fig. 1.1. When the periodicity extends throughout the solid, one speaks of a **single crystal,** or monocrystal, as opposed to a **polycrystal** which consists of small crystals, called **grains,** arranged in random directions and adhering together at their boundaries (Fig. 1.2).

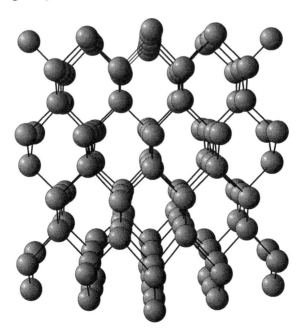

**Fig. 1.1** Model of silicon crystal, seen along the < 110 > direction [1]. (Crystallographic directions are discussed in the following section).

**2**

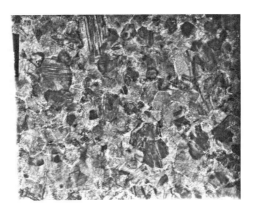

Fig. 1.2 Micrograph of polycrystalline silicon. The crystallites are called grains and the regions between the grains are called grain boundaries. (Courtesy: W. Tice, IBM).

Noncrystalline solids are called **amorphous.** Normally, solids assume the crystalline form because it is the atomic arrangement of minimum energy. In most solids at room temperature atoms occupy, on the average, fixed positions relative to each other. The atoms, however, are in constant vibration about their equilibrium position.

Semiconductor properties are not exclusive to single crystals. Many non-crystalline materials exhibit semiconductor properties similar to those of crystalline semiconductors. In this chapter, however, we restrict ourselves to the discussion of the preparation and properties of single crystal semiconductors, in particular silicon since this is the best understood and most widely used materials in integrated circuits (IC).

## 1.1 Crystals and Crystallographic Orientations

A crystal can be described in terms of a periodic **lattice.** We can attach to each lattice point an atom or a group of atoms called the **basis.** The basis is repeated in space to form the crystal (Fig. 1.3).

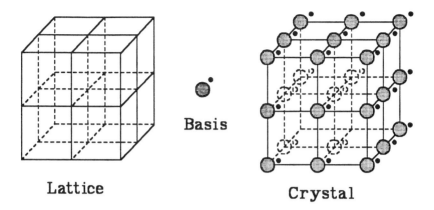

**Basis**

**Lattice**

**Crystal**

**Fig. 1.3** Lattice + basis = crystal structure.

One can think of three intersecting families of parallel crystal planes passing through the lattice and dividing the crystal space in small, identical parallelepipeds. The crystal can then be visualized as a repetition in space of one parallelepiped, called the **unit cell** (Fig. 1.4). The edges of the unit cell define the chosen directions of crystal axes. The lengths OA, OB and OC are called the **lattice constants** a, b and c. In a cubic crystal, a = b = c and the axes are perpendicular.

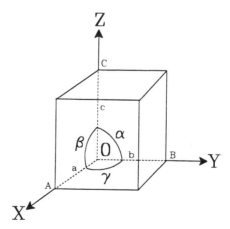

**Fig. 1.4** Unit cell of the most general crystal type, showing crystal axes OX, OY and OZ, and the lattice constants a, b and c. Angles $\angle BOC$, $\angle AOC$ and $\angle AOB$ are $\alpha$, $\beta$ and $\gamma$. In a cubic crystal a = b = c and $\alpha = \beta = \gamma = 90°$.

Crystal planes are best represented by **Miller indices** which are specified as follows:

1. Let the plane intercept the crystal axes at the lattice points $n_1a$, $n_2b$ and $n_3c$, relative to a chosen origin O (Fig. 1.4).

2. The intercepts are expressed as integral multiples of the lattice constants a, b and c. The planes are defined by a set of integers $n_1$, $n_2$ and $n_3$.

3. The reciprocals of the numbers $n_1$, $n_2$ and $n_3$ are usually fractions. Reduce the fractions to the smallest three integers having the same ratio as the fractions. The three integers obtained are called the Miller indices h, k, and l.

The group of integers (hkl) defines a set of equally spaced parallel planes, as illustrated in Fig. 1.5. If a plane crosses the axis on the negative side of the chosen origin, then the corresponding index is negative and is written with a bar over it such as $(h\bar{k}l)$. For cubic crystals of a single element the (100), (010), (001), ($\bar{1}$00), ($0\bar{1}0$) and ($00\bar{1}$) planes are indistinguishable, as are the planes (110), (101) and (011).

A family of equivalent planes is described with the indices enclosed in braces: $\{hkl\}$. These braces define planes which may have different Miller indices but are equivalent by symmetry. For example, in the silicon crystal $\{100\}$ describes the equivalent planes (100), (010), (001), ($\bar{1}$00), ($0\bar{1}0$), and ($00\bar{1}$). A particular direction in the crystal is denoted by a vector which is specified by a group of indices in brackets [uvw]. The indices are integers and have no common factor larger than unity. The direction [uvw] is obtained by moving from the chosen origin over a distance ua along the a-axis, vb along the b-axis and wc along the c-axis. The vector that connects the chosen origin to the point thus obtained is then the direction specified by the indices [uvw]. In a cubic crystal the [100] direction is the x-axis while the (100) plane is perpendicular to the x-axis. In cubic crystals the [uvw] direction is always perpendicular to a plane (hkl) having the same indices. A set of equivalent directions in a crystal is represented as <uvw>.

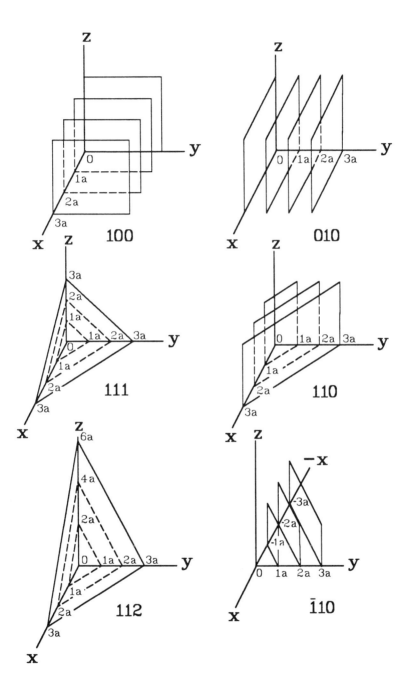

**Fig. 1.5** Miller indices of some important planes in a cubic crystal.

**6**

## 1.2 The Silicon Crystal

The silicon crystal has a diamond structure and belongs to the cubic crystal system. In diamond crystals, pairs of valence electrons with opposite spins are shared between four neighboring atoms and form **covalent bonds.** The 3s and 3p orbitals of the parent atoms are mixed to form a new set of equivalent four bonding orbitals, called *hybridized orbitals,* that are directed toward the four nearest neighbors (Fig. 1.6). In the diamond structure each atom is symmetrically surrounded by four equally spaced atoms; it forms tetrahedral bonds with its four nearest neighbors. This type of bonding is very strong, highly localized, and directional because the distribution of valence electrons around the atom becomes shifted toward the nearest neighbor. The silicon crystal is therefore very hard, and has a high melting point. Because of the directionality, the crystal does not assume the closest packing configuration, but allows a large volume per atom.

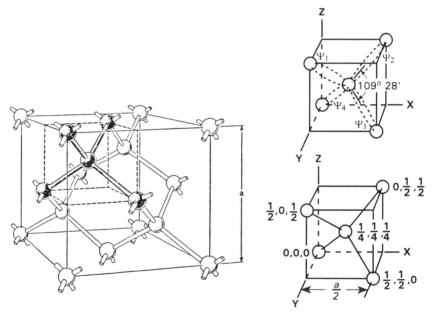

**Fig. 1.6** Diamond structure and subcell illustrating the tetrahedral bonds. Each atom has four nearest neighbors. The hybridized orbitals are $\psi_1$, $\psi_2$, $\psi_3$, and $\psi_4$.

The unit cell configuration in a diamond crystal is best described by projecting the cube onto a two-dimensional plane as shown in Fig. 1.7. The plane of the paper is chosen as one of the cube's planes. The numbers then define the location of the atom centers relative to the plane of the paper in fractions of the cube's edge $a$. For example, atoms located at the centers of cube faces normal to the page are labeled 1/2.

For most discussions, it is convenient to use a simplified two-dimensional picture of the silicon crystal as illustrated in Fig. 1.8. The arrangement shows that when all bonds are in place, each silicon atom has four nearest neighbors. Each bond is represented by two links representing two electrons per bond.

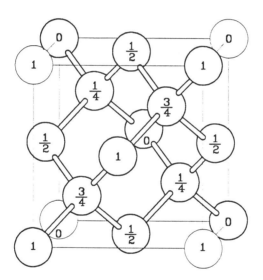

**Fig. 1.7** Three-dimensional representation of the unit cell in a diamond crystal. Numbers show displacement of atoms normal to the plane of the paper in fraction of the cube edge $a$ [2].

When using this simplified scheme, it should be remembered that the structure is three-dimensional, and that the tetrahedral bonds form one continuous three-dimensional chain throughout the solid. The electrons are not constrained to one particular bond. Instead, they can move throughout the crystal, exchanging places with other valence electrons (no net current is

**8**

involved). One must consider the entire system, where the bonding electrons belong to the entire crystal, not to a particular atom. This picture will be of great help when discussing energy-band diagrams and related semiconductor properties.

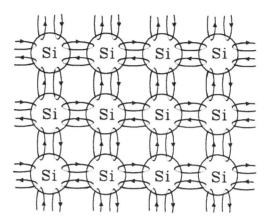

Fig. 1.8 Two-dimensional representation of a silicon crystal.

Important properties of the silicon crystal are given in Table 1.1. The table also contains silicon parameters such as energy gap, electron affinity, effective mass, carrier concentration and dielectric constant. These parameters are discussed in the following chapters.

## 1.2.1 Crystal Growth

The most common technique for growing silicon crystals of commercial dimensions is the Czochralski (CZ) crystal pull method. In this method, polycrystalline silicon **(polysilicon)\*** is melted in a crucible usually made of graphite with fused silica lining (Fig. 1.9). Power is provided by resistance heating or radio-frequency. A seed crystal of known orientation is immersed into the surface of the melt and then slowly withdrawn while rotated.

---

\*Polysilicon is prepared by depositing silicon from gaseous mixtures of trichlorosilene ($SiHCl_3$) or silicon tetrachloride ($SiCl_4$) and hydrogen, (Chap. 3).

**Table 1.1** Some Important Properties of Silicon at 25 °C [3, 4].

| | |
|---|---|
| Symbol | Si |
| Color | Blue-grey |
| Atomic number | 14 |
| Atomic weight | 28.09 g/Mole |
| Crystal structure | Diamond |
| Lattice constant | 5.4307 Å |
| Silicon radius | 1.18 Å |
| Atomic density | $5.02x10^{22}$ atoms/cm$^3$ |
| Density | 2.328 g/cm$^3$ |
| Melting point | 1412 °C |
| Specific heat | 0.7 J/g.K |
| Thermal conductivity | 1.5 W/cm.K |
| Vapor pressure | $10^{-7}$ Torr (1050 °C) |
| Thermal expansion coefficient | $2.6x10^{-6}$K$^{-1}$ |
| Relative dielectric constant | 11.7 |
| Index of refraction | 3.42 |
| Energy gap | 1.12 eV |
| Intrinsic carrier concentration | $1.25x10^{10}$cm$^{-3}$ |
| Electron affinity | 4.05 eV |
| Conduction band density of states | $2.80x10^{19}$cm$^{-3}$ |
| Valence band density of states | $1.04x10^{19}$cm$^{-3}$ |
| Lattice electron mobility | 1450 cm$^2$/V.s |
| Lattice hole mobility | 500 cm$^2$/V.s |
| Average optical phonon energy | 0.063 eV |
| Optical phonon mean-free path | 7.6 nm (e); 5.5 nm (h) |
| Intrinsic Debye length | 24 $\mu$m |
| Intrinsic resistivity | $2.3x10^5$ Ohm-cm |
| Donors | P, As, Sb |
| Acceptors | B |
| Hardness | 7 Mohs |
| Poisson's ratio | 0.42 |
| Tensile Strength in < 111 > | $3.5x10^8$N/m$^2$ |
| Modulus of Elasticity in < 111 > | $1.9x10^{11}$N/m$^2$ |
| Modulus of Elasticity in < 110 > | $1.7x10^{11}$N/m$^2$ |
| Modulus of Elasticity in < 100 > | $1.3x10^{11}$N/m$^2$ |

Both the crucible and the growing crystal are rotated to average variations in temperature and impurity concentration within the melt. The ambient gas is typically argon. The crystal diameter is computer controlled during growth. The crystal is grown oversize, ground to the desired dimension, and then sliced into thin wafers which are used for device fabrication.

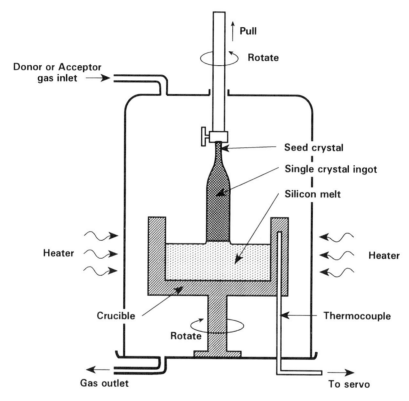

Fig. 1.9 The Czochralski crystal pull method

One of the problems with Czochralski grown crystals is their high impurity content (especially oxygen and carbon). These impurities are introduced unintentionally during crystal growth and originate primarily at the quartz crucible lining. Typical oxygen concentrations range between $5x10^{17}$ cm$^{-3}$ and $10^{18}$ cm$^{-3}$ [5]. Carbon is found in concentrations between $10^{16}$ cm$^{-3}$ and $10^{17}$ cm$^{-3}$ [6]. The presence of *controlled* amounts of oxygen, however, can be advantageous: it can help remove harmful impurities and defects from the surface of the wafer, as discussed in section 1.2.3.

Silicon crystals of high purity can be grown by the floating zone (FZ) method (Fig. 1.10). A polycrystalline rod is clamped in a vertical position in a chuck at the top, and a seed crystal with the desired orientation is clamped at the bottom. The formation of the single crystal begins by melting a small "floating" zone (typically 1.5 cm wide) in contact with the seed. The floating zone is generated by moving a polycrystalline rod relative to a water-cooled induction coil supplied with radio frequency power in the megacycle range. Either the rod is moved downward or the coil is moved upward. In both cases the floating zone starts at the seed and propagates up the rod. The ends of the rod are rotated in opposite directions, resulting in straight, round crystals. As the zone travels from the seed to the top of the polycrystalline rod, the silicon melts and resolidifies into a single crystal which has the same orientation as the seed. The molten zone is held in place by a combination of RF levitation and surface tension and does not come in contact with any crucible. The concentration of impurities is therefore considerably lower in FZ than in CZ crystals. Also, most remaining impurities can be swept to the top of the crystal by making successive passes of the floating zone from seed to top. High purity crystals obtained from the floating-zone technique are used in specific applications, such as high voltage devices, as discussed in the chapters to follow.

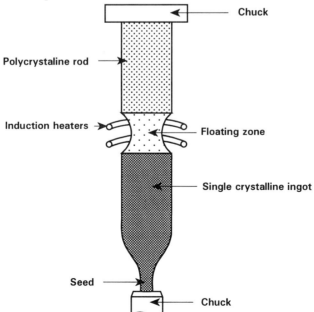

Fig. 1.10 The floating-zone crystal growth technique

Since the molten zone is held together only by surface tension and RF levitation, stability problems limit the grown crystal diameter to $\simeq 10$ cm [7]. Also, the very low oxygen concentration in FZ crystals can become a disadvantage in some applications where intrinsic gettering plays a dominant role in reducing the surface defect density (Sec. 1.2.3).

## 1.2.2  Crystal Doping

The introduction of controlled amounts of selected impurities in lattice sites otherwise occupied by silicon is called **doping.** Lattice sites occupied by impurities are referred to as **substitutional sites.**

In the Czochralski method the crystal is doped by dissolving the desired impurity in the melt (boron for p-type crystals; arsenic, phosphorus, or antimony for n-type crystals). The dopant is added to the melt in the form of a gas, powder, or heavily doped polysilicon. In the floating-zone method the silicon is doped by introducing a controlled amount of the dopant in the argon atmosphere surrounding the crystal during growth. Other techniques for doping the crystal are discussed in chapters 6 and 7.

**Segregation Coefficient**

In the floating-zone technique we found that impurities are swept by the traveling zone to the top of the rod. This is because at a given temperature the concentration of most impurities is higher in the melt than in the solid. At the interface between melt and solid in CZ and FZ grown crystals, the excess dopant in the solid is "thrown-off" into the melt. The difference in impurity concentration between melt and solid is described by a factor called the **segregation coefficient k.** This is the ratio of the impurity concentration (by weight) in the solid to that in the melt, at the interface between the two phases. Equilibrium segregation coefficients of typical impurities in silicon are given in table 1.2. As can be seen, k is less than unity.

**Misfit Factor**

Each silicon atom is situated within a tetrahedron, equidistant from four neighbors (Fig. 1.6). In a hard sphere model it can be assigned a tetrahedral radius $r_o$. Substitutional impurities can also be assigned a "hard sphere" radius $r_i = r_o(1 \pm \Delta)$, where the ratio $\Delta/r_o$ describes the mismatch in size between the impurity and "host" atom, and is called the **misfit factor.** It is indicative of the strain in the lattice caused by introducing impurities into the crystal. Misfit factors of typical dopants in silicon are given in table 1.3. Note that arsenic is the only impurity that has a perfect fit with silicon.

**Table 1.2** Segregation coefficients of typical impurities in silicon

| Element | Segregation Coefficient |
|---------|------------------------|
| B | 0.72 |
| Al | 0.0018 |
| Ga | 0.0072 |
| P | 0.32 |
| As | 0.27 |
| Sb | 0.02 |

**Table 1.3** Misfit factors of typical impurities in Silicon

| Impurity | Radius (Å) | Misfit Factor |
|----------|-----------|---------------|
| B | 0.88 | 0.254 |
| Al | 1.26 | 0.068 |
| Ga | 1.26 | 0.068 |
| In | 1.44 | 0.220 |
| P | 1.10 | 0.068 |
| As | 1.18 | 0.000 |
| Sb | 1.36 | 0.153 |

**Solid Solubility**

There is a limit to the concentration of substitutional impurities that can be incorporated in silicon without seriously disrupting the lattice. This is referred to as the **solid solubility limit** of the impurity. Note that the solubility of typical impurities in silicon is very small and depends on temperature (Fig. 1.11). The highest concentration of boron in silicon for example, is less than 1%. The solubility initially increases with temperature and then

**14**

begins to decrease as the crystal melting temperature is approached. When the maximum solubility is achieved at a certain temperature, the crystal is said to be saturated with the impurity at that temperature. If the crystal is cooled to a lower temperature without removing the excess impurity, a supersaturated condition is created. The crystal may return to the saturated condition by precipitating impurities present above the solubility limit.

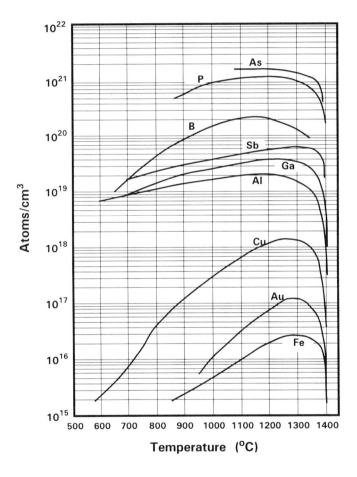

**Fig. 1.11** Solid solubility of impurities in silicon [8, 9].

### 1.2.3 Defects in Silicon Crystals

A model silicon crystal contains only silicon atoms perfectly arranged in a periodic pattern assumed to be infinite in three dimensions (Fig. 1.1). Real crystals, however, terminate at boundaries called **surfaces,** and contain imperfections in the bulk and at surfaces. These imperfections, referred to as **defects,** are inherent to the crystal or created during crystal growth or subsequent processing. Even the best grown crystals contain at least one defect per billion crystal atoms. This is approximately $10^{13}$ defects per cm³. Defects can drastically alter the electrical and physical properties of the crystal. They play an important role in the transport of impurities and carriers in the crystal.

Four types of crystal defects can be distinguished: **point defects, line defects, plane defects** and **volume defects.**

**Substitutional** impurities, **interstitial** atoms, **vacancies** and **vacancy pairs** are examples of point defects. As mentioned, a substitutional impurity is one that occupies a position otherwise occupied by the original crystal atom. An interstitial impurity occupies a position between regular crystal atom sites. A vacancy is a regular site that is not occupied by any atom.

Typical line defects are crystal **dislocations.** For example, a dislocation is formed when ten atoms occupy the space which would be normally occupied by nine atoms.

The most important plane defects are **stacking faults.** These are created when the stack sequence of the crystal atoms is disrupted, as discussed later in this section.

One example of volume imperfections is the oxygen-precipitation-induced defects that can act as "sinks" for heavy metals in large volumes of the crystal.

Pattern and structural imperfections can be introduced in the insulating or conducting layers during processing and handling. These types of defects are discussed in a later chapter in conjunction with process integration and device fabrication.

**Point Defects**

Four types of point defects are shown in Fig. 1.12. A lattice vacancy is shown in Fig. 1.12a. It is created when the four valence bonds of a tetrahedral silicon atom are broken. When the atom which is set free migrates to the crystal surface, the vacancy is called a **Schottky defect** (Fig. 1.12b). When a silicon atom is set free from its lattice site and "eases itself" into an interstitial position, the vacancy-interstitial pair is called a **Frenkel defect** (Fig. 1.12c). Schottky and Frenkel type defects are created by thermal energy: vibrations of lattice atoms interact and may add at a lattice site to displace an atom from its position.

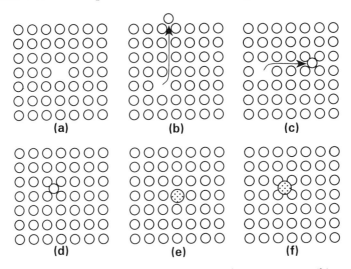

**Fig. 1.12** Point defects. (a) Lattice vacancy. (b) Schottky-type defect. (c) Frenkel-type defect. (d) Interstitial "host" atom. (e) Substitutional impurity. (f) Interstitial impurity.

The density of defects can be determined by the laws of statistical mechanics. The equilibrium density of Schottky defects is found as:

$$n_s = N e^{-E_s/kT} \qquad\qquad 1.1$$

where

$n_s$ = density of Schottky defects (cm$^{-3}$)
N  = atomic density of silicon (cm$^{-3}$)

$E_s$   = energy of formation of a Schottky defect ($\simeq$ 2.3 eV)
k   = Boltzmann constant ($8.62x10^{-5}eV/K$)
T   = absolute temperature (K)

The equilibrium density of Frenkel type defects is given as:

$$n_f = \sqrt{NN'} \; e^{-E_f/2kT} \qquad\qquad 1.2$$

where

$n_f$   = density of Frenkel defects (cm$^{-3}$)
$N'$   = density of available interstitial sites (cm$^{-3}$)
$E_f$   = energy of formation of a Frenkel defect ($\simeq$ 1.1 eV)

Point defects can enhance or retard the motion of impurities in the crystal, as discussed in Chap. 7.

An interstitial "host" atom is a silicon atom which occupies a **void,** that is the space between lattice sites (Fig. 1.12d). A substitutional impurity is shown in Fig. 1.12e, and an interstitial impurity in Fig. 1.12f.

**Line Defects**

Line defects appear in form of dislocations in regions where permanent deformation of the crystal has occurred [10]. The stress producing the dislocation can be mechanical (shear stress), thermal (crystal growth), chemical (misfit of impurity atoms) or radiative (ion bombardment).

Figure 1.13 illustrates one type of dislocation called **edge dislocation.** Here, the dislocation occurs at the termination of an extra plane inserted in the lower half of the crystal. A two-dimensional bubble model of an edge dislocation is shown in Fig. 1.14. When viewed along the indicated arrow, the pattern exhibits an extra line of bubbles on the lower half of the plane [11].

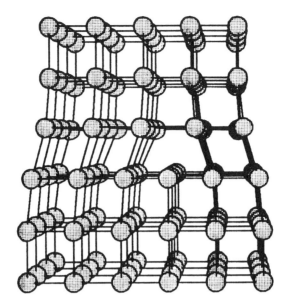

**Fig. 1.13** Edge dislocation. Dislocation occurs at the termination of the extra plane of atoms.

**Fig. 1.14** Two-dimensional bubble model of an edge dislocation. The dislocation can be viewed by turning the page by 30 ° and sighting at a low angle.

The following thought experiment describes the formation of an edge dislocation [12].

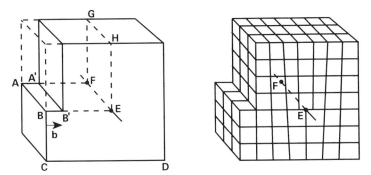

**Fig. 1.15** Thought experiment to illustrate the formation of an edge dislocation. EF represents an edge dislocation line.

Suppose that the crystal in Fig. 1.15 is cut across a plane ABEF, and the upper half displaced so that line A'B' which was initially coincident with line AB is slipped by a distance b, as indicated. If now the two halves are stuck together, an edge dislocation is formed. The crystal plane along which the slip has occurred is known as the slip plane. In this example, the upper half of the crystal is under compression and the lower half is under tension. The termination of the extra plane is an edge dislocation. As in Fig. 1.13, the dislocation line extends out of the paper and is perpendicular to the slip direction.

If the extra plane is above the slip plane, the edge dislocation is positive, below the slip plane it is negative. The slip process resulting from a moving edge dislocation is illustrated in Fig. 1.16.

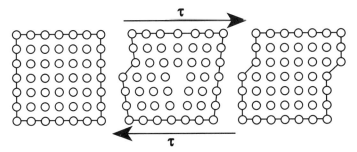

**Fig. 1.16** Movement of dislocation under shearing stress $\tau$, positive edge dislocation moving to the right.

Another type of dislocation is the **screw dislocation** that is shown for a simple cubic lattice in Fig. 1.17.

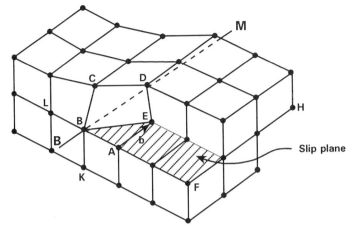

**Fig. 1.17** Schematic representation of a screw dislocation in a simple cubic lattice.

It can be visualized in a thought experiment as follows: Suppose the crystal is cut across the plane BFHM and the upper part displaced in the direction of the vector b. The dislocation line is parallel to b. As one moves around the dislocation line along a path such as AKLCDE, one advances along BM by an amount equal to b for every turn; hence the term "screw dislocation" [1.12].

**Plane Defects**

As mentioned earlier, the most important type of plane defects is the stacking fault [10]. To illustrate a stacking fault, consider the normal stack of silicon atoms in the <111> direction (Fig. 1.18).

The sequence of atoms in the stack can be defined by choosing a reference layer and labelling the atoms in this layer as A-atoms. All atoms in other layers having identical crystallographic positions to those in the reference layer are also named A-atoms and the layers are called A-layers. Layers of atoms in other positions are named B-layers, C layers, and so on. In Fig. 1.18 the sequence of atoms in a stack is ABCABC... When this regular sequence of atoms in the stack is interrupted, a stacking fault is created. A stacking fault appears at the surface of the crystal.

Figure 1.19 shows a disrupted stack sequence. In Fig. 1.19a part of the C layer is missing. This results in a break in the stack sequence and is referred to as an **intrinsic stacking fault.** In Fig. 1.19b an extra A layer is present between a B and a C-layer. The fault created by the extra layer is called an **extrinsic stacking fault.** Figure 1.19c illustrates how a stacking fault propagates to the surface of the crystal.

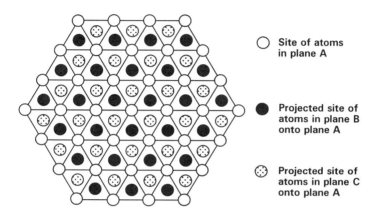

**Fig. 1.18** Normal stack of atoms in silicon, viewed in the <111> direction.

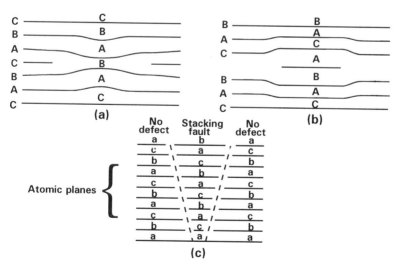

**Fig. 1.19** Stacking fault in silicon. (a) Intrinsic stacking fault. (b) Extrinsic stacking fault. (c) Propagation of stacking fault.

When observed in an optical microscope, stacking faults appear as straight bands at the surface. They lie on (111) planes and intercept the surface along the <110> direction. They appear as equilateral triangles on the surface of (111) oriented silicon, or as squares on the surface of (100) silicon.

In most cases, stacking faults alone do not affect device properties. They can, however, act as nucleation centers for the precipitation of impurities which deteriorate device properties.

**Volume Defects**

Oxygen is one of the main impurities that is unintentionally introduced in silicon during crystal growth in silica crucibles. Its typical concentration in silicon varies from $10^{17}$ - $10^{18}\,cm^{-3}$. Oxygen becomes a major device yield detractor when present at high concentrations in silicon, creating dislocations and sites for heavy-metal precipitation. At concentrations above $10^{17}\,cm^{-3}$, it causes precipitates of silicon-oxygen complexes when the crystal is subjected to high temperatures. This is accompanied by the generation of mobile dislocations around the precipitates [13]. In an interstitial position, oxygen forms a silicon-oxygen bond which exhibits a strong infrared (IR) absorption band at 9.1 $\mu$m. This allows routine measurements of the oxygen content in silicon by infrared spectroscopy.

Along with oxygen, carbon is a major unintentionally introduced impurity in CZ or FZ crystals. It is found in concentrations between $10^{16}$ and $10^{17} cm^{-3}$. At high concentrations carbon can create precipitates and stacking faults during high temperature processing [14]. The effects of carbon are, however, not well understood. As with oxygen, IR spectroscopy is used to measure the concentration of carbon, which exhibits strong absorption at $\simeq 16.6\,\mu m$.

**Process-Induced Defects**

Defects can be introduced into the crystal intentionally and unintentionally. Dopants such as arsenic, phosphorus, antimony, and boron are intentionally introduced to locally change the conductivity of the crystal. Unintentional defects are introduced during crystal growth or during processing. The fabrication of silicon devices requires a certain number of high temperature processing steps. These steps can create unintentional defects, some in conjunction with existing defects in the starting material. Examples of defects introduced during processing are dislocations and stacking faults caused by the misfit of impurities at high concentration; displacement of atoms by ion bombardment (Chap. 6); stress-induced deformations due to different thermal expansion coefficients of superimposed layers; and oxidation-induced stacking faults (Chap. 2).

**Gettering**

There is a favorable mechanism observed in the presence of oxygen at concentrations near $10^{17} cm^{-3}$. As discussed in the following section, the crystal is cut into thin slices, called **wafers,** used for device fabrication. When, prior to device processing, a silicon wafer is heated to $\simeq$ 1000 °C in $N_2$ or HCl atmosphere, most of the oxygen near the surface of the wafer is removed by the atmosphere. Deep in the "bulk" of the crystal, however, oxygen remains at high concentration and precipitates as complexes. The surface of the wafer - where most active devices are fabricated - becomes depleted of oxygen, while a region in the wafer several microns away from the surface becomes rich in defects. The defective region acts as a "sink" that attracts impurities such as heavy metals. As the wafer is heated, the defects are attracted to this sink and thus removed from the surface. This segregation mechanism is called **intrinsic gettering** [15]. The region which is depleted of oxygen is sometimes referred to as the **denuded zone.**

In addition to intrinsic gettering there are treatments which can be applied before or during device fabrication to remove defects and metallic impurities from the regions where devices are fabricated. Gettering can be achieved by intentionally damaging the back of the silicon wafer and then subjecting it to high temperature. The damaged region acts as a sink for impurities, such as heavy metals, which then diffuse from the surface to the back side and deposit there. A common method to create strain in the back side of the wafer is to mechanically damage it, or dope it heavily with impurities such as argon (Chap. 7) [16, 17].

## 1.3 Wafer Preparation

Typical wafers are prepared by grinding the silicon ingot to the desired diameter and then cutting it into single slices of thickness 0.5-1 mm, depending on the wafer diameter. For coarse wafer alignment and future identification, a notch and one or more flats are ground along the length of the crystal before it is sliced. The notch and primary flat are used to position the wafer in the tool and to orient the design with respect to a specified crystallographic direction. The primary flat is typically located on a $(01\bar{1})$ surface and is used (with a secondary flat, when present) to quickly identify the type and crystal orientation of the wafer (Fig. 1.20).

The wafers are then lapped, polished, and cleaned to remove the damage caused by slicing. Wafer cleaning and processing is performed in a "clean-room" environment. A clean room is classified according to the number of particles per unit volume of air that are larger than or equal to a specified diameter. For example, a class 10 room has less than 350 particles per cubic meter that are larger than or equal to 0.3 $\mu$m in diameter. Clean rooms are discussed in more detail in Chap. 5.

The specifications for crystal diameter, type, resistivity, and orientation depend on the anticipated process sequence, the available manufacturing tools, and the required device characteristics. Typical crystal diameters range from 50 mm to 300 mm. For most applications, the wafer is oriented in the < 100 > or < 111 > direction.

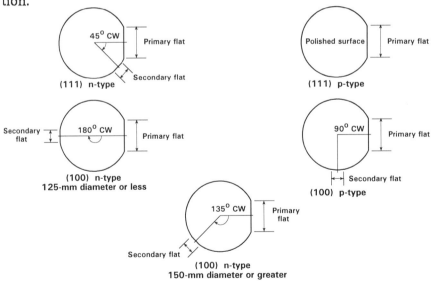

**Fig. 1.20** Convention for silicon wafer identification. Source: Semiconductor Equipment and Materials Institute (SEMI) standard, 1993.

### 1.3.1 Wafer Type and Orientation

The crystal is usually doped p- or n-type while grown, depending on wafer specifications. Only a thin layer of the wafer, approximately 1-$\mu$m to 15-$\mu$m deep, is used to define the device structure. In most cases, the remainder of the wafer acts as a mechanical support.

Several process parameters and device characteristics are sensitive to wafer orientation. For example, the rate of epitaxial growth of silicon depends on its crystallographic orientation. Since the {111} planes have the smallest separation, silicon grows faster along a < 111 > direction than along a < 110 > or < 100 > direction. Also, since (111) planes have the highest density of atoms, the dissolution of silicon in an etching solution is slowest in the < 111 > direction. The rate of thermal oxidation and the diffusivity of impurities are also dependent on the crystal orientation. For example, the oxidation rate of silicon is largest in the < 111 > direction and smallest in the < 100 > direction.

Electronic traps created at the silicon surface increase the surface charge and alter device parameters. The density of these **interface traps** depends on the crystallographic orientation. It is highest in < 111 > oriented wafers and lowest in < 100 > oriented wafers [18]. This is one reason for choosing < 100 > oriented wafers for most device fabrication.

An important step in the manufacture of integrated circuits is dicing the wafer into rectangles called **chips** or **dies.** Typical dicing is performed with a saw that cuts through the wafer along lines that define identical chips (Fig. 1.21). In some cases, however, the chips are separated by initially scribing the wafer along lines and then cleaving the crystal. It is easier to cleave the wafer along (111) planes than in other crystallographic directions, because the tensile strength and modulus of elasticity are largest in the < 111 > direction (Table 1.1). In < 100 > oriented wafers, (111) planes intersect the surface at an angle of 54.74° and in < 110 > directions which are perpendicular to each other. Therefore, for best results, the wafer is cleaved in < 110 > directions (see problems 1.3 and 1.4 at the end of the chapter).

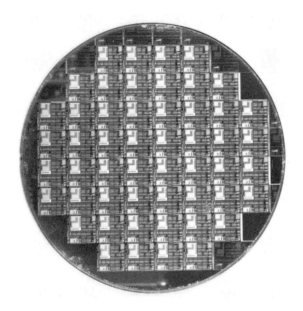

Fig. 1.21 Processed wafer with identical chips.

## 1.3.2 Axial and Radial Variations

Variations in the wafer properties (such as dopant concentration, defect density, scratches) cause variations in device and circuit properties within a chip and from chip to chip. These variations can be induced during crystal growth or wafer preparation. The properties of grown crystals vary axially (along the crystal) and radially from the center to the periphery of the crystal. Axial variations limit the number of wafers that fall within certain specifications. Therefore, wafers are grouped according to the section of the crystal from which they are sliced and those that fall outside the specifications are not used. Of greater importance is the radial variation in the wafer properties. Since processing cost is affected by the number of chips per wafer that pass final test specifications, it is cost effective to eliminate, before processing, those wafers that exhibit large variations in their properties.

## 1.4 Compound Semiconductors

Many important applications, particularly in optoelectronics, are not possible with silicon crystals. Typical examples are light emitting diodes (LEDs), semiconductor lasers, and ultra-high-speed devices. These devices operate best with certain compound semiconductors [19]. Differences in light emission and carrier transport between elemental and compound semiconductors are explained when discussing energy band diagrams of the different types of semiconductors.

The most technologically advanced and widely used compound semiconductor is gallium-arsenide (GaAs). This is a binary compound because it consists of only two different elements - gallium from column III and arsenic from column V of the periodic table. Other binary III-V compound semiconductors of interest are GaP, InP and InSb. Examples of tertiary compounds are AlGaAs and GaAsP. For special applications II-VI compound semiconductors such as CdTe and ZnS can also be used. In this section, however, we only discuss the properties of GaAs crystals.

Gallium has the electronic structure $4s^2 4p^1$, while arsenic has the structure $4s^2 4p^3$. Two silicon-like atoms can be produced from Ga and As by transferring one of the $p$-electrons from As to an empty $p$-level in Ga, so that each atom then has the $4s^2 4p^2$ configuration, similar to that of a group IV atom. The bond between nearest neighbors is thus partially polar, since gallium is left negative and arsenic positive. It is essentially a covalent bond, however, leading to a tetrahedral bond configuration similar to that of silicon. This is because when the bond is formed, the reduction of energy more than compensates for the energy necessary to transfer the electron from As to Ga. Both gallium-arsenide and silicon crystals belong to the cubic lattice family. Compound semiconductor crystals are sometimes said to belong to the **zincblende** structure, which is similar to the diamond structure with one important difference (Fig. 1.22): comparing silicon to gallium-arsenide, for example, we find identical atoms in the silicon crystal (diamond), while in gallium-arsenide (zincblende) each gallium atom is surrounded by four arsenic atoms and vice versa.

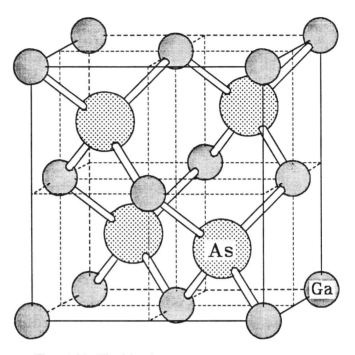

**Fig. 1.22** Zincblende structure, illustrated for gallium-arsenide.

The GaAs structure consists of two interpenetrating face-centered-cubic (fcc) sublattices, each containing atoms of only one type. One atom of the second sublattice is located at one fourth of the distance along a major diagonal of the first sublattice (Fig. 1.23). Gallium-arsenide (100) surfaces contain either only gallium or only arsenic atoms (Fig. 1.24). In either case, each atom has two bonds to atoms in the lower plane. This leaves two dangling bonds per atom at the surface. The preferred cleavage in GaAs is along (110) planes. These intercept < 100 > surfaces at right angles and are orthogonal. When cleaved in these directions the crystal forms parallel plates, which is essential for the preparation of resonant cavities for lasers.

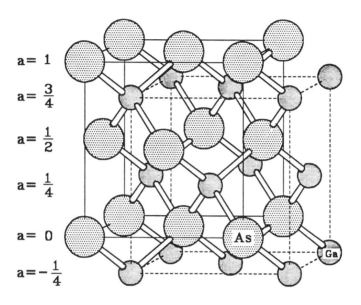

$a = 1$

$a = \dfrac{3}{4}$

$a = \dfrac{1}{2}$

$a = \dfrac{1}{4}$

$a = 0$

$a = -\dfrac{1}{4}$

As

Ga

**Fig. 1.23** Representation of the gallium-arsenide structure as two interpenetrating face-centered-cubic sublattices.

While all (111) planes are identical in silicon, there are two types of (111) planes in GaAs, the (111)Ga and the (111)As plane (Fig. 1.25). All three valence electrons of gallium atoms in the (111)Ga planes are bonded, but only three of the five valence electrons of arsenic atoms in the (111)As planes are bonded. Therefore, As(111) faces exhibit the highest electronic activity since two of five electrons of each arsenic atom are "free". The properties of the two surfaces are thus very different. For example, etching proceeds more rapidly and results in a smoother surface on (111)As than on (111)Ga surfaces. Some important properties of gallium-arsenide are given in Table 1.4.

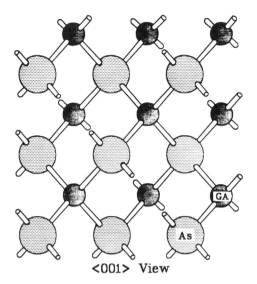

**Fig. 1.24** Gallium-arsenide crystal viewed in the
< 001 > direction.

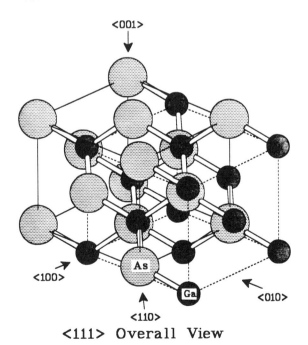

**Fig. 1.25** Gallium-arsenide crystal viewed in the
< 111 > direction

**Table 1.4** Some important properties of gallium-arsenide at 25 °C

| Symbol | GaAs |
|---|---|
| Molecular weight | 144.63 g/Mole |
| Crystal | Zincblende |
| Lattice constant | 5.6533 Å |
| Arsenic radius | 1.18 Å |
| Gallium radius | 1.26 Å |
| Atomic density | $4.42x10^{22}$ atoms/cm$^3$ |
| Density | 5.3176 g/cm$^3$ |
| Melting point | 1238 °C |
| Specific heat | 0.35 J/g.K |
| Linear thermal expansion coefficient | $6.86x10^{-6}$ K$^{-1}$ |
| Thermal conductivity | 0.46 W/cm.K |
| Relative dielectric constant | 13.1 |
| Index of refraction | 3.3 |
| Energy gap | 1.424 |
| Intrinsic carrier concentration | $1.70x10^6$cm$^{-3}$ |
| Electron affinity | 4.07 eV |
| Conduction band density of states | $4.7x10^{17}$cm$^{-3}$ |
| Valence band density of states | $7.0x10^{18}$cm$^{-3}$ |
| Electron mobility | 8500 $cm^2/V.s$ |
| Hole mobility | 360 $cm^2/V.s$ |
| Optical phonon energy | 0.035 eV |
| Donors (As-site) | S, Se, Sn, Te |
| Acceptors (Ga-site) | Zn, Be, Mg, Cd |

## 1.4.1 Crystal Growth

The most common method for preparing GaAs wafers for IC applications is the **Liquid Encapsulated Czochralski (LEC)** technique. Another method of crystal growing, called the **horizontal Bridgman** method, is used primarily to prepare GaAs material for optoelectronic applications. As with silicon, the starting material in the LEC technique is polycrystalline GaAs. It is molten in either a quartz, or boron nitride crucible [20, 21]. In some cases, arsenic and gallium are added separately to form the melt [22]. A cap layer of inert liquid, about 2-cm thick, covers the melt to avoid evaporation of As and the decomposition of GaAs because of the high volatility of arsenic (Fig. 1.26). Boron trioxide ($B_2O_3$) is the most commonly used material for the cap. It floats on the GaAs molten surface, is inert, and is impervious to arsenic diffusion, (provided that the argon or nitrogen ambient pressure is maintained at $\simeq 2$ atm).

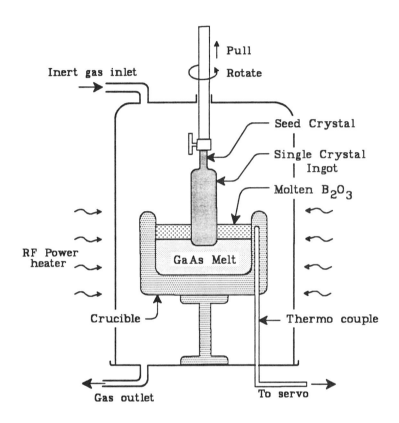

**Fig. 1.26** Principle of the Liquid Encapsulated Czochralski crystal growth method.

A seed having the desired crystallographic orientation is lowered into the GaAs melt through the transparent $B_2O_3$ layer. The seed is cut to allow growth in the $<111>$ or $<100>$ direction. The seed and crucible are rotated in opposite directions to improve crystal uniformity.

The growth rate in the $<111>$ direction is only 0.25-1.50 cm/h. It is limited mainly because of the small thermal conductivity of GaAs (Table 1.4). Growth in the $<100>$ is slower.

The dislocation density in semi-insulating GaAs substrates is higher than in silicon and is of great concern. While large diameter (10-20 cm) silicon crystals can be grown essentially dislocation-free, only very small diameter (less than 1.5 cm) GaAs crystals have been reported with comparable structural perfection [23].

The major cause for dislocation formation in semiconductors is stress induced by temperature gradients during growth and cooling. Under the same growth conditions, gallium-arsenide is more susceptible to dislocation formation for two reasons. First, the bonds in GaAs are not as strong as in Si, so dislocations form more easily. Second, the thermal conductivity of silicon is about three-fold that of GaAs (Tables 1.1, 1.4). Therefore, it is more difficult to reduce thermal gradients in GaAs crystals.

The horizontal Bridgman method consists of loading the material into a long and narrow container, typically made of fused silica of circular cross-section. The container is moved horizontally from a zone of temperature slightly above the material's melting point to a zone of temperature slightly below the melting point. For single-crystal growth, a seed of the desired orientation is placed at one end of the container and partially kept in the zone below the melting point. As the container is slowly moved, the melt freezes onto the seed as a single crystal.

### 1.4.2 Impurities and Crystal Doping

Oxygen and carbon are unavoidable contaminants in GaAs crystals. As a substitutional impurity, oxygen forms a deep donor level which contributes to trapping charge in the bulk of the substrate and reduces its conductivity. Oxygen has a solubility limit of $\simeq 10^{17} cm^{-3}$ in GaAs. Carbon in GaAs creates a shallow acceptor level. It is found in LEC crystals in concentrations of $\simeq 2x10^{15} - 10^{16} cm^{-3}$, depending on the crucible used. The effects of shallow and deep trap levels on conductivity are discussed in the chapters to follow.

Very high resistivity GaAs substrates (termed **semi-insulating,** or SI) are of great importance for the isolation of high speed devices in integrated circuits. Using oxygen as a dopant for making semi-insulating substrates is not practical, since oxygen is highly mobile at device processing temperatures. In some special cases, after all high temperature cycles have been completed, oxygen is introduced locally for device isolation. The electrical properties of undoped LEC GaAs crystals depend strongly on the melt stoichiometry. It is found that the crystal is p-type below a critical arsenic concentration and semi-insulating above it. The semi-insulating property is associated with compensating effects of a deep donor level (referred to as **EL2** ) and shallow carbon acceptors. EL2 levels are created when arsenic atoms occupy gallium positions [24]. Carbon is one of the principal impurities in LEC crystals. It is found at concentrations $2x10^{15}$ cm$^{-3}$ - $10^{16}$ cm$^{-3}$. In some cases, chromium (Cr) is intentionally added to the substrate to create a deep acceptor level and reduce conductivity, leading to SI GaAs. The nature and placement of the Cr level in the GaAs bandgap are, however, not well understood.

**Table 1.5** Dopants in gallium-arsenide.

| Dopant | Segregation coefficient | Dopant type |
|--------|------------------------|-------------|
| Sulfur | 0.3 | n |
| Tellurium | 0.06 | n |
| Tin | 0.08 | n |
| Selenium | 0.3 | n |
| Carbon | 0.8 | n/p |
| Germanium | 0.01 | n/p |
| Silicon | 0.14 | n/p |
| Beryllium | -- | p |
| Magnesium | 0.1 | p |
| Zinc | 0.4 | p |
| Chromium | 0.0006 | SI |

Typical dopants are introduced into GaAs wafers by ion implantation (Chap. 6). Their type and segregation coefficients are given in Table 1.5. Sulfur, selenium, tellurium and tin are shallow donors when they occupy arsenic sites. Beryllium, magnesium and zinc are acceptors when they occupy gallium sites. Carbon, germanium and silicon can be donors or acceptors, depending on whether they occupy arsenic or gallium sites. This can be influenced by the doping conditions which are not detailed here.

# PROBLEMS

**1.1** Assume a hard sphere model and find the ratio of volume occupied to volume available in the following crystals:

a) Simple cubic

b) Body-centered cubic

c) Face-centered cubic

d) Diamond

**1.2** Show that the angle between the tetrahedral bonds of silicon is 109 ° 28′.

**1.3** The plane of the paper is the (100) plane of a silicon wafer. Define the seven remaining directions on the sketch below.

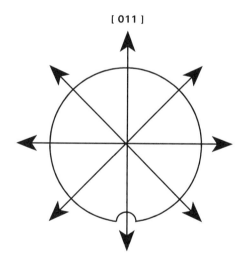

[ 011 ]

**1.4** In problem 1.3 what is the angle between a (111) plane and the plane of the paper?

**1.5** In silicon, the cube edge a = 5.43 Å. Assume a hard sphere model and:

a) Calculate the radius of a silicon atom.

b) Define the coordinates of voids in the cube.

c) Calculate the radius of the largest atom which can fit in a void.

d) Calculate the shortest distance between silicon atoms.

e) Determine the density of silicon atoms in atoms/cm$^3$.

f) Use Avogadro's number to find the density of silicon.

**1.6** Find the number of atoms per square centimeter in a silicon crystal in:

a) (100) plane

b) (111) plane

**1.7** A Czochralski grown crystal is doped with boron. Why is the boron concentration larger at the tail-end of the crystal than at the seed-end?

Do you expect this axial variation to be larger or smaller for arsenic? Why?

Suggest a method to improve the axial uniformity of boron in the floating-zone technique.

**1.8** Assume the energy of formation of a Frenkel-type defect to be 1.1 eV and estimate the defect density at 25 °C and 1000 °C.

**1.9** How many dies of area 2 cm$^2$ can be placed on a 200-mm diameter wafer? Explain your assumptions on die-shape and unused wafer perimeter.

**1.10** Why is the impurity concentration larger in the center of the wafer than at its perimeter?

**1.11** A 1 mm thick silicon wafer having a diameter of 20 cm contains 6.77 mg boron ($_{11}B^5$) uniformly distributed in substitutional sites. Find:

(a) The boron concentration in atoms/cm$^3$.

(b) The average distance between boron atoms.

# References

[1] J. W. Mayer, L. Eriksson, and J. A. Davies, Ion Implantation in Semiconductors, Academic Press, New York, 1970.

[2] C. Kittel, Introduction to Solid State Physics, John Wiley and Sons, New York, p. 27, 1968.

[3] S. M. Sze, Physics of Semiconductor Devices, John Wiley and Sons, New York, 1981.

[4] S. K. Ghandhi, VLSI Fabrication Principles, John Wiley and Sons, New York, 1983.

[5] H. M. Liaw, *Oxygen and Carbon in Czochralski Grown Silicon,* Semicon. International, 2, 71 (1979).

[6] T. Nozaki, Y. Yoshifumi, and N. Akiyama, *Concentration and Behavior of Carbon in Semiconductor Silicon,* J. Electrochem. Soc., 117 (12), 1566-1568, (1970).

[7] W. Dietze, W. Keller, and A. Muehlbauer, *"Float-Zone Grown Silicon,"* in Crystals, J. Grabmaier, Ed., Springer Verlag, New York, 1981.

[8] F. A. Trombore, *"Solid Solubility of Impurity Elements in Germanium and Silicon,"* Bell. Syst. Tech. J., 39, 205-233 (1960).

[9] G. L. Vick and K. M. Whittle, *"Solid Solubility and Diffusion Coefficients of Boron in Silicon,"* J. Electrochem. Soc., 116 (8), 1142-1144 (1969).

[10] D. Hull, Introduction to Dislocations, Pergamon Press, New York, 1965.

[11] W. L. Bragg and J. F. Nye, Proc. Roy. Soc., A190, 474, London (1947); W. L. Bragg and W. M. Lomer, Proc. Roy. Soc., A196, 171, London (1949); C. Kittel, Introduction to Solid State Physics, p. 595, John Wiley and Sons, 1968.

[12] A. J. Dekker, Solid State Physics Prentice-Hall, New Jersey, 1965

[13] J. R. Patel, *"Oxygen in Silicon,"* Semiconductor Silicon, H. R. Huff and E. Sirtl, Eds., Electrochem. Soc., Pennington, New Jersey, 521-545 (1977).

[14] Y. Matsushita, S. Kishino, and M. Kanamori, *"A Study of Thermally Induced Microdefects in Czochralski-Grown Crystals: Dependence on Annealing Temperature and Starting Material,"* Jpn. J. Appl. Phys., 19 (2), L101-L104 (1980).

[15] K. Yamamoto, S. Kishino, Y. Matsushita, and T. Iizuka, *"Lifetime Improvement in Czochralski-Grown Silicon Wafers by the Use of a Two Step Annealing,"* Appl. Phys. Lett., 36 (3), 195 (1980).

[16] G. A. Rozgonti and C. W. Pearse, *"Gettering of Surface and Bulk Impurities in Czochralski Silicon Wafers,"* Appl. Phys. Lett., 32 (11), 747 (1978).

[17] C. W. Pearse, L. E. Katz, and T. E. Seidel, *"Considerations Regarding Gettering in Integrated Circuits,"* Semiconductor Silicon, H. R. Huff and R. J. Kriegler, Eds., Electrochem. Soc., Pennington, New Jersey, 705-723 save81).

[18] P. V. Gray and D. M. Brown, *"Density of $SiO_2$ Interface States,"* Appl. Phys. Lett., 8 (2), 31 (1966).

[19] K.-F. Berggren, *"Quantum Phenomena in Small Semiconductor Structures and Devices,"* 33, 217 (1988).

[20] J. B. Mullen, R. H. Heritage, C. H. Holliday, and B. W. Straughan, *"Liquid Encapsulation Crystal Pulling at High Pressures,"* J. Cryst. Growth, 34, 281 (1968).

[21] T. R. AuCoin, R. L. Ross, M. J. Wade, and R. O. Savage, *"Liquid Encapsulated Compounding and Czochralski Growth of Semi-Insulating GaAs,"* Sol. State Tech., 22, 59 (1979).

38

[22] C. G. Kirkpatrick, R. T. Chen, D. E. Holmes, and K. R. Elliott, *"Growth of Bulk GaAs,"* Gallium Arsenide M. J. Howes and D. V. Morgan, Eds, p. 39, John Wiley and Sons, New York, 1985.

[23] P. F. Lindquist and W. M. Ford, GaAs FET Principles and Technology, J. V. Dilorenzo and D. D. Khandelwal, Eds, p. 1, Artech House, Inc., Massachusetts, 1984.

[24] G. M. Martin, A. Mitonenau, and W. Mircea, *"Electron Traps in Bulk and Epitaxial GaAs Crystals,"* Electronics Lett., 13, 191 (1977).

# Chapter 2

# Thermal Oxidation and Nitridation

## 2.0 Introduction

When silicon is exposed to air at room temperature, it reacts with oxygen to form a very thin silicon dioxide ($SiO_2$) film. This film, sometimes called **native oxide,** is essentially amorphous and approaches a thickness of approximately 0.5 nm after 5 min, 2 nm after 15 h, and 4-5 nm after one year of exposure to air [1]. The reaction between oxygen and silicon occurs at the silicon-oxide interface and, to react with silicon, oxygen must diffuse through the growing oxide and reach the silicon surface. Therefore, as the native oxide grows the rate of penetration of oxygen, and hence the oxidation rate decreases. The oxidation rate can be accelerated by subjecting the wafer to oxygen or water vapor at elevated temperature. The conversion of silicon into silicon dioxide at high temperature is a common processing step in the manufacture of integrated circuits. This step, called **thermal oxidation,** is introduced at different stages of an integrated process technology for various purposes. The grown film is used to properly terminate silicon bonds at the silicon surface, to isolate conductors and semiconductors, or to provide a high-quality dielectric for MOSFET gates, memory-cell nodes, or precision capacitors. Thermal oxidation is also frequently performed to grow a "sacrificial" film that is removed after fulfilling its purpose. The sacrificial film is used, for example, to create a step in silicon for mask alignment (Chap. 4), to remove a certain amount of silicon that has been damaged by a preceding step (such as reactive-ion etching, Chap. 5), or to serve as a screen that blocks the penetration of unwanted impurities during ion implantation (Chap. 6) or diffusion (Chap. 7). Masking against impurities can also be achieved by depositing silicon dioxide rather than growing the film (Chap. 3).

Thermal oxidation and the properties of $SiO_2$ are well established and constitute a large part of this chapter. The conversion of silicon to silicon nitride ($Si_3N_4$), referred to as **thermal nitridation,** is, however, not as well understood as thermal oxidation. Also, the reaction of silicon with nitrogen requires a much higher temperature and is far less frequently used than thermal oxidation. Thermal nitridation is discussed at the end of this chapter in conjunction with local oxidation (LOCOS).

## 2.1 Properties and Structures of $SiO_2$ and $SiO_2$-Si Interface

The unique properties of $SiO_2$ and its interface with silicon have earned this dielectric a dominant role in modern silicon-based technologies. This section summarizes important properties of $SiO_2$ and describes atomic configurations that are suggested for the oxide network and its interface with silicon.

### 2.1.1 Properties and Structure of $SiO_2$

Some properties of silicon dioxide are given in Table 2.1. Their importance to IC manufacturing and operation will become apparent in the following sections and chapters.

**Table 2.1** Important properties of silicon dioxide.

| | |
|---|---|
| Molecular weight | 60.1 g/mole |
| Density (thermal, dry/wet) | 2.27/2.18 g/cm$^3$ |
| Molecules/cm$^3$ | $2.3x10^{22}$/cm$^3$ |
| Melting point | 1700 °C |
| Thermal expansion coefficient | $5.6x10^{-7}$/K |
| Young's modulus | $6.6x10^{10}$N/m$^2$ |
| Poisson's ratio | 0.17 |
| Thermal conductivity | $3.2x10^{-3}$ W/cm.K |
| Relative dielectric constant | 3.7 - 3.9 |
| Dielectric strength | $10^7$ V/cm |
| Energy gap | 8 eV |
| DC resistivity | $\simeq 10^{17}$ Ohm-cm |
| Infrared absorption band | 9.3 $\mu$m |
| Index of refraction | 1.459 |

Oxides grown in a dry atmosphere ($O_2$) exhibit a higher density, implying a lower porosity to impurities than wet oxides (grown in $H_2O$ atmosphere). The thermal expansion coefficient is a measure of stress or strain that the oxide exerts on other materials in contact with it, particularly during high-temperature cycles. Young's modulus and Poisson's ratio describe the mechanical stability of oxide films. The thermal conductivity is an important parameter that affects power dissipation during circuit operation. The stability of $SiO_2$ under high electric fields is expressed as its dielectric strength ($\simeq 10^7$ V/cm). Related to the dielectric strength is the high resistivity ($\simeq 10^{17}$ Ohm-cm), making oxide films very suitable for dielectric isolation. Properties such as interface trap density, energy gap, and dielectric constant are discussed later in conjunction with device operation.

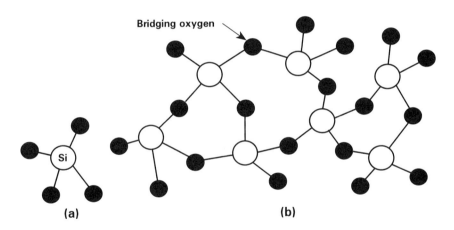

**Fig. 2.1** The silicon dioxide structure. (a) Silicon ion surrounded tetrahedrally by four oxygen ions. (b) Amorphous network of linked tetrahedral elements.

Silicon dioxide can be described as a three-dimensional network constructed from tetrahedral cells, with four oxygen ions surrounding a silicon ion, as shown in a two-dimensional projection in Fig. 2.1. In an ideal network, the vertices of the tetrahedra are joined by a common oxygen ion called a bridging oxygen [2]. The length of a Si-O bond is 0.162 nm and the nearest distance between oxygen ions is 0.227 nm. The free rotation of one tetrahedron with respect to another through the Si-O-Si link and the capability of the Si-O-Si bond angle to vary

42

from 120° to 180° are believed to play an important role in matching amorphous $SiO_2$ with crystalline silicon without breaking bonds [3].

### 2.1.2 Properties and Structure of the $SiO_2$-Si Interface

The properties of the $SiO_2$-Si interface that dominate the behavior of silicon devices are: contamination, interface trap density ($D_{it}$), and surface roughness. Bare silicon surfaces, e.g. those obtained by cleaving the crystal in ultra-high vacuum, exhibit $D_{it}$ levels in the order of $10^{15}\, cm^{-2}$. This corresponds to the density of silicon atoms per unit surface area and is related to silicon dangling bonds created by the termination of the crystal. When grown under controlled conditions in an ultra-clean environment, silicon dioxide properly terminates silicon bonds at the silicon surface, reducing the density of interface traps by five to six orders of magnitude [4,5]. The prime importance of silicon dioxide stems from its ability to properly "passivate" silicon surfaces.

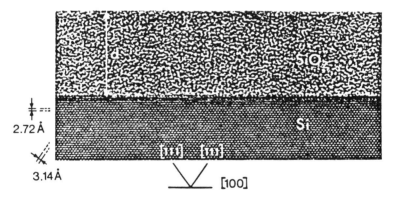

**Fig. 2.2** High resolution transmission electron micrograph of a thermally oxidized silicon surface, illustrating uniform bonding of the oxide to silicon [6].

Surface roughness is defined as the average fluctuation of the silicon surface with respect to a reference plane. The roughness can occur during preparation of the crystal or during subsequent processing. Typical oxidized silicon surfaces of properly prepared wafers exhibit a roughness of less than 0.5 nm, as illustrated in a high-resolution transmission electron micrograph (HRTEM) in Fig. 2.2.

One suggested model for the transition from Si to SiO$_2$ is shown in Fig. 2.3. This model assumes an abrupt transition from single crystal silicon to amorphous SiO$_2$ [7]. There is, however, strong evidence for a few layers of crystalline SiO$_2$ matched to a strained Si crystal, suggesting a crystalline structure near the interface [8]. The strong sensitivity of the interface to the method used to grow the oxide is one of the reasons why a unified microscopic description of the Si-SiO$_2$ interface has not yet been established. Another difficulty in defining the exact atomic configuration at this interface is caused by the non-crystalline character of the SiO$_2$ layer and the inability of measurement tools to determine the atomic structure of disordered systems.

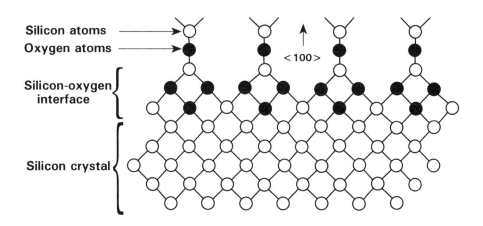

**Fig. 2.3** Proposed model for the atomic configuration at the (100)-oriented Si-SiO$_2$ interface [7].

In a defect-free Si-SiO$_2$ interface, the tetrahedral bonding of crystalline silicon atoms is accommodated by bonding with oxygen atoms of the oxide (Fig. 2.3). A small fraction of surface surface silicon atoms appears, however, not to be bonded to oxygen atoms. Instead, unpaired electrons are localized on the defect silicon atom forming a hybrid orbital in a direction normal to the (111) plane [9]. This "dangling bond", also referred to as a P$_b$ center, can be detected by electron-spin resonance (ESR). It is illustrated for (111) and (100) silicon in Fig. 2.4. There is a good correlation between the density of $P_b$ centers measured by ESR and the electrically measured trap density at the Si-SiO$_2$ interface [10].

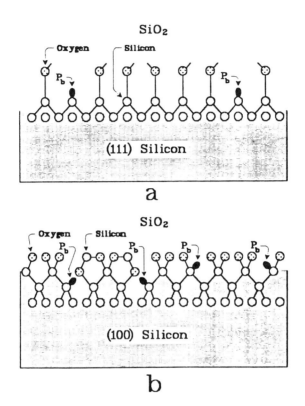

**Fig. 2.4** Illustration of a dangling bond ($P_b$ center). (a) (111) silicon. (b) (100) silicon. [9].

## 2.2 Thermal Oxidation

Thermal oxidation is performed by subjecting the wafer to an oxidizing ambient at elevated temperature. One common objective of an oxidizing system is to obtain a high-quality SiO$_2$ film of uniform thickness, while maintaining a low "thermal budget" (defined as the product of temperature and time). Several methods have been developed to increase the oxidation rate and reduce the oxidation time and temperature. These methods are described in this section after a brief discussion of oxidation reactions, oxidation kinetics, and oxidation systems.

## 2.2.1 Oxidation Reactions

The species used to grow thermal oxide on silicon are dry oxygen and water vapor. For dry oxygen the chemical reaction is

$$Si + O_2 = SiO_2 , \qquad\qquad 2.1$$

and for water vapor the net reaction is

$$Si + 2H_2O = SiO_2 + 2H_2 . \qquad\qquad 2.2$$

In both cases, silicon is consumed and converted into silicon dioxide $SiO_2$ (Fig. 2.5).

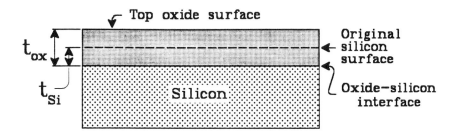

**Fig. 2.5** Consumption of silicon during oxidation.

The thickness of silicon consumed, $t_{si}$, is given by

$$t_{si} = t_{ox} \, \frac{N_{ox}}{N_{si}} \simeq 0.46 t_{ox} \qquad\qquad 2.3$$

where

$t_{ox}$ = thickness of silicon dioxide
$N_{ox}$ = density of oxide molecules ($\simeq 2.3x10^{22} \ cm^{-3}$)
$N_{si}$ = density of silicon atoms ($\simeq 5x10^{22} \ cm^{-3}$)

### 2.2.2 Oxidation Kinetics

For an oxide thickness larger than $\simeq$ 30 nm, the relation between oxide thickness and oxidation time is given by the so-called "Deal-Grove" relation [11]

$$t_{ox}^2 + At_{ox} = B(t + \tau),\qquad\qquad 2.4$$

where A and B are coefficients which depend on temperature, ambient composition and pressure, and crystallographic orientation; $t_{ox}$ is the total oxide thickness; t is the oxidation time; $\tau$ the shift in time to account for an initial oxide thickness.

The relation between $\tau$ and the initial oxide thickness is

$$\tau = \frac{t_i^2 + At_i}{B},\qquad\qquad 2.5$$

where $t_i$ is the initial oxide thickness.

Of particular interest are two limiting forms of the oxide thickness versus time relation. For $t >> \tau$ and $t >> A^2/4B$, Eq. 2.4 reduces to

$$t_{ox}^2 \simeq Bt,\qquad\qquad 2.6$$

which is a simple parabolic expression. The parameter B, referred to as the parabolic rate constant, is proportional to the partial pressure and diffusivity of the oxidizing species in $SiO_2$.

For short oxidation times with $t << A^2/4B$, Eq. 2.4 reduces to the linear form

$$t_{ox} \simeq \frac{B}{A}(t + \tau).\qquad\qquad 2.7$$

The ratio B/A, referred to as the linear rate constant, depends mainly on the surface reaction rate constant. It is the dominant, rate-limiting factor during the initial growth phase.

The two limiting forms of thermal oxidation are shown in Fig. 2.6. The temperature dependence of the parabolic and linear rate constants are shown in Figs. 2.7 and 2.8, respectively. Figures 2.9 and 2.10 compare wet to dry oxidation. The growth rate in wet oxygen is much greater than in dry oxygen, but the higher humidity results in a less dense oxide, as indicated in Table 2.1. The higher growth rate in a wet atmosphere is due to the faster diffusion and higher solubility of water in silicon dioxide. Figure 2.11 compares the diffusivities of oxygen and water in silicon dioxide as a function of temperature.

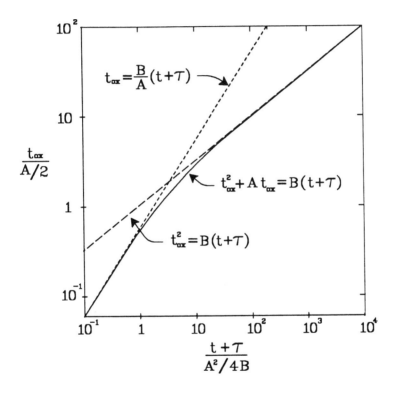

**Fig. 2.6** The general relation for silicon oxidation and its two limiting forms [11].

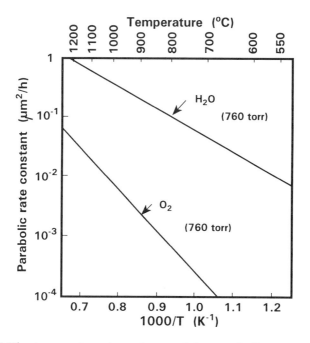

**Fig. 2.7** The temperature dependence of the parabolic rate constant [11].

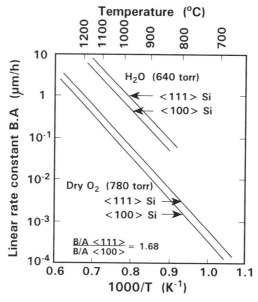

**Fig. 2.8** The temperature dependence of the linear rate constant [11].

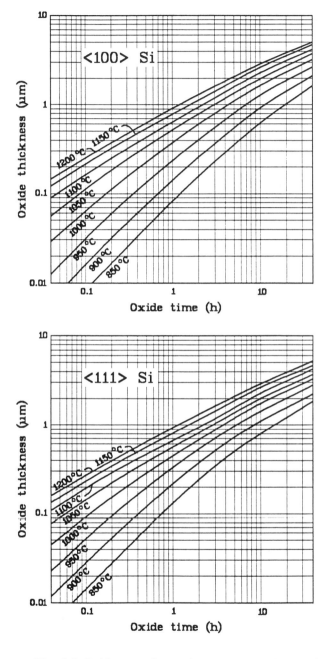

**Fig. 2.9** Oxide growth rate in wet oxygen [11, 12].

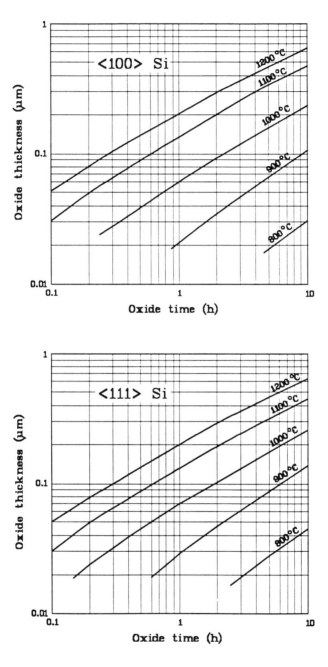

Fig. 2.10 Oxide growth rate in dry oxygen [11, 12].

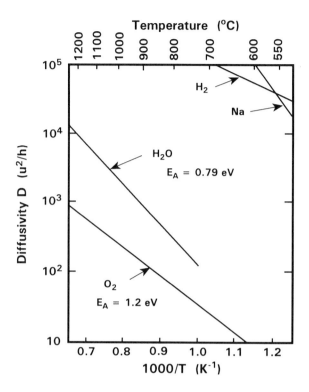

**Fig. 2.11** Temperature dependence of the diffusivity of oxygen and water in silicon dioxide. Also shown are the diffusivities of hydrogen and sodium for later reference [12].

Approximate relations for the parabolic and linear rate constants are given in Table 2.2. When using this table and Eq. 2.4, the time must be defined in hours and the oxide thickness in $\mu$m.

### 2.2.3 Initial Oxidation Stage

While the "Deal-Grove" linear-parabolic relation is followed over a wide oxide thickness and temperature range for wet and steam oxidation, it is found that dry $O_2$ oxidation in the range 0 - 30 nm is faster than predicted by Eq. 2.4 (Fig. 2.12). Since modern technologies emphasize this range of oxide thickness for MOSFETs and capacitors, intensive work has been done to model the initial rapid stage of oxidation [13 – 18].

**Table 2.2** Empirical relations for the parabolic and linear rate constants [12]

| Parabolic: $B = C_1 e^{-E_1/kT}$ $\quad$ Linear : $B/A = C_2 e^{-E_2/kT}$ $C_1, C_2, E_1, E_2$: empirically determined parameters | | | |
|---|---|---|---|
| | <111> Silicon | <100> Silicon | Unit |
| Dry $O_2$ | $C_1 = 7.72 \times 10^2$ | $C_1 = 7.72 \times 10^2$ | $\mu m^2/h$ |
| | $C_2 = 6.23 \times 10^6$ | $C_2 = 3.71 \times 10^6$ | $\mu m/h$ |
| | $E_1 = 1.23$ | $E_1 = 1.23$ | eV |
| | $E_2 = 2.00$ | $E_2 = 2.00$ | eV |
| $H_2O$ (640 torr) | $C_1 = 3.86 \times 10^2$ | $C_1 = 3.86 \times 10^2$ | $\mu m^2/h$ |
| | $C_2 = 1.63 \times 10^8$ | $C_2 = 0.97 \times 10^8$ | $\mu m/h$ |
| | $E_1 = 0.78$ | $E_1 = 0.78$ | eV |
| | $E_2 = 2.05$ | $E_2 = 2.05$ | eV |

One model suggests an oxidation process in which two separate but parallel reactions occur, with a modified linear-parabolic growth law [15 – 18]

$$\frac{dt_{ox}}{dt} = \frac{B_1}{2t_{ox} + A_1} + \frac{B_2}{2t_{ox} + A_2} \qquad 2.8$$

where $B_1$, $B_2$ and $A_1$, $A_2$, are the respective values of the constants in Eq. 2.4 for processes 1 and 2. Initially, for very thin oxides, one of the two parallel processes controls the oxidation rate. The second parallel process which defines the linear regime in Eq. 2.4 takes over rapidly. Eventually, as the oxide thickness approaches "infinity", the kinetics follow the parabolic limit of Eq. 2.4.

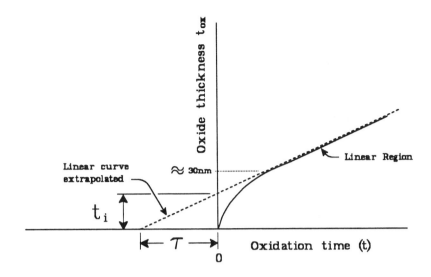

**Fig. 2.12** Rapid, non-linear rate in the initial stage of dry oxidation.

A useful empirical relation for the rapid initial oxidation stage is given as [18]

$$\frac{dt_{ox}}{dt} = Ke^{-t_{ox}/L} + \frac{B}{2t_{ox} + A} , \qquad 2.9$$

where $L \simeq 7$ nm, and $K$ is a fitting parameter approximated as

$$K \simeq K_o e^{-E_A/kT} .$$

Experimentally fitted values for $K_0$ and $E_A$ are given in Table 2.3 [19].

**Table 2.3** Thin oxide growth rate parameters [19].

| Orientation | $<100>$ | $<110>$ | $<111>$ |
|---|---|---|---|
| $K_0$ $(nm/s)$ | $1.10x10^8$ | $8.92x10^8$ | $9.78x10^7$ |
| $E_A$ $(eV)$ | 2.37 | 1.80 | 2.32 |

### 2.2.4 Oxidation Systems

Silicon dioxide layers are typically grown in the temperature range of 400 °C - 1150 °C. This can be performed in resistance-heated furnaces or in rapid-thermal processing chambers with heat provided by, e.g., tungsten-halogen lamps.

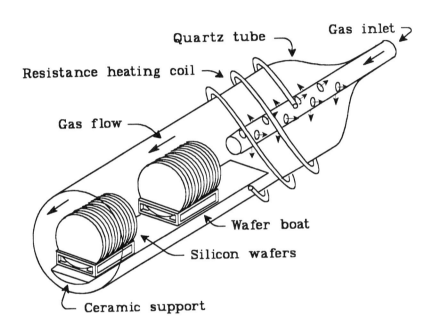

**Fig. 2.13** Typical horizontal furnace tube.

**Furnace Oxidation**

A typical horizontal furnace tube is shown in Fig. 2.13. A batch of wafers is introduced into the furnace in a slow travelling "boat" and heated to the oxidation temperature ("ramp-up"). The wafers are held at this elevated temperature for a specific time and then brought back to a low temperature ("ramp-down"). For dry oxidation, oxygen mixed with an inert carrier gas, such as nitrogen, is passed over the wafers at the elevated temperature, as illustrated in Fig. 2.14a. Wet oxidation is performed by bubbling oxygen through a high purity water bath maintained between 85 °C and

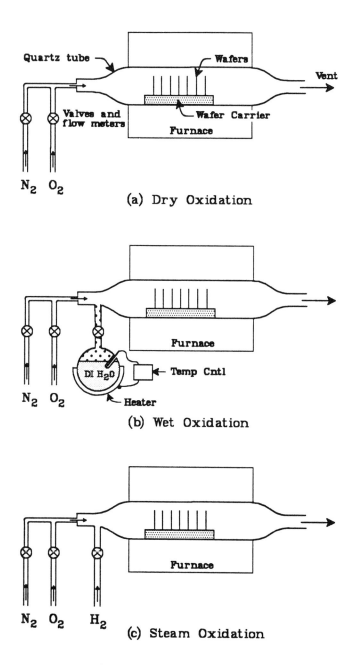

(a) Dry Oxidation

(b) Wet Oxidation

(c) Steam Oxidation

**Fig. 2.14** Typical oxidation systems. (a) Dry oxidation. (b) Wet oxidation. (c) Steam oxidation.

95 °C. The temperature of the bath determines the partial pressure of water in the oxygen gas stream. The mixture is passed over the wafers at the elevated temperature (Fig. 2.14b). In pyrogenic steam oxidation, the oxidizing medium is water vapor formed by a direct reaction of hydrogen with oxygen (Fig. 2.14c).

One important consideration in thermal oxidation is its effect on the distribution of impurities in the bulk of silicon and at the oxide-silicon interface. Since the movement of impurities affects the semiconductor device size and its electrical properties, it is important to control and minimize the effects of oxidation on the impurity profile. This is done by precisely controlling the oxidation temperature and reducing the duration of the heat cycle required to grow an oxide film. The thermal budget required to achieve an oxide film of a certain thickness is considerably smaller for wet oxide than for dry oxide (Figs. 2.9, 2.10). Because of its water content, however, wet oxide films exhibit a lower dielectric strength and more porosity to impurity penetration than dry oxides. Wet oxidation is therefore used when the electrical and chemical properties of the film are not critical. To grow a "high quality" oxide layer while minimizing the oxidation time, it is common practice to begin and end the oxidation process in dry oxygen and to use wet oxidation for the intermediate stage. The intermediate wet oxidation step reduces the thermal budget by increasing the overall oxide growth rate. With this dry-wet-dry process sequence high-quality films are grown on both sides of the oxide layer, so the properties of the triple-layer film become comparable to those of a layer grown by dry oxidation alone.

### Rapid Thermal Oxidation

One limitation of furnace oxidation is its inertia to temperature transitions, resulting in a higher thermal budget than required for oxidation. The thermal budget can be reduced considerably by decreasing the duration of these transitions. This can be achieved by **rapid thermal processing (RTP),** also referred to as rapid isothermal processing. A schematic of an RTP system is shown in Fig. 2.15. The heat source is typically an array of halogen, silicon-carbide, or arc lamps in an optical system. A single wafer is isolated in the chamber and processed in a controlled environment.

During RTP, the wafer is rapidly heated from a low temperature to a high processing temperature. It is held at this elevated temperature for a short time and then brought back rapidly to a low temperature. Typical temperature transition rates range from 10 °C/s to 350 °C/s, compared to about 0.1 °C/s for furnace processing. RTP durations at high processing temperatures vary from 1 s to 5 min. Figure 2.16 compares a furnace oxidation temperature profile to that of an RTP system. As can be seen, RTO reduces the ramp-up and ramp-down durations. Because of the high processing temperature, RTO also reduces the processing time required for oxidation.

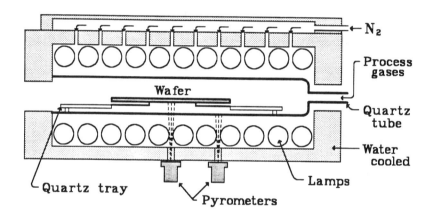

**Fig. 2.15** Cross section view of an RTP system. Courtesy: AG Associates.

The temperature can be measured with an infrared pyrometer from the back side of the wafer, typically at a wavelength of 4-5 $\mu m$. Precise temperature measurement with this method is, however, rather difficult mainly because the "energy reading" depends strongly on the surface conditions of the back side. Accurate temperature measurement is also difficult during transients, where the wafer temperature can change by approximately 1000 °C in a few seconds. Another problem observed with rapid thermal processing is the stress-induced slip at the periphery of the wafer caused by the faster cooling of the wafer edge than its top and bottom surfaces. Methods to overcome the above limitations are being developed and will not be discussed in this chapter. A review of rapid-thermal processes can be found in [20].

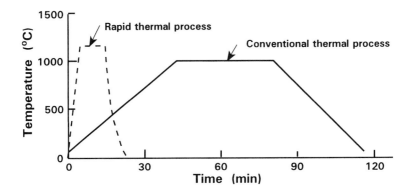

**Fig. 2.16** Comparison of temperature versus time plots for rapid thermal oxidation and conventional furnace oxidation showing the reduced thermal budget with RTO.

Rapid-thermal processing is important where precise thermal control and short high-temperature process times are critical. When used to grow thin oxide films, the process is referred to as rapid thermal oxidation (RTO). Oxide layers with film thickness from 4 nm to 40 nm can be grown in pure oxygen at 900 °C to 1150 °C for a duration of only 15-180 s. In addition, the electrical characteristics of rapid thermal oxides are found to be equivalent to or better than furnace-grown thermal oxides [21 – 26]. RTO grown films on polycrystalline silicon exhibit electrical breakdown fields that approach those of oxides grown on single-crystal silicon [27]. Rapid-thermal processing is also used to grow other insulating films, to activate implanted ions and form shallow junctions (Chap. 6), to alloy contacts and form conducting films, such as titanium-silicide and titanium-nitride, and to reflow glass (Chap. 8).

## High-Pressure Oxidation

At temperatures below 700 °C, atmospheric pressure oxidation is very slow. One of the methods to enhance the oxidation rate at low temperatures is to increase the pressure of oxidizing species, allowing them to penetrate the growing oxide more rapidly. At high partial pressures, the oxidation rate becomes limited by the reaction rate at the silicon-oxide interface rather than by the diffusivity of the oxidants in $SiO_2$. The rate varies directly with pressure P [28,29]. More recent work shows that the linear rate constant B/A for dry $O_2$ is proportional to $P^{0.7}$ rather than P [30,31]. In any case, high-pressure oxidation can be performed at a considerably lower temperature than oxidation at atmospheric pressure, reducing the thermal budget [28 − 32]. For example, under 10 atm of steam, the oxidation rate at 850 °C is comparable to that at about 1200 °C and 1 atm of steam (Fig. 2.17).

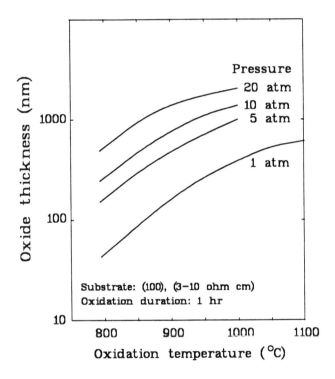

**Fig. 2.17** Dependence of wet oxidation rate on ambient pressure [30].

## Plasma-Enhanced Oxidation

In conventional thermal oxidation, a fraction of molecular oxygen decomposes near the $Si-SiO_2$ interface and provides atomic oxygen for the reaction with silicon. This fraction is, however, negligible below 700 °C, and so is the oxidation rate. Plasma-enhanced oxidation is another method that increases the oxide growth rate at low temperature, reducing the thermal budget [33]. This type of oxidation is typically performed in a pure oxygen plasma which consists of ionized oxygen, electrons, and excited oxygen atoms, created in a high-frequency glow discharge. The wafer to be oxidized is placed in the plasma under an applied electric field that enhances the migration of atomic oxygen toward the silicon surface. It is believed that atomic oxygen created in the plasma plays the dominant role in enhancing the oxide growth rate [34 − 37]. Plasma reactors and their applications are further discussed in Chaps. 3, 5.

## 2.2.5 Second-Order Effects on Oxidation Kinetics

The preceding sections described the dependence of oxidation rate on temperature, time, pressure, oxidizing species and their chemical activity. Other effects that can modify the oxidation rate in an integrated process technology are discussed in this section.

### Stress

Stress is created by the two-dimensional growth of oxide and the resulting volume expansion of the oxidized region. As the oxide grows, the "newly" formed oxide pushes out the "old" oxide which rearranges itself through viscous flow. The viscous stress created by the non-uniform, two-dimensional deformation of the oxide is the fundamental source of retardation in oxide growth [38 − 42]. Stress occurs typically on curved surfaces, as illustrated for the inside and outside corners of a trench in silicon (Fig. 2.18).

Wet oxidation at 900-950 °C, for example, results in oxides that are about 30% thinner at inside corners of a trench than on its sidewalls or top flat surface [38]. The retardation depends on the radius of curvature of the corner. Oxides grown on concave surfaces (compressive stress) are thinner than on convex shapes (tensile stress) [39 − 42]. The effect of stress normal to the oxide surface reduces the surface reaction rate in both concave and convex surfaces by adding to the activation energy the work required to compress or expand the oxide. The diffusivity and solid solubility of the oxidants are, however, decreased by compression and increased by tension. The opposite is true for viscosity. These factors cause the oxide to be thinner on concave surfaces than on convex surfaces.

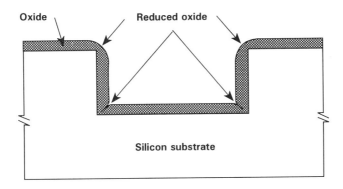

**Fig. 2.18** Oxidation of a silicon trench illustrating the reduced oxidation rate at the corners due to stress.

## Crystal Orientation

As mentioned earlier, the parabolic rate constant depends primarily on the diffusivity of oxidants through the growing oxide, while the linear rate constant depends on the rate of reaction between oxygen and silicon at the silicon surface. Since the density of silicon atoms available for the reaction changes with crystal orientation, the linear rate constant must vary with crystal orientation as (111) > (110) > (100) [43, 44]. For oxides thicker than $\simeq 15$ nm, the linear rate constant is approximately 70% larger in the <111> than in the <100> direction (Fig. 2.10). More recently, it has been found that for very thin oxides and low oxidation temperatures there is a departure from the above orientation order [45,46]. Such a "cross-over" is shown in in Fig. 2.19 where (110) silicon oxidizes faster than (111) at 900 °C in dry $O_2$ when the oxide thickness is is below $\simeq 15$ nm. This cross-over is not well understood.

## Halogens

By adding small amounts of halogen compounds to the ambient gas, the rate of oxidation can be enhanced and the quality of the oxide and oxide-silicon interface can be greatly improved [47, 48]. Chlorine removes harmful impurities, such as sodium ions, from the gas ambient by converting them to their chlorides. Initially, hydrochloric acid (HCl) at a concentration of 3%-7% was used as a chlorine source [47 – 49]. This is now widely replaced by trichloroethane (TCA) or trichloroethylene (TCE), because TCA and TCE are not as corrosive as HCl [50,51]. A model suggests that the reaction between chlorine and silicon promotes the generation of vacancies at the silicon surface. The vacancies then become available to recombine with silicon-interstitials, reducing the density of oxidation-induced stacking faults (discussed in the following section) [52]. The presence of chlorine also enhances the diffusivity of oxygen through the oxide and accelerates its reaction with silicon, increasing the oxide growth rate in dry $O_2$ [53,54]. The mechanisms involved are, however, not well understood. It is suggested that chlorine compounds decompose during oxidation and the fragments diffuse through the oxide, enhancing the oxidation rate [55 – 57].

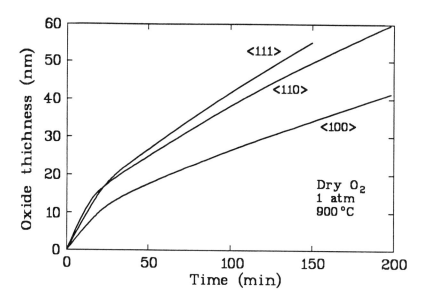

**Fig. 2.19** Effect of crystal orientation on thermal oxidation showing the "cross-over" for very thin oxides [45].

Adding small amounts of fluorine (F) to $O_2$ also enhances the oxidation growth rate and improves the quality of the silicon-oxide interface [58 – 63]. This can be achieved, for example, by introducing diluted (0.011-0.044 vol%) $NF_3$ with $O_2$ during the initial oxidation stage. The presence of fluorine is found to increase the immunity of the oxide against radiation effects, and reduce the surface-state density by "passivating" dangling bonds. Excessive fluorine, however, can degrade the oxide quality by creating non-bridging oxygen in the oxide, and also by displacing oxygen atoms from the silicon surface and creating dangling bonds [64].

### Dopants

Silicon dioxide is not impermeable to dopants. When the oxide is sufficiently thin, impurities such as boron and phosphorus can diffuse through the film. It is therefore necessary to increase the oxide thickness above a certain minimum value when the film is used as a diffusion mask (Chap. 7). The oxide thickness necessary to block the penetration of impurities through an oxide mask is shown in Fig. 2.20 for boron and phosphorus at different diffusion temperatures [65]. When present in the oxide, boron and phosphorus ions ($B^{+3}$, $P^{+5}$) typically assume substitutional sites by replacing silicon in the oxide polyhedra, weakening the oxide network. At high concentration they form borosilicate or phosphosilicate glass (Chap. 3).

The oxidation rate is found to increase with increasing dopant concentration at the silicon surface [66 – 69]. This is shown for boron and phosphorus in Figs. 2.21a, 2.21b. The effect of dopant concentration on oxidation rate is not well understood. One model attributes the enhanced oxidation rate to an increase in the density of vacancies at the oxide-silicon interface [69]. Vacancies increase the linear rate constant and have negligible effect on the parabolic rate constant. Since phosphorus and arsenic have segregation coefficients much less than one, the impurities "pile-up" at the silicon surface during oxidation, while their concentration in the oxide remains low (Chap. 7). It is postulated that the pile-up increases the density of vacancies, explaining the experimentally observed increase in the linear rate constant without appreciably effecting the parabolic rate constant.

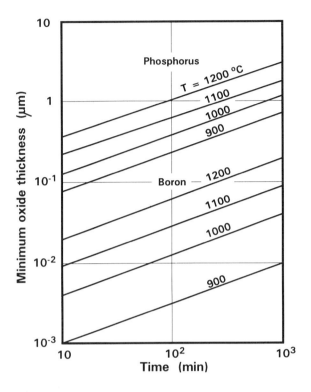

**Fig. 2.20** Required SiO$_2$ mask thickness for boron and phosphorus [65].

The dependence of oxidation rate on the type and concentration of dopants results in a non-uniform film thickness when several regions of the wafer are oxidized simultaneously. This must be taken into account during processing steps subsequent to oxide growth, for example, when etching a pattern in the grown film (Chap. 5), or implanting through the oxide (Chap. 6). Etching a film of non-uniform thickness can result in incomplete removal in the thicker regions or over-etching in the thinner areas.

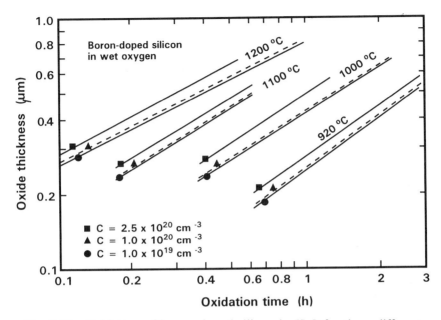

**Fig. 2.21a** Oxidation of boron-doped silicon in $H_2O$ for three different concentrations at four oxidation temperatures [66].

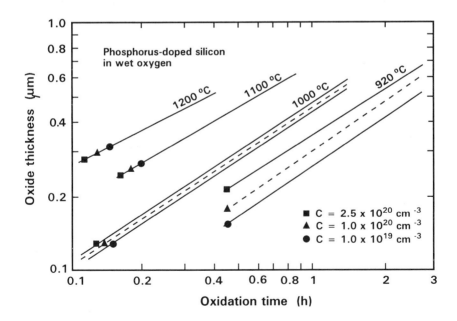

**Fig. 2.21b** Oxidation of phosphorus-doped silicon in wet $O_2$ for three different concentrations at four oxidation temperatures. [66].

**Water**

Water vapor diffuses faster through silicon dioxide than does oxygen (Fig. 2.13), explaining the larger oxidation rate in $H_2O$ than in dry $O_2$ [11, 12, 70]. The presence of water in the oxide, however, weakens the network by reacting with bridging oxygen ions to form non-bridging hydroxyl groups [71]:

$$H_2O + Si\text{-}O\text{-}Si \rightarrow Si\text{-}OH + OH\text{-}Si$$

Weakening the oxide network reduces its breakdown field, and enhances the diffusivity of impurities (such as boron) in the oxide [72].

**Alkali Ions**

Contamination with alkali ions is inevitable during processing and handling, with sodium ions ($Na^+$) being the most common contaminants found in thermal oxides. $Na^+$ can easily migrate through the oxide under the influence of a concentration gradient or an electric field (Fig. 2.13) [73]. Other alkali ions such as potassium and lithium are also mobile in the oxide. The concentration of lithium is, however, negligible when compared to sodium, and the mobility of potassium in the oxide is considerably smaller than sodium [74 − 77]. The drift mobilities of sodium and potassium in the oxide, $\mu_{Na}$, $\mu_K$, are approximated by the following empirical relations [75]

$$\mu_{Na} = e^{-0.66/kT} \quad (cm^2/Vs), \qquad 2.10$$

$$\mu_K = 0.03\, e^{-1.09/kT} \quad (cm^2/Vs). \qquad 2.11$$

The higher mobility of sodium ions is the reason why $Na^+$ is considered as the most important ionic contaminant in thermal oxides.

When present in the oxide, sodium is positively charged and occupies interstitial sites. It is introduced in the form of an oxide which reacts with $SiO_2$ to form non-bridging oxygen ions and create positively charged sodium ions:

$$Na_2O + Si\text{-}O\text{-}Si \rightarrow Si\text{-}O + O\text{-}Si + 2\,Na^+$$

The presence of sodium ions weakens the oxide network and increases the oxide charge, degrading the stability of device parameters. The migration of sodium ions can be inhibited and its charge neutralized by depositing a layer of phosphosilicate glass (PSG) on top of the oxide (Chap. 3). The PSG layer traps and neutralizes sodium in its network, protecting the underlying films from its contamination [78]. Silicon nitride ($Si_3N_4$) is also used extensively as a barrier to impurities (Sec. 2.3). It is superior to PSG because it does not polarize and is not hygroscopic.

**Hydrogen**

Hydrogen permeates very rapidly through silicon dioxide (Fig. 2.13). While this property is essential to effectively anneal an oxidized silicon surface in a hydrogen or forming-gas (10% $H_2$, 90% $N_2$) atmosphere, the presence of excess hydrogen can cause process-integration problems by enhancing the diffusivity of boron in the oxide (Chap. 7). Annealing in hydrogen or forming gas is frequently used to reduce the interface trap density and stabilize the silicon surface. As discussed earlier, interface traps are associated with dangling bonds, referred to as $P_b$ centers. The $P_b$ centers are "passivated" by a reaction with hydrogen. This can be achieved with thermal $H_2$, forming gas, or in a hydrogen plasma. When the oxide is covered with an aluminum film, the reaction of Al with hydroxyl groups releases atomic hydrogen that migrates to the silicon surface at the anneal temperature. For oxides covered with polysilicon (as in modern MOSFET structures), $H_2$ trapped at the oxide-polysilicon interface is an apparent source of hydrogen [79]. In all cases, the reaction results in H-$P_b$ bonds and the interface trap density is reduced.

**2.2.6 Effects of Oxidation on Silicon**

As the oxide boundary moves into silicon, a small excess of silicon atoms are injected into the crystal, increasing the density of silicon interstitials above their thermal-equilibrium level. Excess interstitials are used to explain the growth of stacking faults and the enhanced diffusivity of some dopants in silicon during its oxidation.

**Oxidation-Induced Stacking Faults (OSF)**

Stacking faults and dislocations can be induced during dry and wet oxidation. Oxidation-induced stacking faults (OSF) are caused by the agglomeration of point defects generated during oxidation [80]. Excess silicon remains in the crystal during oxidation because of the incomplete conversion of silicon into silicon dioxide. Some of the excess silicon atoms occupy interstitial sites and are injected as "self-interstitials" from the oxidizing interface into the bulk of the crystal. Self-interstitials nucleate at strain centers in silicon where they generate extrinsic stacking faults [81, 82]. The strained centers can be created by oxygen or other precipitates, or by mechanical damage. Typical OSF growth times are in the range of hours at temperatures above 1100 °C [83]. Therefore, lowering the oxidation temperature and reducing the oxidation time reduces the size of the stacking fault. The OSF density is found to be larger for p-type wafers than for n-type of the same dopant concentration. This is attributed to the larger concentration of bulk defects and more efficient intrinsic gettering in p-type silicon [84].

**Oxidation-Enhanced Diffusion (OED)**

The diffusivity of substitutional impurities, such as boron, is greatly enhanced during oxidation, even at distances greater than 10 $\mu$m away from the oxidized region [80,85]. This is also attributed to excess silicon interstitials that are injected during oxidation. OED is discussed in more detail in Chap. 7.

## 2.3 Silicon Nitride

The importance of silicon nitride ($Si_3N_4$) to modern technologies stems primarily from its impermeability to most impurities and its unique dielectric and etch properties (Chap. 5). This section describes the formation and properties of silicon nitride films and their use for local oxidation and composite insulators. Other applications are discussed in the following chapters.

### 2.3.1 Thermal Nitridation

Silicon nitride films can be directly grown on silicon by reacting nitrogen or a nitrogen compound, such as ammonia ($NH_3$) with surface silicon atoms at elevated temperature, typically 1000 °C - 1300 °C. The reaction with ammonia is

$$3Si + 4NH_3 \rightarrow Si_3N_4 + 6H_2 \,. \qquad\qquad 2.12$$

The imperviousness of the growing $Si_3N_4$ film to nitridizing species, however, limits its thickness to less than 5 nm, even at the very high temperatures [86]. This is one of the reasons why thermal nitridation is not as frequently used as thermal oxidation. Thermal nitridation is only used for special applications that take advantage of the high dielectric constant of the film (resulting in a high capacitance per unit area) and its impermeability to oxygen. For other applications, the nitride film is deposited rather than grown (Chap. 3).

## 2.3.2 Properties of Silicon Nitride

Some important properties of $Si_3N_4$ are given in Table 2.4. One of the unique properties of silicon nitride is its ability to form a barrier against the migration of most impurities. It prevents ionic contaminants (such sodium and potassium) from moving toward the silicon surface where they can cause serious instabilities in device characteristics.

**Table 2.4** Important Properties of Silicon Nitride (25 °C)

| | |
|---|---|
| Molecular weight | 140.28 g/mole |
| Density | 2.9-3.2 g/cm$^3$ |
| Relative dielectric constant | 6-7 |
| Thermal expansion coefficient | $4 \times 10^{-6}$/K |
| Dielectric strength | $10^7$ V/cm |
| Energy gap | 5 eV |
| Resistivity | $10^{16}$ ohm-cm |
| Infrared absorption band | 11.5 $\mu$m |
| Index of refraction | 2.05 |

The properties of the $Si$-$Si_3N_4$ interface are not as well understood as those of $Si$-$SiO_2$. Direct contact of deposited $Si_3N_4$ with silicon is, however, avoided when the wafer is subjected to elevated temperature because of the observed stress-induced damage caused by the thermal mismatch between the two materials.

### 2.3.3 Local Oxidation of Silicon (LOCOS)

Oxygen and water move very slowly through the nitride. Therefore, when patterned over silicon, the nitride film constitutes an efficient oxidation mask which prevents the oxidants from reaching the silicon surface covered by nitride. In addition, the nitride oxidizes very slowly compared to silicon. These properties are used to locally oxidize silicon and form isolation regions in integrated circuits. Local oxidation of silicon (LOCOS) is widely used to isolate regions in silicon, called active areas, where devices are constructed (Fig. 2.22) [87]. A common variant of LOCOS is to grow a thin (10-50 nm) oxide layer on silicon, referred to "pad oxide", and then deposit a thicker (50-250 nm) nitride film on top of the oxide (Fig. 2.22a). The role of the pad oxide is to relieve the stress transmitted to silicon by the nitride at high temperature. The nitride is patterned by lithography and etch techniques discussed in Chaps. 4 and 5 (Fig. 2.22b). The crystal is then subjected to wet oxidation for 2-4 h. The silicon oxidizes where the nitride has been removed but not in regions covered by nitride (Fig. 2.22c). Typical LOCOS thicknesses are in the range 250 nm - 800 nm.

#### Oxidation of Nitride

During oxidation of silicon a thin layer at the top of the nitride film is converted to silicon dioxide. One criterion for defining the nitride mask thickness is that it should be greater than the thickness that will be consumed during all oxidation cycles. At moderate pressure (1-8 atm), this layer is only about 4% of silicon consumed (Fig. 2.23), so that a relatively thin nitride layer is required to block oxidation [88, 89].

#### Lateral Extent ("Bird's Beak")

Some oxidants move laterally at the edges of the nitride and cause the oxide to grow under the nitride, forming a lateral extension of LOCOS into the active area. Due to the volume expansion of the oxidized region, the nitride is lifted at the periphery of the opening

(Fig. 2.22c). Immediately after oxidation, the lateral extension of LOCOS into the active area is 300-500 nm. This extension is frequently called "bird's beak" because of its form. Since the bird's beak increases the effective size of isolation at the cost of the active area, several methods have been conceived to reduce it's extent. The extent of the bird's beak can be reduced by decreasing the thickness of the pad oxide or eliminating the oxide entirely to avoid oxidation under the nitride film via the thin pad oxide. In the presence of a pad oxide, the bird's beak decreases with increasing rigidity of the nitride and hence with increasing nitride thickness.

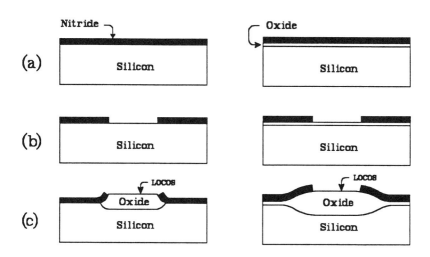

**Fig. 2.22** Local oxidation of silicon (LOCOS). Left: without pad oxide. Right: with pad oxide. (a) Pad nitride deposition. (b) Definition of nitride mask. (c) LOCOS showing the lateral extent of oxidation (bird's beak).

## Sealed Interface

The extent of the bird's beak can be reduced considerably by growing (or depositing) nitride in direct contact with silicon. With complete elimination of any oxide between the nitride film and silicon, a very abrupt transition can be achieved, inhibiting lateral diffusion of oxidants. This "sealed interface local oxidation"

(SILO) can be formed by direct nitridation of silicon in an ammonia ($NH_3$) plasma [90,91].

**Stress and Defects**

As mentioned earlier, direct deposition of nitride on silicon can cause stress-induced defects along the periphery of the local oxide when the structure is subjected to oxidation at elevated temperatures [92 – 96]. The nitride is known to exhibit very high tensile stress which results in a horizontal force along at the periphery of the opening. Defects are created when this force per unit area exceeds the critical stress for dislocation generation.

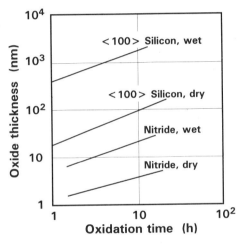

**Fig. 2.23** Comparison of oxidation rates of silicon and silicon nitride [88, 89].

The density of dislocations is considerably reduced by forming a thin pad-oxide film between silicon and nitride. The pad oxide reduces the force transmitted to silicon by relieving the stress through the viscous flow of the oxide. It acts as a buffer which cushions the transition of stress between silicon and nitride. The thicker the pad oxide, the less force is transmitted to the silicon. The oxide layer, however, provides a lateral path for oxidants. As the pad oxide thickness increases, there is more lateral oxidation of silicon, and the nitride loses its effectiveness as a mask. Increasing the nitride thickness reduces the lateral diffusion of oxidants. Therefore, the pad oxide and pad nitride thicknesses must be optimized to minimize the extent of the bird's beak without generating dislocations.

## Polysilicon Buffer LOCOS

Polysilicon buffer LOCOS (PBL) is a modified LOCOS process that uses a stress-relief polysilicon film between the nitride and pad oxide layers [97 – 105]. This allows the deposition of a thick nitride layer (150 - 300 nm) on top of the pad oxide without gener-ating dislocations during oxidation. The polysilicon buffer layer is typically 40-50 nm thick, deposited on a 5 - 10 nm pad oxide layer (Fig. 2.24). Both atmospheric and high-pressure oxidation can be used. PBL results in a lateral extent of only 100 nm, as compared to 300 - 500 nm with conventional LOCOS.

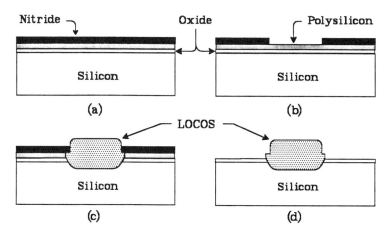

**Fig. 2.24** Polysilicon buffer LOCOS processing steps. (a) Pad oxide growth, deposition of polysilicon buffer layer and pad nitride. (b) Definition of isolation regions. (c) Oxidation. (d) Pad removal [98].

## Limitations of LOCOS

One of the limitations of using LOCOS isolation is the lateral extent discussed in the previous section. The fractional loss of active area due to this "encroachment" increases as the isolation dimensions are reduced. For example, for a minimum isolation spacing of 500 nm, a bird's beak of 100 nm per edge constitutes a

40% increase in isolation area. When the space decreases to 250 nm, the encroachment occupies 80% of the isolation area.

Another important limitation is the sharp decrease in the LOCOS thickness as the isolation spacing is reduced below 1 $\mu$m [106 − 108]. The narrower the opening, the thinner the oxide (Fig. 2.25). The extent of the bird's beak is, however, independent of isolation dimensions [107]. Two mechanisms have been suggested to explain the decrease in the oxidation rate. The first is related to the two-dimensional diffusion of the oxidant and the resulting decrease in its concentration in the narrow nitride window [107]. The second mechanism attributes the oxide thinning to the compressive stress during the two-dimensional oxidation [108]. As the space between isolation regions is reduced, the moving oxide boundaries from the nitride edges push against each other. This, together with the volume expansion during oxidation causes compressive stress, reducing the oxidation rate.

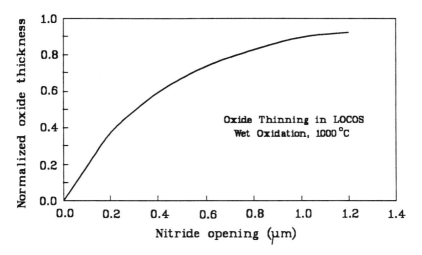

**Fig. 2.25** Dependence of LOCOS thickness on isolation size [108].

The dependence of LOCOS thickness on isolation dimensions can have a great impact on the effectiveness of isolation and on subsequent processing steps in submicrometer and deep submicrometer designs. A more planar isolation technique becomes necessary as the device dimensions are reduced below $\simeq 0.5 \mu m$.

## 2.3.4 Composite Insulators

Thermal oxide is the most common dielectric used for the gate of field-effect transistors and storage capacitors. High quality oxide films can be grown by conventional furnace oxidation, or by rapid thermal oxidation (RTO), as discussed earlier. It is important, however, to define how the quality of an insulator film is measured. In most applications, the insulator must sustain an electric field in the range $5x10^6 - 1.2x10^7$ V/cm. The insulator field-strength is determined by measuring its dielectric breakdown voltage, that is the voltage across the film that results in a specified current density through the structure. The dielectric must also be electrically stable over a long period of time ($\simeq 10$ yr). This requires a high structural uniformity of the film, and allows only little accumulation of trapped charge within the insulator and at its interfaces. The insulator stability is determined by measuring its time-dependent dielectric breakdown (TDDB) and the time-dependent shifts in capacitor and MOSFET characteristics under accelerated stress. Finally, the insulator must be impervious to impurities, such as sodium and boron.

The above requirements are rather difficult to achieve with silicon dioxide alone when the oxide thickness decreases below $\simeq$ 10 nm. Considerable attention is given to composite dielectrics such as "nitrided oxides" (NO) and "oxidized nitrided oxides" (ONO) to form high-quality ultra-thin insulators (< 10 nm). Nitrided oxides (oxynitrides) can be formed by heating thermally oxidized silicon in a nitridizing ambient such as pure ammonia [109 – 111], or in nitrous oxygen ($O_2/N_2O$) [112 – 115]. This produces a very thin nitride layer on top of the oxide. Nitridation occurs predominantly at the oxide surface, and forms a barrier layer which retards further nitridation. Some nitridation can, however, occur at the oxide-silicon interface, as was initially observed in conjunction with LOCOS, resulting in the so-called Kooi effect [116]. Oxynitrides grown in $N_2O$ exhibit similar properties as films grown in ammonia, but are essentially free of hydrogen [112] (one of the deleterious effects of hydrogen is the enhancement of boron diffusivity in silicon dioxide [72]).

Considerable improvement of the insulator quality can be achieved by re-oxidizing the nitrided oxide (ONO). Highly reliable and controllable ONO films can be formed by using rapid thermal processing [113, 117, 118]. As with thermal oxide, RTP allows the growth of composite insulators at elevated temperatures while maintaining a low thermal budget. Typical RTP temperatures range from 1000 °C - 1150 °C, and typical growth times range from 2-30 s.

One side effect of nitridizing a thermal oxide layer in ammonia is the enhanced diffusion of boron and phosphorus in silicon under the nitrided region (Chap. 7). The enhanced diffusion is attributed to the growth of silicon dioxide during nitride formation, resulting in the injection of self-interstitials into the underlying silicon substrate [119 – 122]. It is believed that oxygen is displaced from the $SiO_2$ network by the incoming nitrogen and oxidizes the underlying silicon. Direct nitridation of silicon, however, retards the the diffusion of boron and phosphorus [123]. Both the enhancement and retardation of diffusion must be considered when defining an integrated process technology [124].

The formation of stacked ONO structures for storage capacitors is discussed in the following chapter.

# PROBLEMS

**2.1** A $<100>$ oriented silicon wafer is subjected to 20 min in dry oxygen followed by 20 min in wet oxygen, both at 1100 °C. Find the oxide thickness and the thickness of silicon consumed.

**2.2** The sketch below shows a window in a thin nitride-oxide dual layer defining a region in which silicon is selectively removed to a depth of 300 nm. A wet oxidation step follows at 1000 °C. Estimate the time required for the central part of the oxide in the window to grow to the same level as the flat portions of the nitride surface.

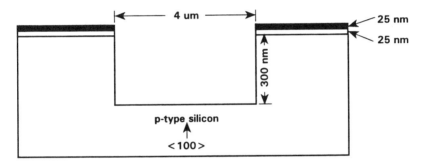

**2.3** An "active" MOSFET area is opened through a $Si_3N_4$ mask and a LOCOS field oxide is grown outside the active region. Then the nitride mask is removed and the wafer receives a thin oxidation step. The field oxide is grown at 1100 °C, and the thin oxide at 900 °C, both in an $H_2O$ ambient. The target thickness is 350 nm for LOCOS and 30 nm for the thin oxide. Assume (100) silicon and $\tau = 0$, and determine:

(a) The minimum nitride thickness required to block field oxidation inside the "active" region.

(b) The duration of the LOCOS step.

(c) The duration of the thin oxide step.

(d) The step height between thin and field oxide.

**2.4** A silicon pillar has the form of a hollow right cylinder with an inner radius of 0.25 $\mu$m and an outer radius of 1 $\mu$m. The top of the cylinder is covered with a 20-nm thick nitride film. Thermal oxide is grown on the surface of the pillar to a thickness of 200 nm. Compare the inner to the outer volume expansion of the pillar. Disregard curvature effects.

**2.5** A wafer is contaminated with one billionth of a gram of sodium ions during the growth of 10 nm thermal oxide on a 200-mm diameter wafer. Assume sodium to be uniformly distributed in the oxide and calculate the sodium ion concentration per unit area and per unit volume of oxide.

**2.6** A hexagonal trench is formed in a $<100>$ oriented wafer. The wafer is then subjected to thermal oxidation that forms a 20-nm thick oxide film on its top surface. How thick is the oxide on the trench surfaces?

# References

[1] S. I. Raider, R. Flitsch, and M. J. Palmer, *"Oxide Growth on Etched Silicon in Air at Room Temperature,"* J. Electrochem. Soc., 122 (3), 413-418 (1975).

[2] W. H. Zachariasen, *"The Atomic Arrangement in Glass,"* J. A. Chem. Soc., 54, 3841 (1932).

[3] C. R. Helms, *"The Atomic and Electronic Structure of the Si-SiO$_2$ Interface,"* The Physics and Chemistry of SiO$_2$ and the Si-SiO$_2$ Interface, C. R. Helms and B. E. Deal, Eds., p. 187, Plenum New York (1988).

[4] M. M. Atalla, E. Tannenbaum and E. J. Scheibner, *"The Silicon - Silicon Dioxide System,"* Bell. System Tech. J., 38, 749 (1959).

[5] P. V. Gray and D. M. Brown, *"Density of SiO$_2$ Interface States,"* Appl. Phys. Lett., 8 (2), 31 (1966).

[6] G. Hollinger, R. Saoudi, P. Ferret, and M. Pitaval, *"The Microstructure of SiO$_2$-Si(100) Interfaces Investigated by XPS and HRTEM,"* in The Physics and Chemistry of SiO$_2$ and the Si-SiO$_2$ Interface, C. R. Helms and B. E. Deal, Eds., p. 211, Plenum Press, New York (1988).

[7] C. R. Helms, *"The Structural and Chemical Properties of the Si-SiO$_2$ Interface,"* Semiconductor Silicon, p. 455 (1981).

[8] A. Ourmazd and J. Bevk, *"The Structure of the Si-SiO$_2$ Interface: A review,"* in The Physics and Chemistry of SiO$_2$ and the Si-SiO$_2$ Interface, C. R. Helms and B. E. Deal, Eds., p. 187, Plenum Press, New York (1988).

[9] E. H. Poindexter, E. R. Ahlstrom, and P. J. Caplan, Proc. Intl. Conf. on the Physics of SiO$_2$ and its Interfaces, S. T. Pantilides, ed., p. 227, Pergamon Press, N. Y. (1978)

[10] N. M. Johnson, D. K. Biegelsen, M. D. Moyer, S. T. Chang, E. H. Poindexter, and P. J. Caplan, *"Electronic Traps and P$_b$ Centers at the Si/SiO$_2$ Interface: Band-Gap Energy Distribution,"* Appl. Phys., 56 (10), 2844-2849 (1984).

[11] B. E. Deal and A. S. Grove, *"General Relationship for the Thermal Oxidation in Silicon,"* J. Appl. Phys. 36 (12), 3770-3778 (1965).

[12] S. K. Ghandhi, VLSI Fabrication Principles, John Wiley and Sons, New Yor (1983).

[13] A. G. Revesz, B. J. Mrstik, H. L. Hughes, and D. McCarthy, *"Structure of SiO$_2$ Film as Revealed by Oxygen Transport,"* J. Electrochem. Soc., 133 (3), 586-592 (1986).

[14] S. M. Hu, *"Thermal oxidation of Silicon: Chemisorption and Linear Rate Constants,"* J. Appl. Phys., 55 (11), 4095-4105 (1984).

[15] Y. J. Van Der Meulen, *"Kinetics of Thermal Growth of Ultra-Thin Layers of SiO$_2$ on Silicon. I. Experiment,"* J. Electrochem. Soc., 119 (4), 530-534 (1972).

[16] M. A. Hopper, R. A. Clarke, and L. Young, *"Thermal Oxidation of Silicon. In Situ Measurement of the Growth Rate Using Ellipsometry,"* J. Electrochem. Soc., 122 (9), 1216-1225 (1975).

[17] C. J. Han and C. R. Helms, *"Parallel Oxidation Mechanism for Si Oxidation in Dry O$_2$,"* J. Electrochem. Soc., 134 (5), 1297-1302 (1987).

[18] H. Z. Massoud, J. D. Plummer, and E. A. Irene, *"Thermal Oxidation of Silicon in Dry Oxygen: Growth Rate Enhancement in Thin Regime. I. Experimental Results,"* J. Electrochem. Soc., 132 (11), 2685-2693 (1985).

[19] P. M. Fahey, P. B. Griffin, and J. D. Plummer, *"Point Defects and Dopant Diffusion in Silicon,"* Reviews of Modern Physics, 61 (2), 289-384 (1989).

[20] R. Singh, *"Rapid Isothermal Processing,"* J. Appl. Phys., 63 (8), R59 (1988).

[21] J. Nulman, J. P. Krusius, and A. Gat, *"Rapid-Thermal Processing of Thin Gate Dielectrics. Oxidation of Silicon,"* IEEE Electron. Dev. Lett., EDL-6 (5), 205 (1985).

[22] J. Nulman, J. Scarpulla, T. Mele, and J. P. Krusius, *"Electrical Characteristics of Thin Gate Implanted MOS Channels Grown by Rapid Thermal Processing,"* IEDM Tech. Dig., p. 376 (1985).

[23] M. M. Moslehi, S. C. Shatas, and K. C. Saraswat, *"Thin SiO$_2$ Insulators Grown by Rpaid Thermal Oxidation of Silicon,"* Appl. Phys. Lett., 47 (12), 1353 (1985).

[24] A. M. Hodge, C. Pickering, A. J. Pidduck, and R. W. Hardeman, *"Silicon Oxidation by Rapid Thermal Processing (RTP),"* in Rapid Thermal Processing, Materials Research Soc. Symp. Proc., T. O. Sedgwick, T. E. Seidel, and B. Y. Tsauer, Eds., 52, 313 (1986).

[25] A. G. Associates, *"Rapid Thermal Oxidation Systems,"* Solid-State Technology, 29 (8), 167-178 (1986).

[26] H. Fukuda, T. Arakawa, and S. Ohno, *"Highly Reliable Ultra-Thin SiO$_2$ Films Formed by Rapid Thermal Processing,"* IEDM Tech. Dig., 451 (1989).

[27] A. Mauri, S. C. Kim, A. Manocha, K. H. Oh, D. Kostelnick, and S. Shive, *"Application of Rapid Thermal Oxidation to the Development of High Dielectric Strength Polyoxides,"* IEDM Tech. Dig., 676 (1986).

[28] C. Su, *"Low Temperature Silicon Processing Techniques for VLSI Fabrication,"* Solid-State Technology, 24 (3), 72-82 (1981).

[29] L. N. Lie, R. R. Razouk, and B. E. Deal, *"High Pressure Oxidation of Silicon in Dry Oxygen,"* J. Electrochem. Soc., 129 (12), 2828-2834 (1982).

[30] R. R. Razouk, L. N. Lie and B. E. Deal, *"Kinetics of High Pressure Oxidation of Silicon in Pyrogenic Steam,"* J. Electrochem. Soc., 128 (10), 2214-2220 (1981).

[31] M. Hirayama, H. Miyoshi, N. Tsubuchi, and H. Abe, *"High Pressure Oxidation for Thin Gate Insulator Process,"* IEEE Trans. Electron Dev. ED-29 (4), 503-507 (1982). Soc., 128, 2214 (1981).

[32] M. L. Reed and J. D. Plummer, *"Si-SiO$_2$ Trap Production by Low-Temperature Thermal Processing,"* Appl. Phys. Lett., 51 (7), 514 (1987).

[33] J. R. Ligenza, *"Silicon Oxidation in an Oxidation Plasma Excited by Microwaves,"* J. Appl. Phys., 36 (9), 2703-2707 (1965).

[34] A. M. Hoff and J. Ruzyllo, *"Atomic Oxygen and the Thermal Oxidation of Silicon,"* Appl. Phys. Lett., 52 (15), 1264 (1988).

[35] R. Ghez and Y. J. Van der Meulen, *"Kinetics of Thermal Growth of Ultra-Thin lyaers of SiO$_2$ on Silicon,"* J. Electrochem. Soc., 119 (8), 1100-1106 (1972)

[36] J. Blanc, *"A Revised Model for the Oxidation of Silicon by Oxygen,"* Appl. Phys. Lett., 33 (5), 424 (1978)

[37] E. A. Irene, *"Evidence for a Parallel Path Oxidation Mechanism at the Si-SiO$_2$ Interface,"* Appl. Phys., Lett., 40 (1), 74 (1982)

[38] R. B. Marcus and T. T. Sheng, *"The Oxidation of Shaped Silicon Surfaces,"* J. Electrochem. Soc., 129 (6), 1278-1282 (1982).

[39] D. J. Chin, S. Y. Oh, S. M. Hu, R. W. Dutton, and J. L. Moll, *"Two-Dimension Oxidation Model,"* IEEE Trans. Electron Dev., ED-30 (7), 744-749, (1983).

[40] D. B. Kao, J. P. McVittie, W. D. Nix, and K. C. Saraswat, *"Two-Dimensional Silicon Oxidation Experiments and Theory,"* IEDM Tech. Dig., 388 (1985).

[41] P. Saturdja and W. G. Oldham, *"Modeling of Stress-Effects in Silicon Oxidation Including the Non-Linear Viscosity of Oxide,"* IEDM Tech. Dig., 264 (1987).

[42] D. B. Kao, J. P. McVittie, W. D. Nix, and K. C. Saraswat, *"Two-Dimensional Thermal Oxidation of Silicon - II. Modeling Stress Effects in Wet Oxides,"* IEEE Trans. Electron Dev., ED-35 (1), 25-37 (1988).

[43] J. R. Ligenza, *"Effect of Crystal Orientation on the Oxidation Rates in High Pressure Steam,"* Phys. Chem. 65, 2011 (1961)

[44] W. A. Pliskin, *"Separation of the Linear and Parabolic Terms in the Steam Oxidation of Silicon,"* IBM J. Res. Dev., 10, 198 (1966).

[45] E. A. Irene, H. Z. Massoud, and E. Thierny, *"Silicon Oxidation Studies: Silicon Orientation Effects on Thermal Oxidation,"* J. Electrochem. Soc., 133 (6), 1253-1260 (1986).

[46] E. A. Lewis and E. A. Irene, *"The Effect of Surface Orientation on Silicon Oxidation Kinetics,"* J. Electrochem. Soc., 134 (9), 2332-2339 (1987).

[47] R. J. Kriegler, Y. C. Cheng and D. R. Colton, *"The Effect of HCl and $Cl_2$ on the Thermal Oxidation of Silicon,"* J. Electrochem. Soc., 119, 388-392 (1962).

[48] Y. J. Van Der Meulen, C. M. Osburn and J. F. Ziegler, *"Properties of $SiO_2$ Grown in the Presence of HCl or $Cl_2$,* J. Electrochem. Soc., 122, 284 (1975).

[49] S. P. Murarka, *"Role of Point-Defects in the Growth of Oxidation-Induced Stacking Faults in Silicon. II. Retrogrowth, Effect of HCl Oxidation and Orientation,"* Phys. Rev. B, 21, 892 (1980).

[50] M. C. Chen and J. W. Hile, *"Oxide Charge Reduction by Chemical Gettering with Trichloroethylene during Thermal Oxidation,"* J. Electrochem. Soc., 119, 223 (1972).

[51] E. J. Janssens and G. J. Declerck, *"The Use of 1,1,1,-Trichloroethane as an Optimized Additive to Improve the Silicon Thermal Oxidation Technology,"* J. Electrochem. Soc., 125 (10), 1696-1703 (1978).

[52] D. W. Hess and B. E. Deal, *"Kinetics of the Thermal Oxidation of Silicon in $O_2/HCl$ Mixtures,"* J. Electrochem. Soc., 124 (5), 735-740 (1977).

[53] J. Monkowski, *"Role of Chlorine in Silicon Oxidation,"* Solid-State Technology, 22 (7), 58-61 (1979).

[54] B. R. Singh and P. Balk, *"Thermal Oxidation of Silicon in $O_2$-Trichloroethylene,"* J. Electrochem. Soc., 126 (7), 1288-1294 (1979).

[55] K. Ehara, K, Sakuma, and K. Ohwada, *"Kinetics and Oxide Properties of Silicon Oxidation in $O_2 - H_2 - HCl$ Mixtures,"* J. Electrochem. Soc., 126 (12), 2249-2254 (1979).

[56] K. Hirabayashi and J. Iwamura, *"Kinetics of Thermal Growth of $HCl$-$O_2$ Oxides on Silicon,"* J. Electrochem. Soc., 120 (11), 1595-1601 (1973).

[57] Z. M. Ling, L. H. Dupas, and K. M. De Meyer, *"Modeling of the Oxide Growth in a Chlorine Ambient,"* in "The Physics and Chemistry of $SiO_2$ and the $Si$-$SiO_2$ Interface," C. R. Helms and B. E. Deal, Eds., p. 53, Plenum Press, New York (1988).

[58] M. Morita, S. Aritome, M. Tsukude, T. Murakawa, and M. Hirose, *"Low-Temperature $SiO_2$ Growth Using Fluorine-Enhanced Thermal Oxidation,"* Appl. Phys. Lett., 47 (3), 253 (1985).

[59] B. R. Weinberger, G. G. Peterson, T. C. Eschrich, and H. A. Krasinski, *"Surface Chemistry of HF Passivated Silicon: X-Ray Photoelectron and Ion Scattering Spectroscopy Results,"* J. Appl. Phys., 60 (9), 3232-3234 (1986).

[60] E. F. Da Silva, Jr., Y. Nishioka, and T. P. Ma, *"Radiation and Hot-Electron Hardened MOS Structures,"* IEDM Tech. Dig., 848 (1987).

[61] Y. Nishioka, E. F. Da Silva, Jr., Y. Wang, and T. P. Ma, *"Dramatic Improvement of Hot-Electron-Induced Interface Degradation in MOS Structures Containing F or Cl in $SiO_2$,"* IEEE Electron Dev. Lett., EDL-9 (1), 38 (1988)

[62] J. Kuehne, W. Ting, G. Q. Lo, T. Y. Hsieh, D. L. Kwong, and C. W. Magee, *"Radiation and Hot-Electron Effects in MOS Structures with Gate Dielectrics Grown by Rapid Thermal Processing in $O_2$ Diluted with $NF_3$,"* Proc. 6th Intnl. Symp. on Silicon Mater. Sci. Techn., Semiconductor Silicon, 364-375 (1990).

[63] U. S. Kim and R. J. Jaccodine, *"Fluorine Enhanced Oxidation of Silicon Related Phenomena,"* Extended Abstracts, The Electrochem. Soc., 179th Meeting, p. 376, May 5-10 (1991).

[64] P. J. Wright and K. C. Saraswat, *"The Effect of Fluorine in Silicon Dioxide Gate Dielectrics,"* IEEE Trans. Electron Dev., ED-36 (5), 879-889 (1989).

[65] M. Ghezzo and D. M. Brown, *"Diffusivity Summary of B, Ga, P, As and Sb in $SiO_2$,"* J. Electrochem. Soc., 120 (1) 146-148 (1973).

[66] B. E. Deal and M. Sklar, *"Thermal Oxidation of Heavily Doped Silicon,"* J. Electrochem. Soc., 112 (4), 430-435 (1965).

[67] C. P. Ho, J. D. Plummer, J. D. Meindl and B. E. Deal, *"Thermal Oxidation of Heavily Phosphorus Doped Silicon,"* J. Electrochem. Soc., 125 (4), 665-671 (1978)

[68] E. A. Irene and D. W. Dong, *"Silicon Oxidation Studies: The Oxidation of Heavily B- and P-Doped Single Crystal Silicon,"* J. Electrochem. Soc., 125 (7), 1146-1151 (1978).

[69] C. P. Ho and J. D. Plummer, *"Si/$SiO_2$ Interface Oxidation Kinetics: A Physical Model for the Influence of High Substrate Doping Levels. I. Theory, II. Comparison with Experiment and Discussion,"* J. Electrochem. Soc., 126 (9), 1516-1528 (1979).

[70] E. A. Irene and R. Ghez, *"Silicon Oxidation Studies: The Role of $H_2O$,"* J. Electrochem. Soc., 124 (11), 1757-1761 (1977).

[71] A. G. Revesz, *"The Defect Structure of Grown Silicon-Dioxide Films,"* IEEE Trans. Electron Dev., ED-12 (3), 97-102 (1965).

[72] J. Y. C. Sun, C. Wong, Y. Taur, C. H. Hsu, *"Study of Boron Penetration Through Thin Oxide with $P^+$ Polygate,"* VLSI Tech. Dig., 17 (1989).

[73] E. H. Snow, A. S. Grove, B. E. Deal, and C. T. Sah, *"Ion Transport Phenomena in Insulating Films,"* J. Appl. Phys., 36 (5), 1664-1673, (1965).

[74] G. F. Derbenwick, *"Mobile Ions in $SiO_2$: Potassium,"* J. Appl. Phys., 48 (3), 1127-1130, (1977).

[75] J. P. Stagg, *"Drift Mobilities of $Na^+$ and $K^+$ in $SiO_2$ Films,"* Appl. Phys. Lett., 31 (8), 532 (1977)

[76] D. P. Kennedy, P. C. Murley, and M. Kleinfelder, *"On the Measurement of Impurity Atom Distributions in Silicon by the Differential Capacitance Technique,"* IBM J. res. Dev., 12, 399 (1968).

[77] A. G. Tangena, N. F. De Rooij, and J. Middelhoek, *"Sensitivity of MOS Structures to Contamination with $H^+$, $Na^+$, and $K^+$ Ions,"* J. Appl. Phys., 49 (11), 5576-5583 (1978).

[78] J. S. Logan and D. R. Kerr, Sol. State Res. Conf., New Jersey (1965).

[79] K. L. Brower, *"Chemical Kinetics of Hydrogen and $P_b$ Centers,"* in The Physics and Chemistry of $SiO_2$ and the Si-$SiO_2$ Interface, C. R. Helms and B. E. Deal, Eds., p. 309 Plenum Press, New York (1988).

[80] S. M. Hu, *"Formation of Stacking Faults and Enhanced Diffusion in the Oxidation of Silicon,"* J. Appl. Phys., 45, 1567 (1974).

[81] R. B. Fair, *"Oxidation, Impurity Diffusion and Defect Growth in Silicon - An Overview,"* J. Electrochem. Soc., 128 (6) 1360-1368 (1981).

[82] S. M. Hu, *"Kinetics of Interstitial Supersaturation and Enhanced Diffusion in Short-Time/Low-Temperature Oxidation of Silicon,"* J. Appl. Phys., 57 (10), 4527-4532 (1985).

[83] R. B. Fair, *"Oxidation-Induced Defects and Effects in Silicon During Low Thermal-Budget Processing,"* in *"The Physics and Chemistry of SiO2 and the Si-SiO2 Interface,"* C. R. Helms and B. E. Deal, Eds., p. 459, Plenum Press, New York (1988).

[84] P. W. Koob, G. K. Fraundorf, and R. A. Craven, *"Reduction of Surface Stacking Faults on N-Type (100) Silicon Wafers,"* J. Electrochem. Soc., 113 (4), 806-810 (1986).

[85] P. S. Dobson, *"The Effect of Oxidation on Anomalous Diffusion in Silicon,"* Philosophical Mag., 24, 567-576 (1971).

[86] T. Ito, S. Hijiya, T. Nozaki, H. Arakawa, M. Shinoda, and Y. Fukukawa, *"Very Thin Silicon Nitride Films Grown by Direct Thermal Reaction with Nitrogen,"* J. Electrochem. Soc., 125 (3), 448-452 (1978).

[87] J. A. Appels, E. Kooi, M. M. Pfaffen, J. J. H. Shototje, and W. H. C. G. Verkuylen, *"Local Oxidation of Silicon and Its Applications in Semiconductor Device Technology,"* Philips Res. Repts., 25, 118 (1970).

[88] I. Fraenz and W. Langheinrich, *"Conversion of Silicon Nitride into Silicon Dioxide Through the Influence of Oxygen,"* Solid-State Electronics, 14, 499 (1971).

[89] M. Miyochi, N. Tsubouchi, and A. Nishimito, *"Selective oxidation of Silicon in High Pressure Steam,"* J. Electrochem. Soc., 125 (11), 1824-1829 (1978).

[90] J. Hui, T. Y. Chiu, S. Wong, and W. G. Oldham, *"Selective Oxidation Technologies for High Density MOS,"* IEEE Electron Dev. Lett., EDL-2 (10), 244 (1981).

[91] J. Hui, T. Y. Chiu, S. Wong, and W. G. Oldham, *"Sealed Interface Local Oxidation Technology,"* IEEE Trans. Electron Dev., ED-29 (4), 554-561 (1982).

[92] E. Bassous, H. N. Yu, and V. Maniscalco, *"Topology of Silicon Structures with Recessed SiO2,"* J. Electrochem. Soc., 123 (11), 1729-1737 (1976).

[93] A. Bohg and A. K. Gaind, *"Influence of Film Stress and Thermal Oxidation on the Generation of Dislocations in Silicon,"* Appl. Phys. Lett., 33 (10), 895 (1978).

[94] P. Deroux-Dauphin and J. P. Gonchon, *"Physical and Electrical Characterization of SILO Isolation Structure,"* Trans. Electron Dev., ED-32 (11), 2392-2398 (1985).

[95] K. Shibata and K. Taniguchi, *"Generation Mechanism of Dislocations in Local Oxidation of Silicon,"* J. Electrochem. Soc., 127 (6), 1383-1387 (1980).

[96] Y. Tamaki, S. Isomae, S. Mizuo, and H. Higuchi, *"Evaluation of Dislocation Generation on Silicon Substrate by Selective Oxidation,"* J. Electrochem. Soc., 130 (11), 2266-2270 (1983).

[97] R. V. Havemann and G. P. Pollack, U.S. Patent No. 4, 541, 167 (1986).

[98] R. A. Chapman, R. A. Haken, D. A. Bell, C. C. Wei, R. V. Havemann, T. E. Tamg, T. C. Holloway, and R. J. Gale, *"An 0.8 μm CMOS Technology for High Performance Logic Applications,"* IEDM Tech. Dig., 362 (1987).

[99] T. Nishihara, K. Tokunaga, and K. Kobayashi, *"A 0.5 μm Isolation Technology Using Advanced Polysilicon Pad LOCOS (APPL),"* IEDM Tech. Dig., 100 (1986).

[100] J. M. Sung, C. Y. Lu, and K. H. Lee, *"The Impact of Poly Removal Techniques on Thin Thermal Oxide Property in Poly Buffer LOCOS Technology,"* IEEE Trans. Electron Dev., ED-38 (8), 1970-1973 (1991).

[101] N. Hoshi, S. Kayama, T. Nishihara, J. Aoyama, T. Komatsu, and T. Shimada, *"1.0 μm CMOS Process for Highly Stable Terra-Ohm Polysilicon Load 1 Mb SRAM,"* IEDM Tech. Dig., 300 (1986).

[102] Y. Han and B. Ma, *"Isolation Process Using Polysilicon Buffer Layer for Scaled MOS/VLSI,"* Extended Abstracts 84-1, Electrochem. Soc., p. 98 (1984).

[103] N. A. H. Wils and A. H. Montree, *"A New Sealed Poly Buffer LOCOS Isolation Scheme,"* Microelectronic Eng. (Netherland), 15, 643 (1991).

[104] N. Hoshi, *"An Improved LOCOS Technology Using Thin Oxide and Polysilicon Buffer Layers,"* Sony Research center Reports, 23, 300 (1984).

[105] J. M. Sung, C. Y. Lu, B. Fritzinger, T. T. Sheng, and K. H. Lee, *"Reverse L-Shape Scaled Poly Buffer LOCOS Technology,"* IEEE Electron Dev. Lett., EDL-11 (11), 549 (1990).

[106] J. Hui, P. V. Voorde, and J. Moll, *"Scaling Limitations of Submicron Local Oxidation technology,"* IEDM Tech. Dig., 392 (1985).

[107] T. Mizuno, S. Sawada, S. Maeda, and S. Shinozaki, *"Oxidation Rate Reduction in the Submicrometer LOCOS Process,"* IEEE Trans. Electron Dev., ED-34 (11), 2255-2259 (1987).

[108] J. W. Lutze, A. H. Parera, and J. P. Krusius, *"Field Oxide Thinning in Poly Buffer LOCOS Isolation with Active Area Spacing to 0.1 μm,"* J. Electrochem. Soc., 137 (6), 1867-1870 (1990).

[109] T. Ito, T. Nakamura, and H. Ishikawa, *"Advantages of Thermal Nitride and Nitroxide Gate Films in VLSI Process,"* IEEE Trans. Electron Devices, ED-29 (4), 498-502 (1982).

[110] N. L. Naiman, F. L. Terry, J. A. Burns, J. I. Raffael, and R. Aucoin, *"Properties of Thin Oxinitride Gate Dielectrics Produced by Thermal Nitridation of Silicon Dioxide,"* IEDM Tech. Dig., 562 (1980).

[111] R. Jayaraman, W. Yang, and C. C. Sodini, *"MOS Electrical Characteristics of Low-Pressure Re-Oxidized Nitrided-Oxides,"* IEDM Tech. Dig., 668 (1986).

[112] A. Uchiyama, H. Fukuda, T. Hayashi, T. Iwabuchi, and S. Ohno, *"High Performance Dual-Gate Sub-Halfmicron CMOSFETs with 6-nm Thick Nitrided $SiO_2$ Films in an $N_2O$ Ambient,"* IEDM Tech. Dig., 425 (1990).

[113] H. Hwang, W. Ting, D. L. Kwong, and J. Lee, *"Electrical and Reliability Characteristics of Ultrathin Oxynitride Gate Dielectric Prepared by Rapid Thermal Processing in $N_2O$,"* IEDM Tech. Dig., 421 (1990).

[114] H. Fukuda, M. Yasuda, T. Iwabuchi, and S. Ohno, *"Novel $N_2O$-Oxynitridation Technology for Forming Highly Reliable EEPROM Tunnel Oxide Films,"* IEEE Electron Device Lett., 12 (11), 587 (1991).

[115] C. Q. Lo, W. Ting, J. Ahn, and D.-L. Kwong *"P-Channel MOSFETs with Ultrathin $N_2O$ Gate Oxide,"* IEEE Electron Device Lett., 13 (2), 111 (1992).

[116] E. Kooi, J. E. VanLierop, and J. A. Appels, *"Formation of Silicon Nitride at an Si-$SiO_2$ Interface during Local Oxidation of Silicon and during Heat Treatment of Oxidized Silicon in $NH_3$ Gas,"* J. Electrochem. Soc., 123 (7), 1117-1120 (1976).

[117] T. Hori and H. Iwasaki, *"Ultra-Thin Re-Oxidized Nitrided-Oxides Prepared by rapid Thermal Processing,"* IEDM Tech. Dig., 570 (1987).

[118] T. Hori and H. Iwasaki, *"The Impact of Ultrathin Nitrided Oxide Gate-Dielectrics on MOS Device Performance Improvement,"* IEDM Tech. Dig., 459 (1989).

[119] S. Mizuo, T. Kusaka, A. Shintani, M. Nanba, and G. Higuchi, *"Effects of Si and SiO₂ Thermal Nitridation on Impurity Diffusion and Oxidation-Induced Stacking Fault Size in Si,"* J. Appl. Phys., 54, (7) 3860-3866 (1983).

[120] S. T. Dunham, *"Interstitial Kinetics Near Oxidizing Silicon Interfaces,"* J. Electrochem. Soc., 136 (1), 250-254 (1989).

[121] M. Moslehi and K. Saraswat, *"Thermal Nitridation of Si and SiO₂ for VLSI,"* IEEE Trans. Electron Devices, ED-32 (2), 106-123 (1985).

[122] P. Fahey and P. Griffin, *"Investigation of the Mechanism of Si Self-Interstitial Injection from Nitridation of SiO₂ Films,"* Extended Abstracts, The Electrochem. Soc., 179th Meeting, p.486, May 5-10 (1991).

[123] P. Fahey, R. W. Dutton, and M. Moslehi, *"Effects of Thermal Nitridation Processes on Boron and Phosphorus Diffusion in (100) Silicon,"* Appl. Phys. Lett., 43 (7), 683 (1983).

[124] M. Moslehi, R. A. Chapman, M. Wong, A. Paranjpe, H. N. Najm, J. Kuehne, R. L. Yeakley, and C. J. Davis, *"Single-Wafer Integrated Semiconductor Device Processing,"* IEEE Trans. Electron Devices, 39 (11), 4-32 (1992).

# Chapter 3

# Thin Film Deposition

## 3.0 Introduction

The controlled deposition of thin organic and inorganic films is an important step in the manufacture of integrated circuits. These films are deposited to remain as an inherent part of the device structure (Fig. 3.1), or to constitute intermediate layers that are used for particular processing steps and then removed. The methods for the deposition of thin films fall into three broad categories: chemical vapor deposition (CVD), physical vapor deposition (PVD), and overlapping techniques which combine both physical and chemical processes [1,2]. This chapter discusses the deposition of single crystal silicon, polysilicon, silicon dioxide, silicon nitride, and gallium arsenide. The deposition of metals and metal compounds is discussed in Chap. 8.

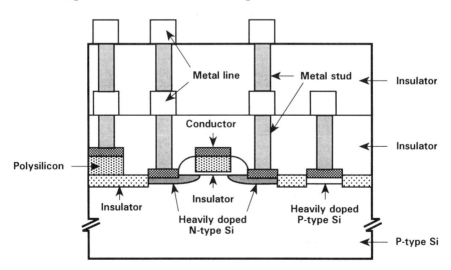

**Fig. 3.1** Schematic cross-section of a typical two-level metal integrated circuit element.

## 3.1 Chemical Vapor Deposition

Chemical vapor deposition is the process in which a film is deposited by a chemical reaction or decomposition of a gas mixture at elevated temperature at the wafer surface or in its vicinity. Typical examples of CVD films are single crystal silicon (epitaxy), polycrystalline silicon, silicon-dioxide, silicon-nitride, composite dielectrics, phosphosilicate glass (PSG), borosilicate glass (BSG), borophosphosilicate glass (BPSG), compound semiconductors (such as gallium arsenide), metals, and metal compounds.

CVD can be performed at atmospheric pressure (APCVD), or at low pressure (LPCVD). LPCVD is rapidly replacing conventional APCVD because of lower cost and superior film properties obtained at low pressure, as discussed later in this section [3]. Three important methods to enhance the deposition rate and reduce the deposition temperature are discussed in section 3.2. These are plasma-enhanced CVD (PECVD), photo-enhanced CVD (photo-CVD) and electron cyclotron resonance plasma (ECR plasma).

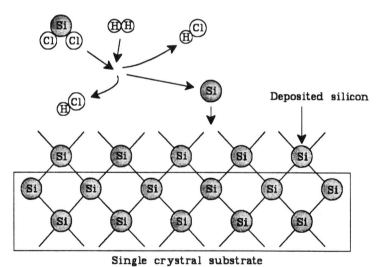

**Fig. 3.2** Schematic description of silicon epitaxial deposition on a silicon substrate.

### 3.1.1 Epitaxy

Epitaxy is a high temperature CVD process where a single crystal layer is deposited on a single crystal substrate. The word *epitaxy* comes from two Greek words: *epi*, which means "upon", and *taxis*, which means "arranged". Epitaxy is thus the arrangement of atoms on an ordered substrate which acts as the seed crystal. The atoms produced by the gas reaction impinge on the substrate surface and "move around" until they find the correct location to bond to the surface atoms, forming a layer of the same crystallographic arrangement as the substrate (Fig. 3.2). The purpose of epitaxy is to deposit a single crystal layer having a dopant type and concentration independent of the dopant type and concentration of the substrate. The layer can be deposited free of oxygen and carbon, the two most common contaminants in Czochralski grown crystals.

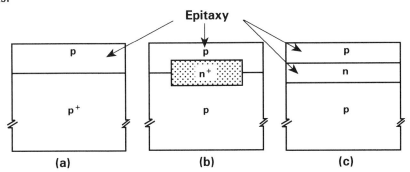

**Fig. 3.3** Three typical applications of epitaxial deposition. (a) Lightly doped layer deposited on a heavily doped substrate. (b) Heavily doped layer buried in a lightly doped region of opposite polarity. (c) Abrupt change in polarity between thin epitaxial layers.

While silicon epitaxy was first developed to enhance the performance of bipolar transistors [4], epitaxial films are now widely used for other device applications in CMOS and mixed bipolar and CMOS (BiCMOS) technologies. Three typical configurations are illustrated in Fig. 3.3. In Fig. 3.3a, a lightly doped epitaxial layer is deposited on a heavily doped substrate. The lightly doped layer is the region where active devices are constructed and the heavily doped substrate constitutes a low resistance circuit path. Figure 3.3b shows how a heavily doped layer is "buried" in a region of lower dopant concentration and of opposite polarity. The heavily doped layer is first defined and formed in the substrate using lithography, etching, and doping techniques discussed in the following chapters. This layer becomes "buried" after a lightly doped epitaxial film is deposited on the substrate. Figure 3.3c shows thin epitaxial layers which change abruptly from one polarity to the other; the precise control of layer thickness and dopant concentration allows the design of very high-speed devices.

Depending on the application and deposition conditions, the epitaxial thickness can vary from 2 nm to over 100 $\mu m$. When the deposited layer is of the same material as the substrate (such as silicon deposited on silicon), one speaks of **homoepitaxy.** When the layer and the substrate are not of the same composition, the deposition is called **heteroepitaxy.** In heteroepitaxy the crystallographic arrangement in the deposited layer and the substrate may be slightly different. An example of heteroepitaxy is the deposition of gallium arsenide (GaAs) over silicon. The deposition of GaAs is, however, complicated by the requirement that more than one element must be co-deposited to form a film of stoichiometric composition.

**Deposition Techniques**

The major techniques to grow epitaxy are liquid-phase epitaxy (LPE), molecular-beam epitaxy (MBE), and vapor phase epitaxy (VPE). In LPE, the growth occurs from a saturated liquid melt which is in contact with the substrate. This technique is widely used for the deposition of GaAs layers for optoelectronic devices [5]. A typical LPE system consists of a graphite holder sliding a sample between melts of different compositions (Fig. 3.4).

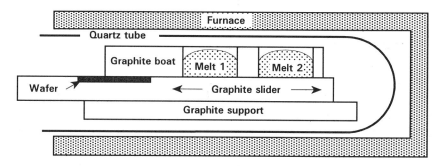

**Fig. 3.4** Graphite slider LPE system.

The epitaxial deposition results from the controlled cooling of the supersaturated melt. The main disadvantage of LPE is the nonuniformity in large-area layers. This method is therefore not considered for large-scale integration. In MBE, the material is evaporated from the melt and deposited on a heated substrate, as further detailed in section 3.2. The most common method for the deposition of silicon and GaAs layers is the vapor phase epitaxy discussed in this section. Silicon VPE is done by passing a gaseous compound of silicon over single crystal silicon wafers that are held at 800 °C - 1150 °C. GaAs VPE uses typically an organometallic gallium source and arsine ($AsH_3$) as reactants.

**Surface Preparation and Epitaxial Defects**

Epitaxial layers essentially replicate the same crystallographic defects as the substrate on which they are grown. Also, the presence of foreign material or oxide residues on the substrate usually creates stacking faults, the most important crystallographic defects in epitaxial films [6 − 8]. Therefore, a defect-free, single-crystal layer can only be deposited on a clean and damage-free substrate. To remove the damage caused by mechanical polishing, the wafer is subjected to a gas mixture of typically 0.1% HCl in $H_2$ above 1150 °C. This ensures a low defect density by removing the damaged silicon layer (at an approximate rate of 1.5 $\mu$m/min) following the reaction:

$$Si + 2HCl = SiCl_2\uparrow + H_2\uparrow \qquad 3.1$$

Since a layer of native oxide always forms on bare silicon exposed to air for a prolonged period, it is important to remove this oxide film prior to epitaxial deposition. This is achieved inside the reaction chamber, just before starting the deposition, by "baking" the wafers in the presence of hydrogen at typically 1150 °C [9]. This converts oxide residues to volatile SiO by the reaction:

$$SiO_2 + H_2 = SiO\uparrow + H_2O\uparrow . \qquad 3.2$$

It is also important to reduce the water concentration to less than one part per million (ppm), since otherwise silicon dioxide can form at the surface when the temperature approaches 900 °C, following the reaction:

$$Si + 2H_2O = SiO_2\downarrow + 2H_2\uparrow, \qquad\qquad 3.3$$

## Reactions

The most common reaction used for silicon epitaxial deposition is the hydrogen reduction of silicon tetrachloride ($SiCl_4$), silane ($SiH_4$), dichlorosilane (DCS, $SiH_2Cl_2$), or trichlorosilane (TCS, $SiHCl_3$). Typical deposition temperatures and rates are shown in Table 3.1. For depositions at temperatures in the range 800-1050 °C, the pyrolitic decomposition of of silane can also be used [10]. The gas mixture is passed above the wafer holder, also called **susceptor,** at a controlled flow rate. A boundary layer of reduced gas velocity, also referred to as the **stagnant layer,** forms above the wafers. It is across this boundary layer that the reactants are transported to the wafer surface; the reaction by-products diffuse back through the boundary and are removed by the main gas stream (Fig. 3.5).

## Hydrogen Reduction of Silicon Compounds

This method introduces a gas mixture of silicon-chlorine compounds and hydrogen into the epitaxial system at elevated temperature. In all cases, the reaction in the gas phase results in the formation of hydrogen ($H_2$), hydrochloric acid (HCl) and silicon dichloride ($SiCl_2$). Silicon dichloride is then adsorbed by the hot silicon surface where it either decomposes to form silicon and silicon tetrachloride, or it is reduced by hydrogen to form silicon and HCl. Beginning with silicon tetrachloride, the chain of reactions can be summarized as [11]:

$$SiCl_4 + H_2 = SiHCl_3 + HCl\uparrow \qquad\qquad 3.4$$

$$SiHCl_3 + H_2 = SiH_2Cl_2 + HCl\uparrow$$

$$SiH_2Cl_2 = SiCl_2 + H_2\uparrow$$

$$SiHCl_3 = SiCl_2 + HCl\uparrow.$$

$$SiCl_2 + H_2 = Si\downarrow + 2HCl\uparrow , \qquad\qquad 3.5$$

or

$$2SiCl_2 = Si\downarrow + SiCl_4\uparrow .$$

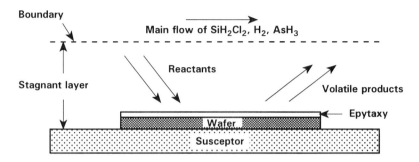

**Fig. 3.5** Transport of reactants and reaction by-products through a boundary layer of reduced velocity (stagnant layer) in an epitaxial system.

The substrate acts as a catalyst for the last two reactions. TCS and DCS are found as intermediate compounds to the overall reaction. The deposition of silicon could therefore start with these compounds rather than with silicon tetrachloride. Growth with DCS has the lowest activation energy, allowing epitaxial deposition at a lower temperature than with other compounds and resulting in a smaller thermal budget (Tab. 3.1) [12].

**Table 3.1** Epitaxial growth by hydrogen reduction of silicon compounds [12 – 14].

| Silicon compound | Growth rate ($\mu$m/min) | Activation energy (eV) | Temperature range (°C) |
|---|---|---|---|
| $SiCl_4$ | 0.4-1.5 | 1.6-1.7 | 1150-1250 |
| $SiHCl_3$ | 0.4-2.0 | 0.8-1.0 | 1100-1200 |
| $SiH_2Cl_2$ | 0.4-3.0 | 0.3-0.6 | 1050-1150 |

## Pyrolitic Decomposition of Silane

When silane ($SiH_4$) is subjected to a temperature above 500 °C, the silicon-hydrogen bonds break and the molecules decompose to give silicon and hydrogen [10,15]. In an epitaxial reactor, silane is carried by a high purity hydrogen stream. When the gas mixture is heated to approximately 500 °C, silane decomposes as:

$$SiH_4 = Si\downarrow + 2H_2\uparrow. \qquad\qquad 3.6$$

Silicon deposits on the wafer surface while hydrogen returns to the gas stream. For single crystal to form at atmospheric pressure, however, the temperature must be typically in the range 800-1050 °C [13]. Lower temperatures have been used at reduced pressure.

## Deposition Rate

The three major steps for CVD epitaxy are: the transport of reactants to the wafer surface, the reaction at or near the wafer surface, and the removal of reaction by-products.

The rate of incorporation of silicon atoms into lattice positions is a function of crystal orientation. The "growth" rate is slowest in a direction normal to (111) planes because these planes have the highest density of atoms per unit area. The higher deposition rate in other crystallographic directions can give rise to "shingle-like" faceted surfaces on (111) wafers or pattern shift (discussed below). Faceting is not observed on (100) oriented surfaces.

Typical deposition rates range from 0.2 - 3.0 $\mu m/min$, with the highest observed for silane and the lowest for silicon tetrachloride. The reaction in Eq. 3.5 with $SiCl_2$ is reversible. At high HCl concentrations more silicon is etched away than deposited, following the reaction in Eq. 3.1. A typical variation of deposition rate with temperature is shown Fig. 3.6 [16]. There is a negative rate (etch) at low and high temperatures. At points X and Y, the net deposition rate is zero.

The temperature dependence of deposition rate is shown for different silicon compounds in Fig. 3.7 [17]. At low temperatures, the deposition rate is reaction controlled, as in the exponential region on the right of the figure. This region is hence characterized as reaction-rate limited. As the temperature increases to the left, the deposition rate varies more slowly with temperature. This is a region where the deposition rate is either limited by the amount of reactants reaching the surface or by the reaction products diffusing away. Most epitaxial depositions are performed in this region because they are less sensitive to temperature variations.

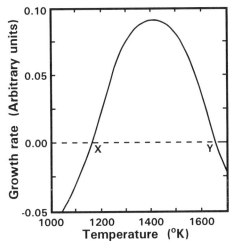

**Fig. 3.6** Deposition rate of CVD silicon versus temperature [16]

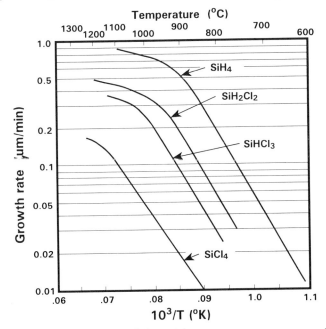

**Fig. 3.7** Dependence of deposition rate on temperature for different silicon compounds [17]

### Doping and Autodoping

The epitaxial layer can be doped while grown by adding controlled amounts of the dopant compounds to the gas stream [18, 19]. Typical dopant sources are hydrides of the impurity, such as phosphine ($PH_3$), arsine ($AsH_3$), antimonine ($SbH_3$), and diborane ($B_2H_6$) [20]. The halides are diluted in hydrogen at a typical concentration of 1-20 ppm. They dissociate at the wafer surface to yield the dopant element.

In addition to intentional doping, unintentional dopants are introduced from the substrate, either by solid-state diffusion (Chap. 7) or by evaporation, an effect which is described in Fig. 3.8 and termed **autodoping** [21]. This is a process at high temperature in which impurity atoms are transferred from a heavily doped region into the vapor, transported within the boundary layer, and then deposited onto the growing epitaxial layer [22 − 31]; regions which are "downstream" are more heavily doped than regions which are "upstream".

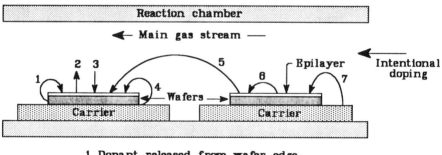

1 Dopant released from wafer edge.
2 Dopant evaporating from wafer surface.
3 Dopant adsorbed at surface.
4 Dopant released from wafer backside.
5 Dopant released from adjacent wafers.
6 Dopant released and adsorbed by wafer.
7 Dopant released from carrier (susceptor).

**Fig. 3.8** Sources of dopants introduced into the epitaxial layer.

When a heavily doped region is exposed, two autodoping mechanisms affect the epitaxial profile: vertical into the epitaxy above the region, and lateral into areas surrounding the region. Because of this dopant redistribution, the impurity concentration does not change abruptly from substrate to epitaxy (Fig. 3.9). The vertical profile can be divided into three regions doped from different sources. Region A is the result of intentional doping of the film while it is deposited. This is typically in the range $10^{15}$ - $10^{17}$ cm $^{-3}$. The second region B the range $10^{17}$-$10^{20}$ cm $^{-3}$ is due to the solid-state diffusion of dopants from the substrate into the epitaxy. Region C is caused by autodoping. The dopants which continue to evaporate from the surface and cause autodoping are supplied by solid-state diffusion from inside the wafer. Therefore, autodoping decreases with time until the intentional doping predominates and the profile becomes flat.

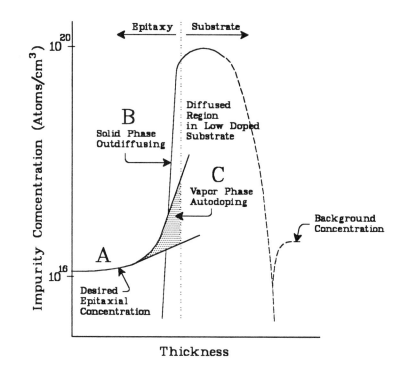

**Fig. 3.9** Impurity profile measured normal to the substrate. A: intentional doping. B: solid-state diffusion. C: Autodoping.

Lateral autodoping can cause a leakage current path or even a short between adjacent regions. Figure 3.10 shows how lateral autodoping can cause a short [32] during the initial stages of epitaxy; the heavily doped region acts as a two-dimensional source of impurities.

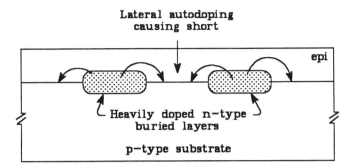

**Fig. 3.10** Short between two adjacent regions caused by lateral autodoping.

Several methods have been implemented to reduce autodoping [33]:

a) Back side seal with a film of polysilicon, silicon dioxide, or silicon nitride to prevent evaporation from the substrate.

b) Extended "prebake" at high temperature to reduce the surface concentration of the heavily doped region by depleting the exposed region from impurities.

c) Use of dopants with lower vapor pressure, such as Sb instead As. (The lower solid solubility limit of Sb in silicon, however, limits its maximum concentration to about $2x10^{19}$ cm$^{-3}$).

d) Deposition of an undoped "cap-film" followed by the deposition of the desired layer (two-step deposition) [23].

e) Reduced pressure deposition, increasing the dopant mean-free path and hence the probability that the species escape from the boundary before being adsorbed at the surface of the crystal (Fig. 3.11) [33 − 36].

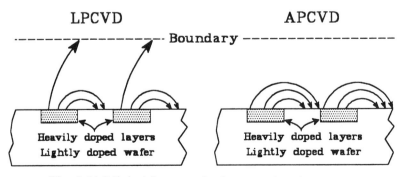

**Fig. 3.11** Minimizing autodoping at reduced pressure epitaxy (LPCVD).

## Pattern Shift and Distortion

In typical bipolar process technologies, a pattern of heavily doped buried layers is defined prior to epitaxial growth (Fig. 3.3b). The pattern is made visible by locally oxidizing silicon where the layer is formed and then removing the oxide. This leaves a sharp, 50-100 nm deep edge at the perimeter of the buried layer marking its location. Since the pattern is used for alignment of subsequent masking steps (Chap. 4), it should maintain its position, shape and "sharpness" as it propagates through the epitaxial layer. Under most deposition conditions, however, the pattern becomes distorted and shifts laterally with respect to its initial position because of the varying growth rate in different crystallographic orientations. Figure 3.12 illustrates how a pattern can shift and be distorted during epitaxial growth [37]. In some cases, a complete "washout" is observed, making alignment very difficult.

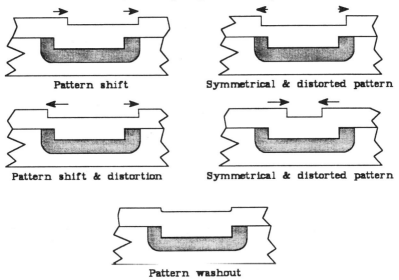

Fig. 3.12 Illustrations of pattern shift and distortion.

Pattern shift and distortion depend on crystal orientation, deposition rate, temperature, and reactants used for deposition. Pattern shift is reduced when the deposition temperature is increased and the deposition rate is reduced [37,38]. It is also found that pattern shift decreases when the ratio of Cl to Si is reduced in the reacting compounds; there is less pattern shift with $SiH_2Cl_2$ than with $SiCl_4$. The use of silane reduces pattern shift but increases pattern distortion for (100) substrates [39]. The effect of silicon source on pattern shift and distortion is not well understood. There is less shift at reduced pressure than at atmospheric pressure epitaxy [40]. It is believed that because the reduced pressure increases the mean free path of silicon atoms, silicon deposits more isotropically and the pattern propagates with less shift.

### Epitaxial Systems

Deposition systems, also called **reactors,** consist of four main elements: a reaction chamber and wafer holder, an inductive or a radiant heat source, a controlled gas-flow system, and an exhaust system to remove the reaction by-products. Epitaxial reactors are operated in a "cold-wall" mode, i.e., the chamber walls are maintained at low temperature to avoid the dissociation of compounds at the walls. When inductive heating is used, the susceptor absorbs energy and conveys it to the wafer by conduction and radiation. This "indirect" heating, however, causes a thermal gradient across the thickness of the wafer, resulting in lattice stress and wafer "warpage". For large wafers, radiant heating is the preferred approach because the problem of stress is avoided by heating the susceptor and wafer at the same rate. Three principle types of commercially used reactors are: the horizontal reactor, the vertical reactor (pancake), and the barrel reactor (Fig. 3.13). The reactor names are associated with the shape of the susceptor and its position with respect to the gas stream.

### Horizontal Reactor

The simplest horizontal reactor consists of an inductively heated quartz tube in which the gas mixture is introduced at one end and the reaction by-products "pushed-out" at the other end (Fig. 3.13a). Since the concentration of silicon compounds decreases "downstream", it is necessary to tilt the susceptor by about 5° to compensate for the depletion of the reactants. The horizontal reactor is simple and has a large throughput (number of process wafers/h), but the deposition rate is difficult to control over the entire length of the reactor.

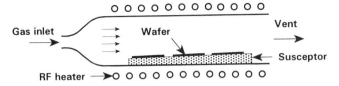

(a) Simple horizontal reactor

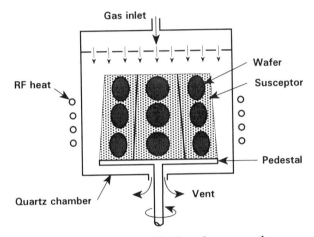

(b) Barrel reactor for batch processing

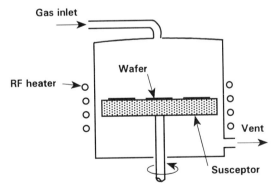

(c) Simple vertical reactor

**Fig. 3.13** Commercial epitaxial reactors. (a) Simple horizontal reactor. (b) Barrel reactor. (c) Vertical (pancake) reactor.

## Barrel Reactor

In a barrel reactor the susceptor is cylindrical and the wafers are held in niches about 3° from the vertical (Fig. 3.13b). The gas flow is roughly parallel to the wafer surface. Improved characteristics are achieved by rotating the susceptor to reduce non-uniformities in the gas flow and epitaxial growth. Heat is provided by induction or radiation.

## Vertical Reactor

In a vertical reactor the gas mixture flows almost normal to the wafer surface. The susceptor is heated inductively by "pancake" shaped elements located beneath it (Fig. 3.13c). Excellent uniformity is achieved, however, at the cost of throughput.

## Metalorganic Chemical Vapor Deposition (MOCVD)

The equipment described in the preceding section can also be used to deposit compound semiconductors (such as GaAs, AlGaAs, GaAsP...) from organometallic sources at very low temperature [41, 42]. Several organometallic sources of group III metals, for example, have sufficient vapor pressure to be introduced into the epitaxial reactor by vapor transport. For gallium arsenide and gallium arsenide phosphide, triethylgallium, $(C_2H_5)_3Ga$, and trimethylgallium, $(CH_3)_3Ga$, are commonly used to react with arsine; the carrier gas is hydrogen ($H_2$). Typical reactions are of the form

$$(CH_3)_3Ga + AsH_3 = GaAs\downarrow + 3CH_4\uparrow . \qquad 3.7$$

The layer thickness varies from 2 nm to 2 $\mu$m and has a uniformity of 1-5%. Deposition temperatures range from 600-850 °C. The layers can be doped with group II, IV or VI elements during deposition [43]. MOCVD is important for the controlled deposition of ultra-thin, doped or undoped semiconductor heterolayers (such as AlGaAs/GaAs and GaAs/Si), primarily used for lasers, light-emitting diodes (LED), and optoelectronic integrated circuits (OEIC) [44].

### 3.1.2 Polysilicon

Polysilicon has found many important applications in integrated circuits. Heavily doped films are used to form the gate conductor in MOSFETs [45], as a dopant source for contacts of very shallow junctions in high-speed bipolar transistors and MOSFETs [46 − 54], as a conductive layer for interconnections, and for "electrically trimmable resistors" [55, 56]. At moderate to low dopant concentrations polysilicon is used for high value resistors in memory cells [57], precision resistors and capacitors in analog designs [58 − 62], and thin-film transistors (TFT) [63,64]. Undoped (intrinsic) polysilicon has a typical resistivity of $10^9$ - $5x10^9$ Ohm-cm, and is used for conformal coating or filling of isolation grooves separating devices on the same chip.

**Deposition and Properties of Polysilicon**

Polysilicon is an aggregate of small crystallites called **grains,** separated by thin regions called **grain boundaries.** Inside each grain the silicon atoms are arranged periodically forming a single crystal region. The grain size depends on the deposition temperature, heat treatment, and dopant concentration and type. The average grain size ranges from 0.3 - 1.0 $\mu$m. Grain boundaries are only 0.5 - 1 nm wide [65].

Polysilicon is typically deposited at low pressure (0.2 - 1.0 torr) at a rate of 10-20 nm/min by the pyrolitic decomposition of silane at 550 °C - 650 °C, following the reaction in Eq. 3.6 [66 − 68]. When deposited below 575 °C, the film is amorphous with no detectable structure, but crystallizes readily on further heat treatment above 600 °C. If deposited at about 620 °C, polysilicon exhibits a columnar structure with a diameter ranging from 30 - 300 nm. The grains increase in size upon heat treatment at higher temperatures [69 − 72]. Grain size and grain boundaries have a significant effect on the electrical and physical properties of polysilicon films [73 − 76]. These properties are discussed later in conjunction with device physics.

**Doping of Polysilicon**

The film can be doped while deposited (in-situ) by adding arsine, phosphine, or diborane to the gas mixture. Doping can also be performed after deposition by ion implantation and diffusion

(Chaps. 6,7) [77]. Impurities diffuse considerably faster in polysilicon than in single-crystal silicon. This is attributed to the enhanced diffusion along grain boundaries and defect-rich regions within the grains [78 − 80]. When the dopant concentration exceeds the solid solubility limit in silicon, the excess dopant deposits at the grain boundaries. The distribution of dopants between grain and grain boundary is, however, a rather complex mechanism which also involves a change in grain size. A considerable fraction of arsenic, phosphorus, and boron segregates at grain boundaries [81, 82]. For example, as-deposited arsenic in-situ doped films with an average grain-size of 20 nm show a segregation of about 30% of the total arsenic concentration to the grain boundaries. After annealing at 1000 °C, the mean grain size increases to $\simeq$ 400 nm and more arsenic diffuses into the grains [81]. By successively subjecting a heavily doped film to 800°C and 1000 °C, phosphorus and arsenic are found to segregate to the grain boundaries at the lower temperature, causing the film resistivity to increase; the dopants diffuse back into the grain at the higher temperature, causing the resistivity to decrease [82]. In some cases, dopant redistribution is accompanied by grain growth [83].

**Reactors**

Reduced pressure, "hot-wall" reactors are widely used to deposit polysilicon, silicon dioxide, and silicon nitride. They are similar in construction to those described in Fig. 3.13. LPCVD reactors exhibit a higher uniformity, and operate at a lower temperature and cost than conventional atmospheric pressure reactors. A horizontal reactor consists of a quartz tube, radiantly heated by a three- or five-zone furnace. The reactor pressure ranges from 0.2-2.0 torr and the temperature from 250-950 °C. The wafers are placed standing on edge. The large molecular mean-free path at low pressure allows the arrangement of densely spaced wafers without substantially impacting film uniformity. The vertical furnace shown in Fig. 3.14a operates essentially in the same fashion as a horizontal furnace, except for the orientation of the wafers. In a vertical furnace, the wafers are loaded in a horizontal position on a quartz cassette which is either lowered or raised into the furnace by an elevator mechanism. The main advantages of a vertical furnace over a horizontal furnace is the reduced area occupied in a clean room ("footprint"), and the reduced contamination during wafer loading.

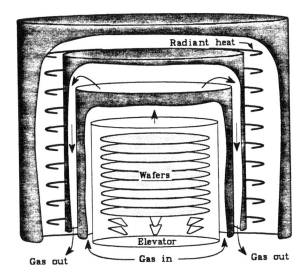

**Fig. 3.14a** Vertical reactor.

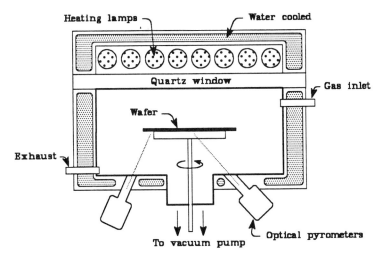

**Fig. 3.14b** Single-wafer LPCVD reactor.

As the wafer size increases (150 mm, 200 mm, 400 mm ...) the "batch-size" and hence the throughput of a conventional furnace decreases. The single-wafer reactor shown in Fig. 3.14b becomes very attractive because it exhibits a large throughput (independent of wafer size) and very high film uniformity. Another attractive feature of the single-wafer reactor is the "clustering" of other processes (such as oxidation, oxide depositions, and nitride deposition) in the same tool [84]. For example, the growth of an oxide film of high quality can be clustered with the deposition of a polysilicon film on top of the oxide in the same chamber to avoid contamination during wafer transport.

## Oxidation of Polysilicon

Polysilicon is thermally oxidized primarily to isolate the film from adjacent conductors. The oxide-growth kinetics that were discussed for single crystal silicon in Chap. 2 are also applicable to polysilicon.

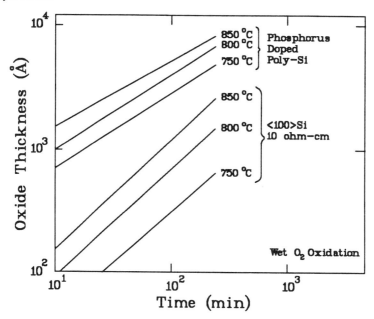

Fig. 3.15 Oxide thickness versus oxidation time for wet oxidation of phosphorus doped polysilicon and single crystal silicon [86].

Since polysilicon exposes different crystallographic orientation to the oxidizing ambient, the growth rate is an average of all grain orientations, intermediate between (111) and (100). The rate of oxidation of heavily doped polysilicon is rather difficult to predict because of grain boundary and segregation effects. For heavily phosphorus doped films, segregation at the grain boundaries can cause considerable increase in the rate of oxidation (Fig. 3.15) [85,86]. This enhanced oxidation rate is less pronounced with arsenic and not observed on boron doped films. At moderate phosphorus concentrations, the oxidation rate is smaller for polysilicon than for single-crystal silicon because at such concentrations phosphorus moves deeper into the polysilicon grains away from the surface where it affects the oxide growth-rate [87].

Because the surface of polysilicon exposes grains of random crystallographic orientations, an oxide film grown on polysilicon is essentially "rough" (Fig. 3.16) [88, 89].

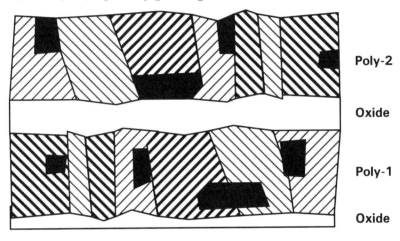

Poly-2

Oxide

Poly-1

Oxide

**Fig. 3.16** Surface roughness and asperities in grown oxide films on polysilicon.

The oxide not only grows at the exposed grain surface but also inside the grain boundaries. Surface roughness causes the electric field to increase at asperities, increasing the leakage current through the oxide and reducing the dielectric breakdown voltage [90,91]. The increase in local field is used in some electrically erasable and programmable read-only memory (EEPROM) designs to enhance the read and write operations. In most cases, however, the increased local field can have a serious impact on the device "life-time". Another reliability consideration is the oxide growth within the grain boundary and subsequent removal of the oxide, causing metal penetration into the film during contact definition (Chap. 8).

### 3.1.3 Selective Epitaxial Growth (SEG)

Selective epitaxial growth is the deposition of a single crystal film in "seed" windows in an insulator mask, without polysilicon nucleation on the insulator (Fig. 3.17a) [92]. Silicon SEG and its derivatives, simultaneous single crystal/polysilicon deposition (SSPD, Fig. 3.17b), and epitaxial lateral overgrowth (ELO, Fig. 3.17c) have initiated the design of important new devices [93, 94].

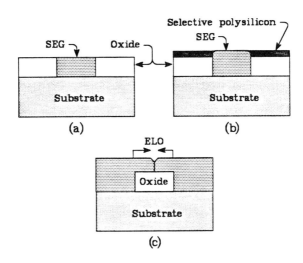

**Fig. 3.17** Selective silicon deposition. (a) Selective epitaxial growth. (b) Simultaneous single crystal/polysilicon deposition. (c) Epitaxial lateral overgrowth. The facets are caused by the varying growth rates with crystallographic orientation.

SEG and ELO films are typically used for active devices (MOSFETs, bipolar transistors) and must therefore be grown defect free. Selective polysilicon films are used for filling trenches or forming contacts and local interconnections.

## Selectivity

While silicon nucleates "instantaneously" over exposed silicon, an incubation period is required before nucleation occurs over silicon dioxide. This period depends on the silicon source gas, deposition temperature and pressure, and on the oxide surface conditions. Silicon nucleation over oxide occurs by adsorption and condensation of silicon atoms at nucleation sites. This process can be inhibited if small amounts ($\simeq$ vol. 2%) of HCl are added to the gas mixture before the incubation period elapses. In this respect, the decomposition of silane is different than the hydrogen reaction with dichlorosilane. Silane decomposes pyrolytically into silicon and hydrogen, while the reaction with DCS results in HCl which becomes free to etch the adsorbed silicon atoms over the oxide. Selectivity can be achieved by controlling the amount of HCl in the gas mixture. With DCS the "process window" for selectivity is much wider at low temperature ($<$ 1000 °C) and pressure ($<$ 80 torr) [95,96]. Deposition with DCS under low-pressure, low-temperature conditions improves the surface planarity and uniformity [97,98], and reduces defects induced in sidewall oxides [99].

Deposition can be performed in a series of growth/etch steps to inhibit formation of polysilicon over the oxide. As the growth of single crystal proceeds, spurious nucleation starts to form on the oxide after a time which exceeds the incubation period. To prevent this, silicon deposition is stopped after a time approximately equal to the nucleation time and the gases are switched to $H_2/HCl$ (2% HCl) mixture which etches polysilicon nuclei over the oxide without appreciably affecting the single-crystal film [100]. In the initial deposition stage the epitaxial film grows vertically in the oxide opening until it reaches the top oxide surface. If deposition continues, the film begins to expand laterally along the surface of the oxide while continuing vertical growth, resulting in ELO. [100, 101].

## Deposition Methods

Most SEG films are deposited at a temperature of 900 - 1000 °C and a pressure less than 80 torr, using a DCS-HCl-$H_2$ mixture in a radiantly heated reactor [95,96,102 – 104]. The most important condition for obtaining high quality SEG films is a damage-free surface, free of tenacious contaminants such as silicon dioxide and carbon-containing films. Also no contamination should occur in the reactor prior to or during epitaxial deposition [105]. The surface is typically cleaned by an in-situ "prebake" in hydrogen at at 900 - 1000 °C, or by etching in an $HCl/H_2$ mixture. The partial pressures of water and oxygen must be maintained below $10^{-10}$ torr and $10^{-13}$ torr, respectively. This can be achieved under LPCVD conditions. Deposition at atmospheric pressure has been achieved with DCS by "ultra-cleaning" the gas mixture entering the reactor [106]. Silane has also been used for SEG at about 900 °C in an ultra-clean environment [107,108]. A higher selectivity is achieved with silane at low temperature when using a borophosphosilicate glass mask [109].

Adding 10-20% germane ($GeH_4$) to the gas mixture increases the growth-rate and incubation period on silicon dioxide, increasing the maximum selective thickness that can be achieved [110]. The most probable reason for improved selectivity is the generation of highly volatile GeO through the reactions [111]:

$$GeH_4 = Ge\downarrow + 2H_2\uparrow \qquad\qquad 3.8$$

$$Ge + SiO_2 = GeO_2\downarrow + Si\downarrow$$

$$GeO_2 + Ge = 2GeO\uparrow$$

Deposition is achieved at 600 °C from pyrolysis of silane and 10% $GeH_4$/90% $H_2$. Selectivity is obtained by growing for a time shorter than the incubation time needed for grain nucleation on $SiO_2$ [112]. Nucleation with silane can also be controlled at a temperature as low as 550 °C, using borophosphosilicate glass as a mask and adding 10-20 % $GeH_4$ to the gas mixture. Under these conditions, the deposition rate on (100) silicon becomes much larger than on (111), reducing faceting [112].

## SEG Applications

The selective deposition of defect-free, single-crystal silicon has initiated the design of a wide range of novel devices. This section gives a brief summary of some of the important applications of SEG in VLSI/ULSI.

## Isolation

One of the limitation of LOCOS isolation discussed in the preceding chapter is the lateral extent of oxidation (bird's beak). The use of SEG provides dielectric isolation without an increase in area by depositing a single-crystal film in thick-oxide windows or in trenches formed in the substrate [113 – 120]. One typical example is shown in Fig. 3.18 where windows in a 1 μm thick thermal oxide layer define regions for active devices. Heavily doped buried layers are defined and implanted in these regions, using techniques described in the following chapters, to locally reduce the substrate resistivity. This is followed by selective deposition of single-crystal films in which devices are constructed. The thick thermal oxide isolates the devices while maintaining minimum dimensions.

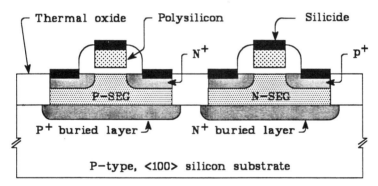

**Fig. 3.18** Device isolation by SEG. The thick thermal oxide regions isolate devices while maintaining their minimum dimensions. Heavily doped buried layers are implanted prior to SEG to reduce the local substrate resistivity (Adapted from [119]).

Another isolation technique is to define trenches in silicon and coat its sidewalls with an insulator (Fig. 3.19). Selective epitaxial growth over the bottom of the trench of p-type silicon provides the trench fill connecting to the substrate. An oxide cap is grown and defined over the trench opening to isolate the trench from crossing conductors [120]. Techniques to etch trenches and insulators are discussed Chap. 5.

## Suppression of Autodoping

Lateral autodoping is suppressed by "capping" the heavily doped buried layers with SEG (Fig. 3.20). After definition and implantation of the heavily doped layers, a 300 nm SEG film is formed over the exposed regions. The rest of the surface is protected by the oxide mask during selective epitaxy. A thick epitaxial layer is deposited after removal of the oxide mask. The SEG film suppresses lateral autodoping during the deposition of the thick epitaxial film [121].

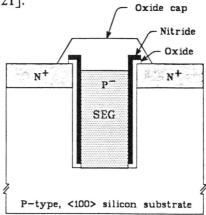

Fig. 3.19 Selective epitaxially refilled trench [120].

## Device Fabrication on ELO

The extension of single crystal silicon over an insulator by ELO is one technique to form silicon on insulator (SOI) structures. MOSFET and bipolar transistors on SOI exhibit a higher performance and lower power dissipation than on conventional substrates [101, 122 − 124]. Also, integration of devices is easier with SOI since isolation becomes simpler. Figure 3.21 illustrates the definition of a MOSFET on SOI formed by ELO. Other methods to form SOI structures and techniques to isolate devices within a structure are discussed in the following chapters.

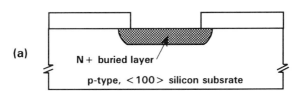

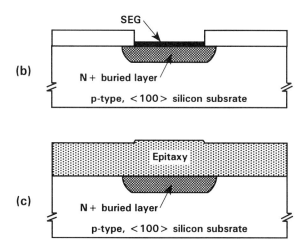

**Fig. 3.20** Suppression of lateral autodoping by SEG capping. (a) Definition and implantation of heavily doped buried layer in oxide window. (b) 300 nm SEG cap. (c) Removal of thick oxide and deposition of thick epitaxial layer. [121].

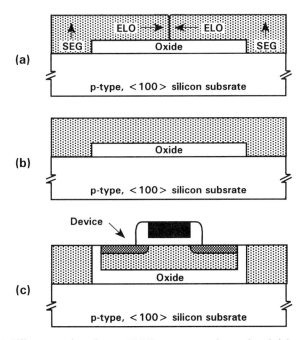

**Fig. 3.21** Silicon on insulator (SOI) structure by epitaxial lateral overgrowth (ELO) [123]. (a) Merger of two ELO regions. (b) ELO after chem-mech polishing. (c) Device definition and isolation.

**114**

### Elevated Junction Contacts

A pn junction, the boundary between regions of opposite dopant
polarities, is the building block of semiconductor devices. It is typically formed by ion implantation and diffusion (Chap. 6). Since the
junction depth below the surface can be reduced to less than 60
nm in ULSI structures, it becomes extremely difficult to directly
form and contact the junction without causing damage to the
structure. This problem can be solved by selective epitaxial growth,
as illustrated for a MOSFET source and drain in Fig. 3.22
[125 − 127].

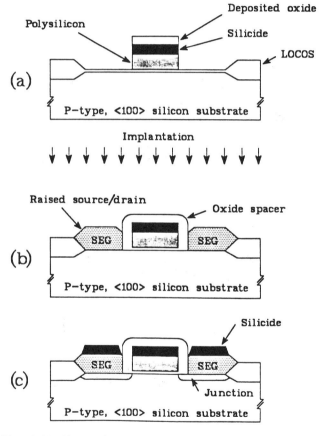

**Fig. 3.22** Elevated source and drain in a MOSFET
by SEG. (a) Definition of gate.   (b) Selective
epitaxial growth and implantation. (c) Contact metallurgy.

After cleaning the exposed silicon surface, a 200-500 nm thick SEG layer is formed and then doped by implantation. The layer becomes then the source of impurities which diffuse below the original silicon surface to form the junction. Elevating the silicon surface places the damage caused by implantation and contact metallurgy away from the junction (Chaps. 6,7).

### Defects

The two most important problems with elevated source/drain by SEG are faceting and sidewall oxide defect formation. Faceting occurs because the deposition rate of silicon is not the same in all crystallographic orientations; it is larger on (100) than on (111) planes. This results in a non-uniform growth normal to the silicon surface (Fig. 3.22). One of the problems caused by faceting is the non-uniform penetration of implanted species. The dopants penetrate deeper in the grooves caused by faceting than in other SEG regions, resulting in deeper junctions under the faceted area. Faceting can be reduced with a low-temperature, low-pressure DCS-HCl-H$_2$ mixture, and by orienting all sides of the junction in the <100> direction.

Defects in the oxide can be caused by the conversion of SiO$_2$ to volatile SiO:

$$Si + SiO_2 = 2SiO . \qquad\qquad 3.9$$

Their formation is accelerated in the presence of a silicon source in an oxygen- and water-deficient ambient atmosphere, a condition needed for good crystalline quality SEG [128]. This situation is similar to a post-oxidation anneal in an oxygen-free or oxygen-deficient atmosphere where a considerable amount of oxide defects have been observed [129,130]. In SEG this can result in oxide consumption, undercut and voids in "weak spots" of the oxide. The water vapor and oxygen free environment necessary to obtain high-quality single crystalline silicon becomes a problem for oxide films during SEG or ELO. The problem can be alleviated by reducing the SEG deposition temperature below 850 °C.

### 3.1.4 Chemical Vapor Deposition of Silicon Dioxide

Silicon dioxide films can be deposited by a chemical vapor reaction at a temperature as low as 400 °C, without consuming silicon in the substrate. The deposition temperature can be considerably reduced by photo- or plasma-assisted CVD (Sec. 3.2). The deposition and annealing conditions depend on how the oxide is used. When used as a thin insulator between conducting layers, the oxide film is deposited undoped and then "densified" at elevated temperature to increase its dielectric breakdown field. In many cases, however, the film constitutes a diffusion or implantation mask, or a "cap" to prevent outdiffusion of dopants from an underlying film (Chap. 7). Doped CVD oxide films have found many applications. Depending on the impurity concentration in the oxide, doped oxide can be used as a diffusion source, getterer, or reflow material. Phosphosilicate glass (PSG), for example, is used as a getterer that neutralizes sodium ions and inhibits their diffusion [131 – 134]. When heated above approximately 900 °C, the PSG film softens and flows, providing a smooth topography which improves the definition of contacts and metal patterns [135]. The softening temperature can be reduced to about 700 °C by adding boron to the film to form a boro-phosphosilicate glass layer (BPSG).

The most common methods to deposit silicon dioxide are oxidizing silane with oxygen at low pressure and low temperature (typically 400-450 °C) and decomposing tetraethylorthosilicate ( $Si(C_2H_5O)_4$ ), also called tetraethoxysilane (TEOS), with or without oxygen at low pressure and at a temperature between 650-750 °C. The reaction of silane with nitrous oxide ($NO_2$) is also used to produce stoichiometric silicon dioxide or silicon-rich oxide by changing the $NO_2/SiH_4$ ratio. Very low temperature (250-400 °C) APCVD oxide can be deposited by reacting TEOS with ozone.

### Oxidation of Silane

When silane reacts with oxygen at temperatures below 500 °C, the final products are silicon dioxide and hydrogen [66, 136, 137]

$$SiH_4 + O_2 \rightarrow SiO_2 + 2H_2 . \qquad 3.10$$

The advantage of low pressure, low temperature oxidation (LPLTO) of silane is that the oxide film can be deposited over metals with a low melting point (such as aluminum and copper). The composition, however, is not stoichiometric and the film exhibits a low dielectric breakdown field ($< 6x10^6$ V/cm as compared to $> 1.2x10^7$ V/cm for thermal oxide), and does not cover steps uniformly (Fig. 3.23). LPLTO is therefore typically used for passivation after metal deposition, and in conjunction with other insulating films (Chap. 8). Stoichiometric oxide can be formed by reacting silane or DCS with nitrous oxide, using a large excess of $NO_2$ at atmospheric pressure [138, 139]. The reaction temperature, however, is in the range 750-900 °C, in excess of the melting temperature of typical metals used for interconnections.

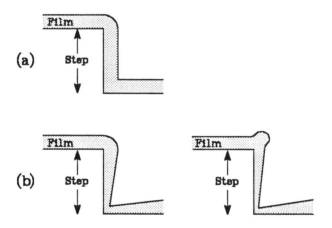

Fig. 3.23 To illustrate step coverage. (a) Uniform step coverage. (b) Non-uniform step coverage. (Adapted from [142]).

## TEOS Oxide

The pyrolysis of TEOS to get an oxide film is similar to the decomposition of silane to form polysilicon. A carrier gas, typically nitrogen, is bubbled through liquid TEOS to provide a gas mixture of controlled TEOS partial pressure in the reaction chamber. TEOS consists of a silicon atom, symmetrically bonded to four

$OC_2H_5$ groups, and its decomposition at 650-750 °C results in silicon dioxide and by-products of organic and organosilicon compounds that cannot be determined predictably [140]. The net reaction is:

$$Si\ (C_2H_5O\ )_4 \rightarrow SiO_2 + \text{by-products.} \qquad 3.11$$

The deposition rate depends on the TEOS partial pressure and temperature. TEOS also reacts with oxygen at low pressure/high temperature (650-750 °C) to produce an oxide film [143, 144]. The overall reaction is:

$$Si\ (C_2H_5O\ )_4 + 12\ O_2 \rightarrow SiO_2 + 8\ CO_2 + 10\ H_2O\ . \qquad 3.12$$

LPCVD TEOS films exhibit excellent uniformity and step coverage. The high decomposition temperature of TEOS, however, precludes its use for the deposition of oxide over aluminum and many silicides (Chap. 8).

**TEOS-Ozone APCVD Oxide**

Oxide films can be deposited at atmospheric pressure and a temperature as low as 250 °C by reacting TEOS with ozone ($O_3$). The films exhibit smooth oxide profiles over steps, excellent filling of high aspect-ratio (ratio of depth to width) gaps between metal lines, and good electrical characteristics. The TEOS/ozone method is most suitable for inter-metal level dielectrics [145 − 150]. The simplified overall thermal reaction between TEOS and ozone is similar to Eq. 3.12:

$$Si(C_2H_5O)_4 + 8\ O_3 \rightarrow SiO_2 + 10\ H_2O + 8\ CO_2\ . \qquad 3.13$$

**Silicon-Rich Oxide (SRO)**

When the ratio of $NO_2$ to $SiH_4$ is reduced, the reaction of silane with nitrous oxide produces an oxide film with excess silicon. For a gas-phase ratio $NO_2/SiH_4 \simeq 3$, there is approximately 14% excess silicon in the oxide film [138, 151]. Silicon rich oxide (SRO) is also called semi-insulating polysilicon (SIPOS) because the material exhibits an electrical conductivity intermediate between $SiO_2$ and intrinsic silicon. SRO films are typically deposited under low pressure and at a temperature between 550-750 °C. The film consists of two distinct phases: silicon islands with the film and CVD oxide [138]. The presence of silicon islands enhances the local electric field by about a factor of 2 over the field in stoichiometric oxide of the same thickness. This increases the electron injection from an electrode into SRO, and then from SRO into an overlaying or underlying stoichiometric oxide film [151]. Such configurations with enhanced electron injection are used in some electrically-alterable read-only-memory (EAROM) designs with floating polysilicon gates [152, 153].

### 3.1.5 Phospho-, Boro- and Borophosphosilicate Glass

As mentioned above, oxide films can be doped by adding small amounts of arsine, phosphine, or diborane to the gas mixture [136]. Doped oxide films typically contain 15-20 % arsenic, phosphorus, or boron when they are used as diffusion sources. The concentration is lower when the film is used for passivation. Phosphosilicate glass can be grown by the simultaneous oxidation of silane and phosphine. The oxidation of phosphine results in phosphorus pentoxide ($P_2O_5$) and hydrogen.

$$4PH_3 + 5O_2 \rightarrow 2P_2O_5 + 6H_2 . \qquad 3.14$$

PSG can be used as a getterer to protect against the migration of sodium ions. For this purpose, only a small percentage (2-3 wt%) of phosphorus is required. The PSG film is also used as a coating which can be smoothed by subjecting the glass to a temperature above 800 °C after deposition. The "flow" of the glass at high temperature tends to round-off sharp edges and form planar surfaces over sharp steps, avoiding discontinuities of interconnecting lines that are defined on the glass surface. The concentration of $P_2O_5$ in a PSG film also affects its the etch rate, thermal expansion coefficient, and electrical polarization. as discussed in the following chapters.

Borosilicate glass (BSG) finds applications similar to those of PSG. A BSG film is formed by simultaneously oxidizing diborane and silane. The reaction of diborane with oxygen is:

$$2B_2H_6 + 3O_2 \rightarrow 2B_2O_3 + 6H_2 . \qquad 3.15$$

The resulting boron trioxide ($B_2O_3$) is incorporated into silicon dioxide while it is deposited.

The thermal expansion coefficient and softening temperature of PSG and BSP films is strongly dependent on their $P_2O_5$ and $B_2O_3$ concentrations. The film can therefore be tailored to provide a suitable coat that flows and matches the underlying layers by co-depositing the correct amount of $B_2O_3$ and $P_2O_5$ with the oxide. Typically, borophosphosilicate glass with about 4 wt% boron and 4 wt% phosphorus flows at a temperature between 800 and 900 °C. An increase in phosphorus concentration above 7-8 wt% causes corrosion problems with aluminum interconnections.

### 3.1.6 CVD Deposition of Silicon Nitride

Silicon nitride is practically impermeable to oxygen, and water [154]. A nitride film of appropriate thickness can be used as a mask to block oxidation of the silicon underneath and will oxidize very slowly at the surface (Chap. 2). Silicon nitride is also an excellent barrier against the migration of sodium and other deleterious contaminants [155]. The film, however, is not deposited directly on silicon because it creates defects induced by stress ($> 10^9$ Pa. Films thicker than about 200 nm sometimes crack because of the high stress.

Extensive use is made of nitride films deposited on oxide, forming an important composite dielectric for MOSFETs and capacitors. The nitride is also widely used as an "etch stop" for selective etching of other films, such as silicon and silicon dioxide (Chap. 5).

A film of stoichiometric composition ($Si_3N_4$) can be deposited by reacting silane with ammonia at 700-900 °C at atmospheric pressure [156, 157]:

$$3SiH_4 + 4NH_3 \rightarrow Si_3N_4 + 12\,H_2 . \qquad 3.16$$

The deposition rate increases rapidly with temperature. While at 700 °C the rate is less than 1 nm/min, it increases to 100-200 nm/min at 900 °C [158].

A typical silicon nitride LPCVD system uses the reaction of dichlorosilane and ammonia at a temperature of 700-800 °C:

$$3SiCl_2H_2 + 4NH_3 \rightarrow Si_3N_4 + 6HCl + 6H_2 . \qquad 3.17$$

The LPCVD deposition rate is 15-20 nm/min and can be controlled easier than with APCVD. While excellent uniformity and high throughput have been achieved, considerable departure from the stoichiometric composition $Si_3N_4$ is observed with the use of LPCVD. Typical silicon to nitrogen ratios range from 0.7-1.1 [136]. The films also contain large amounts of bonded hydrogen in form of Si-H and N-H [159]. The nominal index of refraction of LPCVD nitride films is 2.01, and the typical etch rate in buffered hydrofluoric acid (BHF) is 1 nm/min (Chap. 5). Both properties are used to check the nitride quality. A high index of refraction is caused by a silicon-rich film, and a low index by an increase in hydrogen content [142]. Table 3.2 summarizes the properties of LPCVD and PECVD nitride. The PECVD parameters are added for later reference.

### 3.1.7  Composite Dielectrics

The advantages of composite dielectrics prepared by thermal oxidation and nitridation were discussed in Chap. 2. Stacked oxide and CVD nitride films have also found important applications where they exhibit superior dielectric properties to those of simple oxides. For example, stacked oxide-nitride-oxide (ONO) films are more reliable over shaped or rough silicon surfaces (such as trench or polysilicon surfaces) than simple oxide [160]. Stacked ONO capacitor insulators are typically formed by thermally oxidizing the silicon surface, depositing an LPCVD nitride film on the first oxide, and then oxidizing the nitride to form the top oxide layer [160, 161].

One important consideration when defining composite insulators is the trap density in the bulk of the films and at their interfaces. While stacked ONO films exhibit less leakage current and higher breakdown fields than simple thermal oxide grown over a similar surface, they may not be suitable for MOSFET gates because of their inherent large trap density. For nitride layers thicker than 5 nm electron trapping, causing shifts in the

MOSFET turn-on voltage, is significant. The shifts are practically eliminated if the nitride film thickness is reduced below $\simeq$ 4nm and the composite film subjected to rapid-thermal nitridation (RTN) [162].

**Table 3.2** Physical and chemical properties of LPCVD and PECVD nitride films.

| Property | LPCVD | PECVD | Ref. |
|---|---|---|---|
| Deposition temperature (°C) | 700-900 | 250-350 | |
| Composition | $Si_3N_4(H)$ | $Si_xN_yH_z$ | [136] |
| Atomic % hydrogen | 4-8 | 15-35 | [142, 159] |
| Density (g/cm³) | 2.9-3.1 | 2.4-2.8 | [136, 142] |
| Expansion coefficient ($10^{-6}$/K) | 4 | 4-7 | [136] |
| Refractive index | 2.01 | 1.8-2.5 | [142] |
| Dielectric constant | 6-7 | 6-9 | |
| Energy gap (eV) | 5 | 4-5 | |
| Dielectric strength (V/cm) | $10^7$ | $6x10^6$ | |

There are, however, applications which require a large trap density near the nitride-oxide interface. For example, some electrically erasable and programmable read-only memory (EEPROM) cells rely on charge storage in the nitride traps. For this application, a 40-50 nm thick LPCVD nitride layer is deposited on 1.5-2 nm thermally grown oxide film. The state of the memory-cell is determined by the presence or absence of stored charge in discrete traps near the nitride-oxide interface. The transfer of charge to and from the traps occurs by tunneling through the thin oxide. When properly designed, the structure is able to retain charge for over ten years [163].

## 3.2 Chemical-Physical Deposition Processes

The major disadvantage of the thermally activated CVD processes described in the preceding section is the associated high deposition temperature. The changes in dopant profile and device geometry caused by high temperature processing become less tolerable in ULSI where the device dimensions shrink below 0.5 $\mu$m. Also, the maximum allowable processing temperature after silicide formation and metal deposition precludes the use of conventional CVD processing for inter-metal dielectric deposition and passivation. These limitations have led to the development of alternative low-temperature processes which use physical enhancements to chemical reactions, with the primary focus on photo-, plasma-, and ECR-plasma-enhanced CVD.

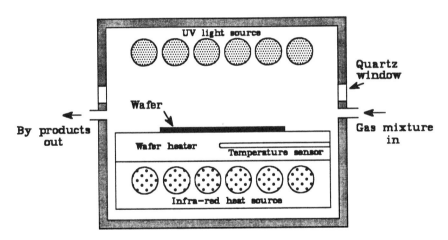

**Fig. 3.24** Schematic of a photo-enhanced CVD deposition system.

## 3.2.1 Photo-CVD Processes

In photo-CVD processes, the reaction between gas molecules is activated or accelerated by electromagnetic radiation, typically ultraviolet light. Photons from lamps or lasers provide energy to dissociate the reactants and enhance their reaction at the wafer surface (Fig. 3.24). The molecules absorb energy either directly, or indirectly via a photosensitiser [164]. In the first case, the photo-wavelength determines which chemical bond of the molecule will be broken. For example, in the reaction between silane and ammonia to produce silicon nitride, both $SiH_4$ and $NH_3$ selectively absorb energy when the wavelength is shorter than about 220 nm [165]. Indirect absorption is achieved by adding a photosensitiser (typically mercury vapor) to the gas mixture. The mercury atoms can be excited with radiation by a quartz mercury lamp (253.7-nm wavelength), and then transfer their kinetic energy to the reactants by collision, producing electrically neutral, free radicals which react at the substrate surface [166, 167].

PHCVD processes have been reported for the deposition of oxide, nitride, epitaxial silicon, metals and amorphous silicon [168]. One advantage of PHCVD is the very low deposition temperature (typically 150 °C), and the greatly reduced radiation-induced damage (compared to PECVD). Another advantage, in particular when using laser-beam CVD (LCVD), is the potential of implementing sequential in situ processing steps without exposing the substrate to the atmosphere, minimizing the risk of contamination and oxidation [169, 170]. For LCVD, UV photons of 193 nm wavelength produced from an excimer laser source (operating on an argon/fluorine mixture) dissociate the reactant molecules; the photo-dissociated products condense on the substrate surface as a solid film. The laser-beam can be directed parallel or perpendicular to the substrate surface. In the parallel beam case the photons do not impinge directly on the film, reducing radiation damage considerably below that of the perpendicular beam arrangement. Since the laser is monoenergetic, the number of dissociation channels is reduced, resulting in improved process repeatability. LCVD also offers the capability of locally depositing films by directing the beam to a particular wafer area. LCVD silicon dioxide and silicon nitride films have been deposited at 380 °C at a rate of 70-80 nm/min under a pressure of 2-6 torr. The gas mixtures are $N_2O/SiH_4/N_2$ for oxide and $NH_3/SiH_4/N_2$ for nitride [169, 170].

## 3.2.2 Plasma-Enhanced Chemical Vapor Deposition (PECVD)

As with photo-assisted CVD, the motivation for using plasma-enhanced deposition processes is that the substrate can be kept at low temperature, typically 300 °C or lower. This is achieved by reacting gases in a glow discharge (plasma) which supplies much of the energy needed for the reaction. The plasma is created by applying a high electric field at a frequency of typically 13.56 MHz across the gas mixture (Fig. 3.25) [136, 171 – 173]. The plasma contains high energy electrons, gas molecules, fragments of gas molecules, and free radicals. A small percentage (0.1-1%) of the gas molecules and fragments will be ionized, and the ion density is in the $10^9 - 10^{12} cm^{-3}$ range [174 – 176]. Because of their light mass, electrons acquire much higher kinetic energy during each cycle than do the ions. The electron energy ranges from 1-20 eV. Therefore, the effective electron temperature can reach $10^4 - 10^5 K$, while the average molecule temperature remains low (typically 100-400 °C). High-energy electrons then transfer energy to the gas molecules by impact and initiate a reaction which would have otherwise not been possible at low temperature. The reaction takes place when the molecules are broken up and the fragments combine at the substrate surface to form the film. The properties of plasma deposited films depend on several variables such as electrode configuration and separation, power level and frequency, gas composition, pressure and flow rate, and substrate temperature [177]. The actual details of intermediate reactions during plasma deposition are not well understood.

Parallel-plate reactors of the type shown in Fig. 3.25 are most commonly used in the production of thin oxide, nitride, and silicon films.

### PECVD Oxide

PECVD oxide films are typically formed by the reaction of silane (or its halides) at $\simeq$ 350 °C with $O_2$, $N_2O$, or $CO_2$ in a glow discharge. A silane and nitrous oxide gas mixture tends to produce oxide films with better uniformity than an oxygen reactant gas [178 – 180].

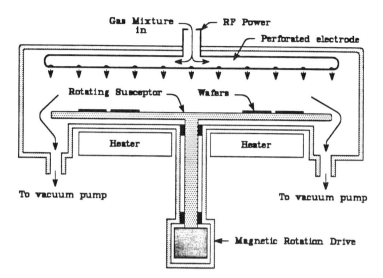

**Fig. 3.25** A typical plasma-assisted parallel plate deposition system [173].

The overall reaction of silane with nitrous oxide is [178]:

$$SiH_4 + 2N_2O \rightarrow SiO_2 + 2N_2 + 2H_2 . \qquad 3.18$$

Plasma enhancement of TEOS oxide (PETEOS) has also been achieved at a temperature as low as 350 °C [181]. Films obtained from the above reactions invariably contain impurities such as bonded hydrogen, and nitrogen [182]. The hydrogen concentration ranges from 5-10 atomic percent and exists in the form of Si-H or Si-O-H. Therefore, these films do not have the stoichiometric composition of thermal oxides and are typically used for passivation and inter-metal level dielectrics, or lithographic masks.

High quality silicon dioxide has been deposited by using a large excess of helium as a diluent (80-99% of the total gas flow). Helium serves as a buffer medium between the plasma and the reactants, limiting undesirable reaction channels. These films exhibit electrical characteristics comparable to those of thermal oxide [183,184].

## Silicon Nitride

PECVD nitride films can be produced from silane (or DCS, $SiCl_2H_2$) and ammonia or nitrogen at 300-400 °C [179, 185]. Unlike $Si_3N_4$ deposited at 750-900 °C, PECVD nitride films are not stoichiometric and depend strongly upon the method and conditions for film deposition. They typically contain excess silicon and large amounts (15-35 atomic %) of bonded hydrogen in form of Si-H, Si-H$_2$, Si-H$_3$, N-H, and N-H$_2$ [186 – 189]. A large fraction of hydrogen is released during thermal annealing and can cause degradation of the film characteristics or the silicon surface, with N-H bonds decomposing at a lower temperature than Si-H bonds [188, 189]. The lower the deposition temperature, the higher the hydrogen content and etch rate (Chap. 5) [187]. PECVD nitride films are commonly denoted SiN or $Si_xN_yH_z$ to distinguish them from stoichiometric nitride.

One important property of nitride films is stress. Nitride films can exhibit tensile or compressive stress, depending on the deposition conditions and subsequent processing [190, 191]. For example, films deposited on silicon at temperatures above 550 °C and at plasma frequencies below 4 MHz exhibit tensile stress ( $\simeq 10^9$ Pa). Compressive stress is observed at plasma frequencies below 4 MHz and deposition temperatures above 550 °C [190]. Ion implantation increases the density of broken bonds and leads to disturbed short-range order of the Si-N structure, expanding the nitride layer [188, 190]. High stress, notably high tensile stress at steps, can cause cracking of the film during further processing [158, 192, 193].

## Silicon Oxynitride

The problems associated with the hydrogen instability and mechanical stress in plasma silicon nitride have led to the development of silicon oxynitride ($SiO_xN_y$) for final passivation and intermetal level dielectrics. Oxynitride films can be formed at 300 °C by introducing oxygen, typically in the form of nitrous oxide, with a silane-ammonia or silane-ammonia-nitrogen gas mixture into a plasma reactor of the type shown in Fig. 3.25 [194 – 196]. Thin oxynitride layers have been deposited under similar conditions to produce "storage films" in read-only memory cells [197]. The film composition can be varied by changing the concentration of nitrous oxide. The incorporation of oxygen in the nitride lattice results in a reduction of mechanical stress by one to two orders of magnitude below the stress observed with PECVD nitride [194, 196].

### Silicon

The most widely used technique for plasma enhanced deposition of amorphous, polycrystalline, and single-crystal silicon films is the decomposition of silane or dichlorosilane.

Hydrogenated amorphous silicon (a-Si:H) can be deposited by plasma-enhanced CVD at low pressure ($\simeq$ 1 torr), using a silane-hydrogen gas mixture, with the substrate held at 25-400 °C [198, 199]. These films are used primarily as the semiconductor material in thin-film transistors (TFT), photovoltaic cells (such as solar cells), and switching elements for liquid-crystal displays. Incorporating hydrogen in the film reduces the density of energy states in the forbidden gap and increases the optical efficiency of solar cells [200].

PECVD films deposited below 600°C are amorphous, while films deposited above 625°C are polycrystalline [201]. All films become polycrystalline after annealing at 1000°C. PECVD polysilicon films are not as temperature sensitive as LPCVD films formed by thermal decomposition of silane. The deposition at low temperature and very low pressure in LPCVD increases the sensitivity of the deposition rate to temperature. Incorporating a plasma into the LPCVD environment reduces this sensitivity by driving the chemical reaction at the silicon surface [201, 202]. The variation of deposition rate from 525 °C to 725°C in PECVD is only 25%, as compared to doubling of the deposition rate in LPCVD when the temperature is increased by 25°C [201].

There is great interest in using PECVD polysilicon rather than amorphous silicon for TFTs because of the higher carrier mobility in polysilicon [203]. Also, semi-insulating polysilicon (SIPOS) formed by PECVD is attractive because the film can be deposited at a temperature as low as 300°C, while APCVD and LPCVD films require temperatures above 600°C [204]. PECVD SIPOS films are used for passivation, high impedance resistors, and to shield electrical fields. The film's conductivity is not too high to cause substantial leakage, but high enough to operate the layer as a field-shield. As with thermal deposition, PECVD SIPOS is formed by reacting silane with nitrous oxide, either at atmospheric pressure or under low pressure.

Deposition of epitaxial silicon at low temperature is important to minimize autodoping and solid-state diffusion. The constraints with respect to contamination and crystalline perfection are, however, much more stringent than with polysilicon and amorphous silicon. The most important step before deposition of low-temperature epitaxy is the removal of native oxide from the silicon surface. This is achieved in PECVD by in-situ surface cleaning by the plasma, for example, by bombarding the silicon surface with argon plasma during deposition at 775°C [205]. The improved quality of PECVD epitaxial films over thermal deposition is attributed to this in-situ cleaning.

PECVD epitaxy was initially formed by the pyrolysis of silane at 1050°C under atmospheric pressure, and at 800°C under low pressure [206]. Improved reactor design and substrate surface quality has allowed epitaxial deposition at a lower substrate temperature ($\leq 700°C$) and at higher rate than conventional CVD ($\simeq 2$ nm/s vs $\simeq 0.8$ nm/s) [207]. The higher rate is attributed to plasma excitation and regeneration of the gas-phase pyrolysis of silane. Adding small amounts of $GeH_4$ to silane at the beginning of plasma deposition to form a Si-Ge buffer layer improves the epitaxial quality and increases the deposition rate to 3-4 nm/s at 750 °C [208].

## Remote Plasma

PECVD has a lower operating temperature than LPCVD and APCVD, and is used wherever low temperature is needed. There are, however, problems which have limited the use of PECVD films. Since PECVD relies on an rf source (such as microwaves) to ionize the reactive gases, the reactants possess high energy which can be transferred to the film and cause damage. While most of the radiation-induced damage can be annealed out at high temperature, this annealing step eliminates the advantage of low temperature processing. Radiation effects can be reduced by using alternate reactor geometries where the plasma is removed from the wafer vicinity, allowing plasma molecules to reach the wafer while preventing high energy electrons and ions from impinging on the film and wafer surface. Another problem associated with PECVD films is their high content of bonded hydrogen and oxygen [180,186,187,189]. These effects can be minimized by isolating the plasma region from the growth surface using a technique known as remote plasma-enhanced CVD (RPECVD) [209 – 219]. This is a variation of the PECVD process in which only one of the constituent gases, the oxygen or nitrogen containing molecule (or a mixture of this gas with a rare gas) is rf excited and then mixed with silane (Fig. 3.26) [211, 212]. The objective of not directly exciting silane is to eliminate gas-phase plasma reactions which generate precursor species that can introduce unwanted bonding groups into the deposited film.

Remote plasma consists of: (a) rf excitation of the gas mixture that contains nitrogen or oxygen, (b) transport of the excited species out of the plasma region, (c) mixing of the transported species with neutral silane, and (d) CVD reaction at the heated substrate to form the film [212]. Besides minimizing plasma-induced surface damage by high-energy particle bombardment, RPECVD allows low-temperature (300-500 °C) deposition of stoichiometric $SiO_2$ and $Si_3N_4$ without introducing bonded hydrogen or oxygen in $Si_3N_4$ or $SiO_2$ [213, 214]. RPECVD has been used to deposit high epitaxial silicon with a low defect density [216, 217]. Excited particles (such as argon) and energetic electrons are mixed with silane, causing its dissociation and increasing the mobility of adsorbed atoms at the silicon surface. This process achieves very low deposition rates (0.1-1.2 nm/min) which is suitable for heterostructures because of the control it provides over layer thickness. The high surface quality is achieved by in-situ remote plasma hydrogen cleaning in ultra-high vacuum (UHV) prior to silicon deposition.

Hydrogen-free oxynitride films of high quality and controlled composition can be deposited at 300 °C with RPECVD by controlling the flow ratio of oxygen and nitrogen source gases [218]. RPECVD has also been used to deposit hydrogenated microcrystal silicon films ($\mu$c-Si:H) at very low temperature (

< 250°C) [219]. These films exhibit higher conductivity and optical bandgap than hydrogenated amorphous silicon (a-Si:H).

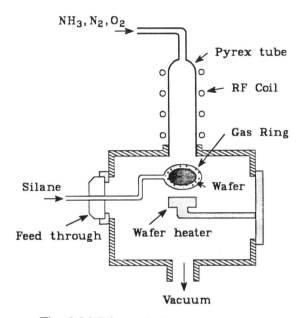

**Fig. 3.26** Schematic Representation of an RPECVD deposition chamber [212].

## Electron-Beam Assisted CVD

In electron-beam assisted CVD (EBCVD), high-energy electrons are injected into the reactant gases to dissociate them through electron-molecule collisions. The dissociated products deposit as a solid film onto the substrate [169]. The electron-beam is created in an "abnormal glow" discharge -- this is the mode in which both the voltage and current density increase when the power fed into the plasma is increased, rapidly increasing the number of secondary electrons [220,221]. The cathodic voltage drop controls the electron-beam energy and the cathode geometry controls the beam shape. As with a laser beam, the electron-beam can be directed parallel or perpendicular to the substrate surface. The advantage of a perpendicular arrangement is the radiation-induced dissociation of bonded hydrogen (Si-H) in the deposited film [222]. The damage caused by impinging electrons can be annealed out at temperatures below 400 °C [223]. Films of silicon nitride and silicon dioxide have been deposited at 350-400 °C using the same gas

**132**

mixture as for the laser-beam and plasma assisted deposition. The deposition rate ranges from 20-50 nm/min [169]. EBCVD oxide and nitride films are, however, inferior to PECVD films with respect to pinhole density and hydrogen content. The excellent uniformity in film thickness and flexibility in deposition make EBCVD suitable for single-wafer processing as the wafer diameter increases above 20 cm.

### 3.2.3 Electron Cyclotron Resonance Plasma Deposition

The purpose of electron cyclotron resonance (ECR) plasma is to increase the path of electrons in the plasma by applying a magnetic field normal to the electron trajectory, enhancing the probability of ionization. Resonance is achieved when the frequency at which energy is fed to an electron circulating in a magnetic field is equal to the characteristic frequency at which the electron circulates. This circulating motion increases the ratio of ionized to non-ionized species in ECR plasmas by three orders of magnitude over that in simple rf plasmas ($10^{-3} - 10^{-1}$ compared to $10^{-6} - 10^{-3}$) [224]. The high efficiency in exciting the reactants in ECR plasmas allows the deposition of films at room temperature without the need for thermal activation.

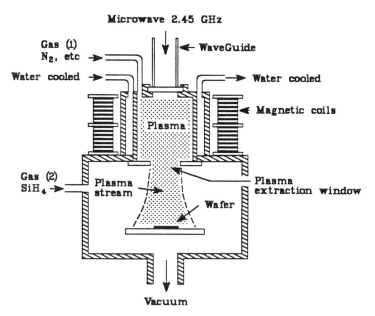

**Fig. 3.27** ECR plasma deposition system [225].

A typical ECR plasma deposition system is shown in Fig. 3.27 [225]. It consists of two chambers, the plasma chamber and the wafer chamber. The reactants are introduced through two separate inlets; oxygen and nitrogen containing molecules are introduced into the plasma chamber, while silane is fed into the wafer chamber. Microwave power is introduced into the plasma chamber through a rectangular waveguide at a frequency of 2.45 GHz (an authorized industrial frequency). The cylindrical plasma chamber operates as a microwave cavity resonator. The electron cyclotron frequency is controlled by magnetic coils arranged at the periphery of the chamber. At the above microwave frequency resonance is achieved at a a magnetic strength of 875 Gauss. A highly activated plasma is then obtained at very low gas pressures ( $10^{-5} - 10^{-3}$ torr) [224]. Ions are extracted from the plasma chamber into the wafer chamber and subjected to a divergent magnetic field that spreads the plasma stream over the entire wafer. The typical ion energy range from 20-30 eV enhances the deposition reaction without appreciably damaging the wafer surface. The wafer temperature is held below 50 °C by cooling. A deposition uniformity of $\pm 5\%$ can be achieved over a 150 mm wafer. In a distributed ECR deposition (DECR), the plasma is generated at the periphery of the cylindrical chamber where microwave rod antennas are distributed in a multipolar magnetic field to satisfy the ECR condition in the vicinity of each antenna (Fig. 3.28) [226, 227].

Silicon nitride is formed by introducing nitrogen into the plasma chamber and silane into the wafer chamber. The deposition rate is about 20 nm/min for a silane and nitrogen flow rate of 10 cc/min and a microwave power of 100 W [224]. When the power is increased to 150 W, the deposition rate increases to 70 nm/min and the refractive index of silicon nitride reaches a minimum of 2.0. ECR plasma-deposited nitride films typically exhibit compressive stress in the range $110^8$-$3x10^8$ Pa. The stress becomes slightly tensile at a microwave power of 150 W, and can be controlled to remain around zero by adjusting the power and gas pressure to 150 W and $\simeq 9x10^{-4}$ torr, respectively [224]. As with remote plasma, the content of bonded hydrogen and oxygen is very small.

To form silicon dioxide films, oxygen is introduced into the plasma chamber. The deposition rate ranges from 20-100 nm/min, depending on microwave power. The refractive index remains in the range 1.46-1.48 and is not strongly affected by power. Also, stress is compressive and remains approximately constant at $2x10^8$ -$3x10^8$ Pa [226 − 229]. Thermal quality oxides have been deposited at 270 °C [228, 229]. High quality oxide films have also been deposited at temperatures below 375 °C by using 2% diluted silane in helium [230].

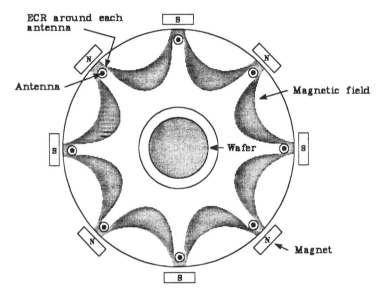

**Fig. 3.28** Schematic of a of distributed ECR deposition system [226].

Silicon can be deposited by introducing an inert gas into the plasma chamber instead of oxygen or nitrogen, resulting in the decomposition of silane in the wafer chamber. Surface cleaning prior to deposition can be achieved at 400 °C by using a hydrogen ECR plasma. The efficient low-temperature cleaning in hydrogen is attributed to an increase in the reaction rate with native oxide [231].

## 3.3 Physical Vapor Deposition

This section describes four important physical vapor deposition (PVD) techniques: vacuum evaporation, sputtering, molecular beam epitaxy, and focused ion beam deposition.

### 3.3.1 Vacuum Evaporation

Vacuum evaporation occurs when a source material (such as aluminum) is heated above its melting point in an evacuated chamber. The evaporated atoms then travel at high velocity in straight-line trajectories and deposit onto wafers which are placed normal to the trajectories (Fig. 3.29) [232]. The source can be molten by resistance heating, rf heating, or with a focused electron beam (electron-gun) (Fig. 3.30).

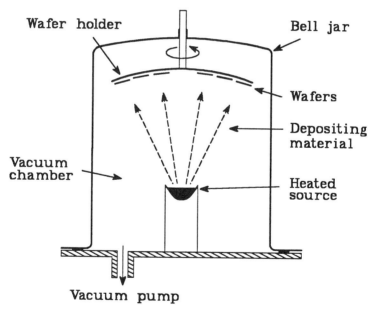

**Fig. 3.29** Schematic of a vacuum evaporation system.

The rate of evaporation depends on the temperature of the source and on the vapor pressure of the evaporant [233]:

$$N_e = (2\pi mkT)^{-1/2} p_e, \qquad\qquad 3.19$$

where $N_e$ is the number of evaporated molecules per unit area and per unit time, $m$ the molecular mass, $k$ the Boltzmann constant, $T$ the absolute surface temperature, and $p_e$ the equilibrium vapor pressure of the evaporant. Therefore, vacuum evaporation is most commonly used for the deposition of metals of high vapor pressure, such as aluminum ($p_e \simeq 0.01$ torr at 1500 K).

Vacuum is necessary to avoid collisions between evaporated source atoms and residual gas molecules. The distance from source to substrate is therefore chosen to be much smaller than the mean-free path of the molecules in the chamber (Eq. 3.20). Vacuum deposition of metals is discussed in more details in Chap. 8.

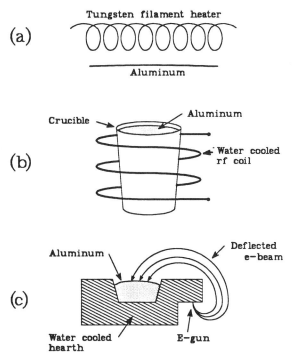

**Fig. 3.30** Typical evaporation sources. (a) Resistance heated. (b) Inductively heated. (c) Electron-beam heated.

### 3.3.2 Sputter Deposition

Sputtering is similar to a billiard-ball event. Ions are accelerated in an electric field toward a target of material to be deposited, where they "knock-off" (sputter) target atoms (Fig. 3.31). The sputtered ions then deposit onto wafers which are conveniently placed facing the target. Argon (Ar $^+$) is typically used for sputtering because it is inert and readily available in a pure form. It is ionized by colliding with high energy electrons in the chamber, and then accelerated in an electric field toward the negatively biased target. The momentum of ions incident on the target is then transferred to surface atoms of the the target material, causing their ejection. Therefore, during sputter deposition, material is removed from the target and deposited onto wafers [234 − 236].

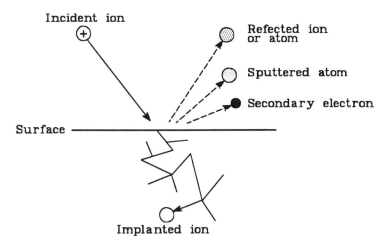

**Fig. 3.31** The sputtering process [235].

The incident ion energy must be large enough to dislodge target atoms, but not too large to cause penetration into the target material (ion implantation). Typical sputtering ion energies range from 500-5000 eV. The number of atoms sputtered from the wafer surface per incident ion is defined as the **sputter yield.** This number varies from $\simeq$ 0.5-1.5, depending the momentum of ions and their angle of incidence.

A simple diode (two electrodes) dc sputtering chamber is shown schematically in Fig. 3.32. Electrons accelerated in the electric field which is created between the metal plate that holds the wafers (anode) and the target (cathode) collide with argon atoms, causing their ionization and generating secondary electrons. The new electrons acquire sufficient energy from the electric field to ionized new argon atoms, forming the plasma. near the target surface. The argon ions are accelerated to the negatively charged target (cathode) where they sputter surface atoms by momentum Typical chamber pressures range from $10^{-3}$ to $10^{-1}$ torr. In this range, the sputtered material travels from target to wafer without considerable collisions, while sufficient collisions between electrons and argon atoms are allowed to maintain the plasma. For insulating targets it is necessary to use an ac power supply. Otherwise, the target will initially be bombarded with positive ions and, since the material is insulating, the target surface will rapidly build-up positive charge repelling incoming positive ions. In an rf sputtering system the target is alternately bombarded by positive ions and then by negative electrons which neutralize the charge [235]. The surface of the target is alternately charged by capacitive coupling at a typical frequency of 13.56 MHz (an allowed industrial frequency). Metals such as Ti, Pt, Mo, W, Au, Ni, and Co can be readily sputtered with either a dc or an rf sputtering system. Sputtering of insulators such as silicon dioxide, silicon nitride, and aluminum oxide requires the use of rf power. While the deposition rate is independent of wafer temperature, the wafer must be heated to about 200 °C for a stable film to form.

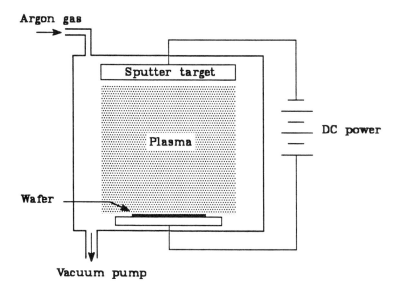

**Fig. 3.32** Schematic of a diode sputtering chamber.

## Ion Beam Sputtering

In an ion beam sputtering system, ions are created outside the chamber rather than in a plasma between the electrodes (Fig. 3.33). A source of ions is accelerated toward the target and impinges on its surface. The sputtered material deposits on a wafer which is placed facing the target. The ion current and energy can be independently modified. Since the target and wafer are placed in a chamber of lower pressure than in any other sputtering system, more target material and less contamination is transferred to the wafer. Ion beam sputter deposition systems, however, do not yet have sufficient throughput to be used in batch manufacturing.

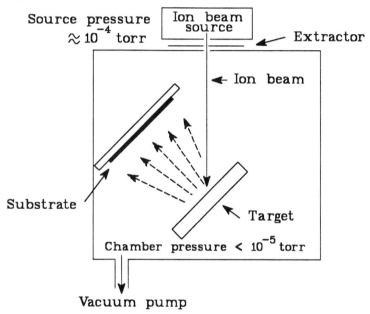

Fig. 3.33 Schematic of an ion beam sputter deposition chamber.

## Magnetron Sputtering

One method to increase the ion density, and hence the sputter-deposition rate, is to use a third electrode (triode sputtering configuration) that provides more electrons for ionization [237]. Another method is to use a magnetic field, as with ECR, to capture and spiral electrons, increasing their ionizing efficiency in the vicinity of the sputtering target. This also reduces the probability of electron-ion recombination at the walls of the chamber. The relatively large mass of the ions essentially prevents their trapping in the magnetic field [238 – 241]. This technique, referred to as **magnetron sputtering,** has found widespread applications for the deposition of aluminum and its alloys at a rate which can approach $1\mu$m/min. The deposition occurs at a much lower voltage than in conventional sputtering systems, thereby reducing the generation of penetrating radiation. Also, the increased deposition rate avoids the formation of a stable oxide film on the aluminum surface by reacting with residual oxidants in the gas, a problem observed at deposition rates below $\simeq 10$ nm/min.

### Reactive Sputter Deposition

When sputtering silicon dioxide or silicon nitride targets, the deposited films does not necessarily have a stoichiometric composition. Silicon nitride may be deficient in nitrogen and silicon dioxide deficient in oxygen. The addition of nitrogen during nitride sputtering or oxygen during oxide sputtering improves the stoichiometry of the film [242, 243]. Another method is to sputter, e.g., silicon with oxygen or nitrogen, or titanium with oxygen, rather than with argon, to form silicon dioxide, silicon nitride, or titanium oxide. This is referred to as **reactive sputter deposition** [244 – 246]. The reaction, however, can take place at the target, in the gas phase, or at the wafer surface. If the reaction occurs at the target, an insulating film forms at the target surface, slowing the the process. Reactions in the gas phase cause an agglomeration of the produced molecules which deposit in clusters, reducing the film quality. The process must therefore be well controlled to induce the reaction at the wafer surface [233].

### Bias Sputter Deposition

By applying rf power to the substrate, ions can be accelerated toward the wafer surface which they bombard. If the energy and density of impinging ions is very large, the film can be sputtered at the same rate as it deposits so that no net deposition occurs. A net deposition can be achieved by controlling the wafer bias. This **bias sputter deposition** technique has several advantages. Bombarding the film with ions of energy in the 5-50 eV range removes contaminants which are loosely trapped in the film, such as inert sputtering gas [234]. Bias sputtering can also be used to deposit conformal films on non-planar surfaces, or to produce planar wafer surfaces prior to metal deposition [247], depending on the bias magnitude. Since without bias most of the sputtered ions travel in straight-line trajectories, sidewalls normal to the substrate surface (such as sidewalls of deep holes or trenches) are not coated at the same rate as planes parallel to the substrate surface. Controlling the substrate bias redirects enough plasma ions toward the growing film so that a fraction of the deposited layer is re-sputtered. Since re-sputtering occurs at a higher rate on parallel "walls" than on normal walls, this method redistributes the deposited material more conformably, however, at the cost of deposition rate. Conformal coverage or planarization are enhanced by increasing the substrate temperature to allow atoms to "move around" on the substrate surface [234].

**Collimated Sputter Deposition**

Holes with large aspect ratios (ratio of depth to diameter) are rather difficult to fill with material, mainly because scattering events cause the top opening of the hole to "seal" before appreciable material has deposited on its floor. This problem can be overcome by collimating the sputtered atoms [248]. This is achieved by placing an array of collimating tubes just above the wafer to restrict the depositing flux to normal $\pm$ 5°. Therefore, atoms whose trajectory is more than 5° from normal are deposited on the inner surface of the collimators.

### 3.3.3 Molecular Beam Epitaxy

In molecular beam epitaxy, a beam of molecules or atoms which emanates from an evaporation or effusion source is transported in an evacuated chamber without collisions, and deposits onto a single-crystal semiconductor surface (Fig. 3.34). For this purpose, the source and wafer are placed in an ultra-high vacuum chamber (UHV, $10^{-10} - 10^{-11}$ torr). At this pressure, the mean-free path of gases is several orders of magnitude larger than the distance between source and substrate (typically 10-20 cm). At very low pressures, the mean-free path can be approximated by [249]:

$$L = \frac{5x10^{-3}}{p}$$ 

where $p$ is the pressure in torr and $L$ the mean-free path in cm.

One of the important features of MBE systems is the use of sophisticated analytical techniques for in-situ process monitoring [250]. These techniques include reflection high energy electron diffraction (RHEED), Auger electron spectroscopy (AES), secondary ion mass spectroscopy (SIMS), electron spectroscopy for chemical analysis (ESCA), and X-ray photoelectron spectroscopy (XPS), as discussed later.

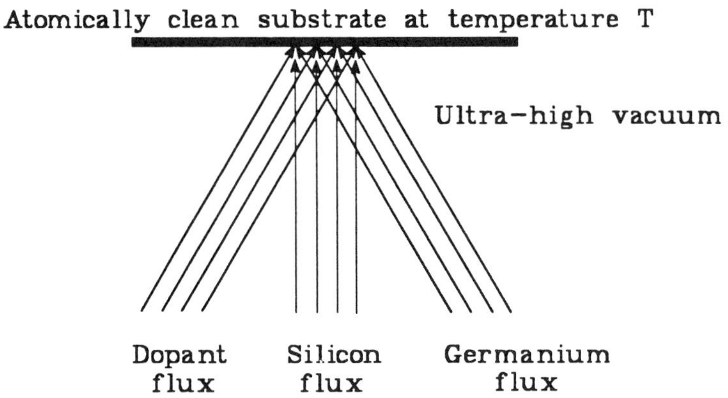

**Fig. 3.34** Schematic representation of MBE deposition of a silicon-based film. The film thickness and composition is defined by controlling the individual fluxes.

Although most of the MBE research was done on elements of groups III and V (such as Ga and As), there is significant work being performed on silicon and germanium, II-VI, IV-VI semiconductors, metals, and oxide superconductors. The main advantages of MBE are: the low epitaxial growth temperature ($450 - 800°C$, which is lower than required for CVD epitaxy), and the low growth rate, independent of wafer temperature, allowing precise control of film thickness. The technique is well suited to form very thin epitaxial layers with abrupt changes in dopant concentrations, allowing the fabrication of digital, microwave, and optoelectronic homo- and heterostructures that are not feasible with other deposition techniques. This section focuses on MBE deposition and doping of silicon, germanium, and III-V compound films.

### Silicon and Germanium Deposition

MBE is not as widely applied to the growth of silicon as it is to III-V compounds, mainly because of the initial difficulty in preparing the silicon surface and controlling the film dopant profile. Recent improvements in surface cleaning and contamination-free ultra-high vacuum environment have, however, allowed the growth and doping of high-quality silicon MBE films [250 – 258].

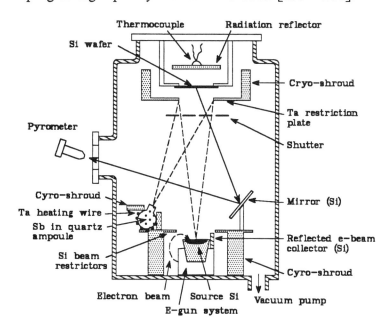

**Fig. 3.35** Schematic of a silicon MBE system [256]. Only a silicon source is shown. A second e-gun source is used for the deposition of germanium with silicon. Antimony doping is performed with an effusion furnace.

A typical silicon or germanium MBE system is shown schematically in Fig. 3.35 [256]. The main components in the chamber are: sources of molecular or atomic beams; a substrate manipulator to heat, translate and rotate the wafer; an ultra-high vacuum system; a cryoshroud surrounding the growth region; shutters to control the deposition rate; a mass analyzer to monitor the gas composition; a thickness monitor; and a pyrometer or thermocouple to control the film temperature. Since silicon and germanium have high melting temperatures, they are best evaporated using electron beam heating (Fig. 3.30). Two separate e-gun sources can be used for the deposition of silicon and germanium. A magnetic field deflects an intense electron beam and directs it into a silicon or germanium "slug" placed in a water-cooled hearth. This creates a molten "pool" from which Si or Ge can evaporate. The hearth is shielded from the source of electrons to avoid direct heating by the tungsten filament, and contamination of the film with tungsten. The wafer is heated to 200-400 °C to increase the silicon and germanium mobilities at the wafer surface. Liquid nitrogen shrouds are used to shield the chamber walls from heat radiation. A pyrometer is used to monitor film temperatures above 750 °C, and a thermocouple for temperatures below below 750 °C. The deposition rate is controlled to the monolayer level by timely actuating the shutters; it is in the order of 10 nm/min, allowing enough time for silicon, germanium, and dopant atoms to find the proper position at the growing film surface. The doping source is typically an effusion furnace, containing, e.g., antimony. Doping is similar to the growth process. A flux of evaporated dopants arrives at the surface and finds appropriate lattice sites in the film. The dopant concentration is controlled by adjusting the dopant flux with respect to the Si or Ge flux.

When depositing germanium on silicon, the mismatch in lattice spacing between the two materials ($\simeq$ 4%) must be considered. A strained superlattice layer can be formed by alternating Ge and Si films, so that the mismatch between the two layers is accommodated by elastic strain between the films. There is, however, a maximum layer thickness (of several nm) above which mismatch induces dislocations in the films.

## Compound Semiconductor Deposition

The epitaxial growth mechanisms of elements from groups III and V of the periodic table has been extensively investigated and reviewed [250, 259 – 262]. As in the case of silicon, the source is the key element of an MBE system. For III and V elements, the standard source is the so-called Knudsen cell (Fig. 3.36). This is an effusion source which consists of a heating element, a surrounding heat shield, a crucible containing the element to be deposited, and a small opening (when compared to the mean-free path of atoms or molecules in the chamber) from which the vapor escapes. Epitaxial growth is performed by directing a beam of elemental sources (such as As, Ga) from the effusion cells onto a GaAs substrate. Separate effusion cells are used for dopants. The film composition is determined by timing the closing and opening of the shutters.

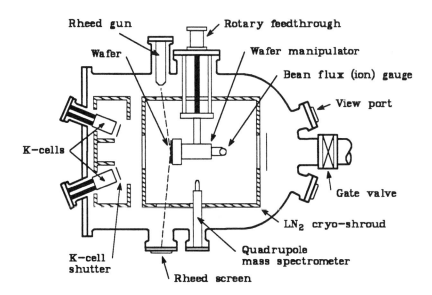

**Fig. 3.36** Schematic cross-section of a typical III-V compound MBE growth chamber using Knudsen (K) source cells [250].

Arsenic and phosphorus evaporate in more than one molecular form (Fig. 3.37). The larger molecules, e.g., $As_4$, typically have a small sticking coefficient (fraction of all atoms incident on the wafer that are adsorbed on the surface). Therefore, the deposition efficiency can be improved by "cracking" the tetramer into two dimers ($As_4 \rightarrow 2As_2$) at the source. This is done by adding a secondary heat zone at the source exit [264].

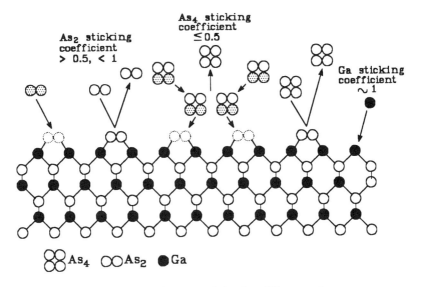

**Fig. 3.37** Schematic to explain the difference between tetramer and dimer sticking coefficients [264].

The MBE technology allows the fabrication of epitaxial films with alternate layers of elemental and compound semiconductors, for example, a heteroepitaxial deposition of GaAs on silicon. The lattice mismatch between GaAs and Si ($\simeq 4.1\%$) is accommodated by depositing buffer layers of $Si_xGe_{1-x}$ composition, where x is the atomic fraction of silicon [265]. The layer thickness is chosen above the critical thickness to segregate misfit dislocations at the interfaces, and distribute the total mismatch in discrete steps. With a final layer of germanium, the residual mismatch with GaAs is reduced to about 0.2 % [265]. A final layer of germanium has also the advantage of more efficiently nucleating GaAs than silicon.

### 3.3.4 Ionized Cluster and Focused Ion Beam Deposition

Research on ionized cluster beam (ICB) and focused ion beam (FIB) deposition has intensified during the last two decade [266 − 268]. A cluster is an aggregate of 100 to 2000 atoms. When it impacts on the wafer surface, the cluster disintegrates into atoms which are then scattered over the surface and deposit to form a film. Focused ion beams are used in a wide range of techniques to locally deposit or remove material. This section briefly describes the ICB and FIB techniques and discusses their applications.

**Ionized Cluster Beam Deposition**

Ion beam epitaxy can be achieved at low substrate temperatures ( $\simeq 80°C$ by imparting low kinetic energies to ions, typically 50-100 eV for silicon, in an ultra-high vacuum [269]. Silicon beams can be formed by evaporating high purity silicon in a boron nitride oven at about 1730 °C and sustaining a plasma in the vapor to create silicon ions. The silicon ion beam is then extracted at low energy and deposited onto a germanium or silicon wafer. With this system, 5-100 nm epitaxial silicon films have been achieved [269].

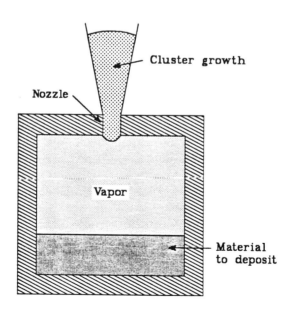

**Fig. 3.38** Formation of a cluster beam [266].

A beam of low energy with a spread of a few eV is, however, rather difficult to produce. Ionized cluster beam deposition avoids this problem by imparting the energy to clusters of 100 to 2000 atoms, resulting in a very small energy per atom [266]. A cluster of atoms is formed by condensation when a supersaturated gas escapes into vacuum through a small nozzle (Fig. 3.38). The cluster is ionized by impact with high energy electrons. The beam of ionized clusters is then accelerated in an electric field to a maximum energy of a few hundred eV. Each atom will then possess a very small energy, comparable to the binding energy of the atom in the film. It is this property of producing a high density beam of low energy atoms that makes ICB attractive for epitaxial growth at very low substrate temperatures.

The basic elements of an ICB film deposition system are: the source, (which consists of a heated crucible containing the element to be deposited), an ionization chamber, and an acceleration column. The ratio of ionized to total cluster flow can be adjusted by controlling the beam of ionizing electrons. When the cluster impacts on the surface, both the ionized and neutral clusters disintegrate into atoms which scatter over the wafer surface. The atoms move along the surface until they find the correct position to rest and form the film (Fig. 3.39). The technique has been used to deposit single crystal metals (e.g., epitaxial aluminum on Si [270]), single crystal oxides and nitrides (e.g., silicon dioxide on Si [267]), and epitaxial gallium arsenide on GaAs [271].

**Focused Ion Beam Deposition**

Focused ion beams (FIB) have found a wide range of applications in restructuring circuits after they are fabricated, mask repairing, micromachining, device cross-sectioning, and nanostructure fabrication (structures with dimensions less than 100 nm). An extensive source of references on FIB can be found in [268]. Since typical beams have a spot size of $\leq$ 100 nm, they must be produced by high intensity sources (1-10 A/cm$^2$). These fall into two categories: liquid metal ion sources and gas field ion sources. Both sources have a needle-type form and rely on field ionization or field evaporation to produce the beam [268]. The beam is then deflected in high vacuum and directed to the desired area without necessitating a masking step (masks are discussed in Chap. 4).

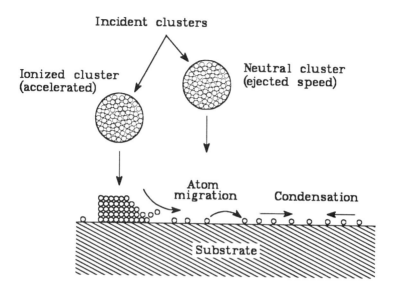

**Fig. 3.39** Illustration of film formation by ionized cluster beam deposition [266].

Focused ion beam lithography (FIBL) uses a beam of ions, rather than photons or electrons to define high resolution patterns on photoresist [272]. The advantages of FIBL are discussed in the next chapter.

Focused ion beam implantation is used to locally introduce dopants into the crystal by accelerating the ions to the appropriate energy (which determines the depth of implanted ions), and deflecting the ion beam to the desired spot of the crystal. Ultra-shallow (< 20 nm) p-type layers of very small dimensions have been fabricated by FIB implantation of Ga $^+$ at 150 keV into n-type silicon [273]. The principle of ion implantation is discussed in Chap. 6. The small ion beam diameter makes the FIB technique suitable for the fabrication of nanostructures. "Bandgap engineering" and "quantum nanostructures" (both discussed later in this series) can be fabricated by locally introducing selected impurities into the crystal [274].

Cutting and attaching techniques are required for circuit repair [275], mask repair [276], and micromachining [277].

Cutting can be performed by locally sputtering the material with a focused ion beam. Attaching is typically achieved by focused ion beam induced deposition (FIB.ID). Tungsten deposition is typically used to connect metal lines by directing a focused beam of gallium ions to the region to be connected in a local atmosphere of hexacarbonyltungsten $(W(CO)_6)$ [278, 279]. The gallium ion induces the decomposition of $W(CO)_6$ and tungsten deposits locally, while the volatile by-products are removed. Local deposition of silicon oxide has been achieved by using a focused beam of $Si^{++}$ ions and a mixture of tetramethoxysilane (TMOS, $Si(OCH_3)_4$) and oxygen [280]. It has been used for local rewiring by etching a hole in the oxide and locally depositing tungsten, without shorting to adjacent conductors.

# PROBLEMS

**3.1** Show that the length L of a stacking fault side measured on the surface of a silicon epitaxial layer of thickness $t_{epi}$ is approximately L = 1.226 $t_{epi}$ for (111) oriented wafers, and L = 1.414 $t_{epi}$ for (100) oriented wafers.

**3.2** The distance between source and wafer in a deposition chamber is 20 cm. Estimate the pressure at which this distance becomes 1% of the mean-free path of source molecules.

**3.3** In an epitaxial deposition process using dichlorosilane, the temperature across a wafer varies from 1050 °C to 1060 °C. Assuming an activation energy of 0.4 eV, estimate the percent variation in deposition rate.

**3.4** If plasma-deposited silicon oxide contains $2x10^{21}$ hydrogen atoms per $cm^3$, find the atomic percent of hydrogen and define an empirical formula for the oxide.

**3.5** Derive the relation between weight percent boron, atomic percent boron, and mole percent $B_2O_3$ for boron-doped silicon dioxide.

**3.6** Assume silicon to be deposited at 1150 °C at a rate of 500 nm/min with an activation energy of 0.6 eV. By how much must the temperature be increased to double this rate?

**3.7** Use the data in Table 3.2 and explain the difference between the densities of LPCVD and PECVD silicon nitride films. Quantify this difference by calculation.

**3.8** A contact hole that must be filled with silicon dioxide has a diameter of 0.5 $\mu$m and a depth of 2 $\mu$m. Which of the following techniques would you use to fill it? (a) Sputtering. (b) Chemical vapor deposition. (c) Thermal oxidation. Explain your choice.

**3.9** How does the average angle of incidence of ions striking on a wafer surface vary with ambient pressure in plasma CVD?

**3.10** A trench in silicon is 4 $\mu$m deep, 0.5 $\mu$ wide, and 20 $\mu$m long. It is filled by successively and conformably depositing 50-nm thick CVD polysilicon films. What is the minimum total film thickness required to fill the trench? What problems could arise when the films are not deposited conformably?

# References

[1] J. L. Vossen and W. Kern, Eds., *Thin Film Processes*, Academic Press, New York, 1978.

[2] K. K. Schuegraf, Ed., *Handbook of Thin-Film Deposition Processes and Techniques*, Noyes Publications, N. J., 1988.

[3] W. Kern and G. L. Schnable, "Low-Pressure Chemical Vapor Deposition *for Very Large-Scale Integration Processing - A Review*," IEEE Trans. Electron Dev. ED-26, 647 (1979).

[4] H. C. Theuerer, J. J. Kleimack, H. H. Loar, and H. Christensen, *"Epitaxial Diffused Transistors,"* Proc. IRE, 48, 1642 (1960).

[5] H. Nelson, *"Epitaxial Growth from the Liquid State and its Application to the Fabrication of Tunnel and Laser Diodes,"* RCA Review, 24, 603 (1963).

[6] I. K. Bonsal, *"Autodoping and Particulate Contaminations during Pre-Diffusion Chemical Cleaning of Silicon Wafers,"* Solid-State Technology, p. 75, July 1986.

[7] G. R. Srinivasan and B. S. Meyerson, *"Current Status of Reduced Temperature Silicon Epitaxy by Chemical Vapor Deposition,"* J. Electrochem. Soc., 134, 1518 (1987).

[8] L. E. Katz and D. W. Hill, *"High Oxygen Czochralski Silicon Crystal Growth Relationship to Epitaxial Stacking Faults,"* J. Electrochem. Soc., Vol. 125, 1151 (1978).

[9] J. Borland, M. Gangani, R. Wise, S. Fong, Y. Oka, and Y. Matsumoto, *"Silicon Epitaxial Growth for Advanced Device Structures,"* Solid State Technology, 111, January 1988.

[10] J. Bloem, *"Silicon Epitaxy from Mixtures of $SiH_4$ and $HCl$,"* J. Electrochem. Soc., 117, 1397 (1970).

[11] J. Nishizawa and M. Saito, *"Growth Mechanisms of Chemical Vapor Deposition of Silicon,"* Proceedings of the 8th International Conference on Chemical Vapor Deposition, Electrochem. Soc., p. 317, 1981.

[12] A. Lekholm, *"Epitaxial Growth of Silicon from Dichlorosilane,"* J. Electrochem. Soc., 120, 1122 (1973).

[13] M. L. Hammond, *"Silicon Epitaxy,"* Solid State Technology, p. 68, Nov. 1978.

[14] C. W. Pearce, *"Epitaxy,"* in S. M. Sze, ed., VLSI Technology, McGraw-Hill, 1983.

[15] S. E. Meyer and D. E. Shea, *"Epitaxial Deposition of Silicon Layers by Pyrolysis of Silane,"* J. Electrochem. Soc., 111, 550 (1964).

[16] E. Sirtl, L. P. Hunt, and D. H. Sawyer, *"High Temperature Reactions in Silicon-Hydrogen-Chlorine Systems,"* J. Electrochem. Soc., 121, 919, (1974).

[17] F. C. Eversteyn, *"Chemical-Reaction Engineering in the Semiconductor Industry,"* Philips Res. Rep., 29, 45 (1974).

[18] McD. Robinson, in F. F. Wang, Ed., *Impurity Doping Processes in Silicon*, North-Holland, Amsterdam, 1981.

[19] R. Reif, T. I. Kamins, and K. C. Saraswat, *"A Model for Dopant Incorporation into Growing Silicon Epitaxial Films,"* J. Electrochem. Soc., 126, 644/653 (1979).

[20] J. Bloem, *"The Effect of Trace Amounts of Water Vapor on Boron Doping in Epitaxially Grown Silicon,"* J. Electrochem. Soc., 118, 1839 (1971).

[21] H. Basseches, R. C. Manz, C. O. Thomas, and S. K. Tung, "Factors Affecting the Resistivity of Silicon Epitaxial Layers," p. 69 in J. B. Schroeder, Ed., Proceedings of the 1961 Semiconductor Metallurgy Conference, Interscience, New York, 1962.

[22] H. B. Pogge, D. W. Bass, and E. Ebert, *Proc. Conf. on Chemical Vapor Deposition, 2nd Intnl. Conf.*, J. M. Blocher and J. C. Withers, Eds., p. 768, Electrochem. Soc., N.Y. 1972.

[23] T. Ishii, K. Takahashi, A. Kondo, and K. Shirakata, *"Silicon Epitaxial Wafer with Abrupt Interface by Two-Step Epitaxial Growth Technique,"* J. Electrochem. Soc., 122, 1523 (1975).

[24] D. Gupta and R. Yee, *"Silicon Epitaxial Layers with Abrupt Interface Impurity Profiles,"* J. Electrochem. Soc., 116, 1561 (1969)

[25] C. O. Thomas, D. Kahng, and R. C. Manz, *"Impurity Distribution in Epitaxial Films,"* J. Electrochem. Soc., 109, 1055 (1962).

[26] A. S. Grove, A. Roder, and C. T. Sah, *"Impurity Distribution in Epitaxial Growth,"* J. Appl. Phys., 36, 802 (1965).

[27] G. R. Srinivasan, *"Autodoping Effects in Silicon Epitaxy,"* J. Electrochem. Soc., 127, 1334 (1980)

[28] B. D. Joyce, J. C. Weaver, and D. J. Maule, *"Impurity Redistribution Processes in Epitaxial Silicon Layers,"* J. Electrochem. Soc., 112, 1100 (1965).

[29] W. H. Shepherd, *"Autodoping of Epitaxial Silicon,"* J. Electrochem. Soc., 115, 652 (1968).

[30] P. H. Langer and J. I. Goldstein, *"Boron Autodoping during Silane Epitaxy,"* J. Electrochem. Soc., 124, 592 (1977).

[31] G. Skelly and A. C. Adams, *"Impurity Atom Transfer During Epitaxial Deposition on Silicon,"* J. Electrochem. Soc., 120, 116 (1973).

[32] G. R. Srinivasan, *"Kinetics of Lateral Autodoping in Silicon Epitaxy,"* J. Electrochem. Soc., 125, 146 (1978).

[33] C. O. Bozler, *"Reduction of Autodoping,"* J. Electrochem. Soc., 122, 1705 (1975).

[34] M. Ogirima, H. Sarda, M. Suzuki, and M. Maki, *"Low Pressure Silicon Epitaxy,"* J. Electrochem. Soc., 124, 903 (1977).

[35] M. J-P. Duchemin, M. M. Bonnet, and M. F. Koelsch, *"Kinetics of Silicon Growth Under Low Hydrogen Pressure,"* J. Electrochem. Soc., 125, 637 (1978).

[36] E. Krullman and W. L. Engel, *"Low Pressure Silicon Epitaxy,"* IEEE Trans. Electron. Dev., ED-29, 491 (1982).

[37] C. M. Drum and C. A. Clark, *"Geometrical Stability of Shallow Surface Depressions During Growth of (111) and (100) Epitaxial Silicon,"* J. Electrochem. Soc., 115, 664 (1968).

[38] C. M. Drum and C. A. Clark, *"Anisotropy of Macrostep Motion and Pattern Edge-Displacement During Growth of Epitaxial Silicon on Silicon Near {100},"* J. Electrochem. Soc., 117, 1401 (1970).

[39] S. P. Weeks, *"Pattern Shift and Pattern Distortion During CVD Epitaxy on (111) and (100) Silicon,"* Solid State Technology, p. 111, Nov. 1981.

[40] R. B. Herring, *"Advances in Reduced Pressure Silicon Epitaxy,"* Solid State Technology, 22, 75 (1979).

[41] P. Rai-Choudhoury, *"Epitaxial Gallium Arsenide from Trimethygallium and Arsine,"* J. Electrochem. Soc., 116, 1745 (1969).

[42] H. M. Manasevit, and W. I. Simpson, *"The Use of Metalorganics in the Preparation of Semiconductor Materials,"* J. Electrochem. Soc., 116, 1725 (1969).

[43] M. L. Ludowise, *"Metalorganic Chemical Vapor Deposition of III-V Semiconductors,"* J. Appl. Phys., 58, R31, Oct. 1985.

[44] G. B. Stringfellow, in *"Organometallic vapor Phase Epitaxy,"* Academic Press, Boston, 1989.

[45] F. Faggin and T. Klein, *"Silicon Gate Technology,"* Solid State Electronics, 13, 1125 (1970).

[46] G. L. Patton, J. C. Bravman, J. D. Plummer, *"Physics, Technology, and Modeling of Polysilicon Emitter Contacts for VLSI Bipolar Transistors,"* IEEE Trans. Electron Devices, ED-33, 1754 (1986).

[47] V. Probst, H. J. Boehm, H. Schaber, H. Oppolzer, and I Weitzel, *"Analysis of Polysilicon Diffusion Sources,"* J. Electrochem. Soc., 135, 671 (1988).

[48] J. Graul, A. Glasl, and H. Murrmann, *"High-Performance Transistors with Arsenic-Implanted Polysil Emitters,"* IEEE J. Solid State Circuits, SC-11, 491 (1976).

[49] J. M. C. Stork, M. Arienzo, and C. Y. Wong, *"Correlation Between the Diffusive and Electrical Barrier Properties of the Interface in Polysilicon Contacted $n^+$-p Junctions,"* IEEE Trans. Electron Devices, ED-32, 1766 (1985).

[50] T. H. Ning and R. D. Isaac, *"Effect of Emitter Contact on Current Gain of Silicon Bipolar Devices,"* IEEE Trans. Electron Devices, ED-27, 2051 (1980).

[51] K. Tsukamoto, Y. Akasaka, and K. Horie, *"Arsenic Implantation into Polycrystalline Silicon and Diffusion to Silicon Substrate,"* J. Appl. Phys., 48, 1815 (1977).

[52] P. Ashburn and B. Soerowirdjo, *"Comparison of Experimental and Theoretical Results on Polysilicon Emitter Bipolar Transistor,"* IEEE Trans. Electron Devices, ED-31, 853 (1984).

[53] A. K. Kapoor and D. J. Roulston, Eds., *Polysilicon Emitter Bipolar Transistors*, IEEE Press, N. Y. 1989.

[54] W. T Lynch and V. Frederick, *"Fabrication of FETs with Source and Drain Contacts Aligned with the Gate Electrode,"* U.S. Patent 4,822,754, April 18, 1989.

[55] K. Kato, T. Ono, and Y. Amemiya, *"A Physical Mechanism of Current-Induced Resistance Decrease in Heavily Doped Polysilicon Resistors,"* IEEE Trans. Electron Devices, ED-29, 1156 (1982).

[56] K. Kato, T. Ono, Y. Amemiya, *"A Monolithic 14 Bit D/A Converter Fabricated with a New Trimming Technique (DOT),"* IEEE J. Solid-State Circuits, SC-19, 802 (1984).

[57] N. Hoshi, S. Kayama, T. Nishihara, J.-I Aoyama, T. Kamatsu, and T. Shihada, *"1.0 μm CMOS Process for Highly Stable Terra-Ohm Polysilicon Load 1Mb SRAM,"* IEDM 1986 Technical Digest, p. 300, 1986.

[58] W. A. Lane and G. T. Wrixon, *"The Design of Thin-Film Polysilicon Resistors for Analog IC Applications,"* IEEE Trans. Electron Devices, ED-36, 738 (1989).

[59] J. E. Suarez, B. E. Johnson, and B. El-Kareh , *"Thermal Stability of Polysilicon Resistors,"* IEEE Trans. Comp. and Manufacturing Technology, 15 (3) June 1992.

[60] C. Kaya, H. Tigelar, J. Paterson, M. de Wit, J. Fattruso, R. Hester, S. Kiriakai, K. S. Tan, and F. Tsay, *"Polycide/Metal Capacitors for High Precision A/D Converters,"* IEDM 1988 Technical Digest, p. 782.

[61] T-I Liou and C-S Teng, *"$n^+$ - Poly-to-$n^+$ -Silicon Capacitor Structures for Single-Poly Analog CMOS and BiCMOS Processes,"* IEEE Trans. Electron Devices, ED-36, 1620 (1989).

[62] S. A. St Onge, S. G. Franz, A. F. Puttlitz, A. Kalinoski, B. E. Johnson, and B. El-Kareh, *"Design of Precision Capacitors for Analog Applications,"* Elect. Comp. and Tech. Confer., San Diego, May 1992.

[63] N. Yamauchi, J-J.J. Hajjar, and R. Reif, *"Drastically Improved Performance in Poly-Si TFTs with Channel Dimensions Comparable to Grain Size,"* IEDM 1989 Technical Digest, p. 353.

[64] R. A. Martin, M. Hack, J. G. Shaw, and M. Shur, *"Intrinsic Capacitance of Amorphous Silicon and Polysilicon Thin Film Transistors,"* IEDM 1989 Technical Digest, p. 361.

[65] C. H. Seager, *"Grain Boundaries in Polycrystalline Silicon,"* Ann. Rev. Mater. Sci., 15, 271 (1985).

[66] R. S. Rosler, *"Low Pressure CVD Production Processes for Poly, Nitride, and Oxide,"* Solid Sate Technology, 20, 63, April 1977.

[67] W. A. Brown and T. I. Kamins, *"An Analysis of LPCVD System Parameters for Polysilicon, Silicon Nitride and Silicon Dioxide Deposition,"* Solid State Technology, 22, 51, July 1979.

[68] M. E. Cowher and T. O. Sedgwick, *"Chemical Vapor Deposited Polycrystalline Silicon,"* J. Electrochem. Soc., 119, 1565 (1972).

[69] T. I. Kamins, M. M. Mandurah, and K. C. Saraswat, *"Structure and Stability of Low Pressure Chemically Vapor-Deposited Silicon Films,"* J. Electrochem. Soc., 125, 927 (1978).

[70] T. I. Kamins, *"Structure and Properties of LPCVD Silicon Films,"* J. Electrochem. Soc., 127, 686 (1980).

[71] S. J. Krause, S. R. Wilson, W. M. Paulson, and R. B. Gegory, *"Grain Growth During Transient Annealing of As-Implanted Polycrystalline Silicon Films,"* Appl. Phys. Lett., 45, 778 (1984).

[72] L. R. Zheng, L. S. Hung, and J. W. Mayer, *"Grain Growth in Arsenic-Implanted Polycrystalline Si,"* Appl. Phys. Lett., 51, 2139 (1987).

[73] A. L. Fripp and L. H. Slack, *"Resistivity of Doped Polycrystalline Silicon Films,"* J. Electrochem. Soc., 120, 145 (1973).

[74] J. Y. W. Seto, *"The Electrical Properties of Polycrystalline Silicon Films,"* J. Appl. Phys., 46, 5247 (1975).

[75] D. M. Kim, A. N. Khondker, S. S. Ahmed, and R. R. Shah, *"Theory of Conduction in Polysilicon: Drift-Diffusion Approach in Crystallin-Amorphous-Crystalline Semiconductor System - Part I: Small Signal Theory,"* IEEE Trans. Electron Devices, ED-31, 480 (1984); *"Part II: General I-V Theory,"* ibid, p. 493.

[76] M. M. Mandurah, K. C. Saraswat and T. I. Kamins, *"A Model for Conduction in Polycrystalline Silicon" - Part I: Theory",* IEEE Trans. Electron Dev. ED-28, 1163 (1981); *"Part II: Comparison of Theory and Experiment,"* ibid, p. 1171.

[77] M. M. Manduhrah, K. C. Saraswat, and T. I. Kamins, *"Phosphorus Doping of Low Pressure Chemically Vapor-Deposited Silicon Films,"* J. Electrochem. Soc., 126, 1019 (1979).

[78] A. G. O'Neill, C. Hill, J. King, and C. Pease, *"A New Model for the Diffusion of Arsenic in Polycrystalline Silicon,"* J. Appl. Phys., 64, 167 (1988).

[79] T. I. Kamins, J. Manoliu, and R. N. Tucker, *"Diffusion of Impurities in Polycrystalline Silicon,"* J. Appl. Phys, 43, 83 (1972).

[80] D. J. Coe, *"The Lateral Diffusion of Boron in Polycrystalline Silicon and its Influence on the Fabrication of Submicron MOSTs,"* Solid State Electronics, 20, 985 (1977).

[81] M. M. Mandurah, K. C. Saraswat, C. R. Helms, and T. I. Kamins, "*Dopant Segregation in Polycrystalline Silicon,*" J. Appl. Phys., 51, 5755 (1980).

[82] C. R. M. Governor, P. E. Batson, D. A. Smith, and C. Wong, "*As Segregation to Grain Boundaries in Si,*" Philosophical Magazine A, 50, 409 (1984).

[83] R. C. Cammarata, C. V. Thompson, and S. M. Garrison, "*Secondary Grain Growth During Rapid Thermal Annealing of Doped Polysilicon Films,*" Mat. Res. Soc. Symp. Proc. 92, 335 (1987).

[84] Z-H Zhou, F. Yu, and R. Reif, "*A Multichamber Single-Wafer Chemical Vapor Deposition Reactor and Electron Cyclotron Resonance Plasma for Flexible Integrated Circuit Manufacturing,*" J. Vac. Sci. Technol., B9, 374 (1991).

[85] M Sternheim, E. Kinsbron, J. Alspector, and P. A. Heimann, "*Properties of Thermal Oxides Grown on Phosphorus In-Situ Doped Polysilicon,*" J. Electrochem. Soc., 130, 1735 (1983).

[86] C. P. Ho and S. E. Hansen, *Technical Report NO SEL 83-001,* Stanford University, Stanford, California.

[87] T. I. Kamins, "*Oxidation of Phosphorus-Doped Low Pressure and Atmospheric Pressure CVD Polycrystalline-Silicon Films,*" J. Electrochem. Soc., 126 838 (1979).

[88] E. A. Irene, E. Tierney and D. W. Dong, "*Silicon Oxidation Studies: Morphological Aspects of the Oxidation of Polycrystalline Silicon,*" J. Electrochem. Soc., 127, 705 (1980).

[89] L. Faraone and G. Harbeke, "*Surface Roughness and Electrical Conduction of Oxide/Polysilicon Interfaces,*" J. Electrochem. Soc., 133, 1410 (1986).

[90] D. J. DiMaria and D. R. Kerr, "*Interface Effects and High Conductivity in Oxide Grown from Polycrystalline Silicon,*" Appl. Phys. Lett., 27 (9), 505 (1975).

[91] R. M. Anderson and D. R. Kerr, "*Evidence of Surface Asperity Mechanism of Conductivity in Oxide Grown on Polycrystalline Silicon,*" J. Appl. Phys., 48 (11), 1834 (1977).

[92] B. D. Joyce and J. A. Baldrey, "*Selective Epitaxial Deposition of Silicon,*" Nature, 195, 485 (1962).

[93] J. O. Borland, "*Novel Device Structures by Selective Epitaxial Growth (SEG),*" IEDM 1987 Technical Digest, p. 12.

[94] R. T. Bates, "*Nanoelectronics,*" Solid State Technology, 32, 101 (1989).

[95] K. Tanno, N. Endo, H. Kitajima, Y. Korugi, and H. Tsuya, "*Selective Silicon Epitaxy Using Reduced Pressure Technique,*" Jap. J. Appl. Phys., 21, L564 (1985).

[96] J. O. Borland and C. I. Drowley, "*Advanced Dielectric Isolation Through Selective Epitaxial Growth Techniques,*" Solid State Technology, 28 (8) 141, (1985).

[97] H. Voss and H. Kurten, "*Device Isolation Technology by Selective Low-Pressure Silicon Epitaxy,*" IEDM 1983 Technical Digest, p. 35, 1983.

[98] H. Pagliaro, J. Corboy, L. Jastrzebski, and R. Soydan, "*Uniformly Thick Selective Epitaxial Silicon,*" J. Electrochem. Soc., 134, 1235 (1987).

[99] S. Nagao, K. Higashitani, Y. Akasaka, and H. Nakata, "*Application of Selective Silicon Epitaxial Growth for CMOS Technology,*" IEEE Trans. Electron Devices, ED-33, 1738 (1986).

[100] L. Jastrzebski, A. C. Ipri, and J. F. Corboy, "*Device Characterization on Monocrystalline Silicon Grown over $SiO_2$ by the (ELO) Epitaxial Lateral Overgrowth Process,*" IEEE Electron Dev. Lett., EDL-4, 32 (1983).

[101] G. W. Neudeck, "*A New Epitaxial Lateral Overgrowth Silicon Bipolar Transistor,*" IEEE Electron Dev. Lett., EDL-8, 492 (1987).

[102] J. L. Regolini, D. Bensahel, E. Scheid, and J. Mercier, "*Selective Epitaxial Silicon Growth in the 650-1100 °C Range in a Reduced Pressure Chemical Vapor Deposition Reactor Using Dichlorosilane,*" Appl. Phys. Lett., 54, 658 (1989).

[103] L. Jastrzebski, J. F. Corboy, J. T. McGinn, and R. Pagliaro, Jr., "*Growth Process of Silicon over $SiO_2$ by CVD: Epitaxial Overgrowth technique,*" J. Electrochem. Soc., 130, 1571 (1983).

[104] J.-C. Lou, C. Galewski, and W. G. Oldham, "*Dichlorosilane Effects on Low-Temperature Selective Silicon Epitaxy,*" Appl. Phys. Lett., 58, 59 (1991).

[105] B. S. Meyerson, "*Low Temperature Silicon Epitaxy by Ultrahigh Vacuum/Chemical Vapor Deposition,*" Phys. Lett., 48, 797 (1986).

[106] T. O. Sedgewick, P. D. Agnello, M. Berkenblit, and T. S. Kuan, "*Growth of Facet-Free Selective Silicon Epitaxy at Low Temperature and Atmodpheric Pressure,* " J. Electrochem. Soc., 138, 3042 (1991).

[107] P. R. Choudhury and D. K. Schroder, "*Selective Growth of Epitaxial Silicon and Gallium Arsenide,*" J. Electrochem. Soc., 118, 106 (1971).

[108] J. Murota, N. Nakamura, M. Kato, and N. Makoshiba, "*Low Temperature Silicon Selective Deposition and Epitaxy on Silicon Using the Thermal Decomposition of Silane Under Ultra-Clean Environment,*" Appl. Phys. Lett., 54, 1007, (1989).

[109] M. Kato, T. Sato, J. Murota, and M. Mikoshiba, "*Nucleation Control of Silicon on Silicon Oxide for Low-Temperature CVD and Silicon Selective Epitaxy,*" J. Crystal Growth, 99, 240 (1990).

[110] M. Racanelli and D. W. Greve, "*Low-Temperature Selective Epitaxy by Ultrahigh Vacuum Chemical Vapor Deposition from $SiH_4$ and $GeH_4/H_2$,*" Appl. Phys. Lett., 58, 2096 (1991).

[111] P. M. Garone, J. C. Sturm, and P. V. Schwartz, "*Silicon Vapor Epitaxial Growth Catalysis by the Presence of Germane,*" Appl. Phys. Lett., 56, 1275 (1990).

[112] M. Kato, C. Iwasaki, J. Murota, N. Mikoshiba, and S. Ono, "*Nucleation Control of Silicon-Germanium on Silicon-Oxide for Selective Epitaxy and Polysilicon Formation in Ultraclean Low-Pressure CVD,*" Extended Abstracts of the 22nd Conf. Solid State Dev. and Mater., Sendai, p. 329, 1990.

[113] K. D. Beyer, V. J. Silvestri, J. S. Markis, and W. Guthrie, "*Trench Isolation by Selective Epitaxy and CVD Oxide Cap,*" J. Electrochem. Soc., 137, 3951 (1990).

[114] L. Jastrzebski, "*SOI by CVD: Epitaxial Lateral Overgrowth (ELO) Process - Review,*" J. Crystal Growth 63, 493 (1983).

[115] T. I. Kamins and D. R. Bradbury, "*Trench-Isolated Transistors in Lateral CVD Epitaxial Silicon-on-Insulator Films,*" IEEE Electron Device Lett., EDL-5, 449 (1984).

[116] A. Ishitani, H. Hitajima, K. Tanno, H. Tsuya, N. Endo, N. Kasai, and Y. Kurogi, "*Selective Silicon Epitaxial Growth for Device-Isolation Technology,*" Microelectronic Eng., 4, 3 (1986)

[117] S. Hine, S. Nagao, N. Tsubouchi, and H. Nakata, "*Solid State Color Imager with Buried Oxide Wall Fabricated by Low Pressure Selective Epitaxy,*" IEDM 1984 Technical Digest, p. 36.

[118] J. Manoliu and J. O. Borland, "*A Submicron Dual Buried Layer Twin Well CMOS SEG Process,*" IEDM 1987 Tech. Dig., p. 20.

[119] P. V. Gilbert, G. W. Neudeck, J. P. Denton, and S. J. Duey, *"Quasi-Dielectrically Isolated Bipolar Junction Transistors with Subcollector Fabricated Using Silicon Selective Epitaxy,"* IEEE Trans. Electron Dev., ED-38, 1660 (1991).

[120] M. Aoki, H. Takato, S. Samata, M. Numano, A. Yagishita, K. Hieda, A. Nitayama, and F. Horiguchi, *"Quarter-Micron Selective-Epitaxial-Silicon Refilled Trench (SRT) Isolation Technology with Substrate Shield,"* IEDM 1991 Tech. Dig., p. 447.

[121] T.-Y Chiu, K. F. Lee, M. Y. Lau, S. N. Finega, M. D. Morris, and A. Voshchenkov, *"Suppression of Lateral Autodoping from Arsenic Buried Layer by Selective Epitaxial Capping,"* IEEE Electron Dev. Lett., 11, 123 (1990).

[122] G. G. Shahidi, D. D. Tang, B. Davari, Y. Taur, P. McFarland, K. Jenkins, D. Danner, M. Rodriguez, A. Megdanis, E. Petrillo, M. Polcari, and T. H. Ning, *"A Novel High-Performance Lateral Bipolar on SOI,"* IEDM 1991 Tech. Dig., p. 663, 1991.

[123] J. Borland, D. Schmidt, and A. Stivers, *"Low Temperature Low Pressure Silicon Epitaxial Growth and its Application to Advanced Dielectric Isolation Technology,"* Extended Abstracts of the 18th Conference on Solid State Devices and Materials, Tokyo, Japan, p. 53, Aug. 1986.

[124] G. Shahidi, B. Davari, Y. Taur, J. Warnock, M. R. Wordeman, P. McFarland, S. Mader, M. Rodriguez, R. Assenza, G. Bronner, B. Ginsberg, T. Lii, M. Polcari, and T. H. Ning, *"Fabrication of CMOS on Ultrathin SOI Obtained by Epitaxial Lateral Overgrowth and Chemical-Mechanical Polishing,"* IEDM 1990 Tech. Dig., p. 587.

[125] H. Shibata, Y. Suizu, S. Samata, T. Matasuno, and K. Hashimoto, *"High-Performance Half-Micron PMOSFETs with 0.1 μm Shallow P⁺N Junction Utilizing Selective Silicon Growth and Rapid Thermal Annealing,"* IEDM 1987 Tech. Digest, p. 590.

[126] T. Yamada, S. Samata, H. Takato, Y. Matsuchita, K. Heida, A. Nitayama, F. Horiguchi, and F. Masuoka, *"A New Cell Structure with a Spread Source/Drain (SSD) MOSFET and a Cylindrical Capacito for 64-Mb DRAMs,"* IEEE Trans. Electron Dev., ED-38, 2481 (1991).

[127] H. Shibata, S. Samata, M. Saitoh, T. Matsuno, H. Sasaki, Y. Matsuchita, K. Hashimoto, and J. Matsunaga, *"Low-Resistive and Selective Silicon Growth as a Self-Aligned Contact-Hole Filler and its Application to 1 M-bit Static RAM,"* VLSI Symp. Techn., 1987 Tech. Digest, p. 75.

[128] J. A. Friedrick and G. W. Neudeck, *"Oxide Degradation During Selective Epitaxial Growth of Silicon,"* J. Appl. Phys., 64, 3538 (1988).

[129] K. Hoffmann, G. W. Rubloff, and R. A. McCorkle, *"Defect Formation in Thermal SiO$_2$ by High-Temperature Annealing,"* Appl. Phys. Lett., 49, 1525 (1986).

[130] K. Hoffmann, G. W. Rubloff, and D. R. Young, *"Role of Oxygen in Defect-Related Breakdown in Thin SiO$_2$ Films on Si,"* J. Appl. Phys., 61, 4584 (1987).

[131] J. M. Eldridge and P. Balk, *"Formation of Phosphosilicate Glass on Silicon Dioxide,"* Trans. Metal. Soc. of AIME, 242, 539 (1968).

[132] K. Massau, R. A. Levy and D. L. Chadwick, *"Modified Phosphosilicate Glasses for VLSI Applications,"* J. Electrochem. Soc., 132, 409 (1985).

[133] D. R. Kerr, J. S. Logan, P. J. Burkhardt and W. A. Pliskin, *"Stabilization of SiO$_2$ Passivation Layers with P$_2$O$_5$,"* IBM J. Res. Dev. 9, 376 (1964).

[134] P. Balk and J. M. Eldridge, *"Phosphosilicate Glass Stabilization of FET Devices,"* Proceedings of the IEEE, 57, 1553 (1969).

[135] A. C. Adams and C. D. Capio, *"Planarization of Phosphorus-Doped Silicon Dioxide,"* J. Electrochem. Soc., 128, 423 (1981).

**160**

[136] W. Kern and R. S. Rosler, *"Advances in Deposition Processes for Passivation Films,"* J. Vac. Sci. Technol., 14, 1082 (1977).

[137] P, J, Tobin, J. B. Price, and L. M. Campbell, *"Gas Phase Composition in the Low Pressure Chemical Vapor Deposition of Silicon Dioxide,"* J. Electrochem. Soc. 127, 2222 (1980)

[138] E. A. Irene, N. J. Chou, D. W. Dong, and E. Tierney, *"On the Nature of CVD Si-Rich SiO_2 and Si_3N_4 Films,"* J. Electrochem. Soc., 127 (11), 2518 (1980).

[139] C. M Giunta, J. D. Chapple-Sokol, and R. G. Gordon, *"Kinetic Modeling of the Chemical Vapor Deposition of Silicon Dioxide from Silane or Disilane and Nitrous Oxide,"* J. Electrochem. Soc., 137 (10), 3237 (1990).

[140] A. C. Adams and C. D. Capio, *"The Deposition of Silicon Dioxide Films at Reduced Pressure,"* J. Electrochem. Soc., 126, 1042 (1979).

[141] H. Huppertz and W. L. Engl, *"Modeling of Low-Pressure Deposition of SiO_2 by Decomposition of TEOS,"* IEEE Trans. Electron Devices, ED-26, 658 (1979).

[142] A. C. Adams, *"Dielectric and Polysilicon Film Deposition,"* in *VLSI Technology*, S. M. Sze, Ed., Mc Graw Hill, New York, 1983.

[143] R. H. Vogel, S. R. Butler, and F. S. Feigel, *"Electrical Properties of Silicon Dioxide Films Fabricated at 770 °C. I: Pyrolysis of Tetraethoxysilane,"* J. Electron. Mater., 14, 329 (1985).

[144] F. S. Becker, D. Pawlik, H. Anzinger, and A. Spitzer, *"Low-Pressure Deposition of High-Quality SiO_2 Films by Pyrolisis of Tetraethylorthosilicate,"* J. Vac. Sci. Technol., B5, 1555 (1987).

[145] Y. Nishimoto, N. Tokumasu, T. Fukuyama, and K, Maeda, *"Low Temperature Chemical Vapor Deposition of Dielectric Films Using Ozone and Organosilane,"* Extended Abstracts, 19th Conference Solid-State Devices and Materials, p. 447 (1987).

[146] H. Kotani, M. Matsuura, A. Fujii, H. Genjou, and S. Nagao, *"Low-Temperature APCVD Oxide Using TEOS-Ozone Chemistry for Multilevel Interconnections,"* IEDM 1989 Technical Digest, p. 669.

[147] S. Nguyen, D. Dobuzinski, D. Harmon, R. Gleason, and S. Fridmann, *"Reaction Mechanisms of Plasma- and Thermal-Assisted Chemical Vapor Deposition of Tetraethylorthosilicate Oxide Films,"* J. Electrochem. Soc., 137 (7), 2209 (1990).

[148] K. Fujino, Y. Nishimoto, N. Tokumasu, and K. Maeda, *"Silicon Dioxide Deposition by Atmospheric Pressure and Low-Temperature CVD Using TEOS and Ozone,"* J. Electrochem. Soc., 137 (9), 2883 (1990).

[149] M. Matsuura, Y. Hayashide, H. Kotani, and H. Abe, *"Film Characteristics of APCVD Oxide using Organic Silicon and Ozone,"* Jpn. J. Appl. Phys., 1530 (1991).

[150] K. Fujino, Y. Nishimoto, N. Tokumasu, and K. Maeda, *"Reaction Mechanism of TEOS and O_3 Atmospheric Pressure CVD,"* 1991 Proceedings, 8th Intnl. IEEE VLSI Multilevel Interconnection Conference, p. 445.

[151] K-T. Chang and K. Rose, *"Enhanced Injection at Silicon-Rich Oxide Interface,"* Appl. Phys. Lett., 49 (14), 868 (1986).

[152] D. J. DiMaria, K. M. DeMeyer, C. M. Serrano, and D. W. Dong, *"Electrically-Alterable Read-Only Memory Using Silicon-Rich SiO_2 Injectors and a Floating Polycrystalline Silicon Storage Layer,"* J. Appl. Phys., 52 (7) 4825 (1981).

[153] D. J. DiMaria, K. M. DeMeyer, and D. W. Dong, *"Dual-Electron-Injector-Structure Electrically-Alterable Read-Only-Memory Modelong Studies,"* IEEE Trans. Electron Devices, ED-28 (9), 1047 (1981).

[154] J. T. Milek, *"Silicon Nitride for Microelectronic Applications,"* Parts 1 and 2, *"Handbook of Electronic Materials,"* Vols. 3 and 6. IFI/Plenum, New York, 1971-1972.

[155] J. V. Dalton and J. Drobek, *"Structure and Sodium Migration in Silicon Nitride Films,"* J. Electrochem. Soc. 115, 865 (1968).

[156] V. Y. Doo, D. R. Nichols, and G. A. Silvey, *"Preparation and Properties of Pyrolytic Silicon Nitride,"* J. Electrochem. Soc., 113, 1279 (1966).

[157] T. Arizumi, T. Nishinaga, and H. Ogawa, *"Thermodynamical Analysis and Experiments for the Preparation of Silicon Nitride,"* Jpn. J. Appl. Phys., 7, 1021 (1968).

[158] K. E. Bean, P. S. Gleim, R. L. Yeakley, and W. R. Runyan, *"Some Properties of Vapor Deposited Silicon Nitride Films Using the $SiH_4 - NH_3 - H_2$ System,"* J. Electrochem. Soc., 114, 733 (1968).

[159] H. J. Stein, B. L. Doyle, and S. T. Picraux, *"Hydrogen Concentration Profiles and Chemical Bonding in Silicon Nitride,"* J. Electronic Material, 8, 11 (1979).

[160] T. Watanabe, N. Goto, N. Yasuhisa, T. Yanase, T. Tanaka, and S. Shinozaki, *"Highly Reliable Trench Capacitor with $SiO_2/Si_3N_4/SiO_2$ Stacked Film,"* IEEE, IRPS 1987 Technical Digest, p. 50.

[161] Y. Ohji, T. Kusaka, I. Yoshida, A. Hiraiwa, K. Yagi, and K. Mukai, *"Reliability of Nano-Meter Thick Multilayer Dielectric Films on Polycrystalline Silicon,"* IEEE IRPS 1987 Technical Digest, p. 55.

[162] H. Iwai, H. S. Momose, T. Morimoto, Y. Ozawa, and K. Yamabe, *"Stacked-Nitride Oxide Gate MISFET with High Hot-Carrier-Immunity,"* IEDM 1990 Technical Digest, p. 235.

[163] H. C. Card and M. I. Elmasry, *"Functional Modeling of Nonvolatile MOS Memory Devices,"* Solid State Electronics, 19, 863 (1976).

[164] R. L. Abber, *"Photochemical Vapor Deposition,"* in Handbook of Thin-Film Deposition Processes and Techniques, K. K. Schuegraf, Ed., Noyes Publications, New Jersey, 1988.

[165] J. W. Peters, F. L. Gebhart, and T. C. Hall, *"Low Temperature Photo-CVD Silicon Nitride - Properties and Applications,"* Solid State Technology, 23 (9) 121 (1980).

[166] C. H. J. Van Der Brekel and P. J. Severin, *"Control of the Deposition of Silicon Nitride by 2573 Å Radiation,"* J. Electrochem. Soc., 119, 372 (1972).

[167] Y. Mishima, M. Hirose, Y. Osaka, K. Nagamine, Y. Ashida, N. Kitagawa, and K. Isogaya, *"Silicon Thin-Film Formation by Direct Photochemical Decomposition of Disilane,"* Jpn. J. Appl. Phys., 22 (1), L46 (1983).

[168] W. I. Milne, F. J. Clough, S. C. Deane, S. D. Baker, and P. A. Robertson, *"Photoenhanced CVD of Hydrogenated Amorphous Silicon Using an Internal Hydrogen Discharge Lamp,"* Appl. Surface Science, 43, 277 (1989).

[169] C. A. Moore, Z-q Yu, L. R. Thompson, and G. J. Collins, *"Laser and Electron Beam Assisted Processing,"* K. K. Schuegraf, Ed., Handbook of Thin-Film Deposition Processes and Techniques, p. 318, Noyes Publications, New Jersey, 1988.

[170] J. J. Rocca, J. Meyer, M. Farrell, and G. J. Collins, *"Glow-Discharge-Created Electron Beams: Cathode Materials, Electron Designs, and Technological Applications,"* J. Appl. Phys., 56 (3), 790 (1984).

[171] J. R. Hollahan and R. S. Rosler, *"Plasma Deposition of Inorganic Films,"* Thin Film Processes, J. L. Vossen and W. Kern, Eds., Academic Press, New York, 1978.

[172] J. L. Vossen and W. Kern, *"Thin-Film Formation,"* Physics Today, 33, 26 (May 1980).

[173] R. S. Rosler, W. C. Benzing, and J. Balod, *"A Production Reactor for Low Temperature Plasma-Enhanced Silicon Nitride Deposition,"* Solid-State Technology, 19 (6), 45 (1976).

[174] S. Rhee, J. Szekely, and O. J. Ilebusi, *"On the Three-Dimensional Transport Phenomena in CVD Processes,"* J. Electrochem. Soc., 134, 2552 (1987).

[175] J. C. Knight, Symposia Proceedings of Material Research Society Fall 1984 Meeting, 38, 371 (1985).

[176] S. Rhee and F. J. Szekely, *"The Analysis of Plasma-Enhanced Chemical Vapor Deposition of Silicon Films,"* J. Electrochem. Soc., 133, 2194 (1986).

[177] M. J. Rand, *"Plasma-Promoted Deposition of Thin Inorganic Films,"* J. Vac. Sci. technol. 16, 420 (1979).

[178] J. R. Hollahan, *"Deposition of Plasma Silicon Oxide Thin Films in a Production Planar Reactor,"* J. Electrochem. Soc., 126, 930 (1979).

[179] H. F. Sterling and R. C. G. Swann, *"Chemical Vapor Deposition Promoted by R.F. Discharge,"* Solid-State Electronics 8, 653 (1965)

[180] A. C. Adams, "Plasma Deposition of Inorganic Films," Solid-State Technology, 24, 135 (1983).

[181] B. L. Chin and E. P. van de Ven, *"Plasma TEOS Process for Interlayer Dielectric Applications,"* Solid-State Technology, 31, 119, April 1988.

[182] A. C. Adams, F. B. Alexander, C. D. Capio, and T. E. Smith, *"Characterization of Plasma-Deposited Silicon Dioxide,"* J. Electrochem. Soc., 128, 1545 (1981).

[183] J. Batey and E. Tierney, *"Low-Temperature Deposition of High-Quality Silicon Dioxide by Plasma-Enhanced Chemical Vapor Deposition,"* J. Appl. Phys., 60, 3136 (1986).

[184] A. A. Bright, J. Batey, and E. Tierney, *"Low-Rate Plasma Oxidation of Si in a Dilute Oxygen/Helium Plasma for Low-Temperature Gate Quality Si/SiO$_2$ Interfaces,"* Appl. Phys. Lett., 58 (6), 619 (1991).

[185] R. G. G. Swann, R. R. Mehta, and T. P. Cauge, *"The Preparation and Properties of Thin-Film Silicon-Nitrogen Compounds Produced by a Radio Frequency Glow Discharge Reaction,"* J. Electrochem. Soc., 114, 713 (1967).

[186] W. A. Lanford and M. J. Rand *"The Hydrogen Content of Plasma Deposited Silicon Nitride,"* J. Appl. Phys., 49, 2473 (1978).

[187] R. Chow, W. A. Lanford, W. Ke-Ming, and R. S. Rosler, *"Hydrogen Content of a Variety of Plasma Deposited Silicon Nitride,"* J. Appl. Phys., 53, 5360 (1982).

[188] H. J. Stein, V. A. Wells, and R. H. Hampy, *"Properties of Plasma-Deposited Silicon Nitrride,"* J. Electrochem. Soc., 126, 1750 (1979).

[189] M. Maeda and H. Nakamura, *"Hydrogen Bonding Configurations in Silicon Nitride Films Prepared by Plasma-Enhanced Deposition,"* J. Appl. Phys., 58 (1), 484 (1985).

[190] V. A. P. Claassen, W. G. J. N. Volkenburg, M. F. C. Willemsen, and M. W. v.d. Wijgert, *"Influence of Deposition Temperature, Gas Pressure, Gas Phase Composition, and Frequency on Composition and Mechanical Stress of Plasma Silicon Nitride Layers,"* J. Electrochem. Soc., 132, 893 (1977). JES 130 1249 (1983).

[191] E. P. EerNisse, *"Stress in Ion-Implanted CVD Si$_3$N$_4$ Films,"* J. Appl. Phys., 48, 3337 (1977).

[192] M. J. Grieco, F. L. Worthing, and B. Schwartz, *"Silicon Nitride Thin Films from SiCl$_4$ Plus NH$_3$: Preparation and Properties,"* J. Electrochem. Soc., 115, 525 (1968).

[193] A. K. Sinha, H. J. Levinstein, and T. E. Smith, *"Thermal Stresses and Cracking Resistance of Dielectric Films ($SiN$, $Si_3N_4$, and $SiO_2$) on Si Substrates,"* J. Appl. Phys., 49 (4), 2423 (1978).

[194] C. M. M. Denisse, K. Z. Troost, J. B. Oude Elferink, F. H. P. M. Habraken, W. F. v.d. Weg, and M. Hendriks, *"Plasma-Enhanced Growth and Composition of Silicon Oxynitride Films,"* J. Appl. Phys., 60 (7), 2536 (1986).

[195] V. S. Nguyen, P. Pan, and S. Burton, *"The Variation of Physical Properties of Plasma-Deposited Silicon Nitride and Oxynitride with Their Compositions,"* J. Electrochem. Soc., 131, 2348 (1984).

[196] W. A. P. Claassen, H. A. J. Th. v. d. Pol, A. H. Goemans, and A. E. T. Kuiper, *"Characterization of Silicon Oxynitride Films Deposited by Plasma-Emhanced CVD,"* J. Electrochem. Soc., 133, 1458 (1986).

[197] Q. A. Shams and W. D. Brown *"Physical and Electrical Properties of Memory Quality PECVD Silicon Oxynitride,"* J. Electrochem. Soc., 137, 1244 (1990).

[198] P. E. Vanier, F. J. Kampas, R. R. Cordeman, and G. Rajeswaran, *"A Study of Hydrogenated Amorphous Silicon Deposited by rf Glow Discharge in Silane-Hydrogen Mixtrure,"* J. Appl. Phys., 56 (6), 1812 (1984).

[199] F. Boulitrop, N. Proust, J. Magarifio, E. Criton, J. F. Peray, and M. Dupre, *"A Study of Hydrogenated Amorphous Silicon Deposited by Hot-Wall Glow Discharge,"* J. Appl. Phys., 58 (9), 3494 (1985).

[200] D. K. Biegelsen, R. A. Street, C. C. Tsai, and J. C. Knights, *"Hydrogen Evolution and Defect Creation in Amorphous Si:H Alloys,"* Phys. Rev. B, 20, 4839 (1979).

[201] T. I. Kamins and K. L. Chiang, *"Properties of Plasma-Enhanced CVD Silicon Films,"* J. Electrochem. Soc., 129, 2326 (1982).

[202] J-J. J. Hajjar and R. Reif, *"Characteristics of Thin-Film Transistors Fabricated in Polysilicon Films Deposited by PLasma Enhanced Chemical Vapor Deposition,"* J. Electronic Materials, 19 (12), 1403 (1990).

[203] D. A. Buchanan, J. Batey, and E. Tierney, *"Thin-Film Transistors Incorporating a Thin, High-Quality PECVD $SiO_2$ Gate Dielectric,"* IEEE Electron Device Lett., 9 (11), 576 (1988).

[204] W. C. Lai, S. S. Ang, W. D. Brown, H. A. Naseem, R. K. Ulrich, and P. V. Dressendorfer, *"Growth Characterization of PECVD Semi-Insulating Polysilicon Films and Resistors,"* J. Electronic Materials, 19 (5), 419 (1990).

[205] J. H. Comfort and R. Reif, *"Chemical Vapor Deposition of Epitaxial Silicon from Silane at Low Temperatures,"* J. Electrochem. Soc., 136 (8), 2398 (1989).

[206] W. G. Townsend and M. E. Uddin, *"Epitaxial Growth of Silicon from $SiI_4$ in the Temperature Range 800-1150 °C,"* Solid-State Electronics, 16, 39 (1973).

[207] T. J. Donahue, W. R. Burger, and R. Reif, *"Low-Temperature Silicon Epitaxy Using Low Pressure Chemical Vapor Deposition with and without Plasma Enhancement,"* Appl. Phys. Lett., 44 (3), 346 (1984).

[208] S. Suzuki and T. Itoh, *"The Effect of Si-Ge Buffer Layer for Low-Temperature Si Epitaxial Growth on Si Substrate by RF Plasma Chemical Vapor Deposition,"* J. Appl. Phys. 54 (3), 1466 (1983).

[209] M. J. Helix, K. V. Vaidayanathan, B. G. Streetman, H. B. Dietrich, and P. K. Chatterjee, *"R.F. Plasma Deposition of Silicon Nitride Layers,"* Thin Solid Films, 55, 143 (1978).

[210] L. G. Meiners, "*Electrical Properties of $SiO_2$ and $Si_3N_4$ Dielectric Layers on InP,*" J. Vac. Sci. Technol., 19, 373 (1981).

[211] P. D. Richard, R. J. Markunas, G. Lucovsky, G. G. Fountain, A. N. Mansour, and D. V. Tsu, "*Remote Plasma Enhanced CVD Deposition of Silicon Nitride and Oxide for Gate Insulators in (In, Ga)As FET Devices,*" J. Vac. Sci. Technol., A 3 (3), 867 (1985).

[212] D. V. Tsu, G. Lucovsky, and M. J. Mantini, "*Local Atomic Structure in Thin Films of Silicon Nitride and Silicon Diimide Produced by Remote Plasma-Enhanced Chemical-Vapor Deposition,*" Phys. Rev. B, 33 (10), 7069 (1986).

[213] G. Lucovsky, P. D. Richard, D. V. Tsu, S. Y. Lin, and R. J. Markunas, "*Deposition of Silicon Dioxide and Silicon Nitride by Remote Plasma Enhanced Chemical Vapor Deposition,*" J. Vac. Sci. Technol., A 4 (3), 681 (1986).

[214] S. S. Kim, D. V. Tsu, and G. Lucovsky, "*Deposition of Device Quality Silicon Dioxide Thin Films by Remote Plasma Enhanced Chemical Vapor Deposition,*" J. Vac. Sci. Technol., A 6 (3), 1740 (1988).

[215] G. Lucovsky, D. V. Tsu, S. S. Kuim, R. J. Markunas, and G. G. Fountain, "*Formation of Thin Film Dielectrics by Remote Plasma-Enhanced Chemical-Vapor Deposition (Remote PECVD),*" Applied Surface Science, 39, 33 (1989).

[216] L. Breaux, B. Anthony, T. Hsu, S. Banerjee, and A. Tasch, "*Homoepitaxial Films Grown on Si(100) at 150 °C Remote Plasma-Enhanced Chemical Vapor Deposition,*" Appl. Phys. Lett., 55 (18), 1885 (1989).

[217] B. Anthony, T. Hsu, L. Breaux, R. Qian, S. Banerjee, and A. Tasch, "*Remote Plasma-Enhanced CVD of Silicon: Reaction Kinetics as a Function of Growth Parameters,*" J. Electronic Materials, 19 (10), 1089 (1990).

[218] T. Fuyuki, T. Saitoh, and H. Matsunami "*Low-Temperature Deposition of Hydrogen-Free Silicon Oxynitride Without Stress by the Remote Plasma Technique,*" Jpn. J. Appl. Phys. Part 1, 29 (10), 2247 (1990).

[219] S. C. Kim, M. H. Jung, and J. Jang, "*Growth of Microcrystal Silicon by Remote Plasma Chemical Vapor Deposition,*" Appl. Phys. Lett., 58 (3), 281 (1991).

[220] J. L. Vossen and J. J. Cuomo, "*Glow Discharge Sputter Deposition,*" in Thin Film Processes, J. L. Vossen and W. Kern, Eds., p. 12, Academic Press, New York, 1978.

[221] B. Chapman, "*Glow Discharge Processes,*" Wiley Interscience, New York (1980).

[222] D. C. Bishop, K. A. Emery, J. J. Rocca, L. R. Thompson, H. Zarnani, and G. J. Collins, "*Silicon Nitride Films Deposited with an Electron Beam Created Plasma,*" Appl. Phys. Lett., 44 (6), 598 (1984).

[223] C. T. Sah, J. Y Sun, and J. J. Tzou, "*Generation-Annealing Kinetics and Atomic Models of a Compensating Donor in the Surface Space-Charge Layer of Oxidized Silicon,*" J. Appl. Phys., 54 (2), 944 (1983).

[224] S. Matsuo, "*Microwave Electron Cyclotron Resonance Plasma Chemical Vapor Deposition,*" K. K. Schuegraf, Handbook of Thin-Film Deposition Processes and Techniques, p. 147, Noyes Publications, New Jersey, 1988.

[225] S. Matsuo and M. Kiuchi, "*Low Temperature Chemical Vapor Deposition Method Utilizing and Electron Cyclotron Resonance Plasma,*" Jpn. J. Appl. Phys., 22, L210 (1983).

[226] M. J. Cooke and N. Sharrock, "*Planarising Silicon Dioxide Layers by Distributed ECR Deposition,*" Proceedings of the Eight Symposium on Plasma Processing, G. S. Mathad and D. W. Hess, Eds., The Electrochem. Soc., 90-14, 538 (1990).

[227] F. Plais, B. Agius, F. Abel, J. Siejka, M. Puech, and P. Alnot, "*Low Temperature Deposition of Silicon Dioxide by Distributed Electron Cyclotron Resonance Plasma Enhanced*

*Chemical Vapor Deposition,"* Proceedings of the Eight Symposium on Plasma Processing, G. S. Mathad and D. W. Hess, Eds., The Electrochem. Soc., 90-14, 544 (1990).

[228] T. T. Chau, S. R. Mejia, and K. C. Kao, *"Electronic Properties of Thin SiO$_2$ Films Deposited at Low Temperatures by New ECR MIcrowave PECVD Process,"* Electronics Lett., 25 (16), 1088 (1989).

[229] S. V. Nguyen and K. Albaugh, *"The Characterization of Electron Cyclotron Resonance Plasma Deposited Silicon Nitride and Silicon Oxide Films,"* J. Electrochem. Soc., 136, 2835 (1989).

[230] R. G. Andosca, W. J. Varhue, and E. Adams, *"Silicon Dioxide Films Deposited by Electron Cyclotron Resonance Plasma Enhanced Chemical Vapor Deposition,"*

[231] T. Shibata, Y. Nanishi, and M. Fujimoto, *"Low-Temperature Si Surface Cleaning by Hydrogen Beam with Electron-Cyclotron-Resonance Plasma Excitation,"* Jpn. J. Appl. Phys. Part 2, 29 (7), 1181 (1990).

[232] J. F. O'Hanlon, *"A User's Guide to Vacuum Technology,"* John Wiley and Sons, New York (1981).

[233] R. Glang, *"Vacuum Evaporation,"* L. I Maissel and R. Glang, Eds., Handbook of Thin Film Technology, p. 11, McGraw-Hill, New York (1970).

[234] J. L. Vossen and J. J. Cuomo, *"Glow Discharge Sputter Deposition,"* J. L. Vossen and W. Kern, Thin Film Processes, p. 11, Academic Press, New York (1978).

[235] B. N. Chapman, *"Glow Discharge Processes: Sputtering and Plasma Etching,"* John Wiley & Sons, New York (1980).

[236] B. N. Chapman and S. Mangano, *"Introduction to Sputtering,"* K. K. Schuegraf, Ed., Handbook of Thin-Film Deposition Processes and Techniques, p. 291, Noyes Publications, New Jersey (1988).

[237] T. C. Tisone, *"Low Voltage Triode Sputtering,"* Solid-State Technology, 18 (12), 34 (1975).

[238] C. F. Powell, J. H. Oxley, and J. M. Blocher, Jr., Eds., *"Vapor Deposition,"* John Wiley & Sons, New York (1966).

[239] T. Van Vorous, *"Planar Magnetron Sputtering: A New Industrial Coating Technique,"* Solid State technology, , 62, Dec. 1976.

[240] A. S. Penfold and J. A. Thornton, *"Electrode Type Glow Discharge Apparatus,"* U.S. Patents 3,884,793 (1975); 3,995,187, 4,030996, 4,031,424, and 4,041,353 (1977).

[241] M. Wright and T. Beardow, *"Design and Advances in the Rotatable Cylindrical Magnetron,"* J. Vac. Sci. and Technol., A 4 (3), 388 (1986).

[242] G. J. Kominiak, *"Silicon Nitride by Direct RF Sputter Deposition,"* J. Electrochem. Soc., 122, 1271 (1975).

[243] S. Suyama et al., *"The Effect of Oxygen-Argon Mixing on Properties of Sputtered Silicon Dioxide Films,"* J. Electrochem. Soc., 134, 2260 (1987).

[244] C. J. Moghab, et al. *"Effect of Reactant Nitrogen Pressure on The Microstructure and Properties of Reactively Sputtered Silicon Nitride Films,"* J. Electrochem. Soc., 122, 815 (1975).

[245] J. Kortland and L. Oosting, *"Deposition and Properties of RF Reactively Sputtered SiO$_2$ Layers,"* Solid-State Technology, , 153, Oct. 1982.

[246] K. G. Geraghty and L. F. Donaghey, *"Kinetics of the Reactive Sputter Deposition of Titanium Oxide,"* J. Electrochem. Soc., 123, 1201 (1976).

[247] T. N. Kennedy, "*Sputtered Insulator Films Contouring over Substrate Topography,*" J. Vac. Sci. and Technol., 13, 1135 (1976).

[248] S. M. Rossnagel, D. Mikalsen, H. Kinoshita, and J. J. Cuomo, "*Collimated Magnetron Sputter Deposition,*" J. Vac. Sci. and Technol. A, Vac. Surf. Films, 9 (2), 261 (1991).

[249] H. Sigiura and M. Yamaguchi, "*Growth of Dislocation-Free Silicon Films by Molecular Beam Epitaxy,*" J. Vac. Sci. and Technol., 19, 157 (1981).

[250] W. S. Knodle and R. Chow, "*Molecular Beam Epitaxy: Equipment and Practice,*" Handbook of Thin-Film Deposition Processes and Techniques, K. K. Schuegraf, Ed., p. 170, Noyes Publications, New Jersey 1988.

[251] B. A. Unvala, "*Epitaxial Growth of Silicon by Vacuum Evaporation,*" Nature, 194, 166 (1962).

[252] K. L. Wang, "*Novel Devices by Si-Based Molecular Beam Epitaxy,*" Solid State Technology, 28 (10), 137 (1985).

[253] S. S. Iyer, R. A. Metzger, and F. G. Allen, "*Sharp Profiles with High and Low Doping Levels in Silicon Grown by Molecular Beam Epitaxy,*" J. Appl. Phys., 52, 5608 (1981).

[254] J. C. Bean, "*Silicon Molecular Beam Epitaxy as a VLSI Processing Technique,*" IEDM Technical Digest, p. 6 (1981).

[255] U. Konig, H. Kibbel, and E. Kasper, "*MBE: Growth and Sb Doping,*" J. Vac. Sci. Technol., 16, 985 (1979).

[256] J. C. Bean and E. A. Sadowiski, "*Silicon MBE Apparatus for Uniform High-Rate Deposition on Standard Format Wafers,*" J. Vac. Sci. Technol., 20, 137 (1982).

[257] Y. Ota, "*Silicon Molecular Beam Epitaxy (n on n+) with Wide Range Doping Control,*" J. Electrochem. Soc., 124, 1795 (1977).

[258] S. S. Iyer, et al., "*Dopant Incorporation Processes in Silicon Grown by Molecular Beam Epitaxy,*" VLSI Sci. and Technol., K. E. Bean and G. A. Rozgonyi, Eds., p. 473, Electrochemical Society, Pennington, New Jersey 1984.

[259] J. C. Bean, S. S. Iyer, and K. L. Wang, Eds., "*Silicon Molecular Beam Epitaxy,*" Material Research Society, Symposium Proceedings, Vol. 220, Pittsburg, Pennsylvania (1991).

[260] J. R. Arthur, "*Interaction of Ga and As$_2$ Molecular Beams with GaAs Surfaces,*" J. Appl. Phys., 39, 4032 (1968).

[261] A. Y. Cho, "*Film Deposition by Molecular Beam Technologies,*" J. Vac. Sci. and Technol. 8, S31 (1971).

[262] L. L. Chang, L. Esaki, W. E. Howard, R. Ludeke, and G. Schul, "*Structures Grown by Molecular Beam Epitaxy,*" J. Vac. Sci. and Technol. 10, 655 (1973).

[263] B. Pamplin, Ed., "*Molecular Beam Epitaxy,*" Pergamon Press, New York 1980.

[264] E. Parker, Ed., "*The Technology and Physics of Molecular Beam Epitaxy,*" Plenum Press, New York 1985.

[265] J. B. Posthill, D. P. Malta, R. Venkatasubramanian, R. R. Sharps, M. L. Timmons, R. J. Markunas, T. P. Humphreys, and N. R. Parikh, "*MBE Growth and Characterization of $Si_xGe_{1-x}$ Multilayer Structures on $Si(100)$ for Use as Substrate for GaAs Heteroepitaxy,*" Silicon Molecular Beam Epitaxy, J. C. Bean, S. S. Iyer, and K. L. Wang, Eds., Material Research Society, Symposium Proceedings, Vol. 220, Pittsburg, Pennsylvania, p. 265 (1991).

[266] I. Yamada, T. Takagi, and P. Younger, *"Ionized Cluster Beam Deposition,"* K. K. Schuegraf, Ed., Handbook of Thin-Film Deposition Processes, p. 344, Noyes Publications, New Jersey (1988).

[267] T. Takagi, I. Yamada, M. Kunori, and S. Kobiyama, *"Vaporized-Metal Cluster Ion Source for Ion Plating,"* Jap. J. Appl. Phys., Suppl. 2, p. 427 (1974).

[268] R. A. D. Mackenzie and G. D. W. Smith, *"Focused Ion Beam Technology: a Bibliography,"* Nanotechnology, 1 (2), 163 (1990).

[269] P. C. Zalm and L. J. Beckers, *"Ion Beam Epitaxy of Silicon on Ge and Si at Temperatures of 400 K,"* Appl. Phys. Lett., 41 (2), 167 (1982).

[270] I. Yamada, H. Inokawa, and T. Takagi, *"Epitaxial Growth of Al on Si(111) by Ionized Cluster Beam,"* J. Appl. Phys. , 56, 2746 (1984).

[271] P. Younger, *"Principle and Application of Ionized Cluster Beam Deposition,"* J. Vac. Sci. Technol., A3, 588 (1985).

[272] R. L. Kubena, J. W. Ward, F. P. Stratton, R. J. Joyce, and G. M. Atkinson, *"A Low Magnification Focused Ion Beam System with 8 nm Spot Size,"* J. Vac. Sci. Technol., B 9 (6), 1937 (1991).

[273] A. J. Steckl, H. C. Mogul, and S. M. Mogren, *"Ultrashallow p +-n Junction Fabrication by Low Energy Focused Ion Beam Implantation,"* J. Vac. Sci. Technol., B 8 (6), 1937 (1990).

[274] P. M. Petroff, Y. J. Li, Z. Xu, W. Beinstingl, S. Sasa, and K. Ensslin, *"Nanostructures Processing by Focused Ion Beam Implantation,"* Focused Ion Beam Implantation," J. Vac. Sci. Technol., B 9 (6), 3074 (1991).

[275] L. R. Harriott, *"Microfocused ion Beam Applications in Microelectronics,"* Appl. Surf. Sci., 36, 432 (1989).

[276] U. Weigmann, H.-C. Petzold, H. Burghause, R. Putzar, and H. Schaffer, *"Repair of Electroplated Gold Masks for X-Ray Lithography,"* J. Vac. Sci. and Technol., B6, 2170 (1988).

[277] T. Ishitani, T. Ohnishi, Y. Madokoro, and Y. Kawanami, *"Focused-Ion-Beam "Cutter" and "Attacher" for Micromachining and Device Transplantation,"* J. Vac. Sci. Technol., B 9 (5), 2633 (1991).

[278] D. K. Stewart, L. A. Stern, and J. C. Morgan,, *"Focused-Ion-Beam Induced Deposition of Metal for Microcircuit Modification,"* Proc. SPIE, Int. Soc. Opt. Eng., 1089, 18 (1989).

[279] Y. Takahashi, Y. Madukoro, and T. Ishitani, *"Focused Ion Beam Induced Deposition in the High Current Density Region,"* Jpn. J. Appl. Phys., 30 (11B), 3233 (1991).

[280] H. Nakamura, H, Komano, K. Norimatu, and Y. Gomei, *"Silicon Oxide Deposition into a Hole Using a Focused Ion Beam,"* Region," Jpn. J. Appl. Phys., 30 (11B), 3238 (1991).

# Chapter 4

# Lithography

## 4.0 Introduction

Lithography comes from two Greek words, *"lithos"* which means stone, and *"graphein"* which means write. Lithography means literally *"writing a pattern in stone"*. In microelectronics the word lithography is commonly used to describe a process in which a pattern is delineated in a layer of material sensitive to photons, electrons or ions. The principle is similar to that of a photo-camera in which an object is imaged on a photo-sensitive emulsion film. After development, the exposed regions of the film are left as metallic silver, while the unexposed regions are removed, resulting in a printed image of the object. While with a photo-camera the *"final product"* is the printed image, the image in microelectronics is typically an intermediate pattern which defines regions where material is deposited or removed. Lithography transforms complex circuit diagrams into patterns which are defined on the wafer in a succession of exposure and processing steps to form a number of superimposed layers of insulator, conductor and semiconductor materials. Typically 8-25 lithography steps and several hundred processing steps between exposures are required to fabricate a packaged semiconductor integrated circuit (IC).

During the past three decades, there has been a steady increase in the number of transistors per chip. This progression is depicted in Fig. 4.1 where the number of bits per memory chip is used as a measure of device density. The figure shows that the bits/chip increases from 1 Kilobit (Kb) in the late 1960s to a projected 1 Gigabit (Gb) by the end of the decade. Both the horizontal and vertical device geometries must shrink steadily to allow such a progression in density. Scaling devices to smaller geometries also increases circuit speed. The minimum feature size, i.e., the minimum line-width or line-to-line separation that can be printed on the surface, controls the number of circuits than can be placed on the chip and has a direct impact on circuit speed. The evolution of integrated circuits is therefore closely linked to the evolution of lithographic tools. Shrinking the vertical dimensions (thickness of insulator, conductor and semiconductor films) is also critical to reducing device size and increasing circuit speed and density.

The pursuit of submicron feature sizes has motivated extensive work in four major lithography technologies: optical (or ultraviolet), electron-beam, X-ray, and ion-beam. While there has been considerable progress in the latter three, optical lithography has remained the dominant technology [1].

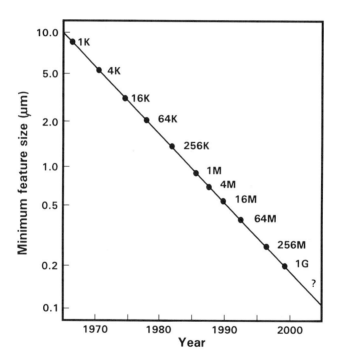

Fig. 4.1 Progress in lithography. Number of bits per memory chip used as "indicator".

## 4.1 Optical Lithography

An optical lithography tool consists of an ultraviolet (UV) light source, a **photomask,** an optical system, and a wafer covered with a photosensitive layer, called **resist** because of its ability to resist chemicals used in subsequent processing. The mask is flooded with

UV light and the mask pattern is imaged onto the resist by the optical system (Fig. 4.2).

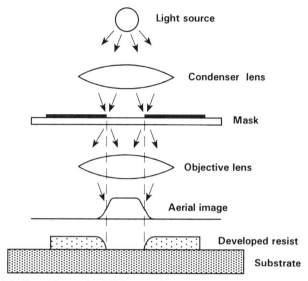

**Fig. 4.2** Generic optical lithography system.

## 4.1.1 Light Sources

The most commonly used ultraviolet light sources for optical lithography are high-pressure arc lamps and laser sources. Three regions of the emitted light spectrum can be distinguished: deep ultraviolet (DUV) in the 100-300 nm range, mid-UV in the 300-360 range, and near-UV in the 360-450 nm range.

### High-Pressure Arc Lamps

A typical arc source is shown schematically in Fig. 4.3. The emitted light (dotted torroid shapes) is reflected by parabolic mirrors into a lens assembly, forming a parallel beam which floods the mask. The source is filled with mercury or a mercury-xenon mixture at a typical pressure of 30-35 atm. The distribution of the emitted spectrum depends on the partial pressures of Hg and Xe and the total pressure of the discharge mixture. Fig. 4.4 shows that for a high-pressure Hg lamp, the minimum emitted wavelength of significant intensity is about 300 nm. This limit can be reduced to about 240 nm by adding xenon to the gas [2]. Heat producing IR wavelengths are typically removed with a beam-splitting mirror, called "cold" or dichroic, that is coated to reflect the wavelength needed for lithography (436 nm, 365 nm, 248 nm) and and transmit the unwanted part of the spectrum to be captured by a "beam dump". The mirror is placed in a location where the geometry requires a bend in the beam.

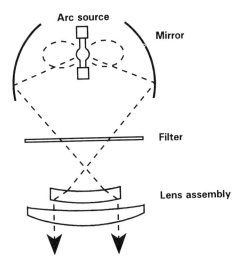

**Fig. 4.3** Typical arc source and surrounding optics.

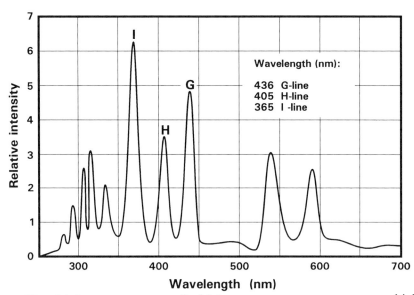

**Fig. 4.4** Output spectrum of a high pressure mercury arc source, highlighting the high-intensity lines used in photolithography.

With a mercury-xenon arc source, the dominant wavelengths are 254 nm (DUV), 313 nm, 365 nm (I-line), 405 nm (H-line), 436 nm (G-line), and longer wavelengths. Since most of the photoresists require photo-energies larger than 2.5 eV for proper exposure (Sec. 4.1.5), only the 436 nm and shorter wavelengths are considered for lithography. For minimum features sizes larger than about 2 $\mu$m, the full emitted spectrum can be used to expose the resist. For smaller feature sizes, the lens is corrected for one or two of the wavelengths, e.g., $\simeq$ 250 nm, and filters are used to remove the rest of the spectrum. This, however, reduces the beam intensity and increases exposure time.

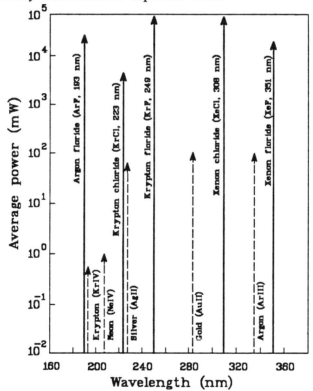

**Fig. 4.5** Comparison of the average output powers from various sources of ultraviolet radiation (adapted from [3])

The G-line is widely used for feature sizes down to $\simeq$ 0.8 $\mu$m, and the I-line for sizes in the range 0.4-0.8 $\mu$m. For feature sizes below 0.4 $\mu$m, shorter exposure wavelengths, such as DUV (248 nm) with very sensitive DUV resist, or I-line exposure with phase-shift mask

**174**

and/or off-axis illumination will be required. Phase-shift masks and off-axis illumination are described in a later section.

## Laser Sources

The most powerful and commonly used laser sources for deep UV photolithography are the **excimer lasers** [3, 4] (Fig. 4.5). The word "excimer" combines the two words "excited" and "dimer". A dimer is a molecule composed of two identical atoms such as $Kr_2$. The word "excimer", however, is also used to describe excited complexes such as rare gas halides, e.g., krypton-fluoride (KrF), xenon-chloride (XeCl), argon fluoride (ArF) [5].

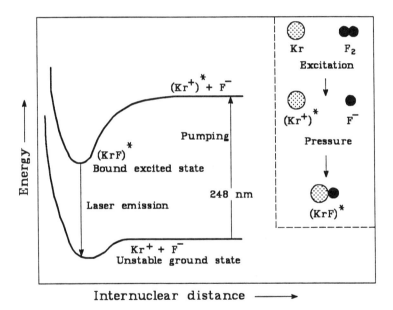

**Fig. 4.6** Excimer laser emissions through transitions from a metastable excited state to an unstable ground state [3].

An excimer is unstable in the ground state but becomes bound or metastable in the excited state. Laser emission occurs through transitions from the excited to the ground state (Fig. 4.6). The dissociation time for the ground state is in the picosecond range ( $10^{-12}$ s), while the lifetime of the excited state is in the nanosecond to microsecond range ($10^{-9} - 10^{-6}$ s) [3]. Therefore, population inversion (higher population of the excited state than the ground state) can be readily achieved by exciting a high pressure mixture of the rare gas and the halogen (or halogen compound) in a high-voltage pulse discharge and producing an excited entity such as:

$$Kr^* + NF_3 \rightarrow (KrF)^* + NF_2,$$

where the asterisk indicates an excited state. Wavelengths at peak emissions from excimer lasers range from 350 nm to 150 nm. The most commonly used excimer for 248-nm exposure is KrF.

The high-power pulses of excimer lasers allows shorter exposure times (10-20 ns) than with "conventional" laser sources, increasing throughput (number of exposed wafers per unit time). Excimer and conventional lasers also differ in the spatial coherence of their emitted beams. Beams are said to be spatially coherent when points on their wavefront remain in phase as the wave propagates (Sec. 4.1.4). Conventional lasers are typically single-mode sources that emit spatially coherent beams, while excimer lasers exhibit multiple modes of emission that result in beams of extremely poor spatial coherence. With spatially coherent beams, any scattering of the optical system causes interference at the wafer surface, resulting in a random pattern of constructive and destructive interference in the resist, called **speckle.** The problem of speckle is almost eliminated with the use excimer laser sources [3].

## 4.1.2 Photomask

A mask for optical lithography consists of a transparent plate, called blank, covered with a patterned film of opaque material. The blank is made of soda lime, borosilicate glass, or fused quartz. The advantages of quartz is that it is transparent to deep UV ($\leq$ 365 nm) and has a very low thermal expansion coefficient. The low expansion coefficient is important when the minimum feature size is less than $\simeq 1.5 \mu m$. Also, distortions related to thermal expansion become more pronounced as the mask size is increased. Therefore, quartz masks are required for exposures onto 150 mm and larger wafers, and for minimum features sizes smaller than 1.5 $\mu m$. The opaque material is typically a very thin ($\leq$ 100 nm) film of chrome, covered with an anti-reflective coating (ARC), such as chrome oxide, to suppress interferences at the wafer surface. High quality photomasks must meet stringent requirements in flatness, accuracy of pattern placement, minimum feature size, linewidth control over the entire area of the mask, and defect density. Variations in flatness can alter the optical path length and cause large distortions on the resist due to defocusing. The coating process is very crucial because any pinholes present in the chrome layer remain as part of the finished mask and are replicated during printing. Temperature variations during mask fabrication and during mask use account for much of the misplacement of the pattern and misregistration between masking levels. For example, a 1° change in temperature in the mask-making system can cause a misplacement of $\simeq 0.1 \mu m$ over 100 mm diameter area of a quartz mask.

**Pattern Generation**

Pattern generation begins by transforming the circuit diagram into geometrical shapes using computer-aided design (CAD) tools. In a typical CAD system, the pattern is designed with a light pen on a cathode-ray tube (CRT). The output of the CAD system is usually in the form of binary data. The data are first translated into machine language, a procedure usually called post-processing. They are then transmitted to an optical, electron-beam, or laser system [6, 7]. To print the pattern on the mask, the chrome film is coated with resist and the system delineates the design shapes in the form of light or electron flashes of different apertures on the resist. The coordinates of exposure on the mask are determined by controlling the position of the x-y stage with a laser interferometer. After developing the resist, the chrome layer is etched away in the regions where the resist is removed.

The most common exposure tool to prepare masks is an electron-beam system. Its advantage over earlier versions of optical mask-making systems stems from the shorter electron wavelength and higher exposure speed compared to UV sources, resulting in higher throughput and tighter dimensional control. Electron-beam systems are described in Sec. 4.3.

A laser pattern generator is shown schematically in Fig. 4.7. This system uses the 364-nm line from an argon-ion laser to create a beam that passes through a beam-splitter and splits into eight parallel beams [8,9]. The beams are independently modulated, converting the binary output from the "rasterizing engine" into writing information as the mask surface is scanned. The beams spot sizes and the printing grid are controlled by the "zoom optics". The beams are then swept across the x-y stage by a rotating polygon mirror. The motion of the stage holding the reticle is controlled by a laser interferometric system, while on-off modulation of the beam exposes the pattern on the mask.

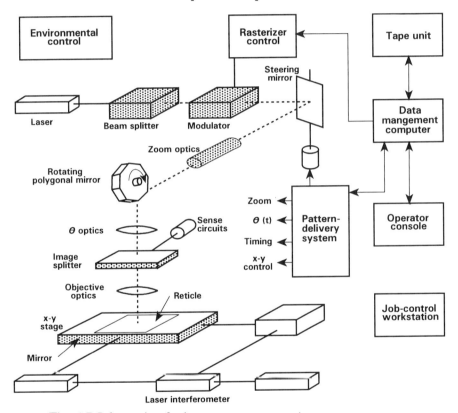

**Fig. 4.7** Schematic of a laser pattern generation system.

The layout of the pattern to be transferred to the mask is typically done on clear-field (correct positive), which means that the image and not its mirror image is seen, and the image as defined will appear opaque on the mask (Fig. 4.8). The image polarity can be reversed during processing. Dark-field (correct negative) masks are preferred over clear-field masks because they

**178**

reduce light scattering effects, and because of the higher probability that particles will fall on an opaque area, where they are not imaged. The first plate which contains one chip pattern or a matrix of several chip patterns is called a **reticle.** It is drawn 1X-5X actual size, with 5X most common for die-size requirements. Typically, a set of 8-25 individual masks is required to fabricate a chip, each mask defining a separate layer on the wafer.

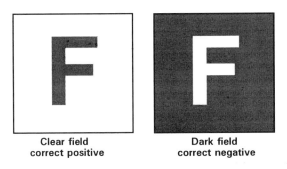

Clear field
correct positive

Dark field
correct negative

**Fig. 4.8** Reticle polarity. (a) Clear-field, or correct positive. (b) Dark-field, or correct negative.

## Mask Defects and Repair

Up to 75% of electron-beam generated masks may be defective and require repair. Automatic mask inspection techniques are available for detecting defects larger than $\simeq 0.1\mu m$ in size. Optical techniques are used to to compare identical adjacent chip patterns on a given mask and, if a difference is detected, it is recorded and categorized as a defect [10]. This method, however, will not detect a systematic error since all chips will have the same error. To detect errors on the reticle, the digitized scan data from the mask is compared to the original digitized layout pattern data. Typical mask defects are missing chrome, pinholes in the chrome layer, or regions of unresolved chrome. Such defects can be usually repaired by locally adding opaque material or removing excess chrome with a laser beam or focused ion beam (FIB). Unresolved chrome is removed by laser evaporation or FIB milling (Chap. 5). Opaque material is added by a local laser- or FIB-induced reaction that deposits, e.g, carbon over the area missing chrome (Chaps. 4,5) [11,12].

Airborne particulates can also deposit on the mask and create defects in the image. Usually, particles on the back of the mask cause no problem since they will be out of focus. One method to reduce the effect of particulates on the front of the mask (chrome-side) is to add a thin protective membrane, called a **pellicle,** over the mask. Particles captured by the pellicle are out of focus and have little effect on the image (Fig. 4.9). Pellicles are typically made of $\simeq 0.7 \mu m$ nitrocellulose (or a teflon-based material) that is transparent to the exposure wavelengths. The pellicle is placed 0.15-0.5 $\mu$m above the mask, depending on the exposure system.

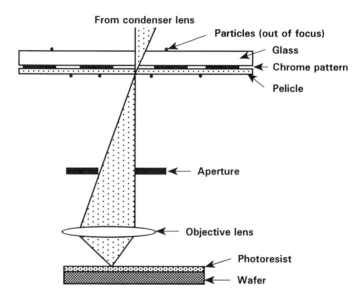

**Fig. 4.9** Protection of a mask with a pellicle. Airborne particulates which deposit on the pellicle and on top of the reticle are out of focus and have little effect on the resist image.

### 4.1.3 Exposure Systems

There are three types of photolithographic exposure techniques: contact, proximity, and projection printing (Fig. 4.10). Contact and proximity printing, also called shadow printing, are the simplest and least expensive methods of imaging. They are, however, limited to minimum feature sizes larger than $\simeq 1$ $\mu$m, and typically used for low-budget laboratory experiments where it is not critical to achieve low defect densities. The predominant exposure method in modern technologies is 1X-5X projection printing.

### Contact and Proximity Printing

Contact and proximity printing are simple exposure techniques in which the mask covers the entire wafer. The mask is oriented with the patterned side facing the resist-coated wafer. This means that the pattern on the mask must contain the mirror image of the pattern on the resist. The mask and wafer-supports allow a vertical movement which brings them in proximity of each other or in intimate contact with each other, and two lateral movement and one rotation about the optical axis for alignment. By opening the shutter, the mask is flooded with UV light so that the entire wafer is illuminated in one exposure, exposing the resist in those regions where the mask is transparent (Fig. 4.11). The main limitations of contact printing are the reduced mask-life and the high defect density in the printed pattern. The repeated contact between mask and wafer causes debris between the films, inducing defects in the resist and the mask, severely reducing the mask lifetime and inducing defects in the resist. This problem is aggravated by non-uniformities in the wafer and mask surfaces.

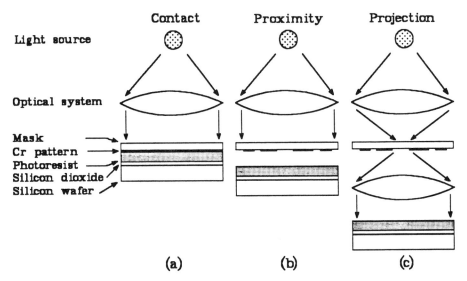

**Fig. 4.10** Schematic of three optical lithographic techniques. (a) Contact. (b) Proximity. (c) Projection.

In proximity printing, a small gap is introduced between mask and resist to eliminate defects caused by direct contact between the surfaces. While the gap increases the mask-life and reduces the defect density below that of contact printing, it also increases diffraction effects and hence the minimum feature that can be printed, as discussed in Sec. 4.1.4.

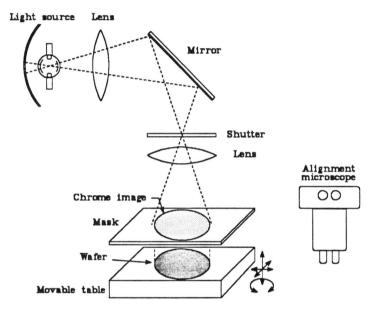

Fig. 4.11 Typical contact or proximity printing system [13].

## Projection Printing

The most commonly used imaging technique is projection printing where lens assemblies are used to focus the mask pattern onto the wafer that is separated from the mask by several centimeters.

## Full-Wafer Scanning Projection Printing

An early version of 1:1 full-wafer scanning (image projected at the same size as object), is shown in Fig. 4.12 [14]. Spherical, reflective mirrors are used to produce the image on the wafer. By using scanning rather than full-field exposure, and keeping the scan direction within the zone of good optical correction ("sweet spots"), aberrations and image distortions are minimized. For this purpose, the mask is illuminated through a slit of a few millimeters in width and imaged onto the resist. The mask and wafer are moved together through this arc of light that covers the entire width of the mask by means of a continuous scanning mechanism. Synchronous movement of wafer and mask reduces effects of vibration on image distortion. A minimum feature size of 1 $\mu$m can be printed with an overlay tolerance of $\pm$ 0.4 $\mu$m using 350-400-nm light. Overlay tolerances are discussed in Sec. 4.1.5.

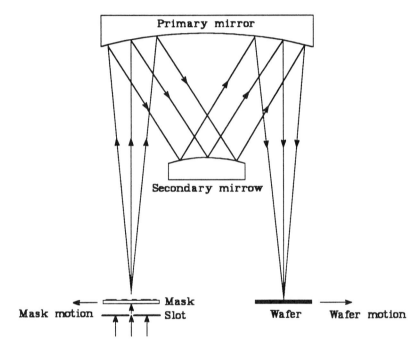

**Fig. 4.12** Schematic of a scanning optical system consisting of two concentric reflecting mirrors [14].

### Step-and-Repeat Projection Systems

The most common printing technique in modern technologies is the step-and-repeat (S/R) projection, (Fig. 4.13). In S/R systems, the image size is projected 1:1 or reduced in size by 2X to 5X (typically 4X). Reducing the image size has the advantage that the features on the reticle do not have to be as small as the final image and are therefore easier to fabricate. Another advantage is that mask defects and imperfections are also reduced in size and hence become less severe. As the chip size increases, however, reductions by factors larger than 5X become difficult because of limitations in mask sizes that can be fabricated with precision. Since it is impossible to design refractive optics to project a mask over an entire 200-mm wafer [15], refractive systems are constructed to project an image over a small portion of the wafer and then step-and-repeat the field over the entire wafer (Fig. 4.13). Typical S/R systems use excimer-laser or high-intensity mercury source, a collimating lens system to focus the illumination on a reticle, and a reduction lens system to image the reticle onto the surface of the wafer. After each exposure, the wafer is mechanically stepped a specified distance, realigned and refocused, and the image projected again. Step-and-repeat systems require several lenses of the highest quality to obtain the needed resolution over the entire projected field.

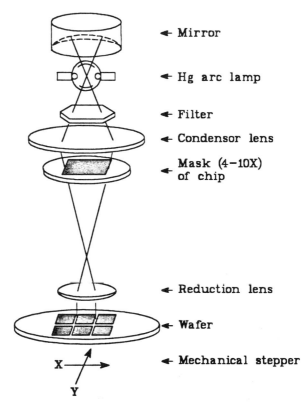

**Fig. 4.13** Schematic of a reduction step-and-repeat projection system.

Step-and-scan systems combine the advantages of "scanners" and "steppers" for large field sizes (20 mm x 32.5 mm and above), increasing dimensional control and throughput. The wafer is stepped, for example, from a double-die field to another and the beam scanned across each field. The reticle stage and wafer stage are both in motion and, for a 4:1 reduction ratio, the reticle moves four times as fast as the wafer.

### 4.1.4 Review of Important Concepts in Optics

The three most important requirements of a lithographic system that must be satisfied over the entire die area are minimum resolution, dimensional control, and alignment. The final wafer dimensions of conductors, insulators, and semiconductors depend, however, not only on the exposure system, but also on the resist properties and subsequent steps used to engrave the pattern on the wafer. Resist properties are discussed in section 4.1.6. Etch processes play the key role in defining and controlling the final image on the wafer and are covered in Chap. 5. The following section discusses the factors that affect the performance and limitations of exposure systems. For this purpose, concepts that are important to the development of an understanding of image projection in optical lithography are briefly discussed. The concepts that are most important to the discussion of exposure systems are spatial coherence, diffraction, resolution, numerical aperture, and depth of focus.

### Spatial Coherence

Beams are said to be spatially coherent when points on their wavefront maintain a fixed phase relationship as the wave propagates. A conventional (single-mode) laser is spatially coherent because separate photons are emitted and amplified in phase. Theoretically, spatial coherence can also be achieved with an infinitely small non-laser source by observing the source from an infinitely remote point where the angle subtended approaches zero. The degree of coherence has a strong effect on image quality. If the light source and condenser lens are such that the light source appears as a point source, the illumination is said to be coherent. If the light source appears to have an infinite extent, an incoherent illumination results. In practice, the light source appears to have a finite extent and the emitted light is said to be partially coherent. It turns out that partial coherence is an important condition for an optimized exposure system, as discussed in the following section.

## Diffraction

Consider a narrow slit of width $a$ that is illuminated by a coherent light source at normal incidence to the plane of the slit (Fig. 14). In geometrical optics, one would draw straight lines casting sharp shadows of the slit-edges on a screen placed behind the slit, defining a boundary with a square-shaped amplitude profile inside the boundary and no light outside it (Fig. 14a). In reality, however, regions outside the boundary are illuminated and the amplitude profile can have a shape as shown in Fig. 14b. Points in the slit at at its edges behave like new sources of wavelets and the light that passes the slit appears to bend around the slit edges, illuminating regions outside the boundary, such as point P in Fig. 15. Bending of of light around obstacles such as a slit-edge is called **diffraction.**

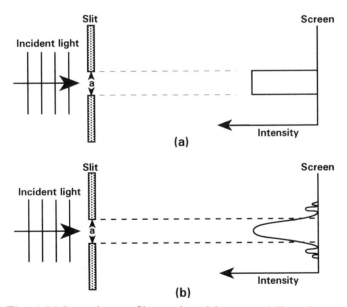

**Fig. 4.14** Intensity profile produced by a spatially coherent beam as it passes by a slit. (a) According to geometrical optics. (b) With diffraction.

The maxima and minima in the profile shown in Fig. 14b are caused by interference of the diffracted light. At the center point $P_0$, the parallel rays have the same optical path length and hence arrive in phase, resulting in a center of maximum intensity. Waves 1 and 2 arriving at point P, for example, can be in phase (constructive interference) or out of phase (destructive interference), depending on the difference in their optical paths.

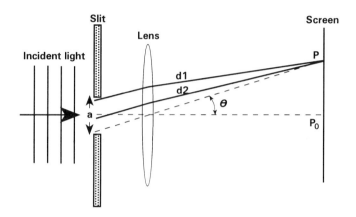

**Fig. 4.15** Single slit of width a illuminated at normal incidence with a parallel beam of coherent light to illustrate diffraction effects.

When the light-source and/or screen are at a finite distance from the diffraction aperture, the wavefronts that are incident on the diffraction aperture and that leave it are not plane and their corresponding rays are not parallel. In this case, one speaks of **Fresnel diffraction.** If the source and imaging screen appear to be at infinite distance from the diffraction aperture, the wavefronts reaching and leaving the aperture can be described by plane waves and their corresponding rays are parallel. This limiting condition, referred to as **Fraunhofer diffraction,** can be established with lenses, as illustrated for a grating in Fig. 4.17 [2]. In this section, only Fraunhofer diffraction is used because it is easier to treat mathematically.

The scale of diffraction depends on the slit-width to wavelength ratio, as illustrated for two different $a/\lambda$ ratios in Fig. 4.16.

Simple analysis shows that minima in intensity occur when the path difference satisfies the condition

$$a \sin \theta = N\lambda,$$

4.1

where N is an integer.

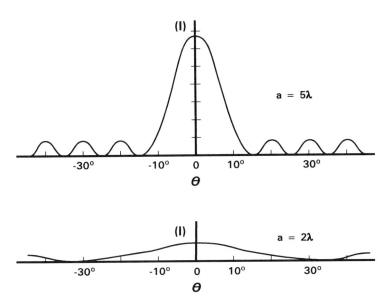

**Fig. 4.16** Diffraction pattern of a long and narrow slit shown for two different slit-width to wavelength ratio.

There is a maximum approximately halfway between each adjacent pair of minima. The phase difference between the top edge and bottom edge of the slit is

$$\phi = \frac{2\pi}{\lambda} a \sin \theta,$$

4.2

where $a.\sin\theta$ is the optical path difference between the top and bottom rays. Defining $\alpha = \phi/2$, the relative intensity at different points of the screen is found as

$$\frac{I_\theta}{I_{max}} = \left( \frac{\sin \alpha}{\alpha} \right)^2,$$

4.3

where $I_\theta$ is the intensity emitted at an angle $\theta$ and $I_{max}$ is the peak intensity at the center.

For a circular aperture of diameter $d$, such as the circular boundary of a converging lens, the scale of diffraction depends on the ratio $d/\lambda$. The diffraction pattern consists of a central maximum disk and circular rings of maxima and minima, with the first minimum ring given by

$$\sin \theta = 1.22 \; \frac{\lambda}{d} \;. \qquad\qquad 4.4$$

The factor 1.22 comes from integration over the circular area.

A diffraction grating consists of a large number of parallel slits of equal width and separation (Fig. 4.17). For example, a mask containing a pattern of equal opaque lines and spaces between the lines constitutes a diffraction grating. The lens arrangement in Fig. 4.17a ensures that parallel rays illuminate the slit at normal incidence and that the parallel rays leaving the grating converge on the screen rather than at infinity. The diffraction pattern for two, three and four narrow slits is illustrated in Fig. 4.17b. The major peaks in the intensity profiles are called principal maxima and occur at the same angle $\theta$ for the three gratings. This angle is given by

$$d \sin \theta = N\lambda , \qquad\qquad 4.5$$

where $d$ is the distance between the centers of two adjacent lines (also called "pitch"), and $N$ an integer called order of diffraction. As the number of slits increases, the positions of the principle maxima remain the same, but their width decreases. When the incident beam makes an angle $\theta_0$ with the normal to the plane, a simple analysis shows that the principle maxima occur at angles $\theta$ such that

$$d ( \sin \theta \pm \sin \theta_0) = N\lambda , \qquad\qquad 4.6$$

where $\theta$ is the angle of the diffracted beam with the normal.

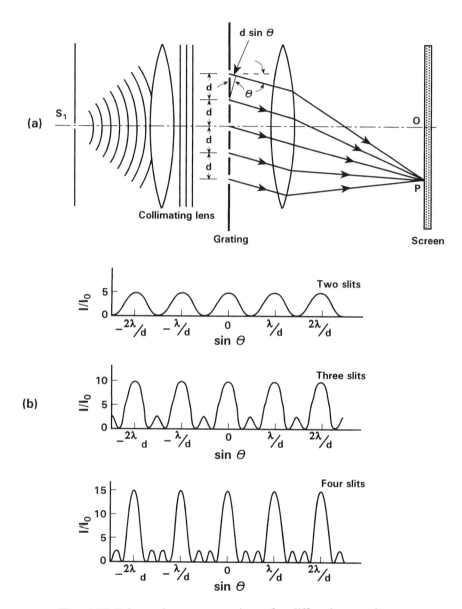

**Fig. 4.17** Schematic representation of a diffraction grating. (a) Arrangement for Fraunhofer diffraction. (b) Intensity profile produced by diffraction gratings with varying number of slits of same interval.

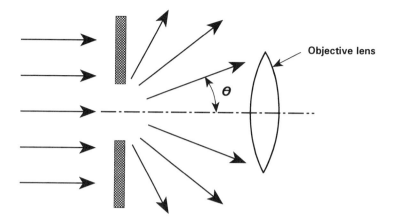

**Fig. 4.18** Definition of the numerical aperture, NA.

## Numerical Aperture

Diffraction in a small opening transforms normally incident radiation into an angular spread (Fig. 4.18). The numerical aperture, $NA$, of a lens is a measure of "how much" of the diffracted light the lens accepts and images. It is defined as

$$NA = n \sin \theta \simeq n \; \frac{\text{Radius of lens}}{\text{Focal length of lens}} \; , \qquad 4.7$$

where $2\theta$ is the maximum cone angle of rays subtended by the lens, and $n$ is the index of refraction of the image medium ($n \simeq 1$ for air). The larger the entrance pupil (large $\theta$), the larger the $NA$ of the projection lens, and the greater the amount of diffracted information that can be collected and imaged. Also, a larger index of refraction reduces the angular spread and increases NA. The numerical aperture determines the number of diffracted orders that can be collected and imaged. It is important to note that the undiffracted zero-order beam consists of a single ray and does not contain pattern information. At least a second ray of the diffracted light is needed to intersect the first ray and reconstruct the image.

## Resolution

The fact that images formed by lenses are diffraction patterns is important when two points of small angular separation must be

distinguished. In optics, resolution is defined by the Rayleigh criterion as the minimum distance at which two points can just be
resolved. This useful (but arbitrary) definition is based on an
angular separation of two points such that the maximum of the
diffraction pattern of one source (point) falls on the first minimum
of the diffraction pattern of the other (Fig. 4.19).

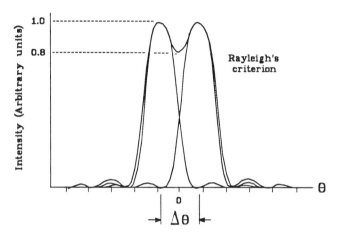

**Fig. 4.19** To define Rayleigh's criterion. The image is
formed in the focal plane of the lens.

Other criteria are sometimes used in lithography to decide
when two points can be resolved. Using Eq. 4.4 and assuming a
very small angle, we find that two objects that are barely resolved
by the Rayleigh criterion must have a separation such that

$$\theta_R \geq 1.22 \frac{\lambda}{d} \, , \qquad 4.8$$

where $d$ is the diameter of the lens. Resolution is therefore
improved by increasing the lens diameter and reducing the wavelength of the illuminating light. It follows that in the absence of
aberrations, two focused images are just resolved when

$$\Delta x = 0.61 f \frac{\lambda}{d} \, , \qquad 4.9$$

or, using the definition of numerical aperture in Eq. 4.7

$$\Delta x = 0.61 \frac{\lambda}{NA} , \qquad\qquad 4.10$$

Therefore, lenses with high numerical apertures also have high resolution. The dependence of resolution on $\lambda/NA$ can be seen from Fig. 4.20. For wavelets emanating from points 1 and 2, destructive interference occurs when $|d_2 - d_1| = \Delta d = \lambda/2$, or $\Delta x = \lambda/(2 \sin \theta)$. For n = 1, this gives $\Delta x = \lambda/2NA$.

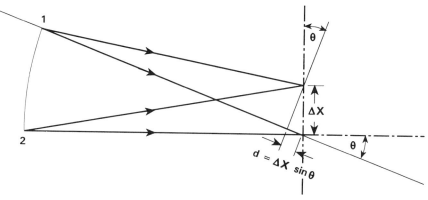

**Fig. 4.20** Dependence of resolution on $\lambda/NA$.

### Depth of Field and Focus

From photography we know that there is a region of "sharpness" situated before and behind the plane of the object to be photographed, and that the depth of this region increases as the pupil-diaphragm is decreased. This region of sharpness is called **depth of field.** The corresponding distance in image space is called **depth of focus (DOF).** The dependence of depth of field on numerical aperture can be seen from a simple analysis in Fig. 4.21. Destructive interference occurs at $\Delta d = \lambda/2$. Therefore, $DOF = \lambda/\sin^2\theta$, or

$$DOF = \frac{\lambda}{(NA)^2} . \qquad\qquad 4.11$$

It follows from the above equations that resolution improves by reducing the wavelength, $\lambda$, and/or by increasing the numerical aperture, NA, of the imaging optics, however, at the cost of reducing the depth of field. Therefore, given a resolution requirement and the imaging wavelength, there is an optimum numerical aperture. If NA is too low, the resolution cannot be achieved, but if NA is too large, the depth of field becomes unacceptable. Varying the imaging wavelength has less impact on the depth of field than varying NA.

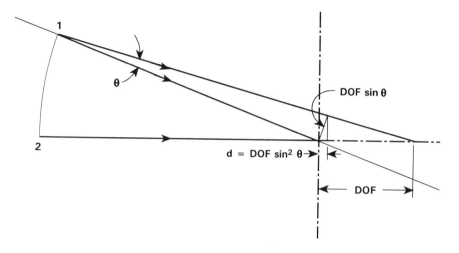

**Fig. 4.21** Dependence of depth of focus on $\lambda/NA^2$.

## 4.1.5 Performance of Optical Exposure Systems

The principle of imaging a mask onto a wafer is rather simple. A light source illuminates the mask with the aid of a condenser lens system, and the light that diffracts from the mask is collected by an objective lens system and imaged onto the wafer (Fig. 4.22). Even with perfect lens or mirror systems, the image quality is limited by diffraction of light. In this case, the image is said to be diffraction-limited. In photolithography, two parameters determine the image quality: resolution, $\Delta x$, and depth of focus, DOF [17, 18]. The definitions of resolution and DOF of an optical system are not unique. In microlithography, minimum resolution is defined as the smallest feature that can be produced in the photoresist with sufficient precision that a subsequent processing step can be performed successfully, under "reasonable manufacturing variations". The pattern to be resolved is chosen as a set of lines of equal width and space, and the ultimate performance of an imaging system is measured by its ability to define the pattern within specified groundrules.

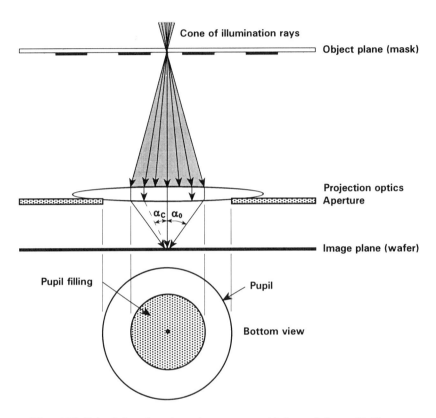

**Fig. 4.22** Principle of an imaging system. (Adapted from [16]).

The surface of the wafer upon which the pattern is imaged is typically non-planar, primarily because of non-uniformities in the starting wafer surface and topographies created by deposition and etching steps during processing. As the plane of image departs from the plane of best focus in the resist, the image quality deteriorates. There is, however, a certain amount of defocus that can be tolerated whereby the image remains within specifications. This tolerance defines the allowable DOF of the system. Depth of focus is closely related to the exposure conditions and, for a given exposure, DOF is typically used as a measure of system performance. Resolution and DOF are determined by the wavelength $\lambda$ of the illuminating light and the numerical aperture, $NA = \sin \alpha_o$, where $\alpha_o$ is the angle of the most oblique ray allowed by the projection optics aperture (Fig. 4.22). Resolution and DOF are best expressed by the simple Rayleigh criteria using "scaling" factors $k_1$ and $k_2$ as

$$\Delta x = k_1 \frac{\lambda}{NA} \, , \qquad\qquad 4.12$$

$$DOF = k_2 \frac{\lambda}{NA^2} \, , \qquad\qquad 4.13.$$

where $k_1$ and $k_2$ are dimensionless factors that depend on the exposure system and resist. $k_1$ typically ranges from 0.57 - 0.87 [19], and $k_2 \simeq 0.5$ - 1.0. In the case of a reduction projector, $\Delta x$ does not refer to the actual mask width but rather to the specified linewidth at the wafer plane.

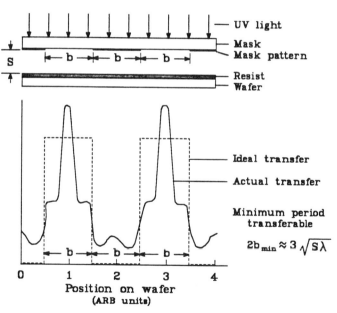

Fig. 4.23 Light distribution profiles on resist surface after light has passed through a mask containing an equal line and space grating [2].

In contact printing, resolution is limited by diffraction between the mask and the bottom of the resist. Even when a perfect optical system is used, the edges of a perfectly delineated chrome pattern in the mask appear rather diffused on the resist because of diffraction effects. Thick resists, or gaps between mask and resist, degrade resolution. Fig. 4.23 shows schematically the effect of diffraction when a mask that consists of a pattern of lines of equal width and space, $\Delta x$, is imaged onto the resist. The "theoretical" resolution limit is approximated by [20]:

$$\Delta x \simeq 1.5 \sqrt{\frac{\lambda t_{PR}}{2}} \qquad\qquad 4.14$$

where $\lambda$ is the wavelength of exposing light and $t_{PR}$ is the resist thickness. For a 400-nm wavelength and a 1-$\mu$m thick resist, this limit is about 700 nm. The theoretical limit is, however, seldom achieved mainly because of non-uniformities in the mask to resist contact. In proximity printing, the distance between mask and resist is increased with spacers and resolution degrades [20]. For a spacer thickness $s$, resolution is approximated by [21]

$$\Delta x \simeq 1.5 \sqrt{\lambda \left( s + \frac{t_{PR}}{2} \right)}, \qquad\qquad 4.15$$

where $s$ typically ranges from 10 $\mu$m. - 20 $\mu$m to ensure that the films do not come in contact with each other. Because of the larger space between mask and resist plane, proximity printing can approach, but not exceed the resolution capability of contact printing. Proximity printing and, to a lesser extent, contact printing also suffer from a shadowing effect called **penumbral blur,** as illustrated in Fig. 4.24. Without diffraction, this shadowing effect limits the resolution to

$$\Delta x \simeq \frac{W(s + \frac{t_R}{2})}{D}, \qquad\qquad 4.16$$

where $W$ is the width of the illuminating source, and $D$ the distance between source and mask pattern.

As mentioned earlier, contact printing is used only to fabricate simple structures that are tolerant to defects, and proximity printing is used to define patterns with minimum feature sizes larger than $\simeq 2 \ \mu m$.

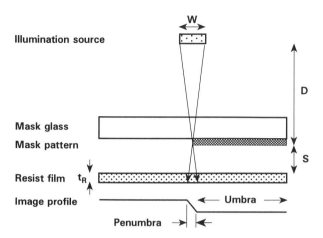

**Fig. 4.24** To define penumbral blur.

**Modulation Transfer Function (MTF)**

Consider an illuminated mask consisting of a grid of opaque lines and clear spaces of equal width $\Delta x$. The intensity of light at the resist plane does not vary abruptly from dark to bright but exhibits gradual transition due to diffraction effects. Even along the center of the opaque regions, there can be considerable intensity. As the grid dimensions are reduced, the light intensity can be approximated by a sinusoidal profile with a maximum $I_{max}$ at the center of the clear field between the lines, and a minimum $I_{min}$ at the center under the opaque regions (Fig. 4.25). $I_{min}$ increases as the "pitch", $p = 2\Delta x$, of the grating decreases. Modulation, $M$ defines the degree to which diffraction effects cause light to fall under the opaque regions between the clear fields. It is defined as

$$M = \frac{I_{max} - I_{min}}{I_{max} + I_{min}} . \qquad 4.17$$

The modulation transfer function (MTF) is the ratio of modulation at the image plane just above the resist (aerial image) to modulation at the mask plane. In most photolithographic applications, the mask consists of opaque chrome features on a glass substrate and $I_{\min} = 0$ under the chrome areas. Therefore, the mask modulation is M = 1.0, regardless of feature size, and MTF is identical to M at the resist plane.

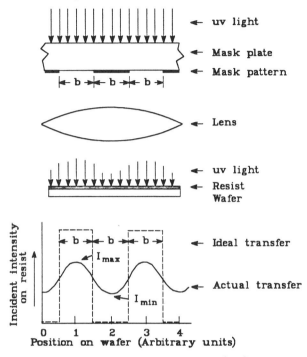

**Fig. 4.25** Definition of the modulation transfer function. Comparison of ideal to actual transfer on the resist plane for a 1:1 projection.

The concept of MTF is useful to describe and predict the the resolution capability of a projection system, independent of resist processing effects. It is a measure of the "fidelity" with which a perfect square-wave pattern is imaged at the resist plane. The modulation transfer function is, however, strictly applicable to incoherent light and sinusoidally varying transmission. Whether this condition is approached with a grid of lines and spaces depends on the illuminating wavelength and NA. In practice, as $\Delta x$ increases above $\simeq 2\mu m$, the MTF loses its meaning since the intensity in the aerial image approaches a binary rather than sinusoidal function. The dependence of MTF on wavelength and minimum feature size is illustrated in Fig. 4.26. From diffraction considerations, we find that the MTF increases with decreasing illuminating wavelength. For a given feature size, this translates into images with steeper edge-slopes, i.e., the higher the value of MTF for a given "pitch", the greater the contrast in the projected image. For typical applications, an MTF of 60% is considered sufficient, but 80% is needed for cases where the image size must be controlled within a tenth of the minimum line width [2].

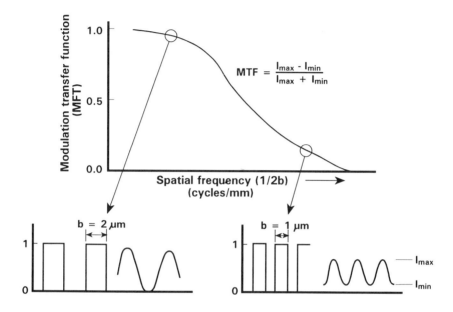

**Fig. 4.26** Effect of minimum feature size and wavelength on MTF.

**Coherence Factor**

Typical projection systems are arranged such that the source rays uniformly fill a circle and the rays emanating from the source are incoherent with respect to each other. The effective size of the source relative to the aperture of the system is described in terms of spatial coherence. Assuming n = 1, the pupil filling factor is defined as

$$\sigma = \frac{NA_{condenser}}{NA_{objective}} = \frac{\sin \alpha_C}{\sin \alpha_O} , \qquad 4.18$$

where $\alpha_C$ and $\alpha_O$ are defined in Fig. 4.27. Smaller values of $\sigma$ correspond to more spatially coherent illumination. When $\sigma = 0$, the source appears as a point source and illumination is fully coherent. For a source of infinite extent, $\sigma = \infty$ and the source is incoherent. In projection systems the aperture is typically under-filled and the values of $\sigma$ range 0.4-0.7, indicating a partial coherent condition [22]. For this reason, $\sigma$ is also referred to as the partial coherence factor. The degree of coherence depends on the source size, its distance from the mask, and the numerical aperture of the optical system.

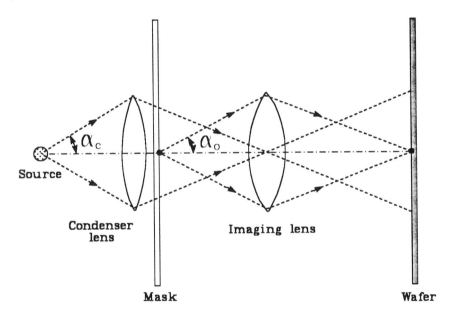

**Fig. 4.27** A refractive lens imaging system using partially coherent light. The condenser NA is $\sin \alpha_C$ and the objective NA is $\sin \alpha_O$.

The degree of coherence affects the modulation transfer function in a rather complex manner. Models are available to simulate the intensity profiles obtained from projection systems of different configurations [23]. The description of these models is, however, beyond the scope of this chapter. Values of $\sigma$ smaller than 0.5 increase the MTF for a given pitch. Increasing the coherence by, e.g., narrowing the illuminating slit is only achieved at the expense of longer exposure time. Also, the image characteristics of isolated edges degrades for $\sigma < 0.7$ because closer points on the mask of the illuminating wavefront interfere and give rise to effects such as ringing [24]. As $\sigma$ increases above 0.7, more detail of the incoherent source is resolved, but it becomes more difficult to maintain linewidth control. Also, the tolerable DOF decreases [2].

### 4.1.6 Photoresists

Photoresists (or resists) are organic compounds whose solubility changes when exposed to ultraviolet light. The regions in the resist that are exposed to light become either more soluble or less soluble in a solvent called **developer.** When the exposed regions become more soluble, a positive image of the mask is defined in the resist, and the compound is called **positive resist.** If the irradiated regions become less soluble in the developer, while the non-irradiated regions are soluble, a negative image of the mask is printed in the resist and the material is called **negative resist.**

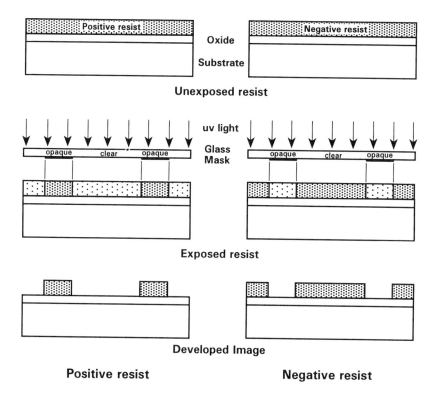

**Fig. 4.28** Mask projection and lithography illustrating the difference between positive and negative resist.

Figure 4.28 illustrates the difference between positive and negative resist. In both cases, the exposed regions must produce a three-dimensional relief image in the resist material that replicates the opaque and transparent regions in the mask. Following exposure and development, the film remaining on the wafer must protect the substrate during subsequent processing such as etching and ion implantation (Chaps. 5,6). While there is continued development of both types of resist, positive resists are the dominant choice for the definition of submicron features. This section discusses some important resist properties, with more emphasis on positive resist.

**Sensitivity and Contrast**

In practice, sensitivity is defined as the minimum light energy that is needed to produce a "good" image. This is closely related to the energy at which the full thickness of negative resist remains, or all of the positive resist is removed, after development (Fig. 4.29). At low-exposure energies, the negative resist remains completely soluble in the developer. As the exposure energy is increased above a threshold energy $E_T$, more resist remains after development until very little of the film is dissolved. For positive resists, there is some solubility, even at zero exposure [25, 26]. As the exposure energy increases, the film solubility increases until all of the film is removed. The incident light energy is defined as

$$E = It,$$

$$4.19$$

where $I$ is the light intensity and $t$ is the exposure time. The light intensity decreases as light passes through the resist, approximately following the Lambert-Beer's relation defined as

$$I(z) = I_0 e^{-\alpha z},$$

$$4.20$$

where $I(z)$ is the intensity of light in the z-direction, $I_0$ the intensity of light at $z = 0$ (typically chosen at the surface of the resist), and $\alpha$ is the absorption coefficient, assumed to be proportional to the concentration of absorbing species in the resist.

High resist sensitivity is desirable because this decreases the exposure time and increases throughput, and also because the resist can then be used for shorter UV wavelengths where the emission of conventional light-sources is reduced [25 − 27]. The dependence of sensitivity on the irradiating wavelength is provided in the form of a spectral response curve. A good match between light source and resist is when the resist is sensitive where the source has strong emission lines, resulting in a shorter exposure time.

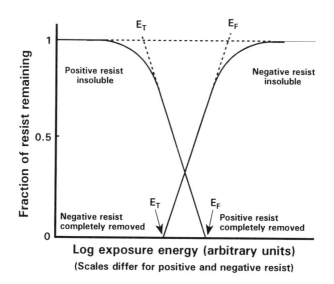

**Fig. 4.29** Sensitivity plot of positive and negative photoresists, showing the resist thickness versus exposure energy (Adapted from [26]).

Another important parameter of a photoresist material is its contrast, $\gamma$, which characterizes the sharpness of the transition from exposure to non-exposure. It is evaluated from the slopes of the linear portions of the curves in Fig. 4.29, and defined for for positive resist as [25 − 28]

$$\gamma = \frac{1}{\log_{10} \dfrac{E_T}{E_F}} \, , \qquad\qquad 4.21$$

where, $E_T$ and $E_F$, defined in Fig. 4.29, are obtained by extrapolating the linear portion of the normalized thickness versus energy plot to 1.0 normalized thickness. Resist contrast affects the slope of the resist image, and hence the image resolution. Other important resist properties are:

**Viscosity,** which affects its flow characteristics and film thickness, and depends on its solid content and temperature.

**Adhesion,** which describes how strongly the film sticks to the substrate. During device manufacture, resists are deposited on different substrate materials (such as oxide, nitride, polysilicon, and metals) and must adhere to these surfaces. Incomplete adhesion can cause a severe distortion or even the loss of a pattern.

**Thermal stability,** i.e., the sensitivity of film and image properties to processing temperature. Most resists are designed to withstand temperatures near 200 °C range, although they typically require additional plasma or UV treatment after development.

**Etch resistance,** which describes how well the film protects the substrate from wet and dry etchants used (Chap. 5).

**Contamination,** especially particulate and metal content. The resist is specified to a very low-level of sodium.

**Shelf-life:** how long the resist can be stored before unacceptable changes in its properties occur.

**Ease of processing:** how difficult to apply, develop and strip.

**Pinhole density:** number of holes per unit area created in the resist because of contaminants and inherent properties. The thinner the resist, the larger the pinhole density.

**Charging,** during plasma deposition, plasma etching (Chap. 5), electron-beam radiation, and ion implantation (Chap. 6). The conductivity of the resist plays an important role in the rate of charging.

### Photoresist Composition

Conventional resists consist of two components: the matrix material, called **resin,** and the **sensitizer** which is the **photoactive compound (PAC).** In positive resists, the sensitizer is also called the **inhibitor** because it inhibits the dissolution of non-exposed regions. The resin is typically insensitive to light. It determines the mechanical and chemical properties of the film, such as adhesion, etch resistance, film thickness, flexibility, and thermal flow stability. The sensitizer is the component of the resist that responds to actinic light (radiation which induces chemical changes). In non-exposed regions, the sensitizer gives the resist its developer resistance. In the exposed regions it gives the film its radiation absorption properties. The sensitizer and resin are diluted in a suitable solvent which keeps the resist in the liquid state until it is applied to the wafer.

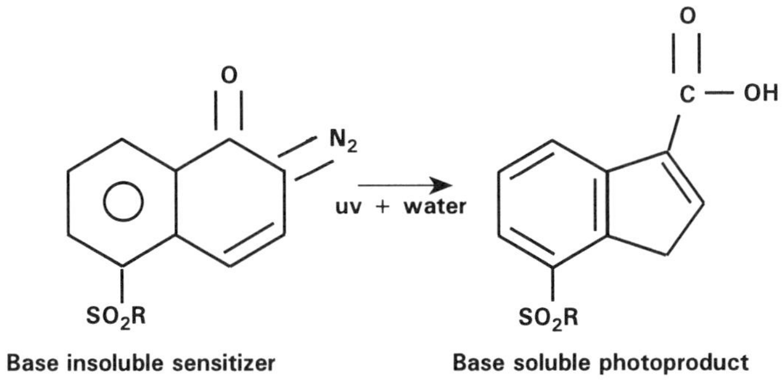

Base insoluble sensitizer          Base soluble photoproduct

**Fig. 4.30** Exposure reaction for a diazoquinone-novolak (DQN) photoresist [25].

### Positive Photoresists for Near-UV Exposure

The most commonly used positive resist consists of diazonaphtoquinone (DQ), which acts as the PAC [25], and novolac (N), a phenolic-formaldehyde resin [29]. The compound DQN is not sensitive to oxygen [30]. The PAC is not soluble in the aqueous base developer. Upon absorption of UV light, the PAC undergoes a structural transformation which is followed by reaction with water to form a base-soluble carboxylic acid (Fig. 4.30). The latter allows dissolution of the resin in alkaline developers and renders exposed resist areas more soluble than unexposed regions. The PAC is "tuned" to be sensitive to the selected illuminating wavelength of the mercury-xenon arc lamp or laser source.

Considerable attention is given to a two-level resist process in which a conventional resist layer is coated with a photobleachable film known as contrast-enhancement layer (CEL). Once exposed, the CEL material becomes transparent in the irradiated regions and opaque in the unexposed areas, forming a contact mask which enhances contrast [31].

## Negative Photoresists

The response of negative resists to actinic light is different than positive resists. Negative photoresists consist of polymers which, before exposure, are not chemically linked to each other. Upon exposure, the resists increase in molecular weight due to polymerization or cross-linking, and become less soluble in the developer. The sensitizer is activated by absorption of light energy in the 200-400 nm range, and then transfers energy to the polymer molecule, enhancing cross-linking.

## Standing Waves

When a thin resist layer is coated on a reflective subsurface and exposed to monochromatic radiation, standing waves are produced in the resist: the reflected wave interferes with the incoming radiation wave and causes the light intensity to vary periodically in a direction normal to the resist (Fig. 4.31a) [17, 32 − 37]. Since oxide, nitride, and polysilicon surfaces are more reflective at deep UV than at 365 nm (I-line) or 436 nm (G-line), standing-wave effects will be more pronounced at shorter wavelengths. Fig. 4.31b shows the simulation of the effect of standing waves on the edge profile of a developed resist image [20, 38]. It can be seen that standing waves cause variations in the development rate along the edges of the resist, and degrade the image resolution. Since the light intensity decreases with increasing depth into the resist (Eq. 4.20), the magnitude of the interference maxima decreases. Standing wave effects that are characteristic of monochromatic sources can be reduced by using broadband illumination, separate anti-reflective coatings (ARC) directly on the substrate surface or on top of the resist, or a bake step between exposure and development of the resist to reduce the extent of "ridges". A typical anti-reflective coating is a 130 nm thick polymer which has a high absorbance at the exposure wavelength, considerably reducing interference due to reflectance from the substrate.

## Resist Processing

Important steps in photoresist processing are shown in Fig. 4.32. Since photoresists are not sensitive to wavelengths larger than 500 nm, the resist is processed in a room which is illuminated with yellow light to prevent unwanted exposure.

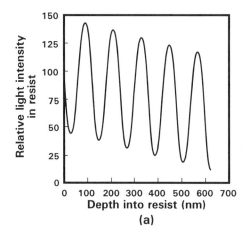

(a)

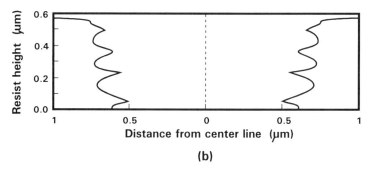

(b)

**Fig. 4.31** Standing wave effect. (a) Periodic variation of light intensity as a function of depth in photoresist on a reflective substrate (Adapted from [32]). (b) Calculated edge profile for a nominal 1 $\mu m$ line in positive photoresist (Adapted from [38]).

## Surface Preparation

Prior to resist application, the wafer surface must be cleaned, dehydrated, and primed to improve the adhesion between resist and substrate. Cleaning steps are required because of the inevitable contamination which occurs during storage and handling between processing steps. Discrete contaminants of concern range in size from about 100-nm in diameter to particles which are visible without magnification, and can cause device failures. Most contaminants require sophisticated techniques to detect, particularly those in the range of 100-300 nm. As the minimum feature size is reduced, the particles of concern will decrease in size from 100 nm to less than 10 nm. Contaminant films may consist of atomic, ionic, or polymer layers, which are sometimes very difficult to detect. Such films can cause gross resist adhesion failure during development or etching processes. Depending on the expected contamination type and level, several surface treatments can be used. In some cases, wet chemical or dry cleaning will be required to remove the film. Contamination control is discussed in the next chapter.

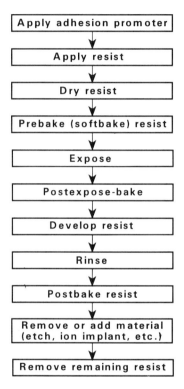

**Fig. 4.32** Important steps in photoresist processing.

### Resist Application

Dehydration and wafer priming are necessary to ensure a hydrophobic wafer surface and good resist adhesion. Dehydration bake is achieved by heating the wafer at about 200 °C for 30-60 minutes to evaporate the water from the surface (and volatilize residual organic contaminants). It is important that the substrate be coated with resist immediately after surface preparation to avoid re-adsorption of water on the substrate.

Surface silicon atoms bond strongly with water and form a monolayer of silanol groups (SiOH) that does not adhere adequately to resist. Adhesion promoters are used to react with silanol and replace the -OH group with an organic radical that adheres well to resist. The most common adhesion promoter is hexamethyldisilazane (HMDS). It is typically applied in the form of vapor at elevated temperature and reduced pressure, improving its uniformity and enhancing its reaction with silanol groups. The most common method of coating the wafer with resist is to deposit a few $cm^3$ of liquid resist with a "moving-arm" dispenser which moves slowly from the wafer center toward its edge while the wafer rotates at slow speed. After dispensing the resist, the wafer is accelerated up to a constant rotational speed, which is held for a specific time (typically 30 s). The resist thickness and defect density must be controlled within tight specifications. Typical thicknesses range from 0.5 $\mu$m - 1.5 $\mu$m and may not vary by more than $\pm$ 15 nm on flat surfaces across the wafer. The thickness increase with the percent solids in the resist (which determines its viscosity), and decreases with increasing spin speed (typically 2500-6500 rpm) [39]. It also depends on the wafer topography. While theoretical models have been developed to predict the resist thickness on flat surfaces, the actual thickness and its dependence on the various parameters must be determined experimentally. For this purpose, calibration curves are prepared by measuring the resist thickness across wafers and from wafer to wafer and projecting an average resist thickness as function of deposition parameters.

### Pre-Expose Bake (Soft-Bake, Prebake)

After coating, the wafer is given a preexpose-bake to remove the resist solvent and increase resist to wafer adhesion. This also causes the resist thickness to decrease and its surface to become less susceptible to particulate contamination. Preexpose bake is typically done at 70 °C-100 °C for about 1 minute with a hot-plate, whereby the wafer is brought in intimate or proximity contact with a heated metal plate. The thermal conductivity of silicon is sufficiently high to allow the resist to reach the plate temperature in seconds.

## Exposure

The mask is aligned to the wafer or, if applicable, to an existing pattern previously defined on the wafer (Sec. 4.1.7), and the mask and wafer are exposed to UV light, as described in the preceding sections. Stringent control of exposure time is required. Underexposure can result in incomplete image formation. Overexposure reduces the resist image size with respect to the mask size in positive resist, but it increases the resist image for negative resist.

## Post-Exposure Bake

Post-exposure bake is introduced primarily to reduce standing wave effects by reducing the extent of "ridges" at the resist edges (Fig. 4.31). It is believed that a bake at 100 °C-110 °C causes the photoactive compounds (PAC) to diffuse between layers of intensity minima, allowing more uniform development of the entire film and reducing non-uniformities along sidewalls.

## Development

Development is a critical step in photoresist processing, playing a key role in defining the shape of the resist profile and controlling the line-width. As mentioned above, a typical developer of DQN photoresist is a solution of tetramethyl-ammmonium hydroxide (TMAH). This solution must have a very low concentration of metal-ions, particularly sodium ions, to reduce contamination of the wafer during development. The most common development techniques are spin, spray and puddle development. In spin development, the developer is dispensed into the rotating wafer, using the same principle as spin coating. Following development, the wafers are rinsed (typically in de-ionized water) to remove the remaining developer, and then dried, while spinning. Spray development uses a a nozzle to create a fine mist of developer over the wafer. This method is believed to reduce the required amount of developer and to result in a more uniform distribution of developer. More recently, static "puddle" development has become widely used. In this technique, the developer is slowly puddled onto the wafer surface, where it remains during the development cycle, typically for 60 s. The newest resists are optimized for puddle processing, resulting in improved control of uniformity and minimal consumption of chemicals.

### Postdevelopment Bake  (Hard-Bake, Postbake)

A postdevelopment bake is typically required to give the remaining resist images the adhesion necessary to withstand subsequent processing steps, such as etching and ion implantation (Chaps. 5,6). Positive resists are baked at 120 °C-140 °C, whereby the resin polymer in the resist crosslinks, making the image more thermally stable Higher bake temperatures may cause the resist to flow and the pattern to deform.

Resist hardening can also be achieved by exposing the resist to deep-UV that enhances crosslinking of the resin in a thin film at the resist surface [40], increasing their thermal stability to $\simeq 200$ °C.

After development, the wafers are measured for pinhole density and uniformity, and the resist images measured as described in Sec. 4.1.7. If the feature sizes are not within a specified range, the resist is stripped and the complete photoresist process repeated.

### Resist Strip

After using the resist pattern for, e.g., etching or ion implantation, the resist is stripped from the surface. Wet-stripping of positive resist typically uses an organic, acid-based solution. The resist removal procedure depends, however, on the nature of the underlying film (oxide, nitride, polysilicon, or metal), and on the process to which the resist pattern has been exposed. For example, a resist which serves as an ion implantation mask is typically very difficult, if not impossible to remove by wet-stripping. Also, even without exposing the resist to a harsh environment, wet-stripping does not always remove the resist completely, and a plasma "descum" becomes necessary to remove the "final monolayer" of resist from the wafer. The most common method to remove positive resist is "plasma ashing".  An oxygen plasma reacts readily with the organic resist but leaves inorganic films (such as oxide, nitride, polysilicon or metal) mostly intact. Plasma ashing is discussed in the following chapter.  One of the major concerns with plasma ashing (as with all plasma electron-beam, and ion-beam technologies) is the effect of radiation and charging on underlying device properties. These device effects are discussed in a later chapter.

### Mid-UV and Deep-UV Resists

Near-UV systems, which includes broadband (no filter), G-line (436 nm), and I-line (365 nm) imaging, is the dominant photolithography technique because of the maturity of its light sources, exposure system and photoresist chemistry. The resolution of photolithographic systems can be increased by extending the tools to shorter wavelengths of exposure radiation [41]. Photoresists used in the near-UV range, however, do not perform well in mid-UV (250-300 nm) and deep-UV (190-250 nm) regions. The problem results from a low sensitivity of the PAC combined with a significant absorption of the resin [3]. The light intensity received near the bottom of the resist is considerably smaller than that received at the top (Eq. 4.20). This invariably produces image profiles with negative sloping walls. To achieve straight-wall images, the resist must absorb only a small percentage of the incident radiation, typically less than 20%. The effect of strong absorption is less pronounced with thinner resists. As the resist thickness is reduced below $\simeq 0.5$ $\mu$m, however, the resist becomes less efficient as a mask during etching, and exhibits strong non-uniformities over steps, and becomes more prone to pinholes.

At deep UV wavelengths, conventional photoresists become practically opaque. Therefore, new positive and negative resists must be developed for deep UV exposure. The major challenge in defining a deep UV resist chemistry is to find a match between the absorption spectrum of the resist and the output spectrum of the exposure tool, with an accompanying redesign of the projection optics [41]. This work is still in progress.

### 4.1.7 Dimensional Control and Alignment

The design of an integrated circuit must follow a set of layout rules that specify the allowable minimum (sometimes maximum) design dimensions, such as linewidth, contact size, spacing between patterns defined by the same mask, and spacings between patterns defined by different masks. Figure 4.33 illustrates some of these dimensions.

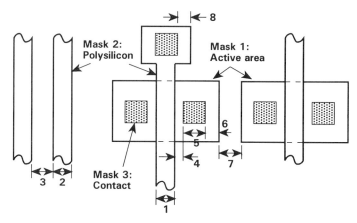

**Fig. 4.33** Examples of dimensions specified in layout rules.

The factors that contribute to the definition of design rules are rather complex. The minimum image size that can be printed on the surface with a specified "fidelity" depends primarily on the resolution of the lithographic tool and resist properties. Other factors, however, such as process-induced fluctuations in pattern dimensions, device reliability, and circuit performance, play important roles in determining design rules. These factors sometimes restrict the patterns to dimensions that are not at the "lithographic limit". For example, etching, oxidation, and lateral diffusion of dopants can cause the final wafer dimensions to deviate considerably from those defined in the resist; reliability concerns, such as electromigration in metals, may limit the metal width or contact size to dimensions larger than the minimum feature size; the minimum MOSFET channel length is frequently designed larger than the allowable minimum feature size to reduce the MOSFET susceptibility to short-channel effects and the impact of channel length fluctuations on circuit performance.

Dimensions that must be rigorously controlled during mask inspection and after imaging on the resist to ensure that circuits operate within specifications are called **critical dimensions (CD).** The imaging process can cause the average linewidth that is measured in the resist to be larger or smaller than on the mask. This difference is sometimes referred to as "bias". Fluctuations in the imaging process cause variations in the resist dimensions. These variations, also called **tolerances,** can be fitted to a Gaussian distribution with a standard deviation $\sigma$, characteristic of the imaging and resist systems. They are taken into account during layout of a circuit by designing devices to be typically tolerant to $3\sigma$ variations. For example, for a standard deviation $\sigma = 30$ nm, the nominal line-width must be chosen so that the circuit will operate reliably and within specification when the width, $W$, varies from a minimum of W - 90 nm to a maximum of W + 90 nm. For more robust designs, the tolerance is extended to $6\sigma$. Tolerances in device dimensions greatly impact die size and circuit performance. Critical dimensions are therefore monitored routinely on the resist and etched wafer patterns.

Alignment refers to the ability of printing images of one mask over images which were printed in a previous masking step with sufficient precision and control (Fig. 4.33). Errors in the placement of one pattern with respect to another are called **overlay errors** and treated statistically. The accuracy of pattern placement on the reticle is crucial to the superposition of successive mask levels during device fabrication. For this purpose, each mask contains specially designed and placed registration marks which allow the alignment of a subsequent mask. Mask image placement is checked during mask inspection, whereby laser interferometric X-Y coordinate measurement tools are used to compare actual mask image positions to the intended design.

For exposure, the reticle is first aligned on its stage using crosshairs or other recognition marks outside the chip design area of the mask glass. By aligning the mask to the tool, reticle placement errors in the x- and y-direction and in the rotational angle $\theta$ are reduced. The wafer is then mechanically pre-aligned, using the notch and flat as references, to position the wafer so that the sensors can detect its markers for the alignment steps that follow. This is done after the wafer is pulled from its container and before the handler places it on the wafer stage.

Alignment is typically done automatically by sensing the light diffracted from alignment marks on the wafer or chip.

Various wavelengths are used for detection, sometimes multiple to minimize phase contrast problems. For coarse (or global) alignment, marks within the kerf outside the chip are used at two positions of the wafer to bring the wafer within a 2-4 $\mu$m of the acquisition range of fine alignment that follows. The fine alignment procedure depends on the exposure system. It uses a separate set of alignment marks, typically placed in the kerf of the die. Some schemes use die-by-die alignment, whereby every die is aligned prior to be exposed. In this process, large errors can occur if alignment targets have poor contrast on a chip. Also, each die's alignment is subject to the errors of the alignment system. Other schemes use two alignment marks on the wafer for global alignment. The wafer is then "free-stepped", using the precision of the X-Y table for good registration. The most widely used method, however, is "enhanced global alignment", whereby a sample of chips (typically 10-15) is measured and the coordinates stored in the system's computer. These positions are then statistically analyzed and a "best fit" stepping pattern generated to drive the stage through the exposure sequence.

Since precision of alignment depends on the contrast of the optical signal obtained from the mark, the mark must exhibit a sufficiently large step and/or change in its refractive index to produce the required contrast. One typical alignment mark is formed by etching a trench in silicon and filling it with silicon dioxide. The first mask, e.g., the active-area mask in Fig. 4.33, can be directly aligned to the mark. The second mask (polysilicon) can also be directly aligned to the initial mark in silicon. Both direct alignments to the mark in silicon are called first-order alignments. If the polysilicon pattern had been aligned to a mark defined by the active area mask rather than the mark initially defined in silicon, alignment of polysilicon to the active area would have been second-order. For a given contrast in alignment marks, second-order alignment tolerances are $\sqrt{2}$ larger than first-order tolerances. There may also be larger orders in the alignment scheme. Typically, first-order overlay tolerances are required to be less than 30% of the minimum feature size.

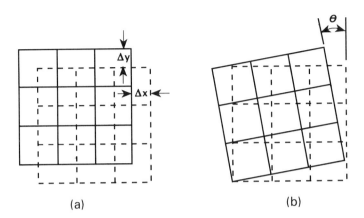

(a)                                    (b)

**Fig. 4.34** Three important types of misregistrations.
(a) X- and Y-misalignments.   (b) Rotations (or $\theta$)
misregistrations.

Figure 4.34 illustrates three major types of misalignments. The first two types are overlay errors in the x- and y- directions (Fig. 4.34a). The third type, sometimes referred to as $\theta$ misalignment, is due to a rotation of one pattern with respect to the other by an angle $\theta$ (Fig. 4.34b).

Another important change in image size is caused by rounding at corners, causing images of square or rectangular shapes to have considerably smaller areas and perimeters than the design dimensions. This is illustrated in Fig. 4.35 for three commonly designed shapes. A designed minimum-size square shape is typically printed as a circle on the wafer (Fig. 4.35a). Small rectangular shapes in the layout can be approximated as ellipses on the wafer (Fig. 4.35b). The effect of corner-rounding on the minimum space between adjacent shapes is illustrated in Fig. 4.35c. The factors that contribute to corner rounding are rather complex. They are related to mask preparation, image projection, and resist development. Processes such as etching and dopant diffusion significantly affect roundings in the final wafer pattern.

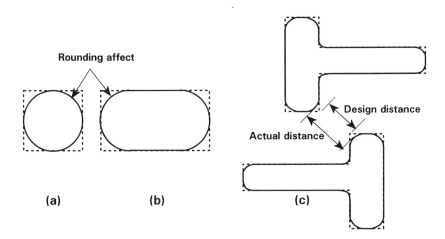

**Fig. 4.35** Rounding at corners. (a) Square shape changes into circle. (b) Rectangular shape changes into ellipse. (c) Distance between adjacent shapes increases.

## 4.2 Resolution Enhancement Techniques

The resolution of an optical imaging system can be improved by increasing the numerical aperture, reducing the wavelength, or reducing the factor $k_1$ in Eq. 4.12. Increasing the numerical aperture and reducing the wavelength, however, decrease the depth of focus, as can be seen from Eq. 4.13. For example, a minimum feature size of $\simeq 400$ nm can be achieved $\lambda = 365$ nm (I-line) and NA $\simeq 0.45$, but the DOF is at best 1 $\mu$m. The minimum printable feature size can be reduced to $\simeq 250$ nm with 248-nm illumination (KrF excimer laser) and NA $\simeq 0.5$, at the cost of further reducing the depth of focus and restricting the flexibility in the design of three-dimensional structures than can be constructed with precision on the wafer [42]. Also, further reduction in wavelength (193 nm, 157 nm) requires the development of new optical systems and resist compositions.

The above limitations have prompted the development of novel techniques that enhance resolution by effectively reducing $k_1$ in Eq. 4.12, and increasing $k_2$ in Eq. 4.13 for a given exposure wavelength and numerical aperture. The methods, however, only improve the minimum feature size and have no impact on overlay tolerances. Optical phase-shifting and off-axis illumination are examples of techniques that are considered to improve resolution and DOF. To simplify the description of their basic principles, coherent illumination is assumed throughout this section.

## 4.2.1 Optical Phase-Shifting

We have seen that with perfect lenses the resolution of the exposure system becomes diffraction limited. That is, because of "bending of light" beyond the edges of the clear field, regions under the opaque areas receive substantial amounts of light, reducing the image contrast in the resist. The image contrast is defined by Eq. 4.17, where $I_{max}$ is the relative intensity in the center of the "ideally" bright area, and $I_{min}$ the relative intensity at the center of the "ideally" opaque region. Fig. 4.36a shows how diffraction can cause substantial exposure of the resist under the opaque regions. As the line/space dimensions decrease, $I_{min}$ increases. Images of two neighboring apertures can be resolved when the intensity under the opaque regions is reduced to a specified fraction of the maximum intensity under the clear field. By introducing phase changes between adjacent images, the light intensity between them can be reduced considerably, allowing closely projected images to be separated. Phase-shifting uses both the image intensity profile and the phase information in the image. A phase-shift mask (PSM) is a two-level mask structure that creates a phase difference between waves from adjacent apertures. The preparation of phase-shifting masks typically requires new mask fabrication techniques and the development of new computer-aided mask design software.

### Alternate Phase-Shifting

Figure 4.36b shows the principle of the first phase-shifting method, initially proposed by Levenson [43] and referred to as alternate phase-shifting or Levenson method.

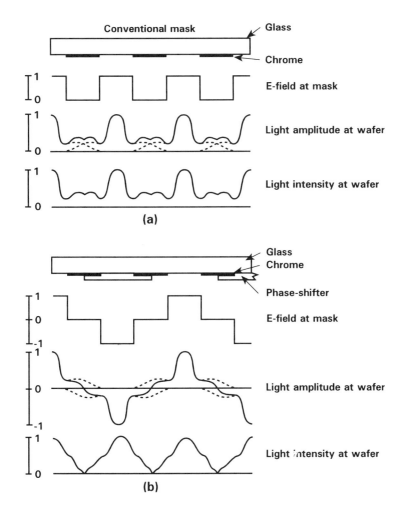

**Fig. 4.36** Alternate phase-shifting mask. (a) Conventional mask showing substantial exposure under the opaque areas. (b) Alternate phase-shifting mask causes destructive interference under the opaque regions, increasing the image contrast [43].

The shifter in Fig. 4.36b can be formed, for example, by depositing and patterning silicon dioxide on chrome, or etching grooves in a quartz blank. Every other transparent region in the periodical design is covered by a shifter film. Mask making consists of two E-beam or laser exposure steps. For example, a chrome film (or other metal film) is first defined and etched, and the phase-shifting film then patterned. The latter exposure is less stringent because the minimum width of the shifter is twice the width of the chrome pattern.

As light passes through the transparent mask-material and phase-shifter, its wavelength is reduced from that in air by the indices of refraction in the two materials. The optical path-difference with and without phase-shifter is $(n-1)d$, where $n$ is the index of refraction of the phase-shifter and $d$ its thickness (Fig. 4.37).

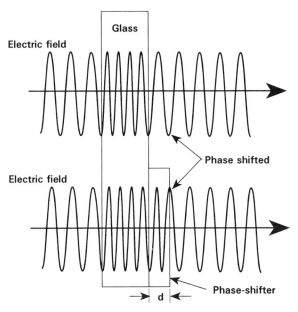

Fig. 4.37 To illustrate the phase difference between light passing through a region coated with a phase-shifter and an uncoated region The sketch shows a phase-difference of 180 ° (Adapted from [44]).

The phase difference is

$$\phi = (n-1)\,\frac{2\pi d}{\lambda}\,.$$

4.22

Phase shifts of 0° and 180° are produced in alternating apertures of the periodic structure by defining the thickness of the shifter as

$$d = \frac{\lambda}{2(n - 1)} \cdot \qquad 4.23$$

The light amplitude and intensity are plotted in Fig. 4.36b. Light amplitudes of opposite phases cancel each other under the opaque regions, considerably improving the image contrast. Since the electric field intensity is forced to pass through zero to -1, resolution is doubled and sharper transitions from exposed to unexposed regions are achieved. The orders of diffraction that pass through the projection optics are compared in Fig. 4.38 for conventional and phase-shift masks.

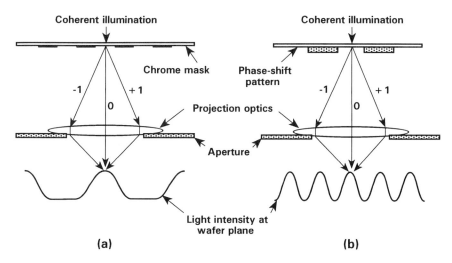

**(a)**          **(b)**

**Fig. 4.38** Diffraction orders passing through the projection lens. (a) Three-beam image formation with conventional transmission mask. (b) Two-beam image formation with phase-shift mask. (Adapted from [16].)

While a periodic grating exhibits 0th, ±1st, ±2nd, ... diffraction orders, only the ±1st order is preserved so that the resolution potential of the imaging lens can be fully utilized. In conventional systems, both the 0th and ±1st order are within the acceptance angle of the projection optics, as shown in Fig. 4.38a. Without at least one of the two 1st-order rays, the image would become a structureless beam because the 0th order contains no information on the pattern. Figure 4.38b illustrates how phase-shifting suppresses the 0th order and enhancing image contrast.

While alternate phase-shifting offers a considerable enhancement in image contrast, it is limited in its applications to periodic grating structures. Consequently, the method is not well suited for isolated features. Some techniques to improve the resolution of isolated openings or lines are described next.

### Rim (or Edge-Enhancement)Phase-Shifting

Rim phase-shifting is applicable to arbitrary mask patterns, enhancing the contrast of isolated opaque features [45,46]. The phase-shifters are placed at the chrome-edges, as shown in Fig. 4.39.

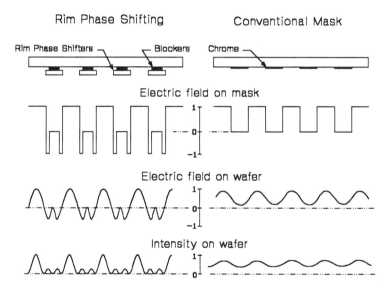

**Fig. 4.39** Rim phase-shifting. Undercutting at the absorber periphery causes phase-shifting to occur only at the rim, increasing image contrast.

To define a rim phase-shift mask, a conventional mask is first prepared with electron-beam or laser exposure. A uniform film of phase-shifting material is then deposited onto the patterned chrome (or other absorber). This is followed by exposure from the glass side of the mask, resulting in self-aligned imaging and delineation of the phase-shifters. The pattern is then used to undercut the chrome by wet-etching (Chap. 5). The width of the shifter is determined by the etch-rate and etch time. Phase-shifting only occurs at the edge of the pattern where it increases the image contrast. Since the phase-shifted regions are smaller than the minimum feature size, the images of those regions are not projected onto the wafer. The phase-shifted regions, however, play a dominant role in reducing the size of the bright regions. By optimizing the width of the shifted regions, a large peak intensity and steep image slope can be obtained. Optimization of the shifter width is, however, only possible for one feature size since all features are typically defined with one mask and exhibit the same undercut.

### Attenuated (or Halftone) Phase-Shifting

An attenuated phase-shift mask is prepared by replacing the opaque part of a conventional mask by an absorbing (half-tone) phase-shifter with a transmittance of $\simeq 0.1$. The halftone material can be made, for example, of a very thin film of chrome having the appropriate thickness to shift the phase by approximately 180° with respect to the clear regions (Fig. 4.40).

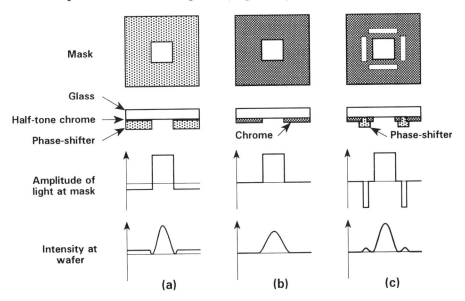

**Fig. 4.40** Attenuated phase-shifting. (a) Phase-shift mask defining an isolated feature and resulting intensity on wafer. (b) Conventional transmission mask and intensity. (c) Rim phase-shifting mask and intensity.

While some light passes through the halftone film, its intensity is too small to expose and develop the resist on the wafer. Since the halftone film also acts as a phase-shifter, only the isolated image is needed for its definition, considerably simplifying mask-making and repair [47].

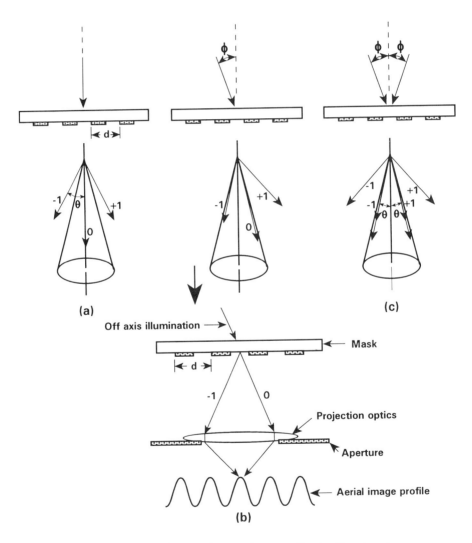

**Fig. 4.41** Comparison of on-axis to off-axis illumination. (a) On-axis illumination. (b) Off-axis illumination, one incident beam. (c) Off-axis illumination, two symmetrically arranged incident beams.

## 4.2.2 Off-Axis (or Oblique) Illumination

The principle of off-axis illumination can be seen from Fig. 4.41, where a mask grating of pitch $d$ is illuminated with coherent light [48 – 51]. In conventional "on-axis" illumination, the 0-order diffracted light travels along the optical axis and the 1st-order diffracted light travels along a direction given by Eq. 4.5 (Fig. 4.41a). As the pitch decreases, the diffracted angle $\theta$ increases and a point is reached where $\theta$ is larger than the acceptance angle of the imaging lens so that the 1st-order beams are suppressed. In this case, only the 0th order passes through the lens, without any information on the pattern. Resolution is approximated for this limit as $\lambda/2NA$. For an optimized oblique illumination, the $\pm$ 1st order beams are kept symmetrical with respect to the 0th order, but one of them (-1 in this case) lies inside the acceptance angle of the lens, allowing the "reconstruction" of the pattern (Fig. 4.41b). The intensity of the 0-order is, however, greater than that of the 1st-order, and the focus latitude with this arrangement is small. Figure 4.41c shows an illumination with two oblique beams, symmetrically arranged with respect to the optical axis. In this case, the 0-order of the left beam coincides with the + 1-order of the right beam and the 0-order of the right beam coincides with the -1-order of the left beam, resulting in a symmetrical intensity profile.

Off-axis illumination reduces the resolution limit and increases the depth of focus. From Eq. 4.6, the resolution limit on the reticle side is found as

$$\Delta x \simeq \frac{\lambda}{2(\sin \theta + \sin \phi)} , \qquad 4.24$$

where $\theta$ and $\phi$ are defined in Fig. 4.41. From the definition of NA in Eq. 4.7 and for 5X projection, the resolution limit at the wafer plane is approximated as

$$\Delta x \simeq \frac{\lambda}{3NA} , \qquad 4.25$$

showing an improvement of 1.5X in resolution over on-axis illumination. The improvement in depth of focus can be seen from a simplified analysis of the optical paths. With conventional on-axis illumination, the optical paths of the 0th and ± 1st order beams are equal when the wafer is placed at the position of best focus. They begin to differ as the wafer is moved away from this position, causing aberrations. With off-axis illumination, the optical paths of the different beams remain equal for a wide range of "defocus" positions, considerably increasing the DOF.

For lines and spaces parallel to the x- and y-direction, a "four-spot" arrangement is chosen, as shown in Fig. 4.42a. This arrangement is, however, not very efficient for oblique patterns, such as lines defined under 45° [48 – 51]. Another illuminating pattern that is considered is the annular aperture stop shown in Fig. 4.42b. Combining off-axis illumination with rim and attenuated phase-shift have been shown to further improve resolution and DOF [50, 51]. Comparison of these techniques is, however, beyond the scope of this chapter.

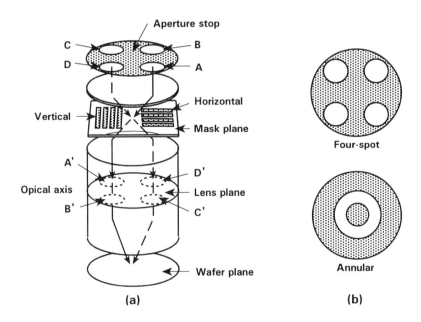

**Fig. 4.42** Illustration of off-axis illumination arrangements. (a) Schematic of projection system. (b) Four-spot and annular aperture stops.

## 4.3 Electron-Beam Lithography

Electron-beam lithography provides a means to reduce image resolution below 100 nm by using accelerated electrons instead of photons [7, 52 – 54]. Electrons, like photons, have particle and wave properties; when accelerated, they acquire a wavelength that can be approximated by

$$\lambda = \frac{1.23}{\sqrt{V}} \, , \qquad\qquad 4.26$$

where $\lambda$ is in nm and $V$ is the accelerating voltage in V. For example, electrons accelerated at 10 keV would have a wavelength of 0.0123 nm, which is several orders of magnitude smaller that the wavelength of ultraviolet light. A beam of electrons can be focused to a few nanometers in diameter and easily deflected by electrostatic and magnetic fields to write a pattern on a mask at a much higher rate than conventional optical reticle generators [55], or to write directly on the wafer without the use of a mask [52]. Other advantages of electron-beam over optical exposure systems are the image placement accuracy (overlay tolerance $\leq$ 50 nm) and greater depth of focus that can be achieved with the extremely small numerical aperture [54, 56]. The main drawbacks of electron-beam systems are the low throughput in the direct-write mode, image proximity effects due to forward and back-scattering of electrons, and high maintenance cost.

Electron-beam lithographic systems are used predominantly in five major areas: fabrication of masks for UV and X-ray lithography; high-resolution patterning of special devices (such as high-speed GaAs circuits); economic custom interconnections of integrated logic circuits, where the cost of fabricating a multitude of masks for small "part-numbers" becomes prohibitively high; research and development, where short turn-around-time for early feasibility demonstration of new device designs is more important than throughput; and in "mix-and-match" schemes, where critical dimensions are printed with electron-beam systems and other patterns with less expensive optical systems.

### 4.3.1 Proximity Effect

While the minimum feature is not diffraction limited, there are parameters other than the wavelength of exposure that limit the resolution of an electron-beam system, notably forward scattering of electrons in the resist and their back-scattering from the substrate. When electrons or ions penetrate a resist and the underlying substrate with a certain energy, they are scattered by collisions with the solid material. Elastic collisions cause electrons to deviate from their incident direction, while inelastic collisions result in energy loss. Therefore, scattering causes the electron-beam to broaden and expose a larger volume of resist than expected (Fig. 4.43).

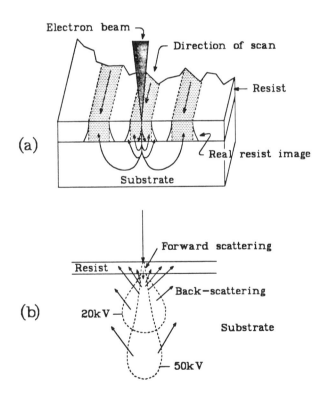

Fig. 4.43 To illustrate the proximity effect. Backscattered electrons cause the resist exposure in adjacent areas. (Adapted from [58].)

Backscattered electrons from the substrate can spread out over distances as large as 1 $\mu$m for 10 keV electrons [2, 26, 57 − 62]. The extent of resist exposure is roughly spherical in shape with a diameter much larger than the original beam-spot size. Consequently, if the features to be printed are close to each other, the exposure of one region will be affected by the exposure in the other. The dependence of exposure at one point on whether or not surrounding points are exposed is known as the **proximity effect.** Software programs can be implemented for proximity corrections. The electron dose for the correct shape and size in the resist image is adjusted as a function of distance from adjacent shapes by typically changing the exposure time [57, 58]. Such corrections, however, require accurate electron scattering and resist exposure models.

## 4.3.2 Electron-Beam Exposure Systems

Electron-beam systems are most commonly used in the scanning exposure mode, whereby the pattern is written sequentially into the resist with a finely focused beam of fixed or variable size. A less frequent mode of exposure is electron-beam projection, where the entire pattern is projected in parallel onto the wafer.

### Scanning Electron-Beam Systems

Figure 4.44 shows a schematic diagram of a scanning electron beam exposure system [52,63,64]. The electron source (gun) consists of typically a thermionic cathode of $TaB_6/W$, $Zr/W$, or $Ti/W$. The beam accelerating voltage can vary from 5-100 keV. The optical column contains electrostatic and magnetic systems for beam-blanking (deflecting off the optical axis), beam-shaping, focusing, and deflection to produce a controlled lateral movement of the beam. The beam shape and size are chosen to optimize performance or throughput. The beam can be shaped to form a fine round "probe" with a Gaussian current distribution and a diameter that is typically one-fourth or one-fifth the minimum feature size. It can also be variable shaped by changing the aperture, as shown in Fig. 4.44. Beam current densities range from 20-100 A/cm². A high-precision motorized mechanical stage (not shown) moves the wafer continuously within one field and in steps from one field to another, allowing exposure of the entire wafer. Machine subsystems, and transfer of pattern data and proximity corrections to the beam deflecting system are computer controlled.

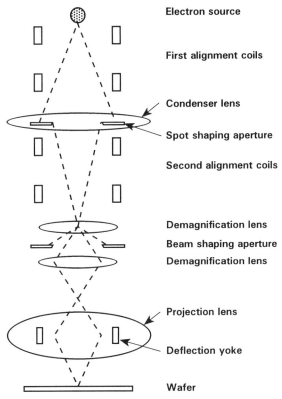

**Fig. 4.44** Imaging concepts in an electron-beam system [52].

There are two basic types of scanning electron-beam systems, raster-scan and vector-scan. Raster-scan exposure systems are very similar to but much more complex than scanning electron microscopes [65]. Since the area of accurate beam deflection is limited, the chip is divided into small subfields, typically about 1 mm on the side, and the stage is moved continuously, while a Gaussian beam of typically 50-200 nm in diameter is scanned serially over each subfield at right angle to the motion of the stage. The beam is deflected off the optical axis (blanked) where no exposure is desired and then unblanked to selectively expose the desired pattern. Overlapping the beam trajectories over the resist ensures that the full shape is exposed. The wafer is then moved mechanically to another subfield and the beam scanned again. Raster-scanning is, however, inefficient for exposure of low-density shapes, such as individual contact openings, since the beam must follow the full serpentine trajectory when travelling from one shape to another (it is blanked off between shapes). In a vector scan system the beam is deflected to only the image to be exposed rather than swept serially over the wafer. This method also uses variable rectangular shaped beam apertures with flat cross-sectional current profiles, and tailors the beam size and shape to fit various pattern elements [66]. By using small shapes to write feature "sleeves" for good edge acuity and larger spots for fast pattern fill-in, the number of spots needed to expose a pattern is reduced, shortening exposure time.

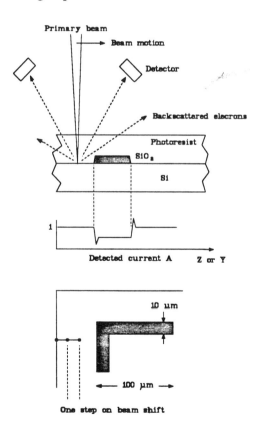

Fig. 4.45 Illustration of alignment mark detection.

**Alignment**

For direct-write, the beam is used as in a backscattering mode to detect features for alignment (registration marks) on the wafer (Fig. 4.45). Registration to a previously defined pattern is accomplished on each chip by scanning the beam across reference marks etched in the substrate and detecting the secondary and backscattered electrons [67]. When printing the pattern on a mask, no registration marks are present, so the system defines its own references on the mask.

## 4.3.3 Electron Resists

The two factors that limit resolution in electron-beam lithography are proximity effects and resist contrast. While electron resists must exhibit high resolution for submicron patterning, they must also be sufficiently sensitive to electron radiation to allow high speed exposure and acceptable throughput. Other factors, such as resistance to plasma etching, play a dominant role in defining the resist composition.

The time required to draw a pattern with an electron-beam is the sum of exposure time, $t_E$, and additional time, $t_{Add}$ required to move and position the stage, transfer data, and so on. For a beam current density, $j$, the exposure time is

$$t_E = k\left(\frac{S}{j}\right)\left(\frac{A}{a}\right),$$
4.27

where $A$ is the total area to be exposed, $a$ the spot area, $S$ the resist sensitivity, and $k$ a pattern filling coefficient. In raster scanning, $k = 1$, and in vector scanning $k$ varies from 0.2-0.4. Increasing the current density is limited by beam instability and resist heating problems. For faster exposure, it is therefore necessary to reduce the resist sensitivity parameter, $S$. Typical mid-UV sensitive resists, such as PAC/novolac, require electron doses in the range of 20-40 $\mu C/cm^2$, resulting in prohibitively long exposure times. These resists are also not applicable to deep UV systems because of their high absorption at the shorter wavelengths and the low DUV flux available for their exposure. The limited sensitivity of optical resists for DUV, electron-beam, and X-ray exposures has prompted the development of new resists that satisfy the sensitivity and resolution requirements of the exposure systems. Most of these resists fall into the "chemically amplified" category [68]. They employ the generation of an acidic species by irradiation with photons or electrons (Fig. 4.46).

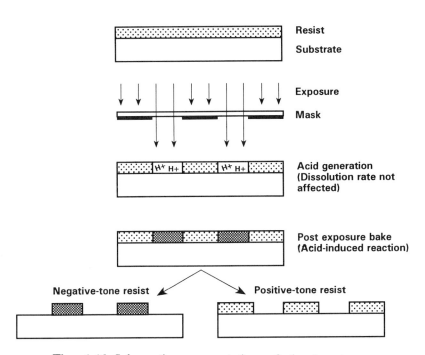

**Fig. 4.46** Schematic representation of the imaging mechanisms in a chemically amplified resist [69, 70].

The high sensitivity of chemically amplified resists is the result of the catalytic action of the acid during post exposure bake [69]. The acid generated in exposed regions acts as a strong catalyst that continues to accelerate chemical changes during post-exposure bake, such as deblocking of a protective group or crosslinking of a matrix resin [69 – 73], hence the label "amplified". Control of temperature, time, and ambient during post exposure bake is more critical than with optical resists. A chemically amplified resist is typically formulated with three main elements: (a) a matrix polymer, (b) a radiation-induced acid generator, and (c) the a capability of affecting differential solubility between exposed and unexposed regions through crosslinking or other molecular transformation. Negative-tone electron resists with sensitivities in the range 1-3 $\mu C/cm^2$ have been obtained using melamine as a cross-linking agent. One commonly used positive-tone electron resist is poly (butene 1 sulfone), or PBS, that has a sensitivity of about 1 $\mu C/cm^2$ [74].

## 4.4 X-Ray Lithography (XRL)

X-ray lithography is a promising technique for integrated circuit manufacturing at dimensions below $\simeq$ 200 nm [75]. When soft X-rays (0.8-2 nm) are used, diffraction effects are negligible down to a linewidth of about 100 nm [76, 77]. X-ray exposure is practically unaffected by small, low atomic mass particles that are frequently "printed" in UV-lithography [78, 79]. Since X-rays pass through many solid materials - including dust and skin flakes - the clean room needed for chip fabrication can be built to specifications much less severe than those needed for ultraviolet lithography, resulting in a greater particle defect tolerance and process latitude [80]. The main disadvantages of X-ray lithography are the fragility and dimensional instability of the mask, and the alignment method that, unless novel alignment schemes are developed, does not offer better accuracy than is achieved with conventional photolithography [53].

### 4.4.1 X-Ray Masks

The most important requirements for an X-ray mask to be used in volume production are [81]:

a) The mask-substrate must be transparent to the exposure radiation in the energy range where the resist is most sensitive and the mask absorber (opaque regions) has the maximum attenuation. Only when the substrate and absorber materials are very different from each other will high-contrast printing be possible.

b) The mask-substrate must exhibit dimensional stability and a very low defect density.

c) Exposure of the resist from X-ray induced photoelectrons should be minimized.

d) When alignment is performed with visible or UV light, the mask must be transparent to this radiation.

Masks for X-rays are more difficult to construct and repair than those used in conventional photolithography, because of the lack of suitable materials. The mask can no longer be made on a quartz plate because the thick plate would absorb the soft X-rays. The fabrication of X-ray masks is complicated by the need for very thin membranes to be used as the transparent substrate, since the attenuation of most materials increases rapidly with thickness. This limits both the durability and dimensional stability of the mask [82,83]. A typical X-ray masks is shown in Fig. 4.47.

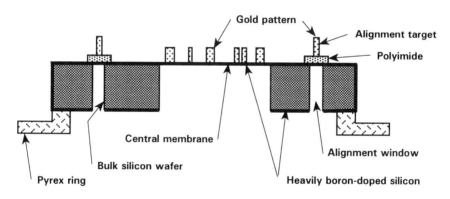

**Fig. 4.47** Cross-section of typical X-ray mask. Courtesy: C. A. Whiting, IBM, Essex Junction, Vermont.

The starting material is a silicon wafer upon which a borosilicate glass (BSG) has been deposited (Chap. 3). The wafer is then subjected to elevated temperature that causes boron to diffuse from the glass and form a $\simeq$ 5-$\mu$m thick, heavily-doped p-type film in silicon (Chap. 7). This film acts as an etch-stop when etching silicon from the backside (Chap. 5). The BSG film is patterned and silicon etched anisotropically from the backside. Etching stops within the heavily doped p-type region, forming a $\simeq$ 3-$\mu$m thick silicon membrane. The structure is then bonded to a pyrex support ring. Polyimide is deposited and patterned over alignment marks, acting as an etch-stop when the thin silicon film is removed from the alignment windows, and as a mechanical support for alignment targets. Because of the low X-ray absorption of most materials, absorber materials in X-ray masks are much thicker (typically 0.5-

$\mu m$ thick) than those of photomasks. They require a high atomic number material, such as gold, to absorb X-rays in the 0.8-2 nm range. A very thin ($\simeq 5$ nm) chrome layer is first deposited for good adhesion. Since the steepness of the gold film (the absorber) affects image sharpness, a very thin ($\leq 30$ nm) gold layer is deposited and used to electroplate a thicker gold through a patterned resist film [84]. Patterning is performed with an electron-beam or laser mask generator [85]. Other heavy metals, such as platinum, tungsten, or palladium, can also be deposited and patterned to form the absorber [86, 87].

The repair of costly X-ray masks is very important. There are two types of mask defects: clear defects (missing absorber structures) and opaque defects (unwanted absorber structures). The requirements on repair accuracy depends on the minimum feature size [88, 89]. For 300-nm minimum features, defects down to 100 nm in size must be repaired. Clear defects can be repaired by photolythic or electron-beam induced tin or tungsten deposition from a tin or tungsten compound atmosphere [90 − 93]. To remove unwanted opaque features, sputter-erosion by means of focused gallium-ion beams is performed [90,92,93].

## 4.4.2 X-Ray Sources

One of the difficulties encountered in XRL was the lack of a bright X-ray source for volume production. A high intensity source is required because typical resists are not very sensitive in the soft X-ray spectral region [94]. The two most promising sources of intense soft X-rays are the synchrotron and the laser-induced plasma radiations [95].

### Synchrotron Radiation (SR)

The electron synchrotron storage ring offers a unique combination of bright short wavelength light with good collimation [96 − 103]. A storage ring consists of a circular vacuum tube in which electrons travel at speeds very close to the speed of light. The electrons are held in their orbit by magnets. Each time the electrons are deflected or "wiggled" by one of the magnets, they emit an intense beam of radiation with a range of wavelengths that extends from visible to hard X-rays. It is well-known that electrons, when accelerated (changing direction is acceleration), emit electromagnetic radiation. While in an antenna the long wavelength radiation is emitted in all directions, relativistic electrons emit the radiation tangent to their trajectories on a curved path. This is similar to a "beam from headlights rounding a curve" [103] (Fig. 4.48).

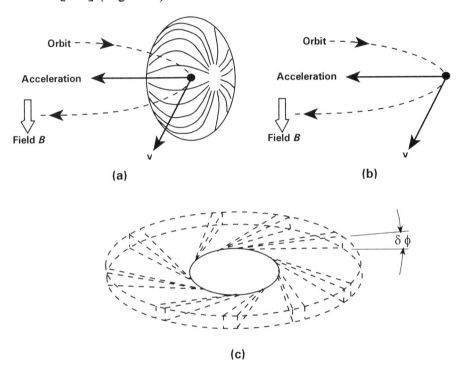

(a)

(b)

(c)

**Fig. 4.48** Synchrotron radiation emission from (a) a single electron at low velocity, (b) a single electron at high velocity, (c) a circulating beam of electrons, producing a continuous radiation fan of vertical opening angle $\Delta\phi$ [96]. Copyright 1993 by International Business Machines, reprinted by permission.

The radiation is in the plane of the storage ring. It is highly collimated and made available for mask illumination at small beam ports spaced at the periphery of the ring. A schematic of a typical synchrotron radiation source is shown in Fig. 4.49. It consists of an electron storage ring which has bending magnets (superconducting in most cases) that deflect the electrons on circular trajectories. The electrons are injected from the linac via the septum magnet, bent into a closed loop by the four dipole magnets, focused by the quadrupole magnets, and accelerated by the rf cavity. Synchrotron radiation is emitted wherever the electron path is bend by the dipole magnets [96].

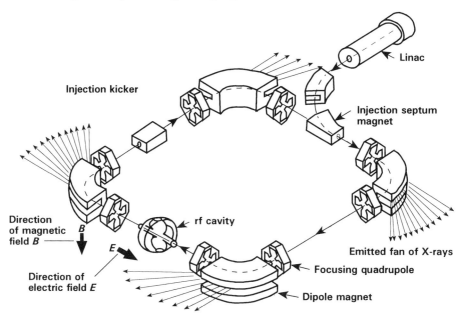

**Fig. 4.49** Schematic of a simple storage ring [96]. Copyright 1993 by International Business Machines, reprinted with permission.

The arc radius of the circular trajectory is (Fig. 4.50)

$$R = 3.34 \frac{E}{B} \quad (m), \tag{4.28}$$

where $E$ is the electron energy in Giga-electron-Volt (GeV), and $B$ the magnetic field strength in Tesla (T). The X-ray beam has a very small width in the vertical direction and will require some sort of scanning. The half-angle of vertical spread of synchrotron radiation is [104]

$$\Delta \phi = \frac{m_o c^2}{E}, \tag{4.29}$$

where $m_o c^2$ is the electron rest-energy. The SR spectrum can be calculated from $E$ and $R$ (or $B$). The wavelengths depend on the electron energy and curvature of the electron path. The peak wavelength of radiation within the angular spread and the emitted power density are given respectively as [104]

$$\lambda_P = 0.235 \frac{R}{E^3} \quad (nm), \qquad P = 14.1 \frac{E^4 I}{R} \quad (mW/cm^2) , \qquad 4.29$$

where $R$ is the radius of curvature (m), $E$ the electron energy (GeV), and $I$ the electron current (mA).

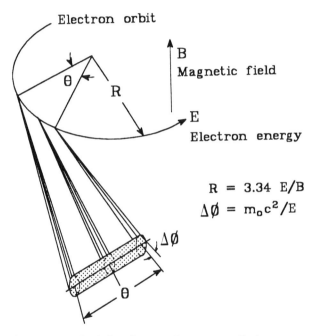

**Fig. 4.50** Principle of a synchrotron radiation source. (Adapted from [104].)

The radiation emitted by synchrotron electrons covers a wide spectral range. To select the desired band incident on the mask, the light is first reflected on a gold mirror at a low angle to "filter out" the high energy rays, and then passed though a $\simeq$ 15-$\mu$m thick beryllium window which absorbs light of wavelength less than $\simeq$ 0.8 nm.

**Laser Sources**

Viable alternatives to synchrotron sources are efficient point sources of soft X-radiation with a high output, consisting of tiny localized high-temperature plasmas [105,106]. They can be created either by a very localized high energy electron discharge, or by focusing a short, high-intensity laser pulse onto a target of high atomic number (such as gold or tin): a small plasma is created where the beam impinges on the metal, and the plasma emits soft X-rays in all directions [107 – 112]. Laser plasma sources exhibit high brightness X-rays which can be controlled in intensity and spectrum by carefully choosing the laser wavelength, target material, and laser pulse width. The source can be used for proximity or projection printing, depending on the availability of collimating and imaging lenses. The systems are compact and sufficiently small to be dedicated to a single processing installation, and offer a viable alternative to the more costly and less flexible synchrotron sources [113]. Two important considerations in the design of laser sources are the efficiency of converting laser irradiation into soft X-rays, and protection of the beryllium window from damage and contamination caused by bombardment with plasma particles.

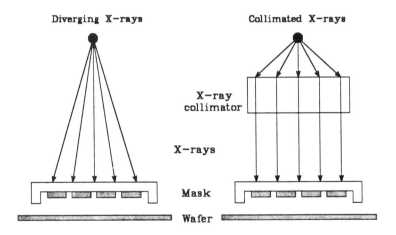

**Fig. 4.51** To illustrate the shadowing effect. (a) Impact on resolution with proximity printing using a point X-ray. (b) Correction with an X-ray collimator. (Adapted from [78].)

### 4.4.3 Imaging

As with UV light, there are two basic configurations for X-ray lithography: proximity and projection. With point X-rays, there is a fundamental trade-off that limits the performance of proximity printing. If the mask and wafer are too far from the source (to ensure a quasi-parallel beam), exposure suffers because of the limited source intensity. If the mask and wafer are too close to the source, the resolution becomes limited by shadowing effects, particularly with beams incident on the edge of the wafer (Fig. 4.51a). For a finite size source, penumbral blur contributes to image distortion (Fig. 4.24). The problem can be solved by collimating the beam (Fig. 4.51b). Refractive optics is not feasible because of the lack of materials to construct lenses to focus X-rays. Most of the lens materials have a refractive index near one, and none of them is sufficiently transparent in the soft X-ray spectral region.

Research continues on materials to construct X-ray collimators and mirrors for X-ray projection printing. The most promising collimator under development in capillary optics is the so-called Kumakhov lens in which hundred of thousands of tiny glass capillaries admit X-rays at different angles of incidence, and then guide them so that they emerge traveling in parallel [114].

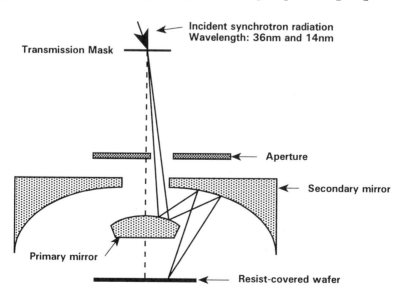

**Fig. 4.52** Schematic of an X-ray reduction system.

**242**

Reflective optics is achieved with special multilayer mirrors coated with thin layers of irridium and chrome, or molybdenum and silicon, depending on the wavelength used. These were originally developed for X-ray astronomy, and are structured in such a way that the reflected waves of X-radiation reinforce each other. The mirrors are shaped like a paraboloid, with the point source placed at the focus, resulting in a collimated beam (Fig. 4.52).

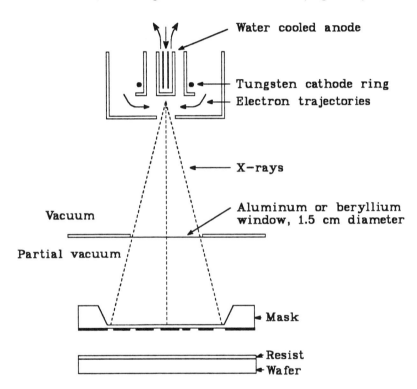

**Fig. 4.53** Principle of a conventional X-ray exposure system [116].

Proximity printing is the most practicable method for using soft X-rays for microlithography [115]. The principle of an early X-ray imaging system is shown schematically in Fig. 4.53 [116]. In this system, a conventional point source of X-rays is produced by a high-intensity 10-20 keV electron-beam which impinges on a metal. The beam of X-rays irradiates the mask and wafer as in a conventional proximity printing system. As mentioned earlier, X-rays must be generated in vacuum. The mask is placed at a distance of typically 30-40 cm from the source in a helium-filled chamber. The source is separated from the mask and wafer by a window, typically beryllium, since it offers the least attenuation in the soft X-ray spectral range. In this wavelength range, diffraction effects are negligible down to a minimum feature of 100-200 nm for a practical mask to wafer distance of 10-40 $\mu m$ [117, 118].

Because of the difficulty in maintaining a membrane stability over a field larger than $\simeq 10$ cm$^2$, typical masks are not designed to be entirely imaged at the same time. Instead, wafer steppers are used to expose only a small portion of the mask at a time. The stepper scans the mask and wafer vertically through a stationary X-ray beam, and steps and aligns the mask prior to each exposure [117 − 122]. While new methods are being developed to use X-rays for alignment with an overlay tolerance $\leq$ 50 nm ($3\sigma$) [120, 123 − 125], present alignment schemes use conventional alignment targets (Fig. 4.47), requiring the same procedure as for optical proximity printing.

### 4.4.4 X-Ray Resists and Resolution

X-ray resists are similar to UV and electron-beam resists. The mechanism of radiation absorption is, however, different. The nature of X-ray interaction with matter depends on the wavelength. Unlike visible and UV radiation, X-ray energy in the wavelength range of 0.6-1.2 nm is negligibly absorbed by valence electrons. Exposure in this range occurs in two stages: first the X-ray photon is absorbed by an atom, and an electron from an interior shell is ejected, generating secondary electrons with energies of a few electron-volts; second the photoelectrons expose the radiation sensitive sites of the resist, by chain scission (positive resist) or cross-linking (negative resist).

Due to the low atomic mass of their constituents, organic materials can be virtually transparent ($\leq$ 10% absorption/$\mu m$) to X-rays, depending on wavelength. Carbon, for example, is three times more absorbant at $\lambda$ = 1.2 nm than at $\lambda$ = 0.8 nm. Therefore, since resists are primarily composed of carbon atoms, their absorbance increases as the exposure wavelength increases [131]. Because the absorption of X-rays in the resist remains low, exposure of the resist is uniform through its entire thickness, resulting in resist images with nearly straight walls.

The resolution with X-ray exposure is typically limited by the electron-beam lithography used to make the mask. Diffraction effects and the migration of photoelectrons excited by the absorbed X-rays can, however, play an important role in determining image resolution The diffraction-limited resolution is approximated by [126, 127]

$$\Delta x \simeq k_x \sqrt{\lambda \left[ s + \frac{t_{PR}}{2} \right]} \, , \qquad\qquad 4.30$$

where $k_x$ = 0.4-1.5, $\lambda$ is the exposure wavelength, $t_{PR}$ the resist thickness, and $s$ the mask to wafer proximity gap (typically 10-40 $\mu m$). The diffraction limited resolution is found as $\Delta x$ = 100 - 200 nm.

The range of electrons generated in the resist is estimated as [128, 129]

$$R_g \simeq 56.2 \, \lambda^{-7/4} \quad (nm), \qquad\qquad 4.31$$

where $\lambda$ is in nm. Consideration of resist scattering and diffraction effects result in an optimum wavelength in the range 0.6-1.2 nm. Shorter wavelengths increase resist scattering while longer wavelengths increase diffraction effects. For beams which are not at normal incidence to the wafer, shadowing effects due to the thick (1 $\mu m$) opaque regions become significant (Fig. 4.51)

While conventional resists, such as DQN and polymethyl metacrylate (PMMA), have been successfully used for X-ray exposure [130], there is a serious concern about the low throughput caused by the limited sensitivity of optical resists to X-ray radiation [131]. As with electron-beam lithography, this has prompted the development of a new class of chemically amplified resists discussed in the preceding section.

## 4.5 Ion-Beam Lithography (IBL)

Ion-beams can be used to expose resists, either through a mask using broad-beams [132, 133], or serially by writing directly on the resist with a finely focused beam, similarly to electron-beam direct write [134, 135]. Because ions have a larger mass than electrons, they transfer their energies more efficiently to the resist and scatter much less (the spread is less than 10 nm) Also, ion-beams produce secondary electrons with much lower energies, and hence shorter range than electron-beams. Consequently, very high resolution and sensitivities can be achieved with ion-beams, without proximity effects that are typically observed with electron-beams lithography. Important advances in ion-beam lithography, a technique still in its early development stage, are reviewed in detail in [136, 137].

### 4.5.1 Ion-Beam Exposure

Columns to focus and deflect ion-beams are similar to those used for electron-beams but more difficult to construct [138]. Masked ion-beam exposure is similar to proximity printing in photo- and electron-beam lithography. It uses a broad-beam (1-2 cm²) which is swept over the mask and exposes the resist. Diffraction effects are negligible, and $\leq$ 100 nm resolution with 100 nm overlay tolerance has been achieved [133]. Focused ion beams (FIB) are used mainly for mask and circuit repair. They are also considered for focused ion-beam lithography (FIBL) to write a pattern with a finely focused beam ($\leq$ 0.1 nm in diameter). Exposure is, however, very slow because of the limited beam intensity. The technique is therefore considered for the preparation of high-resolution 1:1 masks used for X-rays, rather than for direct-writing a pattern on the wafer without using a mask.

### 4.5.2 Ion Sources

The brightness of the ion source limits the maximum intensity that can be focused into a fine beam. Two important sources are considered for lithography: the liquid metal source [136, 138 − 140], and the gas field-ionization source [135, 141, 142]. Liquid metal sources use metals of low melting point and low vapor pressure., typically gallium that melts at 30 °C and has a vapor pressure of $\leq 10^{-12}$ torr. The liquid metal is brought in contact without reacting with a tungsten tip which is held at a high positive potential. The field at the tip is high enough to produce field-evaporation of gallium ions from the melt at currents of $\simeq 1 - 100$ $\mu$A. The gallium ions are extracted, collimated, and scanned electrostatically.

Gas field-ionization sources use tips comparable to those in ion field microscopy. A sharp tungsten tip is held at a very high positive potential. Hydrogen, helium and some rare gases are typical ion sources. Field ionization is achieved by immersing the tungsten tip in a gas, such as $H_2$. As the hydrogen approaches the tip, it is attracted to it by polarization. Electrons tunnel from hydrogen to the tip, creating $H_2^+$ ions which are then repelled by the tip. A beam of such light ions can be formed and focused to very high resolution and current density, and is best suited for directly patterning the resist.

## 4.5.3 Pattern Definition

One of the most attractive applications of focused ion-beams is in ultra-fine lithography with minimum features $\leq 100$ nm. While conventional polymer resists (such as PMMA) and resist processing can be used, other attractive pattern definition techniques exist for ions-beams. In one method, a non-volatile layer is formed at the surface of the resist by implanting a focused beam of gallium ions into the resist, and converting the metal into its oxide [143]. When the film is subjected to an oxygen plasma, regions not covered by gallium oxide are removed by plasma etch, while regions delineated with gallium remain as a fine pattern (etching and implantation are discussed in the following two chapters). In a similar method, silicon ions are implanted into the resist, making the delineated region resistant to plasma etching (Chaps. 5,6) [144]. Another method combines a focused ion-beam with a metal oxide film, which constitutes a high-contrast inorganic resist, to directly define fine metal lines on the substrate [145]. A thin amorphous film of $MoO_3$, $WO_3$, or a combination of both, is deposited onto the substrate. The film is then delineated with a focused beam of $Ga^+$ ions. When the film is subjected to a hydrogen atmosphere at about 800 °C, the oxide is reduced to its metal, defining metal lines as narrow as 50 nm. The above techniques show that focused ion-beam lithography, although in its infancy stage, can be considered as an efficient method to prepare special structures and define masks with minimum features not achievable with other lithographic techniques.

# PROBLEMS

**4.1** An L-shaped line is delineated on resist by proximity printing. Indicate whether the line is defined by leaving resist inside or outside the shape for the following cases:

a) Clear-field mask, negative resist.

b) Clear-field mask, positive resist.

c) Dark-field mask, negative resist

d) Dark-field mask, positive resist.

**4.2** Assume proximity printing in all cases and complete the following table:

| Exposure | Resist sensitivity | Photons/cm² or electrons/cm² |
|---|---|---|
| 365 nm (I-line) | 60 mJ/cm² | |
| 193 nm (DUV | 1000 mJ/cm² | |
| 0.5 nm (X-rays) | 1500 mJ/cm² | |
| Electrons (E-beam) | $5x10^{-6}$ $C/cm^2$ | |
| Ions (Ion-beam) | $10^{-7}$ $C/cm^2$ | |

**4.3** A 1-$\mu$m long and 0.2-$\mu$m wide line is directly exposed with a focused ion-beam. Assuming the sensitivity to be $10^{-7}$ ions/cm², find the total number of ions needed to expose the shape. Comment?

**4.4** A 10 cm x 10 cm square area of a wafer is exposed by direct-write with an electron-beam having a square-shaped beam-spot of 0.125 $\mu$m on the side. The beam energy and current density are 20 keV and 50 A/cm$^2$, respectively. The spot is scanned across the area and blanked where the resist is not exposed. The resist sensitivity is $5x10^{-7}$ C/cm$^2$. Assume the wafer area to be subdivided into picture elements (pixels) of the same size as the beam-spot, and that the duration of spot-blanking to be the same as that of spot-exposure. Find:

a) The wavelength associated with the electron-beam.

b) The number of spots required to expose the wafer field.

c) The total time required to expose one wafer.

**4.5** A mask defining a grating of 0.5 $\mu$m lines and 0.5 $\mu$m spaces is exposed at normal incidence with monochromatic UV-light of wavelength 365 nm.

a) Find the grating's fundamental spatial frequency, N.

b) Show that the phase-difference between two emerging beams at an angle $\theta$ is d = S sin $\theta$, where S is the sum of line-space and line-width.

c) Constructive interference occurs when d = k $\lambda$, where $\lambda$ is the exposing wavelength. Show that constructive interference occurs when sin $\theta$ = k $\lambda$ N, and find all angles that correspond to a maximum of light.

**4.6** Show that for an MTF of 0.6 the normalized electric field amplitude of light swings from 0.5 to 1.

**4.7** The lens system of a camera can be reduced to a thin, converging lens of 50 mm focal distance and a numerical aperture of 0.125. The camera is "focused" to infinity. Assuming that a point is clearly seen if its image size on the film does not exceed 30 $\mu$m in diameter, find the minimum distance of an object that can be printed on the film. Find this distance for NA = 0.25.

**4.8** An aluminum film is covered with silicon dioxide coated with a resist layer. The resist is exposed at normal incidence (z-direction).

with monochromatic light of wavelength $\lambda$. The amplitude of the incident wave is

$$E_i(z) = E_o \sin(\omega k - kz + \phi)$$

and the amplitude of the reflected wave is

$$E_r(z) = E_o \sin[\omega k - k(2d - z) + \phi + \pi],$$

where d is the sum of resist and oxide thickness, $k = 2\pi n/\lambda$, and n is the index of refraction (assumed to be the same for resist and oxide).

a) Derive an expression for the standing wave intensity which can be attributed to the interference between $E_i$ and $E_r$.

b) Derive an expression that predicts the positions of the intensity minima and maxima with respect to the reflecting surface.

**4.9** Consider an X-ray source of diameter $W$ placed at a distance $D$ from a wafer. Show that if we neglect the mask thickness, the width of penumbral shadow can be approximated by $\delta x \simeq Ws/D$, where s is the distance from mask to wafer. For a negligible mask-to-wafer distance and a mask thickness $t_M$, show that the width of mask shadow can be approximated by $\delta x_M \simeq Rt_M/D$, where R is the lateral distance from the center-line of the source-wafer to the particular mask-edge.

**4.10** For a given radiation, the absorption coefficient $\alpha$ of a resist material is $10^4/cm$. Assume $\alpha$ to be uniform and find the resist thickness required to absorb 90% of the irradiating energy.

**4.11** Reliability and yield requirements limit the minimum polysilicon line-width and line-to-line space to 0.5 $\mu m$. The overall $3\sigma$ tolerance (including mask, lithography and etch) on polysilicon line-width is $\pm 0.15 \mu m$.

a) Find the design values of width and space that meet the above wafer requirements.

b) For a grating of polysilicon lines and spaces, what would be the percentile reduction in design "pitch" if the tolerance were "tightened" to $\pm 0.1 \mu m$?

**4.12** MOSFETs are to be constructed with the following four masks in sequence: active area, polysilicon gate, contacts, and metal. The minimum feature size is 0.35 $\mu$m and the tolerance on all image sizes is $\pm$ 80 nm. The masks are independently aligned to a registration mark etched in silicon. The first-order alignment tolerance is $\pm 0.15\mu$m. After processing, nominal dimensions shrink (positive bias) or expand (negative bias) as shown in the table below.

| Pattern | Bias ($\mu$m/edge) |
|---------|--------------------|
| Active area | 0. 25 |
| Polysilicon | -0.03 |
| Contacts | 0.08 |
| Metal | 0.06 |

The following requirements on wafer dimensions must be met:

a) Polysilicon width (MOSFET channel length) $\geq 0.35\mu$m.

b) Polysilicon length over active area (channel width) $\geq 0.6\mu$m.

c) Contacts must remain inside silicon or polysilicon.

d) Contact to polysilicon must remain outside active area.

e) Metal must fully cover contacts.

f) Metal width and space $\geq 0.35\mu$m.

Generate a set of design groundrules that will meet the above requirements.

# References

[1] J. H. McCoy, W. Lee, and G. L. Varnell, *"Optical Lithography Requirements in the Early 1990s,"* Solid-State Technology, 32 (3), 87-92 (1989).

[2] L. F. Thompson and M. J. Bowden, *"The Lithographic Process: The Physics,"* Introduction to Microlithography, L. F. Thompson, M. J. Bowden, and C. Willson, Eds., American Chemical Society, 1983.

[3] K. Jain, Excimer Laser Lithography, SPIE Optical Engineering Press, Bellingham, Washington, 1990.

[4] C. K. Rhodes, Ed., *Excimer Lasers, 2nd ed.,* Springer-Verlag, New York, 1984.

[5] J. J. Ewing, *"Rare Gas Halide Lasers,"* Physics Today, 31, p. 32, May 1978.

[6] F. T. Klostermann, *"A Step-and-Repeat Camera for Making Photomasks for Integrated Circuits,"* Philips Tech. Rev. 30 (3), 57 (1969).

[7] D. R. Herriott, R. J. Collier, D. S. Alles, and J. W. Stafford, *"EBES: A Practical Electron Lithographic System,"* IEEE Trans. Electron Dev., ED-22, 385 (1975).

[8] P. A. Warkentin and J. A. Schoeffel, *"Scanning Laser Technology Applied to High-Speed Reticle Writing,"* Proc. SPIE, 633, 286 (1986).

[9] P. Allen and P. Buck, *"Resolution Performance of a 0.60 NA, 364 nm Laser Direct Writer,"* Proc. SPIE, 1264, 454-465 (1990).

[10] K. Levy and P. Sandland, *"Automated Mask Inspection: What Can it Find, What Can it Miss, and What Can it Do?,"* Kodak Micorelectronics Seminar, p. 84 (1977).

[11] D. Ehlich, R. Osgood, D. Silversmith, and T. Deutsch, *"One-Step Repair of Transparent Defects in Hard-Surface Photolithographic Masks Via Laser Photodeposition,"* IEEE Electron Device Lett., EDL-1, 101 (1980).

[12] P. J. Heard, J. R. A. Cleaver, and H. Ahmed, *"Application of a Focused Ion Beam System to Defect Repair of VLSI Masks,"* J. Vac. Sci. Technol., B3 (1), 87-90 (1985).

[13] B. J. Lin, *"Optical Methods for Fine Line Lithography,"* Fine Line Lithography, R. Newman, Ed., North Holland, Amsterdam 1980.

[14] D. A. Markle, *"A New Projection Printer,"* Solid-State Technology, 17 (6), 50 (1974).

[15] R. E. Tibbetts and J. S. Wilczynski, *"High Performance Reduction Lenses for Microelectronic Circuit Fabrication,"* IBM J. Res. Develop., 13, 192 (1969).

[16] T. Brunner, *"Introduction to Optical Lithography Tools,"* SPIE Symposium on Microlithography, San Jose, California, March 5, 1993.

[17] J. D. Cuthbert, *"Optical Projection printing,"* Solid State Technol., 20 (8), 59 (1977).

[18] J. H. Bruning, *"Optical Imaging for Microfabrication,"* J. Vac. Sci. Technol., 17, 1147-1155 (1980).

[19] B. J. Lin, *"The Optimum Numerical Aperture for Optical Projection Microlithography,"* Proc. SPIE, 1463, 42 (1991).

[20] D. Meyerhofer, *"Simulation of Microlithographic Resist Processing Using the SAMPLE Program,"* RCA Review, 46, 356 (1985).

[21] D. Widmann, K. U. Stein, *"Semiconductor Technologies with Reduced Dimensions, Solid State Circuit,"* Proc. 2nd Europ. Solid State Circuits Conf., 29, p. 1977, 1976.

[22] M. C. King, *"Principles of Optical Lithography,"* VLSI Electronics Micro Structure Science, Vol. 1, N. G. Einspruch, Ed., Academic Press, New York (1981).

[23] C. A. Mack, *"PROLITH: A Comprehensive Optical Lithography Model,"* Opt. Microlith, IV, Proc., SPIE, 538, 207 (1985)

[24] M. M. O'Toole and A. R. Neureuther, *"Influence of Partial Coherence on Projection Printing,"* SPIE Devel. In Semicon. Microlithography IV, 174, 22-27 (1979).

[25] W. M. Moreau, *"Semiconductor Lithography, Principle, Practices, and Materials,"* Plenum Press, New York, 1989.

[26] D. A. McGillis, *"Lithography,"* VLSI Technology, S. M. Sze, Ed., p. 267, McGraw Hill, New York, 1983.

[27] W. R. Runyan and K. E. Bean, Semiconductor Integrated Circuit Processing Technology, New York, 1990.

[28] S. Wolf and R. N. Tauber, Silicon Processing for the VLSI Era, Vol. I, Lattice Press, Sunset Beach, California, 1986.

[29] T. Pampalone, *"Novolac Resins Used in Positive Resist Systems,"* Solid State Technology, p. 115, June 1984.

[30] D. L. Flowers and H. G. Hughes, *"On the Use of Photomask Coatings in Photolithography,"* J. Electrochem. Soc., 124, 1599 (1977).

[31] G. E. Flores and B. Kirkpatrick, *"Optical Lithography Stalls X-Rays,"* IEEE Spctrum, p. 24, Oct. 1991.

[32] D. F. Ilten and K. V. Patel, *"Standing Wave Effects in Photoresist Exposure,"* Image Technology, p. 9, Feb/March, 1971)979)

[33] F. H. Dill, *"Optical Lithography,"* IEEE Trans. Electron Dev., ED-22, 440-444 (1975).

[34] F. H. Dill, A. R. Neureuther, J. A. Tutt, and E. J. Walker, *"Modeling Projection Printing of Positive Resists,"* IEEE Trans. Electron Dev., ED-22, 456-464 (1975).

[35] K. L. Konnerth and F. H. Dill, *"In Situ Measurement of Dielectric Thickness During Etching and Developing Processes,"* IEEE Trans. Electron Devices, ED-22, 452-456 (1975).

[36] S. Middelhoek, *"Projection Masking, Thin Photoresist Layers, and Interference Effects,"* IBM J. Res. Dev., 14, 117 (1970).

[37] M. J. Bowden, *"The Physics and Chemistry of the Lithographic Process,"*

[38] V. Miller and H. L. Stover, *"Submicron Optical Lithography: I-Line Wafer Stepper and Photoresist Technology,"* Solid-State Technology, 127, January 1985. J. Electrochem. Soc., 128 (5), 195C (1981).

[39] B. D. Washo, *"Rheology and Modeling of the Spin Coating Process,"* IBM J. Res. Develop., 21, 190 (1977).

[40] H. Hirahoko and J. Pakanski, *"High-Temperature Flow Resistance of Micron Sized Images in AZ Resists,"* J. Electronchem. Soc., 128 (12), 2645-2647 (1981).

[41] M. J. Bowden, *"A Perspective on Resist Materials for Fine-Line Lithography,"* Materials for Microlithography, L. F. Thompson, C. G. Willson, and J. M. J. Frechet, Eds., American Chemical Society, Washington D.C., 1984.

[42] S. Okazaki, *"Resolution Limits of Optical Lithography,"* J. Vac. Sci. Technol., B 9 (6), 2829-2833 (1991).

[43] M. D. Levenson, N. S. Wiswanathan, and R. A. Simpson, *"Improving Resolution in Photolithography with a Phase-Shifting Mask,"* IEEE Trans. Electron Dev., ED-29 (12), 1892-1901 (1982).

[44] B. J. Lin, *"Phase-Shifting and Other Challenges in Optical Mask Technology,"* SPIE Proceedings, 1486, 54 (1990); *"The Phase-Shifting Mask Technology: Basics, Advanced Topics, and Updates,"* Symposium on Microlithography, SPIE 1992.

[45] T. Terasawa, N. Hasegawa, T. Kurusaki, and T. Tanaka, *"0.30 Micron Optical Lithography Using a Phase-Shifting Mask,"* SPIE Proceedings, 1088, 25 (1989).

[46] Y. Yanagishita, N. Ishiwata, Y. Tabata, K. Nakagawa, and K. Shigematsu, *"Phase-Shifting Photolithography Applicable to Real IC Patterns,"* Proc. SPIE 1463, 207-217 (1991).

[47] T. Terasawa, N. Hasegawa, H. Fukuda, and S. Katagiri, *"Imaging Characteristics of Multi-Phase-Shifting and Halftone Phase-Shifting Masks,"* Jap. J. Appl. Phys., 30 (11B), 2991-2997 (1991).

[48] N. Shiraishi, S. Hirukawa, Y. Takeuchi, and N. Magome, *"New Imaging Technique for 64M-DRAM,"* Proc. SPIE, 1674, 741-752 (1992).

[49] B. J. Lin, *"Off-Axis Illumination - Working Principles and Comparison with Alternating Phase-Shifting Masks,"* Proc. SPIE, 1927, 89-100 (1993).

[50] T. A. Brunner, *"Rim Phase-Shift Mask Combined with Off-Axis Illumination: A Path to 0.5 λ/Numerical Apertures Geometries,"* Optical Eng., 32 (10), 2337-2343 (1993).

[51] K. Ronse, P. Pforrr, K. H. Baik, R. Jonckheere, and L. Van den hove, *"Attenuated Phase Shifting Masks in Combination with Off-Axis Illumination: A Way towards Quarter Micron DUV Lithography for Random Logic Applications,"* Microelectronic Eng., 23, 133-138 (1994).

[52] J. L. Mauer, H. C. Pfeiffer, and W. Stickel, *"Electron-Optics of an Electron-Beam Lithographic System,"* IBM J. Res. Dev., 26, 568 (1977).

[53] A. N. Broers, *"Limits of Thin-Film Microfabrication,"* The Clifford Paterson Lecture, Proc. R. Soc. Lond. A416, 1 (1988).

[54] G. R. Brewer, *"Electron-Beam Technology in Microelectronic Fabrication,"* Academic Press, New York, 1980.

[55] J. P. Ballantyne, *"Electron-Beam Fabrication of Chromium Master Masks,"* J. Vac. Sci. Technol., 12 (6), 1257 (1975).

[56] H. Ahmed and W. C. Nixon, *"Microcircuit Engineering,"* Cambridge Univ. Press, Cambridge, England, 1980.

[57] D. F. Kyser and N. S. Visvanathan, *"Monte Carlo Simulation of Spatially Distributed Beams in Electron-Beam Lithography,"* J. Vac. Sci. Technol., 12, 1305-1308 (1975).

[58] D. F. Kyser and C. H. Ting, *"Voltage Dependence of Proximity Effects in Electron Beam Lithography,"* J. Vac. Sci. Technol., 16 (6), 1759-1762 (1979).

[59] J. S. Greeneich, *"Electron-Beam Processes,"* Electron beam Technology in Microelectronic Fabrication, G. R. Brewer, Ed., Academic Press, New York 1980.

[60] J. S. Greeneich, *"Impact of Electron Scattering on Linewidth Control in Electron-Beam Lithography,"* J. Vac. Sci. Tech., 16, 1749-1754 (1979).

[61] E. Kratschmer, *"Verification of a Proximity Effect Correction Program in Electron-Beam Lithography,"* J. Vac. Sci. Technol., 19, 1264 (1981).

[62] T. H. P. Chang, *"Proximity Effect in Electron-Beam Lithography,"* J. Vac. Sci. Technol., 12 (6), 1271-1275 (1975).

[63] H. C. Pfeiffer, *"Direct Write Electron Beam Lithography - A Production Line Reality,"* Solid-State technology, 223, September 1984.

**256**

[64] H. C. Pfeiffer, R. Butsch, and T. R. Groves, *"EL-3 + Electron Beam Direct Write System,"* Microelectronic Eng., 17, 7-10 (1992).

[65] G. R. Herriott, *"Electron-Beam Lithography Machines,"* Electron-beam Technology in Microelectronic Fabrication, G. R. Brewer, Ed., Academic press, New York, 1980.

[66] H. C. Pfeiffer, *"New Imaging and Deflection Concept for Probe-Forming Microfabrication Systems,"* J. Vac. Sci. Techn., 12, 1170-1275 (1975).

[67] P. Shaw, G. Pollack, R. Miller, G. Varnell, W. Lee, R. Lane, S. Wood, and S. Robbins, *"E-Beam fabrication of 1.25-μm 4K Static Memory,"* J. Vac. Sci. Technol., 19, 905 (1981).

[68] H. Ito, C. G. Wilson, *"Applications of Photoinitiators to the Design of Resists for Semiconductor Manufacturing,"* in Polymers in Electronics, ACS Symp. Series 242, T. Davidson, Ed., 11-23 (1984).

[69] E. Reichmanis, L. F. Thompson, O. Nalamasu, A. Blakeney, and S. Slater, *"Chemically Amplified Resists for Deep-UV Lithography: A New Processing Paradigm,"* Microlithography World, 7-14, Nov./Dec. 1992.

[70] A. A. Lamola, C. R. Szmanda, and J. W. Thackeray, *"Chemically Amplified Resists,"* Solid-State Technol., 34 (8) 53-60, (1991).

[71] J. V. Crivello, *"Applications of Photoinitiated Cationic Polymerization to the Development of New Resists,"* in Polymers in Electronics, ACS Symp. Ser., T. Davidson, Ed., 242, 3-10 (1984).

[72] H. Ito, L. A. Pederson, K. N. Chiong, S. Sonchik, and C. Tsai, *"Sensitive Electron Resist Systems Based on Acid-Catalized Deprotection,"* Proc. SPIE, 1086, 11-21 (1989).

[73] T. X. Neenan, F. M. Houlihan, E. Reichmanis, J. M. Kometani, B. J. Bachman, and L. F. Thompson, *"Chemically Amplified Resists: A Lithographic Comparison of Acid Generating Species,"* Proc. SPIE, 1086, 1-10 (1989).

[74] A. E. Novembre, R. C. Tarascon, L. F. Thompson, W. T. Tange, R. A. Bostic, and D. H. Ahn, *"Initial Manufacturing Performance of an Actively Controlled PBS Resist Development Process,"* Procs. SPIE, 1809, 76-84 (1993).

[75] A. D. Wilson, *"X-Ray Lithography in IBM, 1980-1992, the Development Years,"* IBM J. Res. Dev., 37 (3), 299-318 (1993).

[76] H. I. Smith and M. L. Schattenburg, *"X-Ray Lithography, from 500 to 30 nm: X-Ray Nanolithography,"* IBM J. Res. Dev., 37 (3), 319-329 (1993).

[77] Y. Somemura and K. Deguchi, *"Effects of Fresnel Diffraction on Resolution and Linewidth Control in Synchrotron Radiation Lithography,"* Jpn. J. Appl. Phys. 1, 31 (3), 938, March 1992.

[78] R. A. Della Guardia, D. E. Seeger, and J. L. Mauer, *"X Ray Transmission Through Low Atomic Number Particles,"* Microelectron. Eng. (Netherlands), 9 (1-4), 139, May 1989.

[79] A. Kluwe, H. Lutzke, T. Stelter, K. H. Muller, *"Printability of X Ray Mask Defects at Various Printing Conditions and Critical Dimensions,"* J. Vac. Sci. Technol. B, 8 (6) 1609, Nov./Dec. 1990.

[80] G. Zorpette, *"Rethinking X-Ray Lithography,"* IEEE Spectrum, 29 (6), 33, June 1992.

[81] R. K. Watts, J. R. Mandonado, *"X-Ray Lithography,"* VLSI Electronics, Microstructure Science, Vol. 4, p. 56, Academic Press, New York, 1982.

[82] A. Chiba and K. Okada, *"Dynamic In Plane Thermal Distortion Analysis of an X Ray Mask Membrane for Synchrotron Radiation Lithography,"* J. Vac. Sci. Technol. B, 9 (6), 3279, Nov./Dec. 1991.

[83] H. Braun, *"Lithographie mit Licht-, Elektronen-, und Roentgenstrahlen,"* Physik Unserer Zeit, 10, 68 (1979).

[84] D. Maydan, *"X-Ray Lithography for Microfabrication,"* J. Vac. Sci. Techn., 17, 1164-1168 (1980).

[85] T. R. Groves, J. G. Hartley, H. C. Pfeiffer, D. Puisto, and D. K. Bailey, *"Electron Beam Lithography Tool for Manufacture of X-Ray Masks,"* IBM J. Res. Dev., 37 (3), 411-420 (1993).

[86] M. Gentili, M. Kumar, R. Luciani, L. Grella, D. Plumb, and Q. Leonard, *"0.1 μm X Ray Mask Replication,"* J. Vac. Sci. Technol. B, 9 (6), 3333 (1991).

[87] T. Ohta, Y. Kawazu, and Y. Yamashita, *"Fabrication of X Ray Mask Using W CVD for Forming Absorber Pattern,"* Jpn. J. Appl. Phys. 1, Regul. Pap. Short Notes, 29 (10), 2195, Oct. 1992.

[88] U. Weigmann, H.-C. Petzold, H. Burghause, R. Putzar, and H. Schaffer, *"Repair of Electroplated Gold Masks for X-Ray Lithography,"* J. Vac. Sci. Technol., B6 (6) 2170 (1986).

[89] D. L. Laird and R. L. Engelstad, *"Optimal Design of an X Ray Lithography Mask,"* J. Vac. Sci. Technol. B, 9 (6), 3319, Nov./Dec. 1991.

[90] H. C. Petzold, H. Burghause, R. Putzar, U. Weigmann, N. P. Economou, and L. A. Stern, *"Repair of Clear Defects on X Ray Masks by Ion Induced Metal Deposition,"* Proc. SPIE - Int. Soc. Opt. Eng., 1089, 45 (1989).

[91] W. H. Brunger, *"X Ray Mask Repair by Electron Beam Induced Metal Deposition,"* Microelectron. Eng. (Netherlands), 9 (1-4), 171, May 1989.

[92] P. G. Blauner and J. Mauer, *"X-Ray Mask Repair,"* IBM J. Res. Dev., 37 (3), 421-434 (1993).

[93] T. D. Cambria and N. P. Economou, *"Mask and Circuit Rrepair with Focused Ion Beams,"* Solid-State Technol. 30, 133 (1987).

[94] J. Z. Y. Guo and F. Cerrina, *"Comparison of Plasma Source with Synchrotron Source in XRL,"* Proc. SPIE - Int. Soc. Opt. Eng., 1465, 330 (1991).

[95] S. Hoffman, S. Nash, R. Ritter, and W. Smith, *"Fabrication of a 1 Mbit Dynamic Random Access Memory with Four Levels Using X-Ray Lithography,"* J. Vac. Sci. Technol. B, 9 (6), 3241, Nov./Dec. 1991.

[96] M. N. Wilson, A. I. C. Smith, V. C. Kempson, M. C. Townsend, J. C. Shouten, R. J. Anderson, A. R. Jorden, V. P. Suller, and M. W. Poole, *"The Helios 1 Compact Superconducting Storage Ring X-Ray Source,"* IBM J. Res. Dev., 37 (3), 351-371 (1993).

[97] W. G. Waldo and A. W. Yanof, *"0.25 μm Imaging by SOR X Ray Lithography,"* Solid-State technology, 34 (12), 29, Dec. 1991.

[98] B. Breithaupt, H. H. David, R. U. Ballhorn, E. P. Jacobs, W. Windbracke, and G. Zeicker, *"A 0.25 μm NMOS Transistor Fabricated with X Ray Lithography,"* Microelectron. Eng. (Netherlands), 13 (1-4), 319, March 1991.

[99] J. Warlaumont, *"X Ray Lithography: on the Path to Manufacturing,"* J. Vac. Sci. Technol. B, 7 (6), 634, Nov./Dec. 1989.

[100] A. Heuberger, *"X-Ray Lithography,"* Microelectron. Eng. (Netherlands), 5, 3 (1986).

[101] E. Spiller et al., *"Application of Synchrotron Radiation to X-Ray Lithography,"* J. Appl. Phys., 46, 5450 (1976).

[102] H. Aritome, S. Matsui, K. Mariwaki, and S. Namba, *"X-Ray Lithography by Synchrotron Radiation of the SOR-Ring Storage Ring,"* J. Vac. Sci. Technol. B, 16, 1939 (1979).

**258**

[103] J. Alper, *"Industry's New Magic Lantern,"* High Technology, p. 61, April 1984.

[104] N. Atoda and K. Hoh, *"X-Ray Lithography with Synchrotron Radiation,"* VLSI Symposium (?), p. 48, Year?

[105] V. Jw. Znamenskiy, O. B. Ananjin, Ju. A. Bykovskiy, I. K. Novokov, and A. A. Zhuravlev, *"High Intensity Ultra Soft X Ray Source,"* Proc. SPIE, 1860, 216-220 (1993).

[106] K. Nguyen, D. Attwood, and T. K. Gustafson, *"Source Issues Relevant to X-Ray Lithography,"* OSA Proc. Soft X Ray Projection Lithography, 12, 62-67 (1991).

[107] R. Fedosejevs, R. Babkowski, J. N. Broughton, and B. Harwood, *"keV X Ray Source Based on High Repetition Rate Excimer Laser,"* Proc. SPIE, 1671, 373-382 (1982).

[108] H. Winick, *"Synchrotron Radiation Research,"* Plenum Press, p. 11, New York 1980.

[109] E. Cullman, F. Richter, P. Thompson, and M. Gentili, *"A Radiation Source for X Ray Lithography,"* Microelectron. Eng. (Netherlands), 13 (1-4), 299, March 1991.

[110] K. A. Tanaka, H. Aritome, T. Kanabe, M. Nakatsuka, T. Yamanaka, and S. Nakai, *"Laser Plasma X Ray Source and its Application to Lithography,"* Proc. SPIE - Int. Soc. Opt. Eng., 1140, 350 (1989).

[111] H. A. Hyman, A. Ballantyne, H. W. Friedman, and D. A. Reilly, *"Intense-Pulsed Plasma X-Ray Sources for Lithography: Mask Damage Effects,"* J. Vac. Sci. Technol., 21 (4), 1021 (1982).

[112] R. J. Rosser, R. Feder, A. Ng, F. Adams, P. Celliers, and R. J. Speer, *"Nondestructive Single-Shot Soft X-Ray Lithography and Contact Microscopy Using a Laser-Produced Plasma Source,"* Appl. Opt. 26 (19), 4313 (1987).

[113] M. Richardson, M. Silfvast, W. T. Bender, H. A. Hanzo, A. Yanovsky, J. Feng, and J. Thorpe, *"Characterization and Control of Laser Plasma Flux Parameters for Soft X-Ray Projection Lithography,"* Appl. Opt. 32 (34), 6901-6910 (1993).

[114] L. I. Rudakov, K. A. Bairgarin, Y. G. Kalinin, V. D. Korolev, and M. A. Kumachov, *"Pulsed Plasma Based X Ray Source and New X Ray Optics,"* X-Ray Projection Lithography," Phys. Fluids B, Plasma Physics, 3 (8), 2414-2419 (1991).

[115] D. L. White, J. E. Bjorkholm, J. Bokor, L. Eichner, R. R. Freeman, T. E. Jewell, W. M. Mansfield, A. A. MacDowell, L. H. Szeto, D. W. Taylor, D. M. Tennant, W. K. Waskiewicz, D. L. Windt, and O. R. Wood, *"Soft X-Ray Projection Lithography,"* Solid-State Technology, 34 (7), 37, July 1991.

[116] H. I. Smith and D. C. Flanders, *"X-Ray Lithography,"* Jpn. J. Appl. Phys., Suppl. 16-1, 61-65 (1977).

[117] D. L. Spears and H. I. Smith, *"High-Resolution Pattern Replication Using Soft X Rays,"* Electron Lett., 4, 102 (1972)

[118] S. Yoshida et al., *"X-Ray Lithographic System,"* Proc. Kodak Microelectronics Seminar, San Diego, October 1980.

[119] W. D. Buckley et al., *"The Design and Performance of an Experimental X-Ray Lithography System, Patterning, and Dimensional Stability of Titanium Membrane X-Ray Lithography Mask,"* Proc. Kodak Microelectronics Seminar, San Diego, October 1980.

[120] H. Huber, U. Scheunemann, E. Cullmann, and W. Rohrmoser, *"Application of X-Ray Steppers using Optical Alignment,"* Solid-State technology, 33 (6), 59 (1990).

[121] K. A. Cooper, *"Steppers for X-Ray Microlithography,"* Microelectroni. Manuf. Test., 13 (10), 27 (1990).

[122] S. Ishihara, M. Kanai, A. Une, and M. Suzuki, "*An X Ray Stepper for SOR Lithography,*" NTT Rev., 2 (4), 92 July 1990.

[123] J. Itoh and T. Kanayama, "*A New Interferometric Displacement-Detection Method for Mask-to-Wafer Alignment Using Symmetrically-Arranged Three Gratings,*" Jpn. J. Appl. Phys., 25 (6), L487, (1986).

[124] J. Itoh, T. Kanayama, N. Atoda, and K. Hoh, "*Fine Alignment Exposure System for Synchrotron X Ray Lithography,*" Rev. Sci. Instrum., 60 (7), 1638, July 1989.

[125] H. Huber, U. Scheunemann, W. Rohrmoser, and E. Cullmann, "*Application of X Ray Steppers Using Optical Alignment for Synchrotron Base X Ray Lithography,*" Microelectron. Eng. (Netherlands), 9 (1-4), 151, May 1991.

[126] N. Atoda, H. Kawakatsu, H. Tanino, S. Ichimura, M. Hirata, and K. Hoh, "*Diffraction Effects on Pattern Replication with Synchrotron Radiation,*" J. Vac. Sci. Tech., B1 (4), 1267 (1983).

[127] A. N. Broers, "*Practical and Fundamental Aspects of Lithography,*" Materials for Microlithography, L. F. Thompson, C. G. Willson, J. M. J. Frechet, Eds., American Chemical Society, Washington D.C., 1984.

[128] A. E. Gruen, Z. Naturforschung, 12a, 89 (1957).

[129] J. N. Randall, "*Prospects for Printing Very Large Scale Integrated Circuits with Masked Ion Beam Lithography,*" J. Vac. Sci. Technol. A, p. 777, 1985.

[130] D. Seeger, K. Kwietniak, D. Crockatt, A. Wilson, and J. Warlaumont, "*Fabrication of 0.5 $\mu m$ CMOS Devices by Synchrotron X-Ray Lithography: Resist Materials and Processes,*" Microelectron. Eng., 9, 97 (1989).

[131] D. Seeger, "*Resist Materials and Processes for X-Ray Lithography,*" IBM J. Res. Dev., 37 (3), 435 (1993).

[132] L. Bartlett, "*Masked Ion Beam Lithography: An Emerging Technology,*" Solid State Technol., 29, 215, may 1986.

[133] T. Kato, H. Murimoto, K. Tsukamoto, H. Shinohara, and M. Inuishi, "*Submicron Lithography Using Focused-Ion-Beam Exposure Followed by a Dry Development,*" Proc. 1985 VLSI Symp., p. 72 (1985).

[134] J. A. Doherty, et al. "*Focused Ion beam in Microelectronic Fabrication,*" IEEE Trans. on Components, Hybrids, and Manuf. Technol., CHMT-6, p. 329 (1983).

[135] B. M. Siegel, "*Ion Beam Lithography,*" VLSI Electronics, Microstructure Science, Vol. 16, N. G. Einspruch and R. K. Watts, Eds., Academic Press, New York, 1987.

[136] W. L. Brown, T. venkatesan, and A. Wagner, "*Ion Beam Lithography,*" Solid State Technol., 24, 60, August 1981.

[137] A. B. El-Kareh and J. C. El-Kareh, "*Electron Beam Lenses and Optics,*" Vols., 1 and 2, Academic Press, New York, 1970.

[138] R. Clampitt, K. L. Aitken, and D. K. Jefferies, "*Intense Field-Emission Ion Source of Liquid Metals,*" J. Vac. Sci. Technol., 12 (6), 1208 (1975).

[139] L. W. Swanson, G. A. Schwindt, H. E. Bell, and J. E. Brady, "*Emission Characteristics of Gallium and Bismuth Liquid Metal Field Ion Sources,*" J. Vac. Sci. Technol., 16 (6), 1864 (1979).

[140] K. Gamo, T. Ukewada, Y. Inomoto, Y. Ochia, and S. Namba, "*Liquid Metal Alloy Ion Sources for B, Sb, and Si,*" J. Vac. Sci. Technol., 19 (40), 1182 (1981).

[141] V. E. Krohn and G. R. Ringo, "*Ion Source of High Brightness Using Liquid Metal,*" Appl. Phys. Lett., 27, 479 (1975).

[142] G. R. Hanson and B. M. Siegel, "H$_2$ *and Rare Gas Field Ion Source with High Angular Current,*" J. Vac. Sci. Technol., 16, 1875 (1979).

[143] I. Adesida, J. D. Chin, L. Rathbun, and E. D. Wolf, "*Dry Development of Ion Beam Exposed PMMA Resist,*" J. Vac. Sci. Technol., 21 (2), 666 (1982).

[144] N. Koshida, K. Yoshida, S. Watanuki, M. Komuro, and N. Atoda, "*50-nm Metal Line Fabrication by Focused Ion beam and Oxide Resist,*" Jpn. J. Appl. Phys., 30 (11B), 3246 (1991).

[145] N. Koshida, K. Yoshida, S. Watanuki, M. Komuro, and N. Atoda, "*50-nm Metal Line Fabrication by Focused Ion beam and Oxide Resist,*" Jpn. J. Appl. Phys., 30 (11B), 3246 (1991).

# Chapter 5

# Contamination Control and Etch

## 5.0 Introduction

Clean and etch processes are used to selectively remove organic and inorganic materials from patterned and unpatterned substrate surfaces. When a process is used to remove particulates or unwanted films, it is called cleaning. When a film which was intentionally deposited or grown is removed, the process is called etching. If implemented in a liquid form, these processes are referred to as wet cleaning or etching. Processes that use gases to remove materials by momentum transfer, or by plasma or photo assisted chemical reactions are called dry cleaning or etching. This chapter discusses wet and dry clean and etch processes used in the manufacture of submicron semiconductor devices.

## 5.1 Clean Processes

To successfully manufacture integrated circuits, processes must be carried out in a meticulously clean environment under rigorous control of temperature and humidity. Minute impurity concentrations (less than one part per million) can degrade the performance of a circuit or even cause its failure. Therefore, to achieve a high production yield and device reliability, it is essential to eliminate all sources of contamination. When contamination is unavoidable, cleaning the affected substrate is critical [1,2]. Surface cleaning is particularly important prior to high temperature processes because impurities react and diffuse at much higher rates at elevated temperature. It is also essential to ensure a contamination-free surface after photoresist removal and before critical processing steps, such as MOSFET gate oxidation, low temperature epitaxial growth, bipolar polysilicon-emitter deposition, and contact metallurgy (Chap. 8). One important requirement of any cleaning process is that it minimizes surface roughness to avoid degrading the device performance and reliability. This

**262**

section describes the basic principles of wet and dry cleaning techniques which have become critically important to satisfy stringent requirements on device performance, yield, and reliability in VLSI/ULSI.

### 5.1.1 Contaminants

A large fraction of yield losses is attributed to microcontamination. Contaminants can be organic or inorganic particles, or films of molecular compounds, ionic materials, or atomic species. They can deposit from the surrounding environment, from process equipment such as CVD, plasma reactors and ion implanters (Chap. 6), from chemicals used in processing, or from handling by manufacturing personnel.

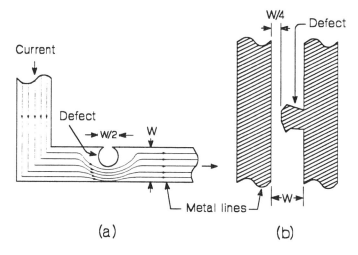

**Fig. 5.1** Illustration of "fatal" defects caused by particle contamination. (a) Material not deposited where it should be. (b) Material deposited where it should not be.

Molecular compounds are particles or films of condensed organic vapors, solvent residues, photoresist, or metal oxides or hydroxides [3]. The most detrimental effects of film contaminants are that they can cause poor adhesion of deposited layers, or inhibit etching of the underlying material. Ionic contaminants can seriously degrade device electrical parameters and stability. Examples ·of ionic contaminants are sodium, potassium, fluorine, and chlorine ions. The most damaging elemental impurities are heavy metals, such as iron, copper, nickel, and gold. These metals have energy levels near the silicon mid-gap and hence directly affect the minority-carrier lifetime and leakage current in silicon devices. The metals can also degrade the gate oxide by, e.g., reducing its break-down field.

Contamination is typically imperceptible to the human eye without magnification. While the human eye can resolve features larger than about 100 $\mu$m in size, several analytical techniques must be used to inspect the wafer for smaller particulate or film contaminants. Among these are bright, dark field, or interference contrast optical microscopy; scanning electron microscopy (SEM); low energy electron diffraction (LEED); photoemission spectroscopy; infrared analysis; Auger spectroscopy; X-ray analysis; and atomic absorption spectroscopy. These and other analytical techniques are discussed in a separate chapter.

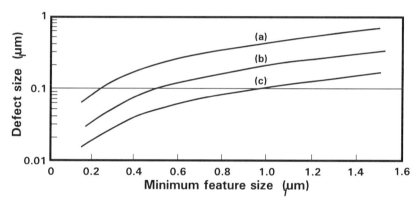

**Fig. 5.2** Rapid decrease in the minimum allowable particle size as the minimum feature size is reduced. (a) Half-rule. (b) Fifth-rule. (c) Tenth-rule.

Particles can cause a device to fail, for example, by improperly defining patterns, creating unpredictable surface topography, inducing leakage current through insulating layers, or accelerating device wearout. It is not always easy to separate defects that detract yield from those that degrade reliability. A particle smaller than the minimum feature size, for example, can create a clear field in a metal line (Fig. 5.1a). The resulting necking of the line does not cause an immediate "open" (line interrupted), but can create an open after a certain duration of current transport through the line. Fig. 5.1b shows how two adjacent conductor lines can be shorted by a particle that falls between the lines and locally inhibits the removal of the conductor film. The short can form immediately or after stressing the conductor for some time. The minimum particle size that can induce a "fatal" defect (defect that causes a device to fail) depends on the minimum feature size and on the region where the particle falls. As a rule of thumb, a particle that exceeds one fifth to one half of the minimum feature size has the potential of causing a fatal defect [3]. A defect of smaller size can be fatal if it falls in a critical area, such as the gate oxide of a MOSFET. Therefore, the size of a fatal defect decreases as horizontal and vertical geometries are reduced. This is illustrated in Fig. 5.2 where the term "fifth-rule", for example, refers to defects that have the size one-fifth of the minimum feature size. To achieve an economically acceptable product yield, the maximum allowable number of fatal particles per unit area must decrease with increasing circuit density, as shown in Fig. 5.3 for dynamic random-access memory (DRAM) chips [4]. From the plots, it follows that a maximum allowable defect density of about $15/m^2$ is projected for the Gigabit era.

Organic vapors in clean rooms are also a serious concern. Highly volatile species, such as methanol and acetone, can be present in clean rooms and deposit on wafers. Their high volatility, however, prevents an accumulation of much beyond a monolayer [5].

The basic steps to achieve contamination-free manufacturing are to filter incoming air to establish a clean-room environment, and to reduce contaminants in liquid and gaseous chemicals to which the wafers are subjected. Since particulates can also emanate from operators, personnel must wear proper garments to protect the wafers.

## Clean Room

All critical process steps are carried out in a clean-room environment. In a clean room, the air is maintained at a controlled temperature and humidity, and is continuously filtered, recirculated, and monitored with respect to particulates. The filtered air flows from ceiling to floor at about 26 m/min which is about the threshold for laminar flow, i.e. uniform velocity of air following parallel flow-lines without turbulence.

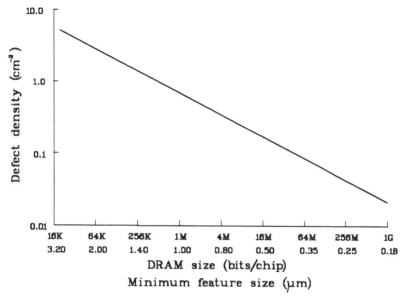

**Fig. 5.3** Decrease in the allowable particle density as a function of DRAM chip progression [4].

Clean-room classes are determined by the maximum number of airborne particulates per unit volume having a particle size larger than 0.5 $\mu$m [6]. A class-M 2.5 (class-10) room, for example, may contain a maximum of 353 such particles per cubic meter (10 per cubic foot). The density of particles of other sizes is also defined for each class (Fig. 5.4). A class-M 1.5 (class-1) environment is required in the highly critical area of lithography.

## Pure Water

Large quantities of water are used during the manufacture of semiconductor devices, especially for frequent cleaning and rinsing. The fundamental requirements on a pure water system is that it provides a continuous supply of ultra-clean water, with very low ionic content. Ionic contaminants in water, such as sodium, iron, or copper, can deposit onto the wafer and cause device degradation. The ionic content is measured by monitoring the water resistivity. A water resistivity of about $1.8 \times 10^7$ Ohm-cm indicates a low ionic content. A pure water system consists of several

**266**

sections which include charcoal filters, electrodialysis units, and a multitude of resin units to demineralize the water [7]. Undissolved solids, bacteria, and other organic materials are removed by a series of filters ranging in pore size from 10 μm down to 0.01 μm, and by circulating the water through an ultraviolet sterilizing unit [7].

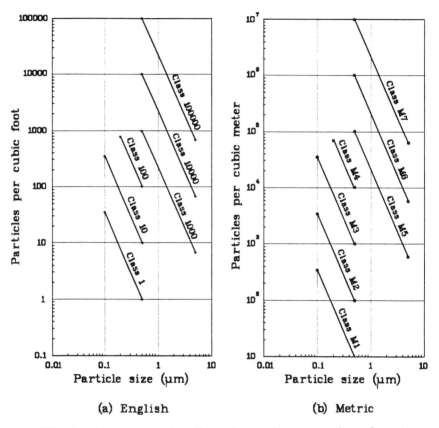

(a) English          (b) Metric

Fig. 5.4 Clean-room class limits in maximum number of particles per unit volume of size equal to or greater than particle size shown. (a) English names and units. (b) SI names and units [6].

## 5.1.2 Wet Cleaning

Wet cleaning consists of immersing a wafer in an appropriate liquid solution, or spraying the wafer with the liquid. This is typically accompanied by agitation and, in the case of particle removal, by scrubbing with a special brush or sonic power.

**Particles Removal**

Particles are removed essentially by momentum transfer. Dislodging particles, however, becomes more difficult as their size decreases. As particle size falls, Van der Waals forces decrease (Problem 5.2), but electrostatic forces increase, and the efficiency of transferring translational momentum to particles decreases rapidly. The relative force to remove a particle is shown as a function of particle size in Table 5.1. The measurements are made on glass beads on a glass slide [8].

**Table 5.1** Dependence of force to remove particle on particle size

| Particle Size ($\mu m$) | Relative force to displace (Gravitational units) |
|---|---|
| 100 | 510 |
| 50 | 2,159 |
| 10 | 57,716 |
| 1 | 674,600 |
| 0.1 | 749,532,300 |

Particle removal methods have evolved from simple immersion, through ultrasonic/megasonic scrubbing, to completely enclosed and automated systems in which cassette-loaded wafers are kept stationary during the entire cleaning, rinsing, and drying process to avoid recontamination when wafers are pulled out of the liquid [3].

Typical mechanical scrubbing consists of rotating a brush near the wafer surface which is sprayed with a jet of high-pressure de-ionized (DI) water (2000-3000 psi). The brush does not come in contact with the wafer surface, but aquaplanes across the surface and transfers momentum to the water. It is the water movement that displaces the particles. One limitation of mechanical scrubbing is that it only translates particles from side to side in openings, such as oxide windows, without removing the particle. Also, as image size decreases, it becomes more difficult for the fluid to reach the particle in openings because of increased surface tension.

The megasonic scrubbing process differs from the conventional ultrasonic techniques in that it uses sonic waves of 850-900 kHz rather than the 20-80 kHz frequency used in ultrasonic scrubbing [9]. The sonic waves are generated by an array of piezoelectric transducers and travel through the liquid, parallel to the wafer surface. The sonic energy rapidly creates and collapses small bubbles in the liquid. The collapsing action generates sonic shock-waves that dislodge particles as small as 0.3 $\mu$m in size from the wafer surface. Particles are then removed from the liquid by continuous filtration to avoid redepositing on the wafer. Sonic scrubbing systems are also combined with standard cleaning solutions described below. Special care must be taken to avoid damaging delicate films, such as gate oxide and metal layers, with excessive sonic energy.

Rinsing and drying are critical steps. Rinsing is typically performed with DI water, preferably in a closed system without removing wafers between steps. Wafer drying is accomplished by physical removal of the water rather than allowing the water to evaporate. Spin drying in the presence of a stream of clean, dry, inert gas is the most widely used technique. In some cases, hot forced-air, capillary, infrared, or solvent vapor drying (typically with isopropyl alcohol or non-flammable azeotropic mixtures) are used to remove adsorbed moisture from the surface.

## Wet Chemical Cleaning

The most commonly used wet chemical cleaning technology is based on hot alkaline or acidic hydrogen peroxide ($H_2O_2$) solutions. These are used to remove chemically bonded films from the wafer surface prior to critical steps.

## RCA Cleans

RCA cleans are based on a two-step process: the Standard Clean-1, referred to as SC-1, followed by Standard Clean 2, SC-2 [10]. Both solutions incorporate the strong oxidizing capability of hydrogen peroxide ($H_2O_2$). SC-1 is an aqueous alkaline solution which removes organic films, while SC-2 is an acidic mixture used to remove alkali ions and cations, and metallic contaminants. SC-1 also etches thermally grown oxide at a rate of about 0.1 nm/min; SC-2 does not attack the oxide. SC-1 is typically a 5:1:1 solution of DI water, "unstabilized" $H_2O_2$ (30%), and ammonium hydroxide ( $NH_4OH$, 27%). This solution is very effective in removing organic contaminants. The $NH_4OH$ dissolves organic films and also complexes Group I and II metals such as copper, silver, cobalt, cadmium, and nickel. SC-2 typically consists of 6 parts $H_2O$, 1 part $H_2O_2$ (30%), and 1 part hydrochloric acid (HCl, 37%) and is effective in removing heavy metals. The processing temperature is 75 °C to 80 °C. Higher temperatures are avoided to minimize excessive decomposition of hydrogen peroxide. To obtain bare silicon surfaces, a 15s immersion in an ultra pure 1% HF-$H_2O$ solution may follow to remove a thin native oxide which can be formed during SC-1. Oxide removal is evidenced by the change from the hydrophilic oxidized surface to the hydrophobic bare silicon surface. This step, however, must be used with great care to avoid recontaminating the wafer [3].

The chemical state of the silicon surface after wet cleans is of great importance. Removal of native oxide from a (100) surface by dipping the wafer in an HF solution, followed by a rinse in DI water results in a hydrophobic surface which contains about 90 % H and 10 % F [11,12]. While hydrogen termination is believed to give a stable silicon surface with controlled oxide growth, the presence of a small fraction of bonded fluorine increases microroughness and degrades the grown oxide quality [13].

Significant reduction in defect density can be achieved by simultaneous removal of particles and contaminant films in one operation combining megasonic with RCA cleans [14].

**Piranha Clean**

Sulfuric acid ($H_2SO_4$) and mixtures of $H_2SO_4$ with other oxidizing agents, such as $H_2O_2$, are widely used to strip photoresist or to clean the wafer after the photoresist has been plasma stripped. The most commonly used mixture is 7 parts $H_2SO_4$ to 3 parts of 30% $H_2O_2$. This solution has earned the name "Piranha" because of its aggressiveness in attacking organic materials. Wafers are first immersed in the Piranha at 125 °C for about 10 minutes and then subjected to a thorough rinse in DI water. This is sometimes followed by SC-1/SC-2 with appropriate DI water rinse to remove inorganic residues.

### 5.1.3 Dry Cleaning

One of the drawbacks of wet cleaning is that surface tension can prevent the liquid from penetrating into high aspect-ratio (deep and narrow) features, leaving part of the surface uncleaned. Another disadvantage is that wet cleaning is not compatible with tool "clustering", i.e., integrating several consecutive operations within the same unit without, for example, interrupting the vacuum [15]. One example of tool clustering is to combine a dry cleaning module with a gate-oxide growth module and a polysilicon deposition module in the same unit to achieve a high degree of contamination control (Fig. 5.5). Also, environmental concerns associated with the use and disposal of large quantities of liquid chemicals and waste produced by wafer cleaning facilities will make the continuous use of wet cleaning less economical. Therefore, substantial cost and environmental benefits can be achieved by replacing conventional wet-chemical cleaning with an equally effective gas-phase, or dry-cleaning process. Dry-cleaning techniques fall in the categories of momentum transfer and gas-phase processes based on thermal, plasma and photo-excited reactions.

### Momentum Transfer

Submicron contaminants can be removed by impact of particles carried by a high-velocity gas stream directed at the surface to be cleaned. The particles can be formed by solidifying either liquid droplets or a gas during rapid cooling.

One method to remove particles and organic films from the wafer surface is to use a jet of fine, ultra-clean ice particles [16]. Ice particles are formed by spraying ultra clean water into cold (-80 °C to -150 °C) nitrogen gas that is formed when liquid nitrogen is sprayed into the freezing chamber. The ice particles are sprayed through a jet nozzle onto the wafer surface. Contaminant particles are removed by the impact and transient melting of ice particles. Melting and re-freezing of ice-particles traps contaminants which are then removed by the impact of subsequent particles. Organic films, such as fingerprints, are also removed by cooling and impact.

Similar systems for momentum transfer use carbon dioxide particles [8, 17], or cryogenic aerosoles of ultra pure argon [18].

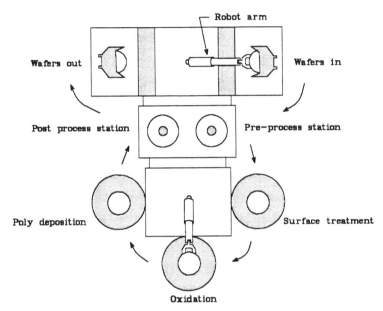

**Fig. 5.5** Illustration of tool clustering. Surface cleaning, gate oxide growth, and polysilicon deposition are performed in separate modules connected in one unit. Wafers are transported by a "wafer handler" without interrupting the vacuum.

## Photo-Enhanced Dry Cleaning

An efficient dry cleaning technique to remove organic impurities from the wafer surface is to expose the wafer with ultraviolet radiation in the presence of oxygen [19]. In such a system, a low pressure mercury vapor lamp emits a spectrum which contains light of wavelength 184.9 nm and 253.7 nm. Oxygen absorbs 184.9-nm radiation to form very active ozone. Ozone dissociates by absorbing 253.7 nm radiation. Therefore, when both wavelengths are present, ozone continuously forms and dissociates. An intermediate product from both the formation and dissociation of ozone is atomic oxygen which is a very strong oxidizing agent. While this technique is suitable to oxidize and volatilize carbon and organic compounds, it is not effective in removing inorganic material and metals.

Removal of metallic impurities from the silicon surface can be stimulated by exposing the surface to UV radiation in a halogen ambient, such as in $Cl_2$ or HCl [20]. Irradiation of the gases with UV produces reactive halogen-radicals in the gas-phase or on the wafer surface. These radicals are chemisorbed at the wafer surface and react with trace metals, such as Fe, Na, Ca, Al, and Mg, to form volatile metal halogen compounds. In the case of Cl, light etching of silicon occurs as the chlorine radicals diffuse through thin native oxide and react with silicon to form volatile silicon chloride compounds, resulting in spontaneous etching of silicon and "lift-off" of the metal contaminants.

## Plasma Assisted Reactions

Plasma energy is also used to create reactive species for cleaning. To avoid radiation damage, the wafer is placed remote from the plasma: the chemically reactive species are removed from the plasma region and supplied to the wafer in form of an afterglow gas (Chap. 3). This remote plasma technique has been used to stimulate chemical reactions in the gas-phase for the removal of organic material with $O_2$, metallic contaminants with HCl/Ar, and thin oxide films in an $NF_3/H_2Ar$ gas mixture [21]. One detrimental side-effect of exposure to the HCl/Ar afterglow is pitting the silicon surface. This can be minimized by adjusting the temperature and time and by forming an ultra thin oxide film in an $O_2$ afterglow prior to HCl/Ar dry cleaning [21].

## 5.2 Etching

Etching is the selective removal of material, either locally where windows are defined, or over the entire wafer without patterning. At least one etching step is performed during the manufacture of semiconductor devices. In Chap. 1 we saw that after mechanically lapping the wafer, several microns of silicon must be etched away to remove the damage caused by lapping. In Chap. 2 we introduced two etching steps, one before and one after local oxidation (LOCOS). Before LOCOS, the nitride film was removed locally, using a resist mask to define regions in silicon where oxidation occurs. After LOCOS, the remaining nitride was removed over the entire wafer without the use of a mask (this is sometimes called a "blanket etch").

The main objective of etching a pattern is to reproduce the image of the mask with a high degree of fidelity. An etched pattern is formed by either a subtractive or additive process (Fig. 5.6). In the subtractive method, the film is first deposited or grown on the substrate and then coated with a photoresist layer. The desired pattern is defined on the resist layer using lithographic methods discussed in Chap. 4. The region covered by resist is protected during dry or wet etching of the film. After etching is complete, the resist is removed in an appropriate solvent or by dry etching as discussed below.

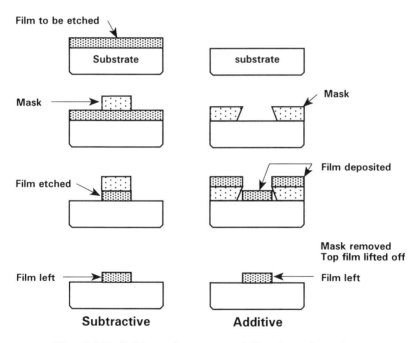

**Fig. 5.6** Definition of a patterned film by subtractive and additive processes.

In the additive process, the resist is coated and defined before the desired film is deposited. The film deposits on top of the photoresist and directly on the substrate surface not covered by resist. The resist is then dissolved away in an appropriate solvent, lifting the unwanted film from the masked area outside the pattern.

This process is also referred to as **lift-off.** For lift-off to occur, the thickness of the deposited film must be smaller than the resist thickness and a gap must remain between the two materials.

Very stringent requirements are imposed on etch processes. Image and sidewall slope control and pattern reproducibility must be achieved across the wafer, wafer to wafer, and run to run. While dry, plasma-assisted etching is best suited to meet the above requirements for submicron technologies, wet etch techniques are still used in some special applications where large tolerances in image dimensions are acceptable.

### Definition of Terms

The film thickness removed per unit time is called the **etch rate.** The etchant must remove the selected material without appreciably affecting the properties of other materials on the surface. The ratio of the etch rate of the material to be removed to the etch rate of another material is called the etch **selectivity.** The most important characteristics of an etchant are the etch rate, selectivity, and dimensional control that can be achieved on the pattern to be etched.

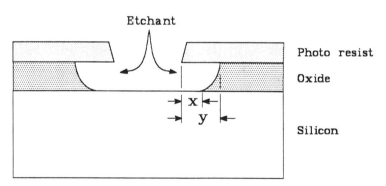

**Fig. 5.7** Definition of two lateral etch dimensions, illustrated for isotropic etching. x is referred to etch bias and y as undercut.

When etching proceeds at the same rate in all directions, it said to be **isotropic** (Fig. 5.7). When the etch rate is considerably larger in one direction than in another, etching is termed **anisotropic.** It is important to distinguish between lateral etching at the top and bottom interfaces of the etched layer when a mask is used to pattern the film. The mask typically consists of a material that is not attacked by the etchant. The top dimension $y$ is commonly referred to as **undercut.** It is approximately equal to the film thickness, until etching reaches the film-substrate boundary. Overetching beyond this point increases the undercut and radius of sidewall curvature. For a long overetch, the sidewalls become essentially vertical (Fig. 5.8). Overetching is necessary to ensure complete removal of the selected film because of non-uniformities in the wafer topography and variations in the film thickness and etch rate. The dimension $x$ in Fig. 5.7 at the film-substrate interface is called **etch bias.**

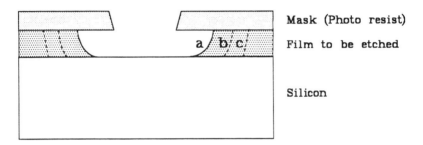

**Fig. 5.8** Effect of isotropic overetching on undercutting and sidewall slope: (a) no overetch, (b) moderate overetch, (c) long overetch.

A comparison between isotropic, moderately anisotropic, and highly anisotropic etching is shown schematically in Fig. 5.9. Vertical anisotropic etching of a film that is deposited conformably in a groove or over a step leaves film "stringers" or "spacers" on the sidewalls, while isotropic etching removes the film entirely (Fig. 5.10).

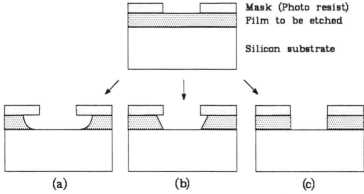

**Fig. 5.9** Etch directionality. (a) Isotropic. (b) Moderately anisotropic (c) Strongly anisotropic (vertical).

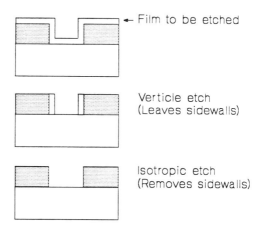

**Fig. 5.10** Effect of etch directionality on sidewall film.
(a) Conformal film deposition over step.
(b) Anisotropic etching leaves sidewall almost intact.
(c) Sidewall fully removed by isotropic etching.

## 5.2.1 Wet Etching

Wet etching is performed by immersing the wafers in an appropriate solution (e.g., in batches of 25 wafers) or by spraying the wafer with the etchant solution.

### The Silicon-Dioxide Etch

Selective removal of silicon-dioxide films from silicon surfaces is the most frequently repeated etching step in device processing. Etching of $SiO_2$ is typically achieved by immersing the wafer in a dilute solution of hydrofluoric acid (HF), buffered with ammonium fluoride ($NH_4F$). This is referred to as a buffered HF solution (BHF), also called Buffered-Oxide Etch (BOE). Buffering HF with $NH_4F$ replenishes the fluoride ions as they are consumed, and results in well controlled etching without appreciably attacking photoresist. Since $SiO_2$ is an amorphous material, its etching in BHF is isotropic. The net reaction is [22]:

$$SiO_2 + 6HF \rightarrow H_2SiF_6 + 2H_2O .$$

$H_2SiF_6$ is a soluble complex which can be removed from the vicinity of the surface by stirring. Spray etching has gradually replaced immersion etching because it greatly increases the etch rate and uniformity by providing fresh etchant at the oxide surface; it also reduces lateral etching.

Table 5.2 summarizes the approximate etch rates of deposited or grown $SiO_2$ in BHF. The loosely structured CVD or sputtered oxide offers a larger area to the etchants, and thus exhibits a faster etch rate than thermally grown oxide.

The presence of impurities in the oxide can also affect the etch rate significantly. A high concentration of phosphorus in the oxide results in a rapid increase in the etch rate. This is attributed to the increased porosity of the phosphosilicate glass (PSG) as HF attacks $SiO_2$, and the rapid subsequent dissolution of $P_2O_5$ in water [23]. With about 8 mole % $P_2O_5$, PSG is removed at a rate of 9-10 nm/s in BHF, compared to 1 nm/min for thermal oxide. The so-called P-etch is an etchant mixture of HF and nitric acid ($HNO_3$) used used to selectively etch PSG [24]. The etch rate of PSG in a P-etch increases rapidly with the phosphorus content of the film. Similarly, the etch rate of borosilicate glass (BSG) increases linearly with the $B_2O_3$ content.

**Table 5.2** Approximate etch rates of oxide in a BHF solution at 25 °C[a]

| Type of $SiO_2$ | Density (g/cm³) | Etch rate (nm/s) |
|---|---|---|
| Dry grown | 2.24 - 2.27 | 1 |
| Wet grown | 2.18 - 2.21 | 1.5 |
| CVD deposited | < 2.00 | 1.5[b] — 5[c] |
| Sputtered | < 2.00 | 10 - 20 |

a) 10 parts of 454 g $NH_4F$ in 680 ml $H_2O$ and one part 48% HF.
b) Annealed at approximately 1000 °C for 10 min
c) Not annealed

The etch rate is also affected by stress in the film. It can be increased by implanting argon at high concentration into the oxide, causing stress in the layer [25]. This property is sometimes used to control the edge-slope of an etched thick oxide layer by gradually increasing the etch rate of the film from bottom to top.

Tapering the edge avoids abrupt steps that can cause thinning or even discontinuities in subsequently deposited films as they pass over the edge. The etch rate and edge profile can also be influenced by varying the $NH_4F/HF$ ratio and temperature [26]. Fig. 5.11 shows how the etch rate of thermal oxide varies with temperature for different $NH_4F/HF$ ratios.

## The Silicon Etch

Silicon etching occurs in two steps: the oxidation of silicon and the removal of the oxide by a complexing agent. The oxidation and the dissolution of the oxide occur simultaneously in the same solution. The most commonly used isotropic wet etchant of single-crystal or polycrystalline silicon is the $HF$-$HNO_3$ solution in acetic acid ($CH_3COOH$, or $HAc$) [27]. Nitric acid oxidizes silicon to form a layer of $SiO_2$, which is then dissolved in $HF$. The oxidation reaction is:

$$Si + 4HNO_3 \rightarrow SiO_2 + 2H_2O + 4NO_2 \,.$$

Acetic acid plays the role of a buffering agent since it reduces the dissociation of nitric acid.

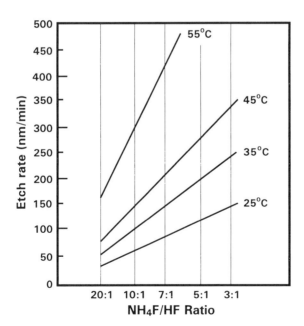

Fig. 5.11 Temperature dependence of oxide etch rate in buffered hydrofluoric acid [26].

Anisotropic etching of single-crystal silicon is used to define V-shaped grooves for special applications. A reaction-limited etch consists of successive dissolution of crystallographic planes. It is slowest on the {111} planes (because of their close spacing) and fastest on the {100} planes. A typical anisotropic etchant of silicon is a dilute solution of potassium hydroxide (KOH) [28]. For example, a solution with 19 wt % KOH in DI water at 80 °C to 82 °C removes (100) crystal at a much higher rate than (110) and (111) planes. The etch ratios are about 400:60:1 for (100):(110):(111) planes. This preferential etch is used to define "self-stopping" V-grooves in < 100> oriented wafers, using an oxide film as a mask (Fig. 5.12). Since (111) planes make an angle of 54.7 ° with the (100) plane, the width ($W$) of the defined image approximately determines the depth ($d$) following the relation

$$d \simeq \frac{W}{2} \tan 54.7 \simeq 0.7W \qquad\qquad 5.1$$

where W is the width of the window at the silicon surface. Another anisotropic etchant of silicon is the hydrazine mixture ($N_2H_4/H_2O$) [29]. A solution of 100 g of $N_2H_4$ and 50 ml DI water at 100 °C results in an etch rate of 50 nm/s on the (100) plane and 5 nm/s in the (111) plane.

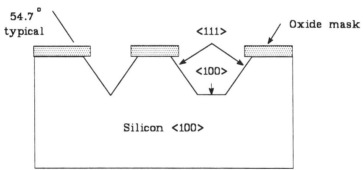

**Fig. 5.12** V-shaped groove formation in < 100 > silicon with anisotropic etching

### The Silicon Nitride Etch

Silicon nitride ($Si_3N_4$) is typically etched in a boiling phosphoric acid solution (85% $H_3PO_4$ at 180 °C). This etchant attacks silicon dioxide very slowly and, to a much smaller extent, silicon [30]. Nitride, oxide, and silicon etch rates in boiling phosphoric acid are about 10 nm/min, 1 nm/min, and 0.2 nm/min, respectively (Fig. 5.13).

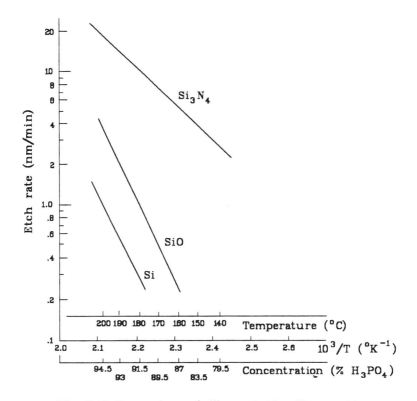

**Fig. 5.13** Comparison of silicon nitride, silicon oxide and silicon etch rates in a hot phosphoric solution [30].

In case photoresist adhesion problems are encountered when etching with hot phosphoric acid, patterning is best achieved by depositing a masking material, such as CVD oxide, on top of the nitride film before coating with resist. The resist pattern is then transferred to the oxide layer, which then serves as a mask for the underlying nitride. A solution which etches nitride and oxide at about the same rate is a mixture of HF and glycerol at 80 °C - 90 °C [31].

## Polysilicon Etch

Etching polysilicon is similar to etching single crystal silicon. In most cases, however, the etch rate is considerably higher because of grain boundary effects, and it is necessary to modify the etch solution to reduce it and make sure that the etchant does not severely attack $SiO_2$, such as the gate oxide in MOSFETs (Fig. 5.14). The $HF/HNO_3$ solution is diluted with water or acetic acid to increase the polysilicon/oxide etch rate ratio. Typical etch mixtures are 10:10:1 $NHO_3:HF:H_2O$, or 20:20:1 $HNO_3:HAc:HF$. Etch rates of 150-170 nm/min can be achieved on undoped polysilicon with these solutions, depending on the grain size. For doped polysilicon, the etch rate typically increases as the dopant concentration increases and the grain size decreases, with phosphorus affecting the etch-rate more rapidly than arsenic and boron.

## Aluminum Etch

Aluminum can be etched in HCl, in a mixture of $H_3PO_4/HNO_3$, in sodium hydroxide (NaOH), or in KOH. The most commonly used wet etchant, however, is a solution of 73% $H_3PO_4$, 4% $HNO_3$, 3.5% $CH_3COOH$, and 19.5% DI water at 30 °C to 80 °C. $HNO_3$ oxidizes aluminum, and $H_3PO_4$ simultaneously dissolves the aluminum oxide. The mixture is agitated during etching to remove gases from the metal surface, resulting in an etch rate of pure aluminum of about 300 nm/min. The etch rate is reduced when copper is added to the aluminum (Chap. 8). As for polysilicon, the aluminum etch rate depends on the grain size.

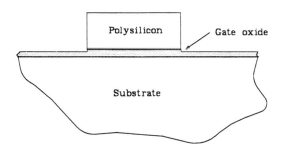

**Fig. 5.14** Selectivity needed when etching the MOSFET polysilicon gate. Etching must stop on the oxide, otherwise single-crystal silicon will be attacked.

## 5.2.2 Dry Etching

Etch processes are judged by their rate, selectivity, uniformity, directionality, surface quality, and reproducibility. While typical wet etch techniques exhibit excellent selectivity and allow processing in large batches, they suffer from the inability to transfer pattern sizes smaller than about 1 $\mu$m with the degree of fidelity and control required for the manufacture of VLSI/ULSI devices. Because wet etching of amorphous and polycrystalline material is isotropic, the horizontal pattern dimensions must be much larger than the thickness of the film to be etched due to the inherent undercutting. These problems are solved by replacing isotropic wet chemical etching with vertical dry etching techniques when defining submicron features. Other advantages of dry over wet etching are the reduced chemical hazard and waste treatment problems, and the ease of process automation and tool clustering [32]. Since most dry etch processes use a low pressure gas in the form of plasma to provide the etchants, dry etching has become synonymous with plasma etching. There are, however, non-plasma etch techniques (such as photo-chemical and vapor etching) that are receiving increased attention and fall in the category of dry etching. A broader discussion of dry etching may therefore be achieved by classifying the processes according to three basic mechanisms: physical, chemical, and chemical-physical [33]. Before discussing these processes, we begin this section with a brief review of basic plasma processes that will help to understand the different plasma-etch techniques.

### Plasma Fundamentals

Plasma processes were introduced in Chap. 4 to describe plasma-enhanced and sputter deposition of thin films. A plasma is a gas which contains equal numbers of positive and negative charges; neutral atoms, radicals, or molecules; and a "gas" of emitted photons. Radicals are molecule fragments with unsaturated bonds. Positive charge carriers are mostly singly ionized atoms, radicals, or molecules created by impact with energetic electrons. The majority of negative charges are free electrons. In the presence of atoms with high electron affinity (such as halogens), negatively charged ions can be created when these atoms capture plasma electrons. Neutral atoms, radicals and molecules can be in the ground or excited state. Photons are emitted when excited species lose energy via spontaneous transitions to lower energy states. This latter process is the basis for the "glow" of the discharge.

A plasma can be created by applying an electric field of sufficient magnitude to a gas. The process can be initiated by an incident electron which gains kinetic energy from the applied electric field. The probability for the electron to collide with and transfer energy to a gas atom or molecule depends on the electron energy, the gas pressure, and the dimensions of the plasma chamber. When a collision occurs, it results in ionization, excitation, or fragmentation of gas molecules. An ionizing collision generates an electron-ion pair. The two new charged particles are accelerated in the electric field and can in turn collide with and ionize other gas particles. As this process continues, the gas breaks down and a plasma is created. The charged particles can be neutralized by recombination within the plasma or at the chamber walls. For a plasma to be sustained, however, the rate of ionization of gas atoms or molecules must be equal to the rate of electron and ion recombination.

Collisions can result in fragmentation of gas molecules into atoms or molecules of smaller size or, for smaller electron energies, in excitation of atoms or molecules to higher energy levels. Some important chemical and physical processes that occur in a plasma are summarized in the Table 5.3 [32]. At thermal equilibrium, the particles can be assumed to move randomly at an average thermal velocity between collisions, approximated by

$$v_{th} \simeq \sqrt{\frac{3kT}{m}} \ ,$$ 5.2

where $k$ is Boltzmann's constant ($k = 8.62x10^{-5}$ eV/K), $T$ the absolute temperature, and $m$ the particle mass. Because of their very small mass, electrons fly at a much higher thermal velocity than gas atoms or molecules. The motion of charged particles under the influence of an electric field is described in terms of a drift velocity, $v_d$, given by

$$v_d = \mu E \, ,$$ 5.3

where $\mu$ is the particle mobility, and $E$ the electric field.

**Table 5.3** Important reactions in a plasma [32].

| Reaction | Example |
|---|---|
| Positive ionization | $Ar + e \rightarrow Ar^+ + 2e$ <br> $O_2 + e \rightarrow O_2^+ + 2e$ |
| Dissociative ionization | $CF_4 + e \rightarrow CF_3^+ + F + 2e$ |
| Fragmentation | $CF_3Cl + e \rightarrow CF_3 + Cle^-$ <br> $C_2F_6 + e \rightarrow 2CF_3 + e$ |
| Dissociative attachment | $CF_4 + e \rightarrow CF_3 + F^-$ |
| Dissociative ionization with attachment | $CF_4 + e \rightarrow CF_3^+ + F^- + e$ |
| Excitation | $O_2 + e \rightarrow O_2^* + e^-$ |
| Photoemission | $O_2^* \rightarrow O_2 + h\nu$ |

$O_2^*$ is the excited state of $O_2$.

Because of their smaller mass, electrons drift at a much higher velocity than ions. As electrons gain kinetic energy from the electric field, their effective temperatures increase above the gas temperature. While the temperature of the atoms and molecules in the gas remains near ambient, electrons can attain high average energies, typically 1-10 eV, corresponding to an effective electron temperature of $10^4 - 10^5 K$ [33 – 35]. This energy is transferred to the gas by collision processes in which ions and highly reactive species are created. It is this property of the plasma that allows high-temperature type reactions to occur at low ambient temperatures and permits the use of temperature sensitive materials such as organic resist masks for etching.

The average distance travelled by particles between collisions, called mean free path $\lambda$, depends on the species and gas pressure following the relation

$$\lambda = \frac{5x10^{-3}}{P} \quad (cm),$$

5.4

where P is the pressure in Torr. In typical plasmas of interest, the chamber pressure ranges from 1 Pa (7.52 mTorr) to 150 Pa (1.13 Torr). This corresponds to a density of gas molecules of $2.7x10^{14} - 4x10^{16}\ cm^{-3}$, and a mean free path of $\simeq 6650\ \mu m$ at the low pressure end and 44 $\mu m$ at the high pressure end. Even at the high pressure end, the mean free path is considerably larger than the depth of typical etched features - an important condition for directional etching. The electron density in such a plasma ranges from $10^9 - 10^{12}\ cm^{-3}$. It follows that the degree of ionization is very small; in typical plasma reactors used for etching, only $10^{-4}$ to $10^{-7}$ of the gas molecules are ionized.

### Sheath Formation

A simple plasma-etch reactor is shown schematically in Fig. 5.15. It consists of a grounded electrode which is typically connected to the chamber walls, a second electrode to which power is applied, and a partially evacuated chamber which contains a low pressure gas of suitable mixture.

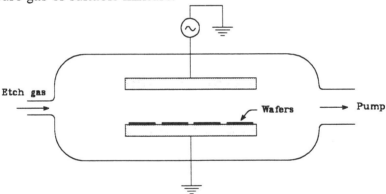

Fig. 5.15 Schematic of a simplified plasma etch reactor.

Most systems use RF rather than DC power to avoid charge accumulation on insulator surfaces. The RF amplitude is in the range 700-1000 V, and its frequency typically 13.56 MHz, an authorized industrial frequency. At such a frequency, most electrons oscillate between the electrodes, increasing the average electron energy and probability of electron-gas collisions. The RF power source is separated from the second electrode by a coupling capacitor to block DC current components. While the plasma chamber as a whole is neutral, recombination of charges at boundary surfaces surrounding the plasma causes charge depletion near these boundaries and the formation of a boundary layer called the **sheath.** The resulting gradient in charges gives rise to a net diffusion of carriers to the boundaries. Since electrons diffuse faster than ions, more electrons than ions leave the plasma initially. Consequently, an excess of positive ions is left in the plasma which now assumes a potential $V_P$ with respect to the grounded walls. The potential $V_P$ gives rise to a drift current component which enhances the motion of holes and retards the motion of electrons to the grounded walls. When steady-state is reached, the electron and ion fluxes are balanced and the sheath is almost depleted of electrons. As a consequence of the reduced electron concentration, the sheath conductivity decreases considerably below that of the plasma region, and the probability for electron-gas collisions with the accompanying decay of excited species by photo-emission is reduced - hence the name "dark space".

The large difference between electron and ion mobilities also creates a sheath near the powered electrode. Since the coupling capacitor suppresses DC current components, electrons can accumulate at the electrode surface which assumes a negative DC voltage superimposed on the time-average AC potential. The powered electrode reaches a "self-bias" negative voltage , $V_{DC}$, with respect to ground [36]. Similarly, when an electrically isolated surface (such as an insulating substrate isolated from ground by an insulating film) is in contact with the plasma, it must receive equal electron and ion fluxes at steady state. Following the same reasoning as above, the isolated surface must acquire a negative potential with respect to the plasma to retard the motion of electrons and enhance the motion of ions so that to equalize the fluxes of both carrier types. The potential of the isolated surface with respect to ground is referred to as its floating potential, $V_F$. As in the case of the grounded walls and the powered electrode, the isolated surface is surrounded by a sheath of reduced electron concentration. The sheaths are typically a few millimeters thick. Therefore, wafers with topographical details of the order of 1 $\mu$m placed on the electrode have a negligible effect on the field distribution in the sheath.

**Potential Distribution**

The potential distribution in a plasma chamber is shown in Fig. 5.16 for a two-electrode parallel-plate reactor, with RF power applied to one of the electrodes. The second electrode, which also includes the chamber walls, is at ground potential.

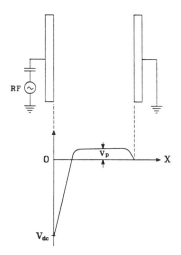

**Fig. 5.16** Potential distribution in a very asymmetric parallel-plate plasma reactor.

The three time-average potentials of importance are the plasma voltage ($V_P$), the "self-bias" voltage of the powered electrode ($V_{DC}$), and the floating voltage ($V_F$). They determine the energies of ions incident on surfaces in the plasma and their effect on etching [36]. For example, the difference between the plasma and floating potential determines the maximum energy with which ions bombard an electrically floating surface. The plasma, because of its degree of ionization and high conductivity, can for practical purposes be regarded as an equipotential volume. It assumes the highest potential of the system, $V_P$. Most of the voltage drops across the sheath because of its high resistance. The sheaths can be represented by parallel-plate capacitors, and the magnitude of $V_P$ approximated from a capacitive voltage dividing network. If the system is perfectly symmetric, i.e. electrodes of equal area and driven by RF signal of equal magnitude, the plasma potential is very large and both electrodes are bombarded with energetic ions to the same extent. Typical planar reactors are, however, asymmetric and the capacitive coupling between electrode and plasma is greater at the larger electrode; the plasma potential is closer to the potential of the larger electrode than to that of the smaller electrode. If the area of the powered electrode is significantly smaller than the grounded area, as is the case for typical planar reactors, the plasma potential is small, in the range $\simeq$20 - 50 V above ground, and a larger sheath potential appears at the smaller electrode than at the large electrode [36]. The average sheath potential of the smaller electrode varies from about 0.25 to 0.5 of the peak to peak voltage of the applied RF signal as the reactor goes from a symmetric to a very asymmetric configuration [32]. In summary, for very asymmetric designs, a large sheath potential difference exists at the smaller electrode causing high-energy ion bombardment of a surface placed on the electrode. The mean voltages across the two sheaths are divided according to the relationship

$$\frac{V_1}{V_2} \simeq \left( \frac{A_2}{A_1} \right)^4 ,$$

5.5

where $A_1$ is the area of the small powered electrode, and $A_2$ the area of the large grounded electrode [37].

## Physical Etch Processes

If a chemically inert gas, such as argon, is ionized and accelerated to impinge on a wafer surface, material can be removed from the surface by momentum transfer, a process similar to sandblasting. This process is used in three distinct modes. When the wafer is immersed in a plasma and subjected to bombardment with high-energy plasma ions which are accelerated across the sheath, the process is termed **sputter etching** [38,39]. When the wafer is physically separated from the plasma and a broad ion beam extracted from the plasma, collimated and accelerated to impinge on the wafer surface in a definite direction with respect to the feature to be etched, the process is called **ion-beam milling** [40 – 42]. The third mode is to spatially selective etch the material with a beam of ions (typically Ga +), focused to submicron diameter (Chap. 3). This technique is referred to as **focused ion**

**beam (FIB) etching** or milling. Although FIB milling has not emerged for dry etching of silicon patterns, it has become an indispensable tool for restructuring a pattern on a mask or integrated circuit (unwanted opaque pattern milled-off), or for diagnostic cross-sectioning of microstructures [43].

Sputter etching and broad ion-beam milling use high-energy ($\geq$ 500 eV), inert gas ions (typically Ar $^+$) to dislodge material from the wafer surface - a highly anisotropic etch process. The principle of sputter etching is shown schematically in Fig. 5.17. The wafers are mounted on the powered electrode, immersed in a glow discharge at a pressure of 1-10 mTorr. Noble gas ions are created in the plasma and accelerated toward the wafer surface. The electric field profile and gas density around the electrode is such that most of the accelerated ions impinge on the wafer surface at normal incidence. If the ion energy is above a threshold value, atoms, molecules or ions are ejected from the wafer surface and vertical etch profiles can be achieved.

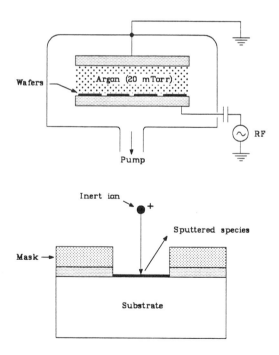

**Fig. 5.17** Principle of sputter etching.

In ion-beam etching (or ion-beam milling), a magnetically confined DC plasma is used to generate ions (Fig. 5.18). The purpose of the magnetic field is to force electrons to follow a helical path between collisions rather than pass directly from cathode to anode, enhancing the electron path length and ionization efficiency at low pressure. A set of grids used in the confinement is biased to extract ions from the source and direct an ion beam to the wafer surface [40 − 42]. The well-collimated ion beam of line-of-sight nature also allows etching of tapered profiles by substrate tilting and rotation. The etch flexibility is increased by the ability to separately control the energy and flux of ions.

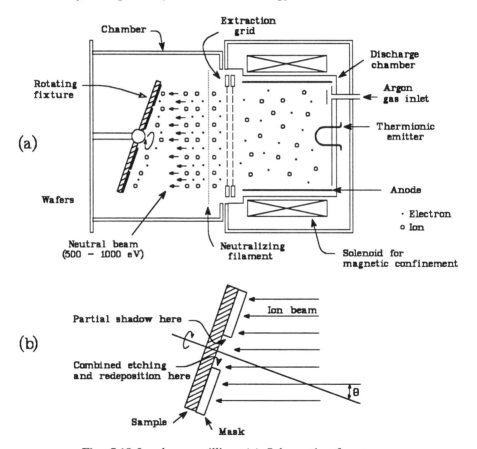

Fig. 5.18 Ion beam milling: (a) Schematic of system, (b) Details of wafer orientation.

The ability to etch virtually any material opened up a variety of applications for ion-beam milling and sputtering, particularly when etching several layers of different chemical composition is required. The inherent poor selectivity and slow etch rate of these purely physical processes, however, severely limit their use in the transfer of submicron patterns. The etch rate, $R$, is directly proportional to the sputter yield - the number of sputtered atoms or molecules per incident ion, given by

$$R = \frac{6.22sjW}{\rho} \quad (nm/\min) \qquad 5.6$$

where $s$ is the sputter yield, $j$ the ion flux (mA/cm$^2$), $W$ the molecular weight of etched material (g/mole), and $\rho$ the density of material to be etched (g/cm$^3$).

The sputter yield depends on the ion energy and its angle of incidence with respect to the surface of the material to be etched. While etching typically occurs at normal incidence during sputtering, tilting and rotation of the substrate during ion beam milling add the flexibility of modifying the sputter yield. Sputter yields of important materials are shown in Table 5.4 for 500 eV argon ions at normal incidence. The corresponding etch rate is found from Eq. 5.6. For example, an etch rate of $\simeq$ 34 nm/min is found for silicon ($s = 0.45$, $W = 28$, $\rho = 2.3$) at $j = 1$ mA/cm$^2$ ( $6.25x10^{15}$ ion/cm$^2$s). Selectivity depends only upon sputter yield differences between materials which, for most semiconductor materials, are within a factor of three of each other (Table 5.4). Therefore, there is little selectivity between sputtering a thin film of the material and sputtering the mask with which windows in the material are defined. For example, a photoresist mask is removed at about the same rate as an underlying silicon dioxide film.

One important concern with sputtering is that the sputtered material is typically non-volatile and tends to re-deposit onto the wafer surface and elsewhere in the system. Other concerns with all processes which involve ion bombardment are the charge build-up on insulated surfaces and resulting damage to the underlying film and semiconductor surface, and the damage to semiconductor surfaces directly exposed to ion bombardment. If the beam strikes a conducting grounded surface, sufficient secondary electrons are generated to balance the space charge of the beam and external neutralization is not necessary. If ions impinge on an insulated surface, however, positive charge can build-up on the surface, damaging the underlying insulator and semiconductor surface. In ion-beam milling, this is avoided by supplying low-energy electrons by a hot filament that is placed in the beam path to neutralize the ions at the wafer surface (Fig. 5.18). Plasma effects on insulators and semiconductor surfaces are discussed at the end of the chapter.

**Table 5.4** Typical sputtering yields of materials for 500 eV argon ions at normal incidence [38]

| Material | Sputtering Yield | Material | Sputtering Yield |
|:---:|:---:|:---:|:---:|
| Al | 1.05 | Au | 2.83 |
| Cu | 2.83 | Si | 0.45 |
| Ti | 0.51 | W | 0.63 |
| Co | 1.22 | SiO$_2$ | 0.25 |
| SiO$_2$ | 0.25 | GaAs | 1.20 |

## Dry Chemical Etching

Dry chemical etching occurs when a chemical reaction takes place between a gas and the wafer surface, with or without a plasma, and the resulting volatile product is pumped off. This process is typically selective and non directional. In plasma-assisted chemical etching, the only role of the plasma is to maintain a supply of reactive species in form of free radicals and excited neutrals. Non-plasma reactions occur when the injected gas is inherently reactive or when the reaction is stimulated by, e.g., photons. Photo-assisted chemical etching and spontaneous dry reactions, such as the chemical etching of silicon in a $ClF_3$ gas or the removal of oxide in anhydrous HF gas or in $HF/H_2O$ vapor are discussed in the following sections.

### Plasma-Assisted Etching

Selective removal of material by a reactive gas created near or within a glow discharge from an initially non reactive gas mixture is called plasma-assisted chemical etching. In most systems, the injected gas itself rarely reacts with the surface being etched and free radicals are believed to be the major reactant species. The gas-mixture is chosen to produce reactive species by molecular dissociation into radicals and excitation of neutrals in the plasma. In purely chemical etching, reactive species migrate to the wafer surface while ion-bombardment is suppressed by increasing the gas pressure and placing the wafers on the grounded electrode in a very asymmetrical reactor, where the voltage across the sheath is small ($\simeq V_P$). Ion bombardment can be totally eliminated by mounting the wafers remote from the plasma, as in the downstream etch reactor described below. Because of the absence of physical enhancement, plasma-assisted chemical etching is essentially isotropic and very selective. For typical plasma gas pressures, the neutral species are incident on the wafer from all directions but, on the average, they make their last gas phase collisions well above the surface relative to the depth of etched features.

The key requirement for successful plasma-assisted etching is the volatility of (or ability to evaporate) etch products. The number of molecules evaporating from the surface per unit area and unit time, $n_e$ is given by [44,45]

$$n_e = 3.51 \times 10^{22} \, P_v \left( \frac{1}{TM} \right)^{1/2} \quad (cm^{-2}s^{-1}) , \qquad 5.7$$

where $P_v$ is the the vapor pressure in Torr, $T$ the temperature in K, and $M$ the molecular weight in g. The vapor pressure is given as [46]

$$\log P_v = C' - 0.2185 \, \frac{\Delta H_v}{T} , \qquad 5.8$$

where $C'$ is a material constant, and $\Delta H_v$ is the heat of evaporation in cal/g.mol, both published for various materials in [47].

The reactive species must be continuously created and supplied to the wafer surface. The generation of reactive species is critical to plasma etching because many of the injected gases used to etch thin films do not react spontaneously with the film material. Etchant species such as radicals created in the plasma have large sticking coefficients compared to "inert" gas molecules, so adsorption occurs easily on the film surface - another condition for plasma etching. Finally, the reaction by-products must spontaneously desorb from the wafer surface and be pumped out of the plasma gas environment. If any of these individual steps does not occur, the overall etch cycle terminates - or never starts.

Because of the lack of directionality, plasma-assisted chemical etching is unsuitable for patterning VLSI/ULSI structures where tolerances on patterned and etched critical features may not exceed 10-15% of the feature size. Also, in isotropic etch processes it is difficult to control the degree of overetching beyond the end point because the rate of undercut may accelerate at the end point where control is needed most, an effect related to "loading" discussed below [34]. Plasma-assisted chemical etching is, however, widely used when isotropic etching is required or when dimensional control is not critical. The most common example of plasma-assisted chemical etching is the removal of photoresist in an oxygen plasma - a cheaper alternative to wet solvent resist stripping, sometimes referred to as **plasma ashing** [48]. Another application is the leveling ("planarizing") of insulator surfaces for multilevel interconnections (Chap. 8). Plasma ashing and insulator leveling are discussed further below in the "Applications" section.

An example where isotropic etching may be required is illustrated in Fig. 5.19. When a conducting film (such as doped polysilicon) is deposited conformably over vertical steps, the film thickness along the vertical edges, as seen in the direction of the ion beam, exceeds the film thickness over the flat field area by approximately the step height. Therefore, if the film is etched directionally to end point over flat areas, material will be left along the step in the form of "stringers" which can cause shorts between adjacent conducting lines. Overetching for a duration based on the step height would eventually clear the stringers, but only at the expense of attacking the underlying film in the flat areas (typically a thin oxide). A short isotropic etch following directional etch to end point over flat areas removes the stringers and causes little sidewall etch in the patterned layers.

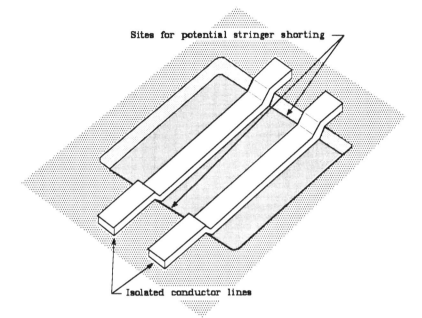

**Fig. 5.19** Formation of "stringers" in directional etching and need for isotropic plasma etching to remove the stringers.

Plasma-assisted etch conditions can be altered from being isotropic to very directional by varying input power, gas pressure, reactor geometry, gas mixture, and gas flow. The control of excited neutrals and ion fluxes is critical to optimizing plasma processes. The main difficulty, however, is the limited understanding of the processes taking place in the bulk of the plasma and at surfaces in contact with it. There is no simple relationship between etch objectives and tool/process variables, leading to a practical problem in scaling up processes for production. In practice one arrives at the correct process "recipe" for each tool by systematically evaluating the response surface of the process. One of the most important variables in this investigation is the choice of gas mixture, also referred to as "etch chemistry" [49].

## Physical-Chemical Etch Techniques

The reaction-rate between gas and wafer surface can be greatly increased by combining chemical and physical processes at or near the surface. Anisotropic etching can be achieved by impinging high-energy ions, electrons, or photons onto the surface, enhancing the reaction between the gas and horizontal surfaces with little attack on vertical sidewalls.

### Reactive Ion Etching (RIE)

The most common etch technique in the manufacture of VLSI/ULSI silicon devices is the so-called reactive-ion etch (RIE) [50]. Positive plasma ions in a parallel-plate RF reactor are used to provide a source of energetic particle bombardment for the etched surface, producing vertical edges in the etched film with negligible undercutting (Fig. 5.20). Ion bombardment increases the reaction rate of spontaneously occurring processes and/or prompts reactions which do not occur without radiation.

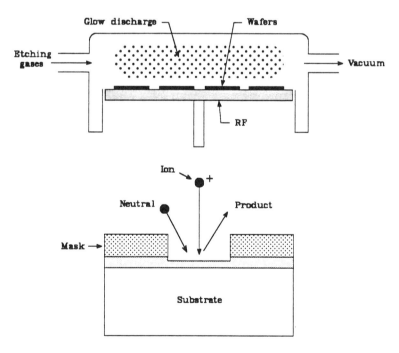

**Fig. 5.20** Reactive-ion etch system. The wafers are placed on the powered electrode of a parallel-plate RF reactor. Horizontal surfaces are subjected to both reactant species and impinging ions, while vertical sidewalls are only subjected to reactive species.

In RIE, the gas pressure is typically about 100 mTorr, and the wafers are placed on the powered electrode. As with sputtering, the low pressure ensures that the transport of ions through the sheath is without collisions, so that ions acquire high kinetic energies and impinge on the wafer at normal incidence - a primary factor in etch directionality [36].

If the wafers were placed on the large (grounded) electrode, as in plasma-assisted chemical etching, the maximum accelerating potential available for positive ions would be approximately the plasma potential, $V_P$, which ranges from only 20-30 V. This would result in negligible ion bombardment and isotropic etch characteristics. In RIE, the wafers are mounted on the powered electrode and the maximum energy available for positive ions is $e(|V_{DC}| + V_P)$, where $V_{DC}$ is the DC self-bias of the powered electrode, which can be as high as 500 eV [36]. The impinging high-energy ions enhance or initiate reactions between the wafer surface and reactive species that diffuse from the plasma and are adsorbed on the wafer surface. This physical-chemical process leads to much larger vertical than lateral etch rates, because horizontal surfaces of patterned features are in the "line-of-sight" of both energetic ions and chemical species, while sidewalls are exposed to only chemical species. The effect of ion bombardment on the reaction rate is illustrated for silicon etching in Fig. 5.21 [51]. A silicon surface in vacuum is initially exposed to a low flux of of $XeF_2$ molecules which spontaneously etches silicon at a low rate. When a flux of 450-eV $Ar^+$ ions is turned on, the etch rate increases dramatically. Finally, when the $XeF_2$ flux is shut off, silicon is again etched at a low rate by physical sputtering of the argon ion beam.

The mechanisms by which ion bombardment enhances or promotes the reaction between an active gas and the wafer surface vary with the gas/solid system and etching conditions [51,52]. They can be categorized as: (1) chemically enhanced physical sputtering, where chemically weakened bonds increase the sputter yield [53], (2) damage enhanced chemical reactivity [54,55], and (3) chemical sputtering, where ion bombardment provides energy to drive the chemical reaction by increasing the mobility of species in the reaction layer, forming volatile products more rapidly which subsequently desorb [56]. The increased anisotropy of RIE is gained at the expense of reduced selectivity (because of ion bombardment), and reduced etch rate (because of the reduced flux density of reactive species at lower pressures). Batch processing is therefore required to overcome the low etch rate and increase throughput. This is done, however, at the expense of uniformity and reproducibility.

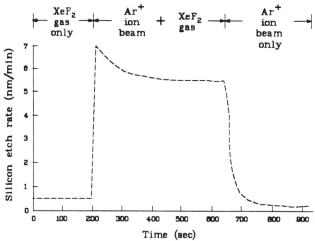

Fig. 5.21 Illustration of ion-enhanced etching of silicon using Ar$^+$ and XeF$_2$. The energy of the Ar$^+$ ions is 450 eV, and the XeF$_2$ flow rate is $2x10^{15}$ molecules/s. Adapted from [49,51]

The term ion assisted etching appears to be more accurate and descriptive than reactive ion etching because in all cases free radicals and not ions are the primary etch species. The role of energetic ions is mainly to enhance the reaction.

**Sidewall Passivation**

Etch directionality can be enhanced by a mechanism known as sidewall passivation (or inhibitor-driven anisotropy). By adjusting the etchant gas composition and reactor parameters, an etch-inhibiting film can be formed on vertical sidewalls, slowing down or completely stopping lateral attack while etching of horizontal surfaces proceeds [57 – 59]. Fig. 5.22 illustrates this mechanism for silicon etching. By adding O$_2$ to a Cl$_2$ plasma, for example, an oxide film can be grown on the sidewalls (not exposed to ion bombardment), while oxide growth is prevented on horizontal surfaces because they are exposed to ion bombardment [49]. A similar mechanism can be induced by choosing a greater elemental ratio of carbon to fluorine in fluorocarbon plasmas. This favors the deposition of involatile polymer films on sidewalls, forming a coating that blocks chemical attack. While the polymer film also deposits on horizontal surfaces, it is removed by ion bombardment, allowing etching of horizontal surfaces to continue.

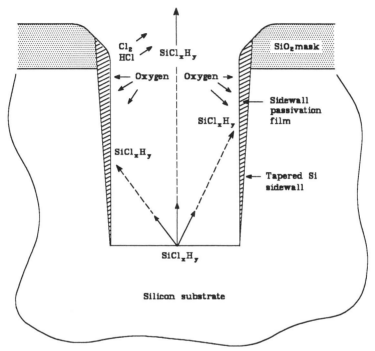

**Fig. 5.22** Formation of an inhibiting film on sidewalls enhances directionality [49].

One problem which is solved by sidewall passivation is the enhanced etch rate of heavily and non-uniformly doped silicon or polysilicon, causing a nonuniform etch profile and loss of dimensional control. While the purely chemical reaction between chlorine and intrinsic silicon is slow at ordinary temperatures, heavily doped n-type silicon or polysilicon can react rapidly with Cl atoms, causing these regions to etch at a higher rate than intrinsic silicon. The sidewall is protected by an inhibitor film to preserve linewidth control when highly doped polysilicon is etched by isotropic etchants such as fluorine or chlorine atoms. Fig. 5.23 illustrates the effects of adding a film-former, trichlorofluoride ( $CCl_3F$), to the plasma gas when etching silicon in $SF_6$. The undercut is plotted as a function of overetch time (time after the polysilicon film was clear from the planar wafer area). The upper dashed curve shows that polysilicon is severely undercut by reactive F atoms created in a pure $SF_6$ plasma. When 12% $CCl_3F$ is added to the gas, however, there is barely an undercut, even after substantial overetching, because an etch-inhibiting film forms on the sidewalls of the polysilicon feature [34].

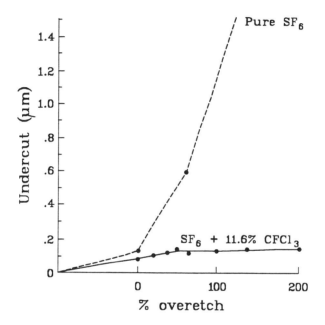

**Fig. 5.23** Undercut of a heavily doped polysilicon film with and without the addition of 11.6% ($CCl_3F$). The formation of a protective sidewall film with $CCl_3F$ inhibits lateral attack by F-atoms. Adapted from [34].

### Reactive Ion-Beam Etching.

In reactive ion-beam etching (RIBE) the configuration is similar to ion-beam milling, except that a reactive rather than inert gas is used, improving selectivity [60,61]. Since the ion energy can be controlled independent of the neutral flux by means of extraction grids (Fig. 5.18), RIBE offers a flexibility not available with RIE. Also, since the wafers are placed remote from the glow discharge, plasma damage is negligible when compared to RIE where the wafers are immersed within the plasma.

A variant of RIBE is the the chemically assisted ion beam etching (CAIBE). Here, a reactive gas is introduced in the vicinity of the wafer and irradiated with a beam of inert ions (typically Ar +). This method allows independent control over ion flux, ion energy, and neutral flux, considerably increasing flexibility in etching [62].

## Focused Ion-Beam Assisted Etching

In focused ion beam assisted etching, a beam of ions (typically Ga +) is incident on the surface and can be deflected to directly produce patterns. A gas such as $Cl_2$ is introduced into the chamber at about 30 mTorr. The vacuum outside the chamber where the FIB is generated is maintained at $10^{-6} - 10^{-8}$ Torr. With this system, an increase in the etch rate of both Si and GaAs has been achieved [63,64]. For example, the etch rate of GaAs in $FIB/Cl_2$ is about ten times larger than the milling rate that would be obtained in the absence of the $Cl_2$ gas. Trenches 6 $\mu$m deep and 0.1 $\mu$m wide at the bottom have been etched in silicon. The maximum etch rate observed corresponds to the removal of 10 Si atoms per incident Ga + ion (35 keV). As with RIE, chemical reactions are induced or accelerated by the impact of a focused beam of energetic ions. The spatial resolution of this process is ultimately limited by the focal beam spot size that can be achieved.

## Photo-Chemical Etching

Photo-chemical etching is a non-plasma process in which photon energy, typically from a laser source, is used to excite gas-phase reactants or adsorbed species at the wafer surface [65]. Directional etching with high selectivity can be achieved by initi-ating a chemical reaction with impinging photons at normal inci-dence to the wafer surface (Fig. 5.24). In photo-assisted etching, radiation effects on devices are reduced or eliminated because bombardment with energetic ions is avoided. The mechanisms involved depend, however, upon the wavelength and power density chosen. For example, in a direct ablative process, the light is absorbed by a surface layer of the wafer and material is removed by thermal evaporation - a physical process. Alternatively, adsorbed species at the surface can be excited in a manner analo-gous to ion-enhanced processes, or gas-phase reactions can be induced to produce chemically reactive species. The effect of irradiation on devices also depends on the wavelength of the laser beam used. For example, traps are created in the thin gate oxide of MOSFETs when the polysilicon-gate film is etched using a ArF (193 nm, 6.4 eV) or KrF (249, 5 eV) excimer laser beam, but not with XeCl (308 nm, 4eV) excimer-laser induced etching [66].

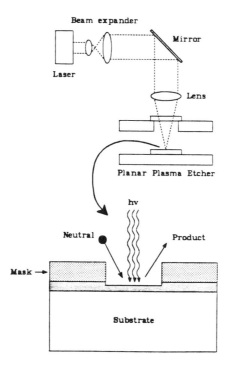

**Fig. 5.24** Principle of photo-chemical etching.

## Reactor Designs

A plasma reactor consists of three basis functions: a vacuum chamber and an evacuation system to maintain reduced pressures ranging from 1 mTorr - 10 Torr, a power supply to create the plasma (glow discharge) in an electric and/or magnetic field, and a controlled source of gases of appropriate mixture. Typical reactors use an RF electric field to transfer power to the discharge, but frequencies ranging from DC to the microwave region have been used and significantly influence the process. The three most widely used reactor configurations are the cylindrical (barrel), the parallel-plate (planar), and the downstream etchers (afterglow or remote plasma).

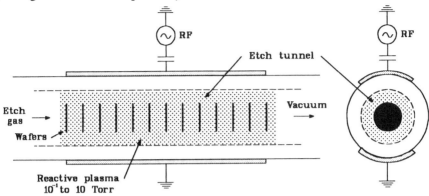

Fig. 5.25 Typical barrel reactor configuration.

## Barrel Reactor

A simple barrel reactor is shown schematically in Fig. 5.25. It consists of a cylindrical dielectric vessel with gas admission at one end and an exhaust pump at the other. The RF power is applied by placing metal electrodes on opposite sides of the vessel. The wafers to be etched are placed vertically in a batch holder in the center of the chamber, with small separation between the wafers. This is why this type of reactor is also called a "volume-loaded system". Concerns about plasma induced radiation damage led to the introduction of a cylindrical mesh to shield the wafers from energetic ions and electrons in the plasma. The protective mesh is shown in Fig. 5.25 as an "etch tunnel". It confines the plasma to the region between the tunnel and the vessel walls, so that the wafers are not subjected to ion bombardment. Etching proceeds by the chemical reaction of long-lived excited neutral species which migrate to the wafer surface. Since this process is essentially isotropic, barrel reactors are used primarily for ashing resist, leveling ("planarizing") insulators, or for etching "non-critical" features where undercutting can be tolerated.

## Planar Reactor

The need for a directional etch process led to the use of planar geometry, also called a parallel-plate or "surface loaded" system (Fig. 5.26). As discussed earlier, these systems are used in two distinct modes. When the wafers are mounted on the grounded electrode opposite to the RF powered electrode (Fig. 5.26a), little ion bombardment occurs and the process is predominantly an isotropic chemical reaction between highly reactive species and the surface material, with little enhancement by ion bombardment. When the wafers are placed directly on the RF powered electrode (Fig. 5.26b), they are in direct contact with the plasma, and therefore exposed not only to reactants, but also to energetic ions. This can result in sputtering, reactive ion etching (RIE), or reactive sputter etching, depending on the gas pressure, species, and ion energy.

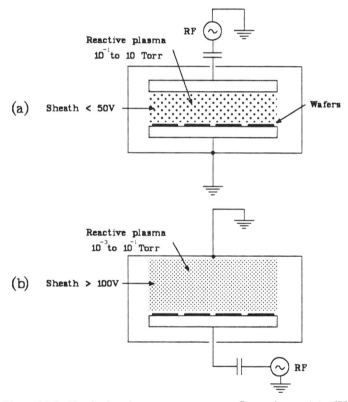

**Fig. 5.26** Typical planar reactor configuration. (a) Wafer mounted on grounded electrode with negligible ion bombardment. (b) Wafer mounted on powered electrode where high-energy ion bombardment enhances the etch rate.

One variant of the parallel-plate configuration is the hexode or Hex reactor which increases the wafer capacity by folding the electrodes in concentric cylinders and expanding the reactor in the vertical dimension (Fig. 5.27). Six wafers are placed vertically in each level. This arrangement also reduces particle deposition on the wafer surface. The planar reactor and its variations are the most common type in use. They are available in single-wafer or batch processing.

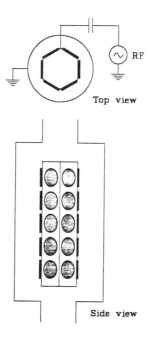

**Fig. 5.27** Hexode reactor configuration as a variant of the parallel-plate reactor. Six wafers are placed vertically at each level.

### Downstream Etchers

To avoid subjecting wafers to high-energy ion bombardment, the wafers can be placed downstream from the plasma, outside the discharge (Fig. 5.28). Since there are no ions to induce directional etching, downstream reactors are isotropic and primarily limited to removing resist or other layers of material where patterning is non-critical. Downstream etching configurations offer several advantages over barrel and planar systems. By using RF or microwave radiation, long lived active species are generated and then transported to the wafer located remote from the plasma. Temperature control problems and radiation damage are

**306**

therefore reduced or eliminated [67, 68]. The wafer holder can be heated to a precise temperature to increase the chemical reaction rate, independent of the plasma. In contrast, it is difficult to precisely control the temperature in barrel or planar reactors, and radiation damage is a concern because the wafers are immersed within the plasma. Downstream etchers are typically single wafer cassette-to-cassette reactor chambers.

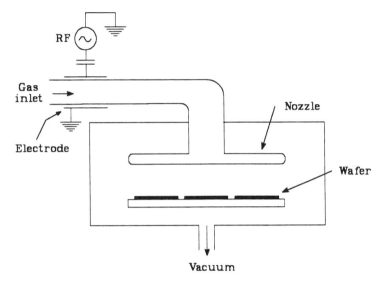

**Fig. 5.28** Schematic of a downstream etcher. (Adapted from [67]).

## Low-Pressure Etch Reactors

It is well-known that the generation of reactive species within the plasma can be accelerated by increasing the plasma-gas pressure (0.1- 10 Torr). The high pressure enhances electron-neutral collisions by decreasing the mean free path of electrons, and also helps cooling the wafers by gas thermal conduction. As the pressure is increased, however, reactive species are subjected to more lateral scattering, increasing the isotropic etch component. The factor of importance in determining the degree of isotropy is the ratio $R_I$ of ion thermal energy ($kT_i$) to the average energy gained from the sheath electric field between collisions ($eE\lambda$) [69]:

$$R_I = \frac{kT_i}{eE\lambda} \, ,$$

5.9

where $k$ is the Boltzmann constant, $T_i$ the average ion temperature, $E$ the average sheath electric field, and $\lambda$ the mean free path for ion-neutral collisions. If $R_I$ is larger than 1, the random thermal ion energy is larger than the energy of accelerated ions gained by the directional electric field, and isotropic etching is enhanced. To increase anisotropic etching, we must have $eE\lambda >> kT_i$. Anisotropy is therefore increased by either increasing the accelerating field, or reducing the gas pressure (increasing $\lambda$), or reducing the average ion temperature. An increase in electric field not only causes more damage from ion bombardment of the wafer surface, but also increases the energy with which ions bombard the grounded electrode causing its sputtering and re-deposition of electrode material on the wafer surface. A more appropriate means of achieving a higher rate of anisotropy is to maintain a low plasma pressure and separate the process of ionization and generation of reactive species from the ion transport process. This can be accomplished by making use of an auxiliary ionization/excitation mechanism in a low-pressure plasma, such as a triode or magnetically enhanced discharge, or by electron-cyclotron resonance (ECR).

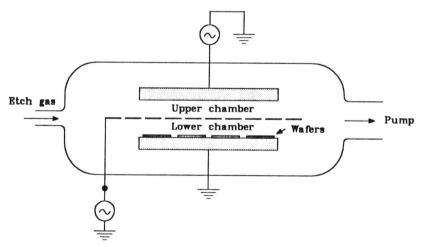

**Fig. 5.29** RF Triode etcher (Adapted from [34]).

### Triode Reactor

In a triode reactor RF power is applied to a third electrode in such a way that the sheath voltage can be controlled almost independently of the generation of reactive species (Fig. 5.29). This allows some degree of decoupling to be gained over the physical and chemical processes [70]. The objective of this system is to achieve a high generation rate of reactive species in the upper zone that is partly shielded from the wafers, and control ion bombardment energies in the lower zone where the wafers are located.

### Magnetically Enhanced Reactive Ion Etching (MERIE)

As the minimum lithographic dimension decreases and the circuit density increases, a lower plasma gas pressure is needed to maintain etch directionality and uniformity in RIE systems, particularly when etching high aspect-ratio (depth to width) features is required. At a lower pressure, the desorption and volatility of etch by-products is more efficient and the probability for ion-neutral collisions in the sheath is reduced, resulting in less lateral attack. As the pressure is reduced in RIE systems, however, the ion density is reduced and the etch rate decreases considerably.

By confining the plasma in a magnetic field, the probability for electron-neutral collisions can be increased, creating a large flux of ions and enhancing the etch rate at low plasma-gas pressure. In such a system, referred to as a magnetically enhanced RIE reactor (MERIE), a magnetic field is produced normal to the electric field between the electrodes by a pair of Helmholz coils (Fig. 5.30). The field can be rotated electrically by modulating the DC current supplied to the coils. It confines electrons to cycloidal trajectories near the powered electrode as illustrated in Fig. 5.31 [49, 70]. This confinement increases the frequency of electron-neutral collisions and improves the ionization efficiency at low pressure. The ratio of ion to neutrals can be increased in MERIE systems by a factor as high as 50 above that in conventional RIE systems at the same gas pressure [49, 71]. Since the magnetic confinement reduces the rate at which electrons impinge on the powered electrode, the self-bias voltage in MERIE systems is typically reduced to about 100 V. This is considerably lower than in comparable RIE systems [72]. With separate adjustment of magnetic field and discharge power, some control over plasma and etchant densities can be achieved independent of sheath potential and DC bias.

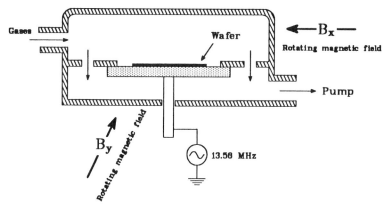

**Fig. 5.30** Schematic of an MERIE system.

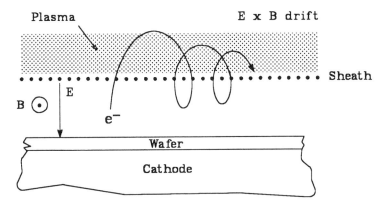

**Fig. 5.31** Trajectory of an electron under the influence of an electric and magnetic field near the surface of a wafer (Adapted from [49]).

Since a high etch rate at low bias can be achieved, there is no need for a high flux of high-energy ion bombardment and substrate damage is reduced [73 – 75]. MERIE systems have been successfully used to etch trenches in silicon, aluminum films, silicon dioxide, and gallium arsenide with negligible damage to the substrate [76 – 79]. When etching silicon or aluminum films, directionality is typically enhanced with sidewall passivation. One disadvantage of these systems, however, is the degraded etch uniformity because of inhomogenities in the magnetic field and ion density.

### Electron Cyclotron Resonance (ECR) Etching

Electron Cyclotron Resonance (ECR) systems were described in Chap. 3 in conjunction with thin-film deposition. ECR reactors combine microwave power with a static magnetic field to force electrons to circulate around the magnetic field lines at an angular frequency equal to the microwave frequency - a condition called cyclotron resonance. This resonance considerably increases the probability for electron-neutral collisions, allowing the generation of very dense plasmas even at pressures as low as $10^{-4}$ Torr [80, 81].

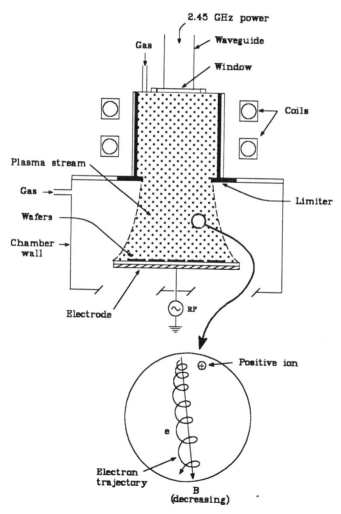

**Fig. 5.32** Schematic of an ECR plasma etcher [80].

For a magnetic flux **B** the angular frequency, $\omega_e$, is

$$\omega_e = \frac{e\mathbf{B}}{m_e} \, , \qquad\qquad 5.10$$

where $e$ is the electronic charge, and $m_e$ the electron mass. The microwave frequency is typically 2.45 GHz, an allowed industrial frequency. From Eq. 5.10 it follows that resonance between the giratory frequency of electrons and incident microwave frequency can be achieved when the magnetic field intensity is **B** = 875 Gauss. The ECR source is mounted above the processing chamber (Fig. 5.32) [82]. Microwave power is coupled via a wave guide through a ceramic window into the ECR source region. Magnetic coils provide a magnetic field **B** that starts at a magnitude of $\simeq$ 1 kGauss and decreases gradually as the distance from the coils increases, following a gradient. At some distance from the coils, the magnetic field drops to 875 Gauss and resonance is achieved. The species diffuse out of the resonant region into the lower chamber where the wafer is located. The wafer can be RF or DC-biased to control the energy of impinging ions.

Important reactor configurations discussed in this section are summarized schematically in Fig. 5.33.

### Loading and Microloading Effects

In typical plasma reactors the etch rate is proportional to the concentration of reactive species which, on the average, is determined by the difference between the rates of generation and loss of species. In typical plasma reactors, the main loss mechanism of etchants is their consumption by reaction with the material being etched. Therefore, more reactive neutrals are depleted as the etchable surface area is increased. Since the the generation of reactive species is essentially independent of the amount of etchable material present, there is a net loss of reactive species which increases as more etchable substrate is exposed. The result is a decrease in the etch rate as the exposed surface area is increased, a phenomenon referred to as **loading effect** [35,53,83 − 85]. Assuming a simplified case of a single etchant (such as fluorine), the dependence of etch rate $R$ on etchable surface area $\Phi$ is found as [83]

$$R = \frac{\beta \tau G}{1 + K\beta\tau\Phi} \ , \qquad\qquad 5.11$$

where $\beta$ is a reaction rate constant, $\tau$ is the lifetime of active species in the absence of etchable material, $G$ is the generation rate of reactive species, and K a constant for a given material and reactor geometry.

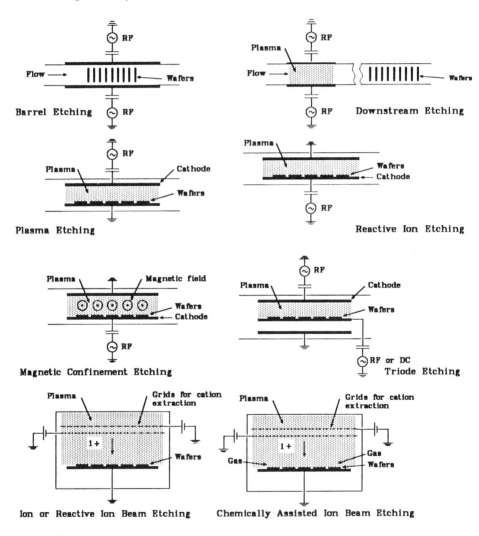

**Fig. 5.33** Schematics of important reactor configurations used in plasma etching [33] (Courtesy: Integrated Circuit Engineering Corp., [101]).

The main point in Eq. 5.11 is that there is negligible loading as long as $K\beta\tau\Phi << 1$, i.e. when the exposed area ($\Phi$) is very small and/or the lifetime ($\tau$) of active species in the absence of etchable material is small. Therefore, loading effects can be reduced by employing plasmas in which the predominant etchant loss mechanism is insensitive to the consumption by reaction. Loading effects are also minimized when the etch rate is controlled by the flux of ion bombardment rather than the etchant supply [86].

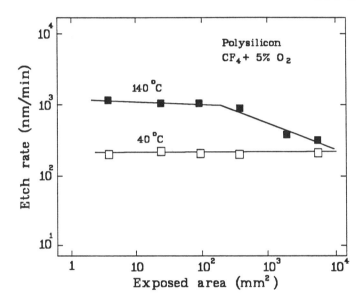

**Fig. 5.34** Etch rate of polysilicon in $CF_4/5\%$ $O_2$ plasma as a function of exposed area at two different temperatures. The loading effect is observed at 140 °C but not at 40 °C [87].

The chemical reaction between neutrals and etchable surface increases exponentially with temperature, while the generation of reactive species is mainly limited by electron collisions and essentially insensitive to wafer temperature. Since ion-induced etching is less temperature dependent than neutral etching, wafer temperature during RIE can influence both loading and anisotropy by changing the rate of etching by neutral reactions [32]. Fig. 5.34 illustrates how temperature can influence the loading effect. The etch rate of polysilicon typically decreases as the polysilicon area exposed to a $CF_2/O_2$ plasma is increased [87]. While a strong loading effect is observed at 140 °C, the etch rate at 40 °C appears to be essentially constant within the indicated range of areas. This can be attributed to the Arrhenius temperature dependence of the rate constant $\beta$ in Eq. 5.8. The loading effect would be observed at 40 °C if still a larger area were exposed.

Not all isotropic dry etchants exhibit the loading effect. Silicon etching in fluorine, for example, shows a strong dependence on etchable surface area because the chemical reaction dominates the loss of fluorine. In most chlorine or bromine containing plasmas, however, the etchant concentration tends to be limited by atom-atom recombination so that the etch rate is rather insensitive to load-size [88]. The ratio of etch rate in an empty reactor (no wafers) to etch rate with $m$ wafers is shown in Fig. 5.35 for three different chemistries. The loading effect is pronounced for a fluorine-based plasma but not for a chlorine- or bromine-based plasma.

The most serious concern caused by the loading effect is the loss of etch control when nearing the the end point. Ideally, as the termination of an etch process is approached, the etch rate should decrease to allow stopping the process at the correct time and minimize overetching. With the loading effect, however, less etchable material is exposed near the end point and the etch rate increases rapidly, so that overetching is carried out at a higher rate than nominal. This makes linewidth control extremely difficult, since accelerated etching occurs on clearing.

One useful application of the loading effect is the ability to signal an increase in etch rate when clearing of the etched film is approached. The effect is sometimes used for end-point detection.

A microscopic loading effect, in which the etch rate is influenced by the size and density of features, is referred to as **microloading.** This is the consequence of localized concentration gradients of etchant species, which are caused by differing rates of reaction with the patterned surface. For feature sizes below $\simeq 1$ $\mu$m and aspect ratios much greater than 1, etching rates are observed to depend on pattern density and aspect ratio. Such dependencies tend to increase the cost of manufacturing because even small changes in device design rules, cell design, or chip configuration can result in time consuming, new plasma process development [89]. The mechanisms underlying pattern and aspect ratio dependent etching are not well understood. The term "microloading" is commonly used to describe the dependence of etch rate on the exposed etchable area defined by the pattern. Microscopic effects have also been attributed to non-uniformities in magnetic field and field gradient, and ion-density and energy in MERIE and ECR etchers [90, 91].

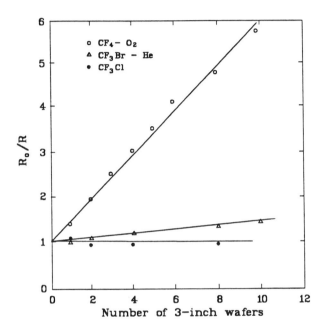

**Fig. 5.35** Etch-rate dependence on exposed surface area for three plasma gas mixtures. The ratio of etch rate with zero wafer loading ($R_o$) to etch rate with wafer loading ($R$) is plotted as a function of the number of loaded wafers. The loading effect is observed with fluorine but not with chlorine or bromine [83].

### Single Wafer Versus Batch Processing

As wafer size increases and critical dimensions decrease, more stringent requirements are imposed on etching. Because of tight process control requirements, it is desirable to have every wafer processed under identical conditions. Etch uniformity and reproducibility are superior when every wafer is exposed in the constant environment of a single wafer etcher rather than in batch processing where several wafers are etched in the same chamber. This permits in situ monitoring of each wafer's processing conditions. The choice between single wafer and batch processing is a trade-off between wafer size, cost, and etch control and uniformity. The drawback with single wafer processing has been reduced throughput. To attain the same throughput as in batch reactors, the etch rate in single wafer processing must be increased. This requires a more intense plasma at higher power density, causing more damage from ion bombardment and sputtering. As the wafer size increases, however, the effective electrode area in batch processing occupied by wafers decreases, increasing the cost of batch processing. Fig. 5.36 compares the cost of single wafer and batch processing as a function of wafer size. A "cross-over" point is observed at a wafer diameter of about 200 mm.

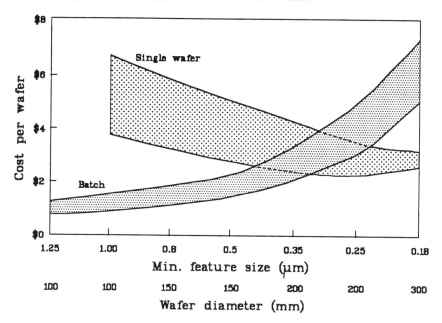

**Fig. 5.36** Cost comparison of single wafer to batch processing as a function of wafer size. A "cross-over" is observed at a wafer diameter of about 200 mm. (Courtesy: L. Larson, SEMATECH).

## Applications

Dry etching has rapidly evolved from a simple resist stripping technique to an indispensable tool for the definition of submicron structures, replacing wet chemical etching for most of the materials used in semiconductor manufacturing. By proper choice of the gas mixture and reactor conditions, specific reactions can be promoted at the wafer surface. This section describes important isotropic and directional dry etch processes and their applications for shaping horizontal and geometries.

### Silicon Etching

The main objective of directional plasma etching of silicon is to form trenches in the silicon substrate for device isolation or for cell capacitor structures in DRAM designs (Fig. 5.37). The aspect ratio of these trenches typically ranges from 1:1 to 15:1, depending on their application.

Several gas combinations are used in RIE reactors to produce vertical trench sidewalls in silicon. While fluorine-based plasmas exhibit high silicon etch rates and rather isotropic etch profiles (barreling), chlorine-based plasmas result in moderate etch rates but improved control of the trench profile. In both cases, grown or deposited $SiO_2$ is typically used as a mask because of the high silicon to oxide selectivity (> 15:1). Ion bombardment plays an important role in obtaining high vertical etch rates, while polymer coating or oxide passivation of the trench sidewall contributes to sidewall angle control [49].

### Fluorine-Based Plasmas

Etching of silicon in a glow discharge containing F atoms is spontaneous. Therefore, anisotropy is not possible without sidewall passivation [86]. Since etching is reaction-limited, a fluorine-based plasma etch exhibits a strong loading effect. Typical gas mixtures used are $CF_4/O_2$ and $SF_6/O_2$; the volatile by-product is $SiF_4$. The presence of oxygen enhances the silicon etch rate and improves selectivity. For example, the etch rate for undoped single crystal silicon at 100 °C is about 30 nm/min for $CF_4$ and 300-450 nm/min for $CF_4/O_2$ [35,67,83]. Fluorine atoms etch undoped (100) Si at a rate given by [92]

$$R = 2.91x10^{11}T^{1/2}n_F\,e^{-0.108/kT}\ (nm/\,\text{min})\,, \qquad 5.12$$

where $n_F$ is the concentration of F atoms (cm$^{-3}$), $T$ the temperature (K), and $k$ Boltzmann constant ($8.62x10^{-5}$ eV/K). Other fluorine-based gas mixtures, such as $SiF_4/O_2$, $NF_3$ and $ClF_3$, have been successfully used in etching silicon [86].

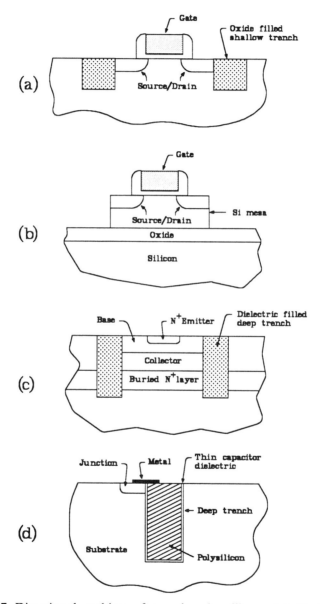

**Fig. 5.37** Directional etching of trenches in silicon. (a) Shallow-trench MOSFET isolation. (b) Isolated silicon-on-insulator (SOI) mesas. (c) Deep-trench isolation in bipolar designs. (d) Deep-trench capacitor for a DRAM cell.

**Chlorine- and Bromine-Based Plasmas**

Chlorine- and bromine-containing plasmas can be used to etch silicon. While the two halogens have similar etching characteristics, chlorine is more widely used because it has a higher vapor pressure and is less corrosive [86]. Unlike fluorine, Cl and $Cl_2$ etch undoped silicon very slowly and require ion-bombardment to achieve a high etch rate [93]. Since etching is dominated by ion bombardment, a high degree of anisotropy can be controlled and only a small loading effect is observed. The most commonly used gases are $Cl_2$, $CCl_4/O_2$, and $SiCl_4/O_2$ with a high etch rate (50-650 nm/min), and high $Si:SiO_2$ selectivity (10-50:1) [86]. The volatile by-product of the reaction with chlorine is $SiCl_4$.

**The Doping Effect**

The etch rate of silicon in a halogen plasma is affected by electrically active dopants, a phenomenon called doping effect. Without sidewall passivation, etching through silicon layers of varying dopant types and concentrations results in different lateral etching of these layers, making the control of etched profiles rather difficult [49]. In a fluorine-based plasma, etching of p-type silicon proceeds at a slightly slower rate than in intrinsic silicon [94], while etching is enhanced in highly doped n-type regions ( $> 10^{19} \, cm^{-3}$) [95]. The doping effect is much more pronounced in chlorine-based plasmas. Etching of heavily doped n-type silicon (or polysilicon) films in Cl proceeds 15-25 times faster than undoped substrates [96,97]. This enhancement is related to the concentration of active n-type carriers rather than the chemical "identity" of the dopant [86]. Electrically inactive dopants have a negligible influence on etching [86,98].

**Polysilicon Etching**

Defining the polysilicon gate is one of the most critical etching steps in the manufacture of MOSFETs. The etch process must be repeatable, uniform, with excellent dimensional and edge-slope control. A very high selectivity to silicon dioxide is required because etching must stop on the thin gate dielectric without causing damage to the underlying silicon surface (Fig. 5.38).

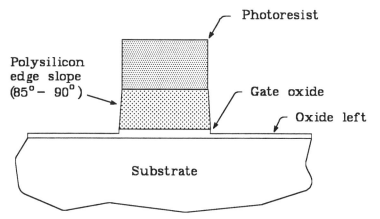

**Fig. 5.38** Definition of the polysilicon gate in MOSFETs. The control of linewidth and edge-slope, and a high selectivity to the underlying gate oxide are critical etch process parameters.

As with silicon etching, there is a trade-off between selectivity and directionality with a fluorine-based plasma. Good anisotropy and selectivity over $SiO_2$ is obtained with chlorine. Also, in the presence of photoresist masking, chlorine-based plasmas tend to deposit a passivating polymer on the polysilicon sidewall, enhancing anisotropy. Enhanced directional etching of polysilicon is also obtained with $SF_6/O_2$, because an etch-inhibiting oxide film is deposited on the sidewall during the process [49]; an etch rate of about 400-680 nm/min with a 20:1 selectivity over $SiO_2$ can be achieved.

**Photoresist**

The first application of plasma etching was "plasma ashing". This is an isotropic etch process of organic photoresists in an oxygen glow discharge where atomic oxygen is produced following the reaction

$$e + O_2 \rightarrow 2O + e \rightarrow O^* + O^* + e.$$

Oxygen atoms react with organic material to form the volatile products CO, $CO_2$, and $H_2O$ [99]. Since most of the underlying film materials are not appreciably attacked by an oxygen plasma, overetching to ensure that all resist residues are removed presents no problem.

A barrel reactor is commonly used for ashing. Typically, a batch of wafers coated with 1 $\mu$m resist can be processed in 5-10 min. The reactor is also used for "descumming" of wafers, i.e. removal of residual layers of photoresist or other organic material following resist development and hard bake. These residues are typically removed in a 1-min exposure to an oxygen plasma. Plasma ashing or descumming can cause device damage due to bombardment by ions (not as intensive as with RIE) in the form of radiation-induced traps in insulators and semiconductor/insulator interfaces. The plasma can even cause electrostatic breakdown of thin gate dielectrics due to charge build-up. Downstream plasma strippers are increasingly used to circumvent these problems.

While plasma ashing has been introduced primarily to reduce cost and hazards of wet chemical waste, it has become an important etching step to remove photoresist that has become insoluble in wet solvents. This occurs when the resist is exposed to fluorine or chlorine dry-etch environments (e.g. during polysilicon or oxide etching) or to high-current ion implantation (Chap. 6). After removing the "hardened" top surface layer of the resist, plasma etching or wet chemical etching can be used to remove the remainder of the resist. Plasma ashing is typically followed by a cleaning step to remove ions and heavy metals that are not volatilized by the oxygen plasma process.

Resist ashing is also used to define polysilicon linewidths smaller than what can be achieved with available lithographic tools. For example, with an oxygen plasma, an effective MOSFET channel length of 0.15 $\mu$m can be achieved in a conventional g-line (436 nm) lithography [100]. Gates with drawn length of 0.7 $\mu$m are initially defined with conventional lithographic methods. Plasma etching is used to partially ash the photoresist (Fig. 5.39). Polysilicon is then defined by directional etching in a $CCl_4$ plasma using the ashed photoresist as a mask. A final polysilicon linewidth of about 0.2 $\mu$m can be obtained with good control.

322

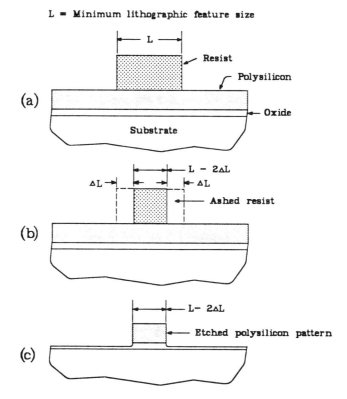

L = Minimum lithographic feature size

(a)

Resist

Polysilicon

Oxide

Substrate

(b)

L − 2ΔL

ΔL

ΔL

Ashed resist

(c)

L − 2ΔL

Etched polysilicon pattern

**Fig. 5.39** Definition of the polysilicon linewidth below the lithographic capability.

Directional, ion-assisted, resist etching is also in wide use in multilevel resist processing. In a typical "trilevel" process, a thick ( $> 1\mu m$) conventional resist is deposited and covered by a very thin (10-20 nm) $SiO_2$ film (Chap. 4). The resist and oxide layers are then coated with a thin conventional photoresist. The pattern is first defined on the top thin resist (of uniform thickness) and then transferred to the $SiO_2$, using the top resist as a mask to etch the oxide (Fig. 5.40). The delineated oxide film is then used as a stencil to etch high aspect ratio images anisotropically in the thick conventional resist or polyimide with an oxygen plasma. Ion bombardment enhances the etch rate of the polymer in the vertical direction. The patterned thick resist layer becomes the final mask for further plasma etching.

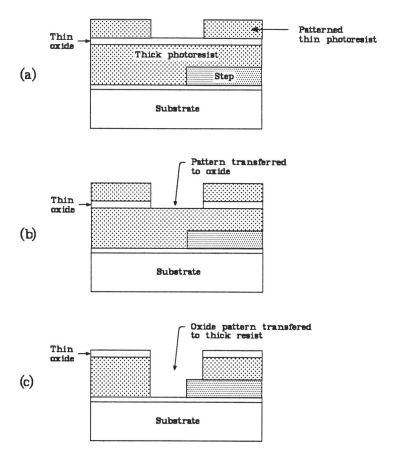

**Fig. 5.40** Etch processes in trilevel resist.

Isotropic etching of resist can be used to level a top insulator surface upon which metal is deposited and delineated (Fig. 5.41). The resulting "planarized" surface is critical to the definition of submicron metal patterns (Chap. 8). The process begins by depositing a dielectric film over the patterned polysilicon layer. This film may consist of a "reflowed" PSG or BPSG upon which an $SiO_2$ layer is deposited (Chap. 3). The high-temperature reflow produces rounded contours of the dielectric film over sharp edges. The film is then coated with a photoresist layer which produces a nearly planar top surface. The films are isotropically etched under conditions that remove the resist and and dielectric at the same rate. This can be achieved by adding oxygen to a fluorine-based plasma such as $CF_4$, or by using a strong oxidizing agent such as

**324**

NF₃ diluted with helium or argon. A leveled ("planarized") dielectric surface is obtained after removal of the resist, allowing the definition of very small contacts with high aspect ratio and metal linewidths and spaces of minimum feature sizes. There is typically no etch stop to this process. Therefore, etching must be performed by proper timing or by in situ monitoring the film thickness, e.g. with a laser interferometer.

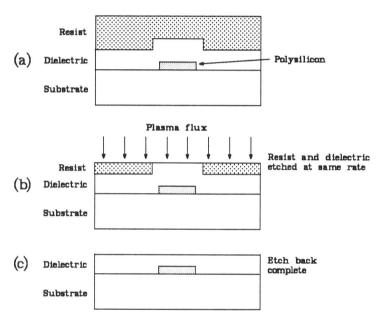

Fig. 5.41 Plasma etch for leveling the top dielectric film prior to contact etch and metal deposition and definition [101].

**Silicon Dioxide Etching**

Silicon dioxide can be etched in plasmas that produce fluorine atoms or by using gas mixtures that generate unsaturated-rich fluorocarbon species [86]. Isotropic etching of $SiO_2$ is also achieved in a hydrogen plasma [102], in anhydrous HF [103], or $HF/H_2O$ vapor [104].

Several processing steps in the manufacture of silicon structures require etching $SiO_2$, with a high selectivity over doped and undoped silicon or polysilicon. Dry etching of high aspect-ratio contact openings in SiO₂ over polysilicon and silicon surfaces is illustrated in Fig. 5.42. For leveled top surfaces the oxide height over polysilicon is smaller than over silicon and a very high selectivity is required when both contacts are etched simultaneously. Otherwise, the polysilicon film would be partially removed as etching proceeds toward the single-crystal silicon surface.

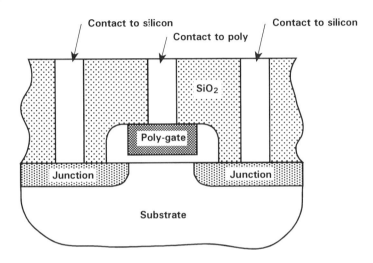

**Fig. 5.42** Directional etching of contact openings in *SiO₂* over polysilicon and silicon surfaces. A high etch selectivity is required when both contacts are etched simultaneously.

High selectivity is also required when etching oxide "spacers" in MOSFETs (Fig. 5.43). Spacers are necessary to keep silicided junctions at a safe lateral distance from the polysilicon gate and source/drain junction edges, and to achieve the correct drain profile (Chaps. 6,7). The spacers are formed by depositing an oxide film conformably over patterned polysilicon with vertical sidewalls and then etching the oxide in RIE. The oxide must be removed from horizontal surfaces without appreciably attacking silicon and polysilicon surfaces.

Since fluorine atoms etch silicon faster than silicon dioxide, special fluorine-deficient fluorocarbon gas mixtures must be used to pattern oxide on silicon or polysilicon surfaces. The etch rates and selectivities for Si and SiO₂ can be controlled by adding oxygen or hydrogen to CF₄. While adding O₂ to a CF₄ gas increases the etch rate of both Si and $SiO_2$ significantly, the dependence on the $O_2$ concentration is different for both materials (Fig. 5.44). From Fig. 5.44 it can be seen that the etch rate of silicon increases until a maximum is reached at 12% $O_2$, while the etch rate maximum of $SiO_2$ occurs at about 20% $O_2$ [105]. The increase in etch rate is attributed to the reaction of oxygen with carbon, releasing additional F atoms for etching.

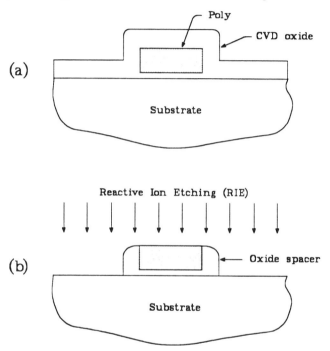

**Fig. 5.43** Formation of oxide spacers in MOSFETs. (a) Conformal deposition of an oxide film over patterned polysilicon with vertical sidewalls. (b) Directional etching of oxide to form spacers. Etching must be completed without appreciable attacking silicon or polysilicon.

At higher $O_2$ concentrations, the etch rate decreases more rapidly for Si than for $SiO_2$. This allows more selective etching of $SiO_2$ over silicon surfaces. By adding $H_2$ to the gas mixture, the etch rate of silicon decreases monotonically until it reaches almost zero for a hydrogen concentration $\geq$ 40% [106, 107]. In contrast, the RIE etch rate of $SiO_2$ remains unaffected at hydrogen concentrations below 40%. This different behavior allows an increase in the oxide to silicon selectivity ($\geq$ 40:1), at an oxide etch rate of about 500 nm/min. The increased selectivity is attributed to polymer formation on silicon but not on $SiO_2$. This is explained by the reaction of carbon from $CF_x$ radicals with oxygen from $SiO_2$ to form volatile CO and $CO_2$, allowing continued etching of $SiO_2$, while polymer deposition on silicon surfaces inhibit their etching (i.e., blocks access to silicon for F) [107].

Another gas mixture used for etching $SiO_2$ is $CHF_3/O_2$. When used in a triode reactor, an oxide to silicon etch-rate ratio of 30:1 can be achieved.

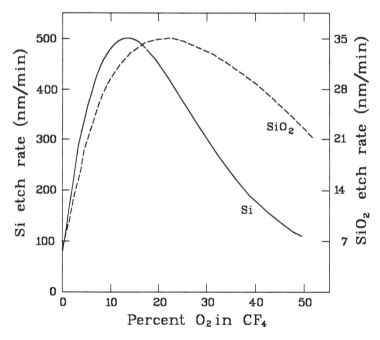

**Fig. 5.44** Dependence of silicon and silicon dioxide etch rates on the concentration of $O_2$ in a $CF_4/O_2$ gas mixture [105].

Selective etching of native oxide is required just before critical steps, such as gate oxide growth or deposition for MOSFETs, emitter polysilicon deposition for bipolar structures, low temperature epitaxial growth, or formation of contact metallurgy (Chap. 8). One method to remove the thin native oxide is to use a low-energy hydrogen plasma [102, 108]. To minimize radiation effects and ion bombardment, a downstream microwave reactor is used which is configured to allow only atomic hydrogen to reach the wafer surface. An independent local discharge is produced immediately above the wafer to control a low flux of ion bombardment. While the flux of neutrals alone easily removes hydrocarbon contaminants, low-energy ion bombardment is required to etch the native oxide. By in-situ monitoring the etch/clean process, optimized conditions can be defined to remove the native oxide with negligible damage to the wafer surface [102]. Native oxide can also be removed in anhydrous HF gas at room temperature and normal pressure, or in HF/H₂O vapor, both without plasma. Selective etching to thermal and CVD oxide is achieved by strictly controlling the concentration of HF in an ultraclean nitrogen carrier gas, and the moisture content in the etch chamber [103]. Selectivity is attributed to the different "critical" HF concentrations needed to etch native, thermal, and CVD oxide. Native oxide etches at a much lower critical HF concentration than thermal and CVD oxides. The proposed etch mechanism is that HF molecules react with water traces of water adsorbed at the oxide surface, resulting in $HF_2^-$:

$$2HF + H_2O \rightarrow H_3O^+ + HF_2^-.$$

The $HF_2^-$ ions reacts with the oxide producing water, enhancing the reaction - a positive feedback:

$$SiO_2 + 2H_3O^+ + 2HF_2^- \rightarrow SiF_4 + 4H_2O$$

As with plasma etch, the by-product is $SiF_4$ which can be monitored by IR transmittance to detect the etch start and end points. The fluorine termination of the bare silicon surface has a detrimental effect on subsequently grown or deposited oxides; it is removed by exposure to irradiation from a Xe-lamp in ultra-high vacuum [103].

### Silicon Nitride Etch

The replacement of hot phosphoric acid by dry etch processes to remove nitride is one of the earliest applications of plasma etching. Wet etching of nitride in hot phosphoric acid has several serious drawbacks. Since photoresist does not adhere properly during the process, a silicon dioxide film must be first deposited and patterned on the nitride. The oxide pattern then serves as a stencil to wet etch the nitride, a complicated and expensive isotropic process which also presents a health hazard of working with a hot acid. These problems are circumvented by using plasma etching. A barrel or planar reactor can be used to plasma etch nitride. Typical etch gas mixtures are $CF_4/O_2$, $SF_4/O_2$, and $C_2F_6/O_2$. An etch rate of 30-50 nm/min can be achieved, depending on the method the nitride was deposited. For example, the etch rate is higher for PECVD than LPCVD nitride. Unfortunately, the nitride cannot be etched selectively over oxide because the gas mixtures used to etch nitride attack the oxide at a higher rate.

**Aluminum Etching**

Aluminum has remained the interconnect metal of choice because of its low resistivity ($\simeq 3\mu$ Ohm-cm at 20 °C), and its compatibility with other semiconductor materials and processing despite its low melting point (660 °C). Aluminum exhibits a high etch selectivity to $SiO_2$ and acts as an etch-stop for contacts in $SiO_2$.

The reaction of aluminum with fluorine results in involatile $AlF_3$ which has a vapor pressure of only 1 Torr at 1240 °C. Hence, the vapor pressure is extremely low at an ambient temperature and the etch rate in a fluorine-based plasma is seriously limited. This is why aluminum is typically etched in chlorine-containing plasmas where the principal etch product is $AlCl_3$. Since the chloride has a vapor pressure of just under 100 Torr at 150 °C (sublimation point 176 °C), it can be easily removed from the plasma chamber [86, 109, 110]. Exposure of aluminum to chlorine molecules at room temperature results in little reaction. In the presence of a plasma, however, active species are created that will readily attack aluminum. With $CCl_4$, the plasma reactions are believed to be of the type:

$$e + CCl_4 \rightarrow CCl_3 + Cl + e \rightarrow CCl_2 + Cl_2 + e \rightarrow CCl_2 + 2Cl + e$$

Before etching of aluminum begins, the thin native oxide, that forms almost instantly on the aluminum surface upon exposure to air, must be removed since it inhibits the reaction with Cl or $Cl_2$. While chlorine atoms etch aluminum spontaneously, they do not attack the $Al_2O_3$ native oxide unless subjected to ion bombardment. A $CCl_4$ plasma with ion bombardment provides sufficient Cl to etch aluminum and its native oxide. In the presence of moisture or $O_2$ in the chamber, however, aluminum oxide is formed as fast as it is removed. Control of water vapor and oxygen content is therefore critical to the etching process. This can be achieved, for example, with single-wafer processing in vacuum loadlocks to isolate the wafer from atmospheric contaminants and minimize the introduction of wafer and oxygen during handling. Since $BCl_3$ is very effective in gettering oxygen and water, a plasma based on a $BCl_3/Cl_2$ mixture has become the standard for etching aluminum. With this gas mixture, an etch rate ratio of about 20:1 can be obtained over $SiO_2$ and 4:1 over photoresist.

Directional etching of aluminum in pure $Cl_2$ is not possible without adding polymerizing agents, such as $CHF_3$, to passivate aluminum sidewalls and prevent undercutting [109]. Sidewall passivation occurs by forming polymers that inhibit aluminum etching on surfaces that are not subjected to ion bombardment [110,111].

In VLSI/ULSI technologies, aluminum is typically alloyed with copper and silicon. Alloying with $\simeq 1\%$ Si is necessary to minimize spiking at contact interfaces due to the solubility of Si in Al. Copper is added at a concentration which ranges from 0.4% for CMOS to 4% for bipolar structures, to reduce electromigration at high current densities (Chap. 8). Adding silicon to aluminum presents no problem to etching since $SiCl_4$ is a volatile etch product at room temperature. Etching of the Cu alloy, however, is more difficult because $CuCl_2$ and $CuCl$ are not volatile at temperatures below 175 °C [44]. Elevated temperature and high-energy bombardment are therefore required to remove the copper.

Another problem in etching modern metal interconnects is the presence of barrier layers and anti-reflective coats (ARC, Fig. 5.45). The barrier and ARC typically consists of a 10-20-nm TiN or TiW film. The lower barrier is needed to prevent spiking (Chap. 8), and the ARC is required to avoid standing waves during exposure that cause necking and notching (Chap. 4). The presence of barriers and ARC complicate the task of etching. To remove Ti-W barrier or ARC films, the gas mixture must be changed to $CCl_4/O_2$, $SF_6$ in Ar or $N_2$, or $SF_6/Cl_2$. TiN films can be removed in $CCl_4/N_2$, $Cl_2/Ar$, or $BCl_3/Cl_2/CHF_3$ [109]. The $SF_6/Cl_2$ gas mixture is also used to "etch-back" tungsten and remove the underlying barrier (Fig. 5.46).

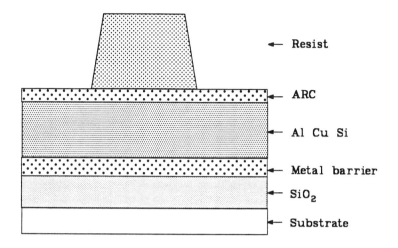

**Fig. 5.45** Composite metallization for VLSI/ULSI circuits [109].

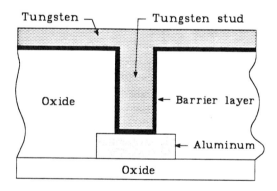

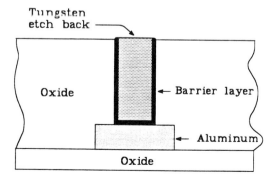

**Fig. 5.46** Tungsten etch-back and barrier removal over "studs".

### A Note on Selectivity

As mentioned earlier, selectivity is a critical etch parameter which refers to the relative etch rate of the film to be etched to the etch-rate of another material exposed to the etchant. Since etching by ion bombardment is rather insensitive to the composition of the material being removed, sputtering exhibits poor selectivity between materials; selectivity in sputtering is related to the difference in sputter yield between materials and decreases with increasing ion energy. In contrast, chemical etch rates vary significantly with the material composition; therefore, chemical etching reactions have the potential of being very selective [32]. In wet etching, the chemical selectivity for the etching of the film with respect to the underlying substrate or the mask material can be infinitely high. In terms of selectivity, ion-enhanced etching processes typically fall somewhere between sputter etching and chemical etching processes.

There are two types of selectivity, one with respect to the masking material and the other with respect to an underlying film. High selectivity with respect to both layers is needed to produce the required pattern resolution with minimal erosion of the mask and attack of the underlying film and other exposed materials.

Typical etch rates are proportional to $e^{-E_a/kT}$, where $E_a$ is the activation energy which ranges from 0.1-2.0 eV. Therefore, changing the wafer temperature can strongly affect selectivity between materials since different materials have different activation energies and hence different etch-rate dependencies upon temperature.

### Endpoint Detection

Because of the limited selectivity associated with plasma etching, endpoint detection is necessary to minimize overetch times. Several methods are available for endpoint detection and plasma monitoring. Some methods measure changes in the gas-phase reactions while others directly monitor the film being etched.

The presence of a reaction product or the absence of a reacting species can be monitored and used to determine when film etching is complete. Detection in these cases is either optical or by mass spectroscopy. The most widely used endpoint detectors are optical emission spectrometers which use a narrow bandpass filter and a photodiode to continuously monitor the intensity of wavelengths associated with the reaction product or reactive species. The point where the intensity of certain spectra reaches or falls below a preset level can be determined, indicating the end of the process. Optical emission spectroscopy can be easily implemented and gives information on both the etching species and etch products. For example, since silicon dioxide and photoresist removal both produce substantial amounts of CO, one of the CO emission bands (184 nm or 297.7 nm) is used to monitor the process. The F concentration can be monitored during etching of silicon or tungsten in a fluorine contained plasma by measuring the emission intensity at 703.7 nm. A depression in the fluorine concentration indicates the endpoint. Etching of aluminum in $CCl_4$ produces AlCl which has a characteristic wavelength of 261.4 nm and is widely used for monitoring the etch process. For Al, Ti, and W, the spectral region around 400 nm is particularly strong emission at 396 nm (Al), 400 nm (Ti), and 401 nm (W) can be used to monitor the progress of etching the metals [109].

Mass spectroscopy requires rapid sampling of the exit gas stream and a high vacuum ($< 10^{-5}$ Torr) which is provided by a separate vacuum chamber and pump.

A laser interferometer typically uses a helium-neon laser beam (632.8 nm) to detect changes in the reflected or refracted light from the film being etched. If the film is opaque, (such as aluminum) the disappearance of the film is detected by a sharp change in the reflected laser light intensity. Polysilicon, silicon dioxide, and silicon nitride have a wavelength region where they are transparent. In this case, an interference pattern can be obtained which is characteristic of the wavelength of the monochromatic light and the film thickness. Interference maxima and minima occur at normal incidence of the laser beam when the following condition is satisfied:

$$2\Delta d = \frac{\lambda}{n} , \qquad\qquad 5.13$$

where $\Delta d$ is the change in film thickness between adjacent maxima or minima, $n$ the film index of refraction, and $\lambda$ the wavelength. A photodiode is used to record the interferometry signal. Laser interferometry is particularly effective when the refractive index of the film being etched is distinctly different from that of the underlying material, e.g., $SiO_2$ on Si. The method is not suitable, for example to monitor trench etching in a silicon substrate. Another limitation is related to the small spot size needed ($\simeq 10^{-5}$ cm in diameter) to obtain a distinct interferogram, introducing a sampling error (particularly in batch processing).

A novel method to monitor etch uniformity over the entire wafer is described in [112]. This technique uses a charge-coupled device (CCD) camera to measure the optical emission interferometry across the wafer, allowing for analysis at one point, along a line, over a chip, or over an entire wafer (Fig. 5.47). This method can also be used to determine when endpoint is reached at a number of locations across the wafer, instead of observing an average endpoint time, as is done with optical emission and a photodiode.

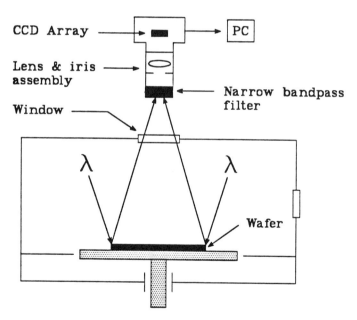

CCD Array

PC

Lens & iris assembly

Window

Narrow bandpass filter

Wafer

Fig. 5.47 CCD system used to measure the etch rate uniformity in situ. Plasma optical emission is gathered from across the entire wafer, using a bandpass filter. A CCD camera is used as a detector so that the etching rate can be monitored at any point on the wafer (Courtesy: T. J. Dalton, W. T. Conner, and H. H. Sawin, MIT [112]).

## Plasma Damage

Plasma damage can be caused by direct irradiation with high-energy ions, electrons, and photons, or indirectly by charging conductive layers exposed to the plasma. In most cases, the damage can be reversed by annealing at elevated temperature. When the accumulated surface charge causes excessive current to pass through an insulator, however, the damage to the dielectric can be permanent.

Plasma damage is particularly important when etching is performed after all hot processes are completed and the MOSFET thin gate dielectric is grown or deposited. In regions where the semiconductor is directly bombarded by ions, material modification of the substrate surface can occur. The damage is manifest as bond rupture and the generation of defects such as dislocations, vacancies, and interstitials. The damaged lattice region may be several tens of nm deep, which is comparable to the thickness of gate oxide and the depth of shallow junctions (Chap. 6). Plasma damage to the silicon surface can change the diffusion profile of implanted boron (Chap. 6) [113]. It can speed-up silicide growth [114], accelerate metal penetration (Chap. 8) [115], or cause defects in a epitaxial silicon [116]. Formation of plasma-induced defects in epitaxial layers is attributed to the surface and subsurface damage by ion bombardment and the RIE-induced contaminants on exposed silicon surfaces. Silicon roughening is of particular importance in a chlorine-based plasma where the formation of "black-silicon", i.e., a highly damaged, non-reflective surface can occur. This has serious implications with respect to the integrity and reliability of an oxide grown on the silicon surface. Thermal oxide grown on a silicon wafer which has been exposed to dry etching will typically exhibit reduced breakdown field. This is a consequence of increased surface roughness leading to enhanced fields in the grown oxide at silicon asperities. The damaged region can be removed by direct etching of the irradiated film, without a plasma, or by growing a "sacrificial" or "recovery" oxide on the damaged region and then etching the oxide without a plasma.

The gate insulator and oxide-silicon interface are the crucial elements of MOSFETs. When $SiO_2$ is subjected to ion bombardment, atomic displacement damage can occur, leading to trapping sites for electrons and holes. Ionic and photon bombardment can also generate neutral traps which can capture charges during device operation. In typical structures, however, the gate oxide is etched with the gate electrode present, and the electrode protects the underlying oxide from direct particle bombardment.

The primary cause of oxide damage during plasma etching is charge buildup on conductors [117]. Plasma non-uniformities across the wafer surface play a major role in this damage. These non-uniformities cause an imbalance in local electron and ion currents which results in charging exposed surfaces, damaging underlying insulators [118]. There are two important currents that affect charging, the ion current which is nearly constant in time and depends linearly on the plasma density, and the electron flux which flows briefly in every RF cycle to balance the ions lost from the central plasma region. Ideally, the ion and electron conduction currents locally balance each other over the RF cycle and charging is not a problem. In a non-uniform plasma, however, ions and electrons do not locally balance, although the system as a whole is neutral. When the plasma potential is minimum, the electron current can be larger than the ion current, and when the plasma potential is maximum, the ion current exceeds the electron current [117]. In this case, wafer charging occurs, resulting in increased voltage across the oxide and decreased voltage across the sheath. The charge build-up continues until the net current is zero or the oxide begins to conduct. The feedback mechanism is caused by the exponential dependence of electron current on sheath voltage.

Early in the process, the gate electrode completely covers the top, sides and bottom of the wafer and is in direct contact with the electrode. Therefore, surface currents balance the local non-uniformities and no charge builds up on the gate. As the end-point is approached, however, the conductive gate material becomes discontinuous and surface current paths are interrupted. Isolated gate electrode islands begin to form and accumulation is observed. The charging rate is determined by the net local current and the gate area in the exposed islands. The increase in oxide field gives rise to Fowler-Nordheim tunneling conduction, typically at weak points in the oxide. In this case, tunneling is the precursor to irreversible damage. When the endpoint is approached, only a small area at the gate edge is exposed to the plasma and charging decreases rapidly.

From the above discussion, it follows that charging is most serious when islands of conducting polysilicon begin to form. Damage of the oxide may therefore scale with the polysilicon gate perimeter rather than with its area when polysilicon is defined [119]. Since plasma ashing is performed after the polysilicon islands are formed, oxide damage caused by ashing tends to increase with gate area rather than gate perimeter [120]. Plasma damage can be avoided by connecting the polysilicon film to junctions in the regions where the islands are formed (this requires one additional masking and implantation step). In this case, the junction current is adequate to dissipate the charge and no oxide damage occurs. Plasma damage caused during etching of contacts and metals, oxide leveling, or resist ashing can be serious and must be considered when defining a process sequence and designing a circuit.

# PROBLEMS

**5.1** The defect-limited yield can be estimated from

$$Y = \frac{1 - e^{-AD_o}}{AD_o} \, ,$$

where $A$ is the "vulnerable area" and $D_o$ the "fatal" defect density per unit area. Assume that a 256 Mbit DRAM chip having a total area of 3.25 $cm^2$ contains $3.5x10^8$ thin oxide gates, each of area 0.25 x 0.5 $\mu m^2$, and find the "fatal" defect density in $cm^{-2}$ that will result in a gate-oxide yield of 99%.

**5.2** Consider a spherical particle of density 2.4 g/cm³ and radius R resting on a flat wafer surface. While the time-average electric moment of the particle is zero, the motion of electrons gives rise to an instantaneous dipole moment $p_1$ on the particle which produces an electric field and induces a dipole moment $p_2$ on the wafer. This creates an attractive force between the particle and the wafer, referred to as the van der Waals force, that can be approximated as [121]

$$F = \frac{hR}{8\pi z^2} \, ,$$

where $h$ represents an energy related to the particle and the surface on which it rests and $z$ is the adhesion distance. Assume $h = 8$ eV and $z = 0.4$ nm. The wafer is turned upside down. At what value of $R$ does the particle begin to fall off the wafer?

**5.3** A doubly ionized, positively charged spherical particle of radius $R$ approaches the surface of a heavily doped semiconductor.

(a) Assume the semiconductor surface to be of infinite dimensions and sketch the distribution of the field emanating from the charged particle and terminating on the semiconductor surface.

(b) Find the force of attraction exerted on the particle immediately before it touches the semiconductor surface for $R = 0.1\mu m$ and $R = 0.5\mu m$.

**5.4** A 0.5 $\mu m$ silicon dioxide layer grown on a flat silicon surface is covered with photoresist in which a 2 $\mu m$ wide window is defined. The wafer is subjected to an isotropic etchant that removes silicon dioxide at a rate of 1.2 nm/s without attacking the resist nor the underlying silicon substrate. Draw the etch profile after 5 minutes, 14 minutes, and 28 minutes of isotropic etching.

**5.5** A multilayer structure consists of 0.1 $\mu m$ thermally grown oxide on silicon, a 1 $\mu m$ PSG layer, and a 0.2 $\mu m$ CVD oxide film coated with photoresist. A 1 $\mu m$ x 1 $\mu m$ contact window is defined in the photoresist and the wafer is subjected to a BHF etch. The etch rates of thermal oxide, CVD oxide and PSG are 1.2 nm/s, 2.5 nm/s, and 10 nm/s, respectively.

a) What is the minimum etch time required to open contacts to silicon?

b) Draw the two-dimensional etch profile when etching is just completed.

c) At what minimum distance should another identical contact be placed from the first, if the material left between the two contacts must have a minimum thickness of 0.5 $\mu m$?

d) Assume the tolerance on film thickness to be $\pm$ 10% for all oxide film types and find the minimum etch time required to ensure that the contacts are open. What should the minimum contact-to-contact space be in this case?

**5.6** A < 100 > oriented silicon crystal is etched in a KOH solution through a 2 $\mu m$ x 2 $\mu m$ window defined in silicon dioxide. The etch rate normal to (100) planes is 67 nm/s. The etch rate ratios are 400:60:1 for the (100):(110):(111) planes. Show the etched profile after 15 seconds, 30 seconds, and 60 seconds.

**5.7** A 0.8 $\mu m$ thick aluminum film is patterned by subtractive isotropic wet etching. Find the undercut and etch bias for a 50% overetch, assuming that the patterning mask is not attacked by the etchant.

**5.8** An active silicon area is defined by a LOCOS boundary of thickness 0.5 $\mu m$. The LOCOS "bird's beak" is approximated by two planes intersecting at the silicon surface and making an angle of 60 ° with respect to the normal. A 25 nm thick sacrificial oxide is grown and then isotropically removed with 50% overetch. What is the extent of the bird's beak after etch?

**5.9** The etch rate of Si and $SiO_2$ in a fluorine plasma can be approximated as

$$\text{Etch Rate} = C n_F T^{0.5} e^{-E_A/kT},$$

where $C$ is a "pre-exponential" factor which is assumed constant, $n_F$ is the concentration of fluorine atoms in the gas, $E_A$ is the activation energy, $T$ is the absolute temperature, and $k$ is the Boltzmann constant ($k = 8.62x10^{-5}$ eV/K).

a) Plot the etch rates of Si and $SiO_2$ in nm/s, together with the Si to $SiO_2$ etch-rate ratio, as a function of $1/T$ in the range 250 K - 450 K for $n_F = 3x10^{15}$ $cm^{-3}$ and the values of $C$ and $E_A$ given in the table below.

| Film | $C$ $(cm^4 s^{-1} K^{-0.5})$ | $E_A$ (eV) |
|---|---|---|
| Si | $4.767x10^{-22}$ | 0.1075 |
| $SiO_2$ | $1.023x10^{-22}$ | 0.1631 |

b) A 10 $\mu m$ deep trench is etched vertically in silicon under the above plasma conditions, using a silicon dioxide mask (directionality is achieved with sidewall passivation). What is the minimum oxide-mask thickness required to ensure that the top silicon surface is not attacked at 298 K? 400 K?

**5.10** A 0.6 $\mu m$ aluminum film is deposited over a flat field oxide region and patterned with photoresist. The metal is then directionally etched at room temperature in a a $BCl_3/Cl_2$ plasma. Assuming a 50 % overetch, what is the minimum photoresist thickness required to ensure that the top metal surface is not attacked? What is the step height created in the filed oxide?

**5.11** A 0.2 $\mu$m polysilicon film is deposited over 10 nm thick gate oxide in active areas, and over 460 nm thick oxide in field areas. The LOCOS bird's beak, which defines the boundary between active and field areas, consists of two planes making an angle of 50 ° with respect to the normal and intersecting at 50 nm above the silicon surface. The polysilicon is then patterned in photoresist and the exposed film etched vertically at a rate of 400 nm/min. Under these conditions, the poly to SiO$_2$ etch rate ratio of 20:1.

a) Assuming a $\pm$ 20 % variation in the polysilicon film thickness, what is the etch time required to remove the polysilicon film completely in the areas exposed to RIE? How thick is the remaining oxide film in the active regions?

b) A 0.15 $\mu$m thick CVD oxide is deposited conformably over polysilicon and then etched directionally with 50% overetch to form spacers along the polysilicon sidewalls. Assuming an oxide to silicon/polysilicon selectivity of 40:1, find the extent of the bird's beak after spacer etch and draw the contour of the spacer showing its position with respect to the silicon surface and polysilicon boundary.

**5.12** Thermal oxide breaks down typically at a field intensity of $1.2x10^7$ V/cm. What is the maximum charge density that can accumulate on the oxide surface before breakdown occurs?

**5.13** A polysilicon film covers a total field oxide area of 400 $\mu m^2$ and a thin gate oxide area of 2 $\mu m^2$. The field and gate oxide thicknesses are 400 and 10 nm, respectively. The structure is subjected to a plasma which exhibits an imbalance in ion and electron fluxes to the polysilicon surface, causing the polysilicon to charge up. Assumptions:

a) The oxide breaks down at a field intensity of $1.2x10^7$ V/cm,

b) The substrate is grounded,

c) The voltage difference between poly and substrate drops fully across the oxide.

d) Oxide breakdown occurs before the fluxes equalize

Find the maximum voltage which the polysilicon film can reach and the polysilicon charge density over the thin and thick oxides.

**5.14** At what frequency does an argon ion circulate in an ECR plasma? In an ECR plasma, an electron circulates at a radius of 0.5 cm. How many revolutions does it make before colliding with a neutral at a gas pressure of 0.1 mTorr?

# References

[1] P. S. Burggraaf, "Wafer Cleaning: Brush and High-Pressure Scrubbers," Semiconductor International, 4 (7), 71 (1981).

[2] T. Hattori, "Contamination-Control Engineering in Wafer Processing: Problems and Prospects," Technical Proceedings SIMCON/Japan 1989, pp. 244-255.

[3] W. Kern, "The Evolution of Silicon Wafer Cleaning Technology," J. Electrochem. Soc., 137 (6), 1887 (1990).

[4] V. Ramakrishna and J. Harrigan, "Defect Learning Requirements," Solid State Technology, (1), 103-105 (1989).

[5] A. J. Muller, L. A. Psota-Kelty, J. D. Sinclair, and P. W. Morrisson, "Concentrations of Organic Vapors and their Surface Arrival Rates at Surrogate Wafers During Processing in Clean Rooms," Proceedings of the First International Symposium of Cleaning Technology in Semiconductor Device Manufacturing, Vol. 90-9, J. Ryzyllo and R. E. Novak, Eds., pp. 204-211, The Electrochem. Soc., Inc., New Jersey (1989).

[6] Federal Standard "Airborne Particulate Cleanliness Classes in Clean Rooms and Clean Zones," FED-STD-209E, September 11, 1992.

[7] G. E. Helmke, "Anatomy of a Pure Water System," Semiconductor International, p. 119, Aug. 1981.

[8] J. McHardy, "Particulate Removal with Dense $CO_2$ Fluids," Third International Workshop on Solvent Substitution, Phoenix, AZ, Dec. 8-11, 1992.

[9] A. Mayer and S. Schwartzman, "Megasonic Cleaning: A New Cleaning and Drying System for Use in Semiconductor Processing," J. Electron. Mat., 8, 885 (1979).

[10] W. Kern and D. Puotinen, "Cleaning Solutions Based on Hydrogen Peroxide for Use in Silicon Semiconductor Technology," RCA Review, 31, 187-206, (1970).

[11] Y. J. Ghabal, G. S. Higashi, K. Raghavachari, and V. A. Borrows, "Infrared Spectroscopy of Si(111) and Si(100) Surface after HF Treatment: Hydrogen Termination and Surface Morphology," J. Vac. Sci. & Technol. A 7, 2104 (1989).

[12] P. O. Hahn, M. Grundner, A. Schnegg, and H. Jacob, "The Si-SiO2 Interface Roughness: Causes and Effects," in the Physics and Chemistry of SiO2 and the Si-SiO2 Interface, C. R. Helms and B. E. Deal, Eds., pp. 401-411, Proceedings of the 173rd Meeting of the Electrochem. Soc., New York (1988).

[13] M. Hirose, T. Yasaka, K. Kanda, M. Takakura, and S. Miyasaki, "Behavior of Hydrogen and Fluorine Bonds on Chemically Cleaned Silicon Surfaces," Proceedings of the Second International Symposium of Cleaning Technology in Semiconductor Device Manufacturing, Vol. 92-12, pp. 1-9, J. Ruzyllo and R. E. Novak, Eds., The Electrochem. Soc., Inc., New Jersey.

[14] W. A. Syverson, M. J. Fleming, and P. J. Schubring, "The Benefits of SC-1/SC-2 Megasonic Wafer Cleaning," Proceedings of the Second International Symposium of Cleaning Technology in Semiconductor Device Manufacturing, Vol. 92-12, pp. 1-17, J. Ruzyllo and R. E. Novak, Eds., The Electrochem. Soc., Inc., New Jersey.

[15] J. Ruzyllo, "Evaluating the Feasibility of Dry Cleaning of Silicon Wafers," Microcontamination, 3 (3), 39 (1988).

[16] T. Ohmuri, T. Fukumoto, and T. Kato, "Ultra Clean Ice Scrubber Cleaning with Jetting Fine Ice Particles," Proceedings of the First International Symposium on Cleaning Technology in Semiconductor Device Manufacturing, 1989, The Electrochemical Society, 182-191, 90-9 (1990).

[17] S. A. Hoenig, *"Fine Particles on Semiconductor Surfaces: Sources, Removal, and Impact on the Semiconductor Industry,"* in Particles on Surfaces, 1: Detection, Adhesion, and Removal, K. L. Mittal, Ed., Plenum Press, New York, pp. 3-16 (1988).

[18] W. T. McDermott, R. C. Ockvic, J. J. Wu, and R. J. Miller, *"Removing Submicron Surface Particles Using a Cryogenic Argon-Aerosol Technique,"* Microcontamination, (10), 33, (1991).

[19] J. R. Vig, *"UV/Ozone Cleaning of Surfaces,"* in Treatise on Clean Surface Technology, Vol. 1, K. L. Mittal, Ed., Plenum Press, New York, pp. 1-26 (1987).

[20] R. Sugino, M. Okuno, M. Shigeno, Y. Sato, A. Ohsawa, T. Ito, and Y. Okui *"UV-Excited Dry Cleaning of Silicon Surfaces Contaminated with Iron and Aluminum,"* Proc. of the 2nd Intern. Symp. on Cleaning Technology in Semiconductor Device Manufacturing, J. Ryzyllo and R. E. Novak, Eds., Vol. 92-12, pp. 72-79, The Electrochem. Soc., New Jersey, (1992).

[21] J. Ruzyllo, D. C. Frystak, and R. A. Bowling, *"Dry Cleaning Procedure for Silicon IC Fabrication,"* IEDM 1990 Technical Digest, 409-412 (1990).

[22] J. S. Judge, *"A Study of the Dissolution of $SiO_2$ in Acidic Fluorine Solutions,"* J. Electronchem. Soc., 118, 1772 (1971).

[23] A. S. Tenney and M. Ghezzo, *"Etch Rates of Doped Oxides in Solutions of Buffered HF,"* J. Electrochem. Soc. 120, 1091 (1973).

[24] W. A. Pliskin and R. P. Gnall, *"Evidence for Oxidation growth at the Oxide-Silicon Interface from Controlled Etch Studies,"* J. Electrochem. Soc. 113, 263 (1966).

[25] A. B. Glaser and G. E. Subak-Sharpe, *Integrated Circuit Engineering,* Addisson-Wessley, New York, 1979.

[26] G. I. Parisi, S. E. Haszko, and G. A. Rozgonyi, *"Tapered Windows in $SiO_2$: The Effect of $NH_4F/HF$ Dilution and Etching Temperature,"* J. Electrochem. Soc., 124 (6), 917 ...

[27] H. Robbin and B. Schwartz, *"Chemical Etching of Silicon, II. The System HF, $HNO_3$ and $H_2O$,"* J. Electrochem. Soc., 106, 505 (1960).

[28] D. L. Kendall, *"On Etching Very Narrow Grooves in Silicon,"* Appl. Phys. Lett. 26, 195 (1975).

[29] M. Declercq, L. Gerzberg, and J. Meindl, *"Optimization of the Hydrazine-Water Solution for Anisotropic Etching of Silicon in Integrated Circuit Technology,"* J. Electrochem. Soc., 122 (4), 545 (1975).

[30] W. van Gelder and V. E. Hauser, *"The Etching of Silicon Nitride in Phosphoric Acid with Silicon Dioxide as a mask,"* J. Electrochem. Soc., 124, 869 (1977).

[31] C. A. Deckert, *"Pattern Etching of CVD $Si_4O_3/SiO_2$ Compositions in HF/Glycerol Mixtures,"* J. Electrochem. Soc., 127, 2433 (1980).

[32] H. H. Sawin, *"A Review of Plasma Processing Fundamentals,"* Solid State Technology, 28 (4),211-216, (1985).

[33] S. J. Fonash *"Advances in Dry Etching- A Review,"* Solid State Technology, 28 (1), 150-158, (1985).

[34] D. L. Flamm and G. K. Herb, *"Plasma Etching Technology - An Overview,"* in Plasma Etching, D. M. Manos and D. L. Flamm, Eds., Academic Press, p.14, New York (1989).

[35] C. J. Mogab, *"Dry Etching,"* in VLSI Technology, S. M. Sze, Ed., McGraw Hill, New York, p. 303 (1983).

[36] J. W Coburn, Plasma Etching and Reactive Ion Etching, American Vacuum Society, New York, 1982.

[37] H. R. Koenig and L. I. Maissel, *"Application of RF Discharges to Sputtering,"* IBM J. Res. Dev., 14 (1), 168 (1970).

[38] J. L. Vossen and J. J. Cuomo, *"Glow Discharge Sputter Deposition,"* in Thin Film Processes, J. L. Vossen and W. Kern, Eds., p. 11, Academic Press, New York, 1978.

[39] J. W. Coburn and H. F. Winters, *"Plasma-Assisted Etching in Microfabrication,"* Ann. Rev. Mater. Sci., 13, 91 (1983).

[40] D. T. Hawkins, *"Ion Milling (Ion Beam Etching), 1975-1978: A Bibliography,"* J. Vac. Sci. Technology, 16, 1051 (1979).

[41] D. Bollinger and R. Fink, *"A New Production Technique: Ion Milling,"* Solid-State Technology, 23 (11), 79-84 (1980).

[42] J. M. E. Harper, *"Ion Beam Etching,"* in Plasma Etching, D. M. Manos and D. L. Flamm, Eds., p. 391, Academic Press, New York (1989).

[43] J. Melngallis, *"Focused Ion Beam Technology and Applications,"* J. Vac. Sci.Technology. B5 (2), 469 (1987).

[44] S. Broydo, *"Important Considerations in Selecting Anisotropic Plasma Etching Equipment,"* Solid State Technology, -- (4), 159 (1983).

[45] R. W. Berry, P. M. Hall, M. T. Harris, Thin Film Technology, p. 24, Van Nostrand, New Jersey (1968).

[46] F. Daniels and R. A. Alberty, Physical Chemistry, p. 126, John Wiley & Sons, New York (1966)

[47] R. C. Weast, Handbook of Chemistry and Physics, p. B-88, The Chemical Rubber Co., Cleveland (1971).

[48] S. M. Irving, *"A Plasma Oxidation Process for Removing Photoresist Films,"* Solid State Technology, 14 (6), 47 (1971).

[49] G. S. Oehrlein and J. F. Rembetski, *"Plasma-Based Dry Etching Techniques in the Silicon Integrated Circuit Technology,"* IBM J. Res. Dev., 36 (2), 140 (1992).

[50] J. A. Bondur, *"Dry Process Technology,"* J. Vac. Sci. Technol., 13, 1023 (1976).

[51] J. W. Coburn and H. F. Winters, *"Ion- and Electron-Assisted Gas-Surface Chemistry - An Important Effect in Plasma Etching,"* J. Appl. Phys. 50 (5), 3189-3196 (1979).

[52] H. F. Winters, J. W. Coburn, and T. J. Chuang, *"Surface Processes in Plasma-Assited Etching Environments,"* J. Vac. Sci. Technol., B1, 469 (1983).

[53] J. L. Mauer, J. S. Logan, L. B. Zielinski, and G. C. Schwartz, *"Mechanism of Silicon Etching by a $CF_4$ Plasma,"* J. Vac. Sci. Technol., 15, 1734 (1978).

[54] D. L. Flamm and V. M. Donnelly, *"The Design of Plasma Etchants,"* Plasma Chemistry and Plasma Processing, 1 (4), 317 (1981).

[55] H. F. Winters and J. W. Coburn, *"Etching Reactions at Solid Surfaces,"* Mater. Res. Soc. Symp. Proc. 38, 189-200 (1985).

[56] T. J. Tu, T. J. Chuang, and H. F. Winters, *"Chemical Sputtering of Fluorinated Silicon,"* Phys. Rev. B, 23, 823-835 (1981).

[57] M. Sato and Y Arita, *"Etched Shape Control of Single-Crystal Silicon in Reactive Ion Etching Using Chlorine,"* J. Electrochem. Soc., 134 (11), 2856-2862 (1987).

[58] R. N. Carlile, V. C. Liang, O. A. Palusinski, and M. M. Smadi, *"Trench Etches in Silicon with Controllable Sidewall Angles,"* J. Electrochem. Soc., 135 (8), 2058 (1988).

**348**

[59] G. K. Herb, D. J. Rieger, and K. Shields, *"Silicon Trench Etch in a Hex Reactor,"* Solid State Technology, 30 (10), 109-115 (1987).

[60] R. A. Powell and D. F. Downey, *"Reactive Ion Beam Etching,"* in Dry Etching for Microelectronics, R. A. Powell, Ed., North Holland Physics Publishing, p. 115, New York (1984).

[61] H. C. Scheer *"Ion Sources for Dry Etching: Aspects of Reactive Ion Beam Etching for Si Technology,"* Rev. Sci. Instrum. 63(5), pp. 3050 3057 (1992).

[62] D. J. Chinn, I. A. Adesida, and E. D. Wolf, *"Profile Formation in CAIBE,"* Solid State Technology 27 (5), 123-129 (1984).

[63] M. Komuro, N. Watanabe, and H. Hiroshima, *"Focused Ga Ion Beam Etching of Si in Chlorine Gas,"* Jap. J. Appl. Phys., 29 (10), 2288-2291 (1990).

[64] J. A. Skidmore, G. D. Spiers, J. H. English, Z. Xu, C. B. Prater, L. A. Coldren, E. L. Hu, and P. M. Petroff, *"Low Damage Anisotropic Radical-Beam Ion-Beam Etching and Selective Chemical Etching of Focused Ion Beam-Damaged GaAs Substrates,"* SPIE, 1671, 268-279 (1992).

[65] P. D. Brewer, G. M Reksten, and R. M. Osgood, Jr., *"Laser-Assisted Etching,"* Solid State Technology, 28 (4), 273-278 (1985).

[66] M. Sekine, H. Okano, K. Yamabe, N. Hayaska, and Y. Horrike, *"Radiation Damage Evaluation in an Excimer Laser Etching,"* Digest VLSI Symposium, p. 82 (1985).

[67] J. A. Mucha and D. W. Hess, *"Plasma Etching,"* in Introduction to Microlithography: Theory, Materials, and Processing, L. F. Thompson, C. G. Willson, and M. J. Bowden, Eds., American Chem. Soc. Symp. Series, 219, 215-285, (1983).

[68] Y. Horiike, M. Shibagaki, *"A Dry Etching technology Using Long-Lived Active Species Excited by Microwave,"* in Semiconductor Silicon, H. R. Huff and E. Sirtl, Eds., The Electrochem. Soc., 77-2, 1071 (1977).

[69] C. B. Zarowin and R. S. Horowath, Proceedings of the Third Symposium on Plasma Etching, The Electrochemical Society, Vol 82-6, p. 50 (1982).

[70] E. Bogle-Rohwer, D. Gates, L. Hayler, H. Kurasaki, and B. Richardson, *"Wall profile Control in a Triode Etcher,"* Solid State Technology, 28 (4), 251-255 (1985).

[71] J. A. Thornton, *"Magnetron Sputtering: Basic Physics and Application to Cylindrical Magnetrons,"* J. Vac. Sci. Technology, 15 (2), 171-177 (1978).

[72] M. Engelhardt, *"Evaluation of Dry Etching Processes with Thermal Waves,"* Solid State Technology, 33 (4), 151-156 (1990).

[73] A. A. Bright, S. Kaushik, and G. S. Oehrlein, *"Plasma Chemical Aspects of Magnetron Ion Etching with $CF_4/O_2$* J. Appl. Phys., 62 (6), 2518-2522 (1987).

[74] G. S. Oerlein, A. A. Bright, and S. W. Robey, *"X-Ray Photoemission Spectroscopy Characterization of Silicon Surfaces after $CF_4/H_2$ Magnetron Ion Etching: Comparisons to Reactive Ion Etching,"* J. Vac. Sci. Technol., A6 (3), 1989-1993 (1988).

[75] O. S. Nakawaga, S. Ashok, and J. K. Kruger, *"A Schottky Barrier Study of HBr Magnetron Enhanced Reactive Ion Etching Damage in Silicon,"* J. Appl. Phys.. 69 (4), 2057-2061 (1991).

[76] C. P. D'Emic, K. K. Chan, and J. Blum, *"Deep Trench Plasma Etching of Single Crystal Silicon using $SF_6/O_2$ Gas Mixtures,"* J. Vac. Sci. Technol. B, 10 (3), 1105-1110 (1992).

[77] C. W. Fu, R. Hsu, and V. Malba, *"Magnetron Enhanced Reactive Ion Etching of Al 1% Si 2% Cu Alloy,"* Low Energy Ion Beam and Plasma Modification of Materials Symposium Anaheim, California, April/May 1991, pp. 385-388 (1991).

[78] M. Sato, D. Takehara, K. Uda, K. Sakiyama, and T. Hara, "*Suppression of Micro-loading Effect by Low Temperature* $SiO_2$ *Etching*," Jap. J. Appl. Phys., Part 1, 31 (12B), 4370-4375 (1992).

[79] M. Meyyappan, "*Magnetron Reactive Ion Etching of GaAs in* $SiCl_4$," J. Vac. Sci. Technology B, 10 (3), 1215 (1992).

[80] T. Hara, J. Hiyoshi, H. Hamanaka, M. Sasaki, F. Kobayashi, K. Ukai, and T. Okada, "*Damage Formed by Electron Cyclotron Resonance Plasma Etching on a Gallium Arsenide Surface*," J. Appl. Phys. 67 (6), 2836-2839 (1989)

[81] K. Suzuki, S. Okudaira, N. Sakudo, and I, Kanomata, "*Microwave Plasma Etching*," Jap. J. Appl. Phys. 16 (11), 1979-1984 (1977)

[82] S. Matsuo, "*Selective Etching of Silicon Relative to* $SiO_2$ *Without Undercutting by* $CBrF_3$ *Plasma*," Appl. Phys. Lett., 36 (9), 768-770 (1980).

[83] C. J. Mogab, "*The Loading Effect*," J. Electrochem. Soc., 124, 1262 (1977).

[84] P. M. Schaible, W. C. Metzger, and J. P. Anderson, "*Reactive Ion Etching of Aluminum and Aluminum Alloys in an RF Plasma Containing Halogen Species*," J. Vac. Sci. Technol., 15, 334-337 (1978).

[85] Y. Horiike and M. Shibagaki, "*A New Dry Chemical Etching*," Jpn. J. Appl. Phys., Suppl. 15, 13-18 (1976).

[86] D. L. Flamm, "*Introduction to Plasma Chemistry*," in Plasma Etching, D. M. Manos and D. L. Flamm, Eds., Academic Press, p.91, New York (1989).

[87] T. Enomoto, M. Denda, A. Yasuoka, and H. Nakata, "*Loading Effect and Temperature Dependence of Etch Rate in* $CF_4$ *Plasma*," Jpn. J. Appl. Phys., 18, 155 (1979).

[88] C. J. Mogab and H. L. Levinstein, "*Anisotropic Plasma Etching of Polycrystalline Silicon*," J. Vac. Sci. Technol., 17, 721 (1980).

[89] R. A. Gottscho, C. W. Jurgensen, and D. J. Vitkavage, "*Microscopic Uniformity in Plasma Etching*," J. Vac. Sci. Technol. B, 10 (5), 2133-2147 (1992).

[90] N. Fujiwara, H. Sawai, M. Yoneda, K. Nishioka, K. Horie, K. Nakamoto, and H. Abe, "*High Performance Electron Cyclotron Resonance Plasma Etching with Control of Magnetic Field Gradient*," Japn. J. Appl. Phys. 1, 30 (11B), 3142-3146 (1991).

[91] K. Koller, H. P. Erb, and H. Korner, "*Tungsten Plug Formation by an Optimized Tungsten Etch Back Process in Non Fully Planarized Topology*," Appl. Surf. Sci., 53, 54-61 (1991).

[92] D. L. Flamm, V. M. Donnelly, and J. A. Mucha "*The Reaction of Fluorine Atoms with Silicon*," J. Appl. Phys., 52 (5), 3633-3639 (1981).

[93] A. W. Kolfschotten, R. A. Haring, A. Haring, and A. E. de Vries, "*Argon-Ion Assisted Etching of Silicon by Molecular Chlorine*," J. Appl. Phys., 55 (10), 3813-3818 (1984).

[94] Y. H. Lee, M. M. Chen, and A. A. Bright, "*Doping Effects in Reactive Plasma Etching of Heavily Doped Silicon*," Appl. Phys. Lett., 46 (3), 260-262 (1985).

[95] L. Baldi and D. Beardo, "*Effect of Doping on Polysilicon Etch Rate in a Fluorine-Containing Plasma*," J. Appl. Phys., 57 (6), 2221-2225 (1985).

[96] G. C. Schwartz and P. M. Schaible, "*Reactive Ion Etching of Silicon*," J. Vac. Sci. Technol., 16 (2), 410-413 (1979).

[97] N. Awaya and Y. Arita, Proc. 6th. Symp. on Dry Processes, IEE, Tokyo, pp. 98-103 (1984).

[98] S. Berg, C. Nender, R. Buchta, and H. Norstroem, *"Dry Etching of N- and P-Type Polysilicon: Parameters Affecting the Etch Rate,"* J. Vac. Sci. Technol., A5 (4), 1600-1603 (1987).

[99] S. E. Bernacki and B. B. Kisicki, *"Controlled Film Formation During CCl₄ Plasma Etching,"* J. Electrochem. Soc., 131 (8), 1926-1931 (1984).

[100] J. Chung, M. C. Jeng, J. E. Moon, A.T. Wu, T. Y. Chan, P. K. Ko and C. Hu, *"Deep Submicrometer MOS Device Fabrication Using Photoresist Ashing Technique,"* IEEE Electron Device Letters, EDL 9 (4), 186 (1988).

[101] *"Dry Etching,"* Practical VLSI Fabrication for the 90s, R. Bowman, G. Fry, J. Griffin, R. Potter, and R. Skinner, Eds., Integrated Circuit Engineering Corp., Arizona (1990).

[102] Z.-H. Zhou, E. S. Aydil, R. A. Gottscho, Y. J. Chabal, and R.. Reif, *"Real-Time, in-situ Monitoring of Room Temperature Silicon Surface Cleaning Using Hydrogen and Ammonia Plasma,"* J. Electrochem. Soc., in Press (1993).

[103] M. Miki, H. Hikuyama, I. Kawanabe, M. Miyashita, and T. Ohmi, *"Gas-Phase Selective Etching of Native Oxide,"* IEEE Trans. Electron Devices, ED-37 (1), 107 (1990).

[104] B. Witowski, J. Chacon, and V. Menon, *"Characterization of an Anhydrous HF Pre Gate Oxidation Etching Process,"* in Cleaning Technology in Semiconductor Manufacturing, J. Ruzyllo and R. E. Novak, Eds., p. 372, Electronchem Soc., Vol. 92 12, New Jersey

[105] C. J. Mogab, A. C. Adams, and D. L. Flamm, *"Plasma Etching of Si and SiO₂ - The Effect of Oxygen Additions to a CF₄ Plasma,"* J. Appl. Phys., 49, 3769 (1978).

[106] L. M. Ephrath, *"Selective Etching of Silicon Dioxide Using Reactive Ion Etching with CF₄/H₂,"* J. Electrochem. Soc., 126, 1419 (1979).

[107] K. Hirata, Y. Ozaki, M. Oda, and M. Kimizuka, *"Dry Etching Technology for 1-μm VLSI Fabrication,"* IEEE Trans. Electron Dev., ED-28, 1323 (1981).

[108] M. Delfino, S. Salimian, D. Hodul, A. Ellingboe, and W. Tsai, *"Plasma Cleaned Si Analyzed in situ by X-Ray Photoelectron Spectroscopy, and Actinometry,"* J. Appl. Phys., 71 (2), 1001-1009 (1992).

[109] P. E. Riley, S. S. Peng, and L. Fang, *"Plasma Etching of Aluminum for VLSI,"* Solid State Technology, , 47-52 (1993).

[110] D. W. Hess and R. H. Bruce, *"Plasma-Assisted Etching of Aluminum and Aluminum Alloys,"* in Dry Etching for Microelectronics, R. A. Powell, Ed., North Holland Physics Publishing, p. 115, New York (1984).

[111] N. Selamoglu, C. N. Bredbenner, T. A. Giniecki, and H. J. Stocker, *"Tapered Etching of Aluminum with CHF₄/BCl₃ and its Impact on Step Coverage of Plasma-Deposited Silicon Oxide from Tetraethoxysilane,"* J. Vac. Sci. technol., B9 (5), 2530-2535 (1991).

[112] T. J. Dalton, W. T. Conner, and H. H. Sawin, *"Interferometric Real-Time Measurement of Uniformity from Plasma Etching,"* J. Appl. Phys., submitted (1993).

[113] K. Shenai, *"Diffusion Profiles of Boron Implanted into Plasma-Etched Silicon Surfaces,"* IEEE Trans. Electron Devices, ED-39 (5), 1242-1245 (1992).

[114] S. W. Pang, D. D. Rathman, D. J. Silversmith, R. W. Mountain, and P. D. DeGraff, *"Damage Induced in Si by Ion Milling or Reactive Ion Etching,"* J. Appl. Phys., 54 (6), 3272-3277 (1983).

[115] L. J. Brillson, M. L. Slade, A. D. Kadnani, M. Kelly, and G. Margaritondo, *"Reduction of Silicon-Aluminum Interdiffusion by Improved Semiconductor Surface Ordering,"* Appl. Phys. Lett, 44 (1), 110-112 (1984).

[116] L. Jen Chung and W. G. Oldham, *"Plasma Etch Effects on Low Temperature Selective Epitaxial Growth of Silicon,"* J. Appl. Phys., 71 (7), 3225 30 (1992).

[117] C. T. Gabriel and J. P. McVittie, *"How Plasma Etching Damages Thin Gate Oxides,"* Solid State Technology, 35 (6), 81-87 (1992).

[118] M. Kubota, K. Hatafuji, A. Misaka, A. Yamano, H. Nakagawa, and N. Nomura, *"Simulational Study for Gate Oxide Breakdown Mechanism due to Non-Uniform Electron Current Flow,"* IEDM 1991 Tech. Digest, p. 891 (1991).

[119] H. Shin, C.-C. King, T. Horiuchi, and C. Hu, *"Thin Oxide Charging Current During Plasma Etching of Aluminum,"* IEEE Electron Dev. Lett., 12 (8), 404 (1991).

[120] S. Fang, A. M. McCarthy, and J. P. McVittie, *"Charge Sharing Antenna Effects for` Gate Oxide Damage During Plasma Processing,"* Proc. 3rd Intl. Symp. on ULSI, Electrochem. Soc., Pennington, New Jersey, Vol. 91-11 (5), p. 473 (1991).

[121] R. A. Bowling, *"An Analysis of Particle Adhesion on Semiconductor Surfaces,"* J. Electrochem. Soc., 132 (9), 2208-2214 (1985).

# Chapter 6

# Ion Implantation

## 6.0 Introduction

The electrical properties of semiconductor crystals can be modified predictably by introducing controllable amounts of dopant impurities into substitutional sites of the crystal. In silicon, for example, the most electronically active substitutional impurities are elements from group III (boron) and group V (arsenic, phosphorus, antimony) of the periodic table. When boron is in excess, holes are created for conduction and the crystal is said to be p-type. For excess arsenic, phosphorus, or antimony, free electrons are created for conduction and the crystal is called n-type. The most commonly used methods to introduce impurities into a semiconductor are doping during crystal or epitaxial growth (Chaps. 1,3), **ion implantation,** and **diffusion.** The basic principle of ion implantation in semiconductor technologies is described by Shockley [1]. It is a low-temperature process in which ionized dopants are accelerated to energies high enough so that when they impact on a target wafer's surface they penetrate to a certain depth. Solid-state diffusion describes the method of introducing impurities into the wafer and/or redistributing them within the solid material by activating their motion at elevated temperature.

In earlier phases of semiconductor technologies, diffusion was the prevalent method used to introduce dopants into the wafer and redistribute them to achieve the desired impurity concentration versus distance, called **profile.** During the past two decades, ion implantation has progressed steadily from its initial use as a complement to diffusion to its present central role in the manufacture of semiconductor circuits. It is now the preferred means of doping semiconductors because of the better control and reproducibility of concentration and depth that can be achieved. The most important feature of ion implantation, however, is its flexibility and ability to achieve impurity profiles which are not possible with simple diffusion. Buried layers with profiles that drop in concentration from the bulk to the surface of the wafer cannot be obtained by diffusion from the surface of the crystal. They are best formed by implanting impurities at one or multiple energies deep into the crystal to achieve a retrograde profile. Since ion implantation is essentially a low-temperature process, it allows the use of temperature-sensitive implantation masks, such as photoresist, to selectively expose regions where dopants are introduced, blocking implantation in those regions covered by the resist. Also, implantation can be performed through films, such as silicon nitride and silicon dioxide, that are impervious to impurities introduced by diffusion, increasing the flexibility in doping the crystal at different stages of the process sequence. Ion implantation is used almost exclusively to dope gallium arsenide, because substantial decomposition of GaAs can occur when subjecting the semiconductor to elevated diffusion temperatures.

Ion implantation is not only applied to dope semiconductors. Silicon-on-insulator (SOI) wafers, for example, can be prepared by implanting oxygen or nitrogen deep in silicon and then subjecting the wafers to elevated temperatures to form a buried-oxide or -nitride layer that separates the top and bottom portions of the silicon wafer. Implantation is also used to introduce gettering damage into the crystal, e.g., by bombarding the back of the wafer with energetic argon ions, or by placing a heavily doped buried layer deep in the crystal. Finally, by controlling heavily doped shallow impurity profiles, it is possible to achieve very high dopant gradients used in special structures, such as IMPATT diodes and microwave transistors.

Ion implantation, however, has some drawbacks. The collision of energetic "guest" ions with target "host" atoms results in crystal damage, commonly referred to as **radiation damage.** The extent of this damage depends on the ion energy and the mass and atomic number of the host and guest atoms. Also, collisions of energetic ions with metal surfaces in the vicinity of the wafers, such as the wafer holder, can cause metals to sputter and contaminate the wafer. Another problem is that implanted atoms do not initially rest on crystal lattice sites and are hence not electrically active. A high temperature treatment for a period of time in a suitable ambient, called **annealing,** is required to activate the dopants and remove the damage. Also, several problems are encountered when forming ultra-shallow impurity profiles. Point defects created during implantation enhance the diffusion of most dopant species during high temperature treatments, resulting in deeper impurity profiles than predicted by simple diffusion theory (Chap. 7). While a low thermal budget can reduce the movement of impurities in the crystal, the low anneal temperature is typically inadequate to remove all crystal damage, leaving residual defects near junctions and causing an increase in junction leakage. Finally, implantation in crystalline targets is not isotropic. Some crystallographic directions allow the ions to penetrate deeper than others. This effect, known as **channeling,** can be reduced, but not fully eliminated, by controlling the angle of incidence of the ion beam, and/or by forming an amorphous film at the crystal surface.

## 6.1 Principle of Operation

The main elements of an ion implantation machine are the ion source, the mass analyzer, the accelerating column, the two-dimensional scanning system, and the wafer chamber. The design of the machine components depends on the anticipated perform-ance of the implanter and its specific doping requirements, including implanted flux, beam current, and power density. A sim-plified schematic of a typical implanter is shown in Fig. 6.1.

### 6.1.1 Ion Source

The purpose of an ion source is to produce and maintain a high-density of ions in a chamber of suitable geometry, so that a focussed beam of ions can be extracted from the source and trans-ported through the implanter system to the target. Impurities are first introduced from a solid, liquid, or gaseous source into the chamber as an unionized gas of elements or compounds. Typical source materials are given in Table 6.1. While arsine ($AsH_3$) and phosphine ($PH_3$) can be used as diluted gas sources for arsenic and phosphorus, solid As and P are preferred because of the toxicity of the gases. Solid or liquid sources with high vapor pressure are thermally vaporized by resistive or inductive heating in or near the source chamber with a thermocouple feedback. The maximum oven temperature is typically 1000 °C to 1100 °C. In case the vapor pressure of the source material is not sufficiently high, a

356

sputtering system is incorporated into the chamber to provide a gas of the species to be ionized. Electrons emitted from an incandescent filament (cathode) at $\simeq 2500$ °C gain energy by acceleration in the field created by the potential difference between the filament and the chamber walls (30-100 V). They are made to oscillate at high energy in an electromagnetic field in the chamber and to collide with and ionize the gas which is maintained at a typical pressure of $10^{-4} - 10^{-3}$ Torr. For a given gas pressure, the ionization rate is proportional to the electron temperature and path length in the chamber.

**Table 6.1** Typical ion-source materials [2 − 5].

| Element | Atomic mass of most frequent | Typical source Material | Melting point (°C) |
|---|---|---|---|
| Antimony, Sb | 121, 123 | Solid Sb<br>Solid Sb$_2$O$_3$ | 630<br>600 |
| Argon, Ar | 40 | Ar gas | -189 |
| Arsenic, As | 75 | Solid As<br>Arsine | 817<br>-116 |
| Beryllium, Be | 9 | Solid Be | 1284 |
| Boron, B | 10,11 | BF$_3$ gas | -128 |
| Gallium, Ga | 69,71 | Solid Ga | 30 |
| Germanium, Ge | 70,72,74 | Solid Ge<br>Germane, gas | 936<br>-165 |
| Nitrogen, N$_2$ | 14 | N$_2$ gas | -210 |
| Oxygen, O$_2$ | 69,71 | O$_2$ gas | -219 |
| Phosphorus, P | 31 | Solid P<br>Phosphine | 44<br>-133 |
| Silicon, Si | 28,29 | SiF$_4$ gas | -90 |

To create a positive ion, at least one electron is removed from the atom. The energy required to remove an electron, called **ionization potential,** increases rapidly when successive electrons are removed to create multiply charged ions. The plasma obtained contains the desired ionized species together with other ionized fragments created in the plasma. Positive ions are extracted from the plasma through an opening in the chamber by applying a positive potential (typically 40-70 kV) between the chamber and the extraction electrode.

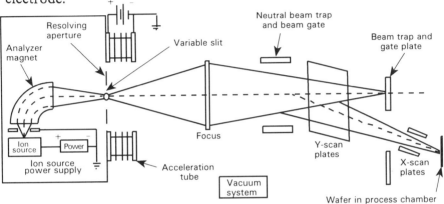

**Fig. 6.1** Schematic of a typical medium current implantation system.

The most abundant species extracted from the chamber are singly ionized dopants, such as $_{31}P^+$, $_{75}As^+$, $_{11}B^+$, $_{121}Sb^+$, and $_{49}BF_2^+$. A much smaller number of these species are present in the extracted beam as double-charged (up to $\simeq 10\%$) and triple-charged (up to $\simeq 1\%$) ions. The generation of these multiply charged ions is desirable to extend the energy range of the accelerator to higher values and obtain deeper impurity profiles. The accelerator's energy capability is extended by a factor of 2 for double-charged ($E = 2qV$) and a factor of 3 for triple-charged ions ($E = 3qV$), however, at the cost of machine throughput. For example, a 400 keV range distribution can be obtained by accelerating double-charged phosphorus ions $P^{++}$ in a 200 keV accelerator.

One important consideration with multiply charged species is the potential charge exchange with residual gas molecules along the ion trajectory resulting in beam "contamination" with ions of lower charge multiplicity that can distort the profile, as illustrated in Fig. 6.2. The same species arrive at the target at different energies and ionized states. A typical profile with the singly ionized "contamination" is shown in Fig. 6.3.

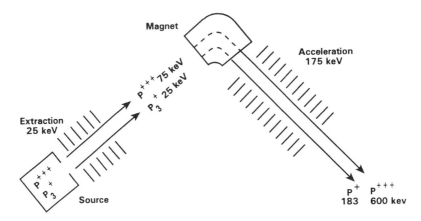

**Fig. 6.2** Contamination of a triple-ionized phosphorus beam with singly ionized phosphorus caused by charge exchange before and during acceleration in a medium energy implanter which is adjusted for a 600 keV $P^{+++}$ ion beam (200 keV accelerating potential) with the beam filter off [6,7].

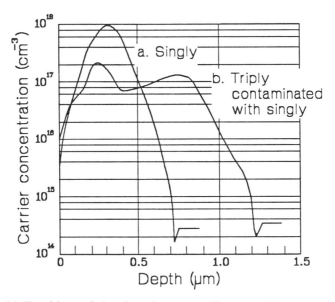

**Fig. 6.3** Double peak in phosphorus profile caused by contamination of triple-ionized beam with singly ionized phosphorus [6].

The presence of singly ionized contaminants causes errors in depth, surface concentration and total dose in silicon [6,7]. The $P^+$ contamination caused by charge exchange before analysis can be eliminated by using a beam filter. In the absence of a beam filter these contaminants reach the target surface with an energy of only 183.5 keV, as indicated in Fig. 6.2. For double-charged phosphorus, the interaction of the $P^{++}$ ions with residual gas molecules in the accelerating tube, causes some charge exchange, reducing a fraction of the $P^{++}$ ions to $P^+$ ions. The $P^+$ contaminants reach the target surface with an energy between 200 keV and 400 keV, depending on where in the accelerating potential the interaction took place [6].

In medium energy implanters, the beam current decreases rapidly as the energy drops below the extraction potential, which is the energy range used to implant shallow and ultra-shallow impurity profiles. To extend the energy range to values lower than the extraction potential without reducing the beam current, molecular species such as $As_2^+$, $As_3^+$, and $BF_2^+$ are extracted and selected. For example, since the mass ratio of $^{49}BF_2^+$ to $^{11}B^+$ is 49/11, the energy required to achieve the same range distribution is approximately 4.46 times larger for $^{49}BF_2^+$ than for single boron ions. This places the implant energy in a range where the beam current is higher and the system is more stable. $BF_2^+$ and $B^+$ ions are obtained in the ionizing chamber from the dissociation of $BF_3$ molecules. Typical systems can supply a $BF_2^+$ beam current which is more than ten times larger than the available $B^+$ beam current, thus considerably reducing the time for high-dose implantation at low energies [2]. Upon impact on the target, $BF_2^+$ dissociates into boron and fluorine. Boron is is incorporated into the crystal while most of the fluorine escapes as a gas. In the presence of an oxide layer, the fluorine rapidly diffuses into the oxide without causing its degradation [8]. Another advantage with $BF_2^+$ ions is the capability of amorphizing the surface to reduced channeling and improve the annealing efficiency, (Sec. 6.2.2) [9]. One potential problem with molecular-ion implantation is the dissociation of the ions after mass analysis or during final acceleration. In the case of $BF_2^+$ the most important dissociation is:

$$BF_2^+ \rightarrow B^+ + F_2$$

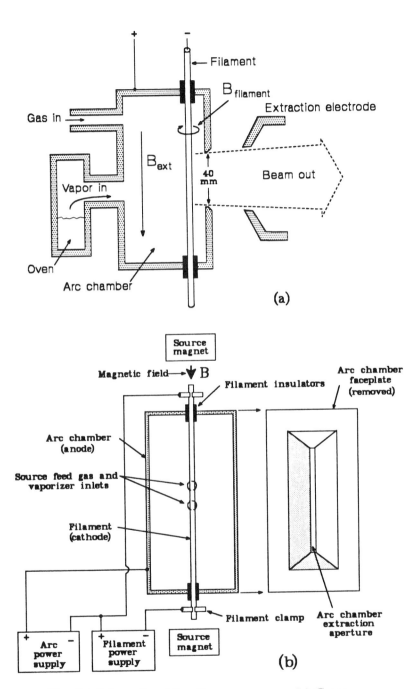

Fig. 6.4 Schematic of the Freeman source. (a) Cross-section. (b) Top view. (Adapted from [3]).

Since $B^+$ is accelerated by the same voltage as $BF_2^+$, it penetrates deeper into the target, thus distorting the projected profile. This problem can be avoided by implementing beam filters at different stages of the beam trajectory.

The beam current intensity is limited primarily by the ionization and extraction efficiencies and, in some high-energy machines, by the charge exchange efficiency. The beam current ranges from 1 $\mu$A - 5 mA for medium-current implanters, and from 5-30 mA for high-current implanters. Higher beam currents (50-200 mA) are used in specific machines, e.g., for oxygen implantation to form buried oxide layers.

State of the art ion sources are described in detail in [2 − 5, 10 − 14]. Of the various types, the Freeman and Bernas sources, both belonging to the hot-filament ion source category, are the most commonly used because of their flexibility in ionizing a wide variety of species and the high beam current that can be achieved [15,16]. The sources are equipped with an oven so that solids can be vaporized. In the Freeman source, the beam is extracted from a slit-shaped aperture in the shape of a wedge about 40 mm long and 2-4 mm wide (Fig. 6.4). Because of the large extraction slit, beam current with several tens of mA are possible.

The main feature of the Freeman source is that the filament ($\simeq$ 2 mm in diameter) is mounted in the most effective position, parallel and extremely close to the ion extraction slit, as shown in Fig. 6.4. An external magnet is used to created a magnetic field parallel to the direction of the filament, but only a weak magnetic field ($\simeq 10^{-2}$ T = $10^{-6} G$) is needed to maintain a stable beam in this configuration. This low magnetic field eliminates the magnetically induced electrical breakdown and undesirable deflection of the ion beam observed with earlier configurations which use a much higher magnetic field. The external magnetic field is normal to that generated by the current heating the filament ($\simeq$ 150 A, $2x10^{-3}$ T). The two magnetic fields and the electric field between filament and chamber walls cause electrons to spiral around the cathode in complex orbits near the filament, resulting in an increase in the electron path-length and ionization efficiency.

The Bernas source is similar to the Freeman source, except for the filament arrangement at one end of the chamber and the

**362**

presence of an electron reflector at the other end facing the filament (Fig. 6.5) [3]. The winding of the filament is such that the
magnetic field created by the heating current is in the same direction as the external magnetic field. This source arrangement
increases the path length of electrons, hence the ionization rate
and beam current. Also, corrosion and sputtering of the filament
by plasma ions is reduced, increasing the filament lifetime.

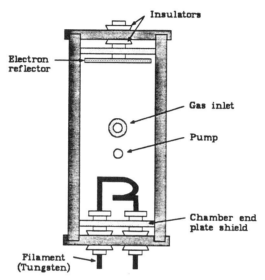

**Fig. 6.5** Bernas source arc chamber schematic (view
from top, front plate removed).

The arrangement of the extraction electrode is shown in
more detail Fig. 6.6. The extraction voltage and gap are typically
adjustable and fundamentally affect the shape of the plasma
extraction surface, and hence the beam current, focussing and
acceleration [3]. Best results are obtained when the ion beam is
extracted from a concave plasma surface, resulting in very small
beam divergence. Since electrons are repelled by the extraction
electrode potential, the beam is not neutralized and the current is
essentially space-charge limited, approximated by

$$J = 4\varepsilon_0 \sqrt{\frac{2q}{M}} \; \frac{V^{3/2}}{d^2} \; ,$$

6.1

where $V$ is the extraction voltage, $d$ is the extraction gap, $\varepsilon_0$ is the permittivity of free space ($\varepsilon_0 = 8.86x10^{-14}$ F/cm), $q$ and $M$ are the ion charge and mass, respectively. Beyond the extraction electrode, the beam is mostly neutralized by electrons trapped in the potential well of the ion beam [3]. Space-charge neutralization is essential to prevent excessive beam divergence, as illustrated in Fig. 6.7.

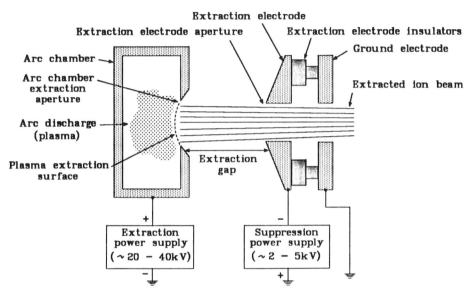

**Fig. 6.6** Schematic of extraction electrode arrangement.

Microwave ion sources have the advantage of not using a filament to maintain the discharge, increasing the source lifetime [13, 14]. As with ECR plasmas (Chap. 5), microwaves at 2.45 GHz are introduced, from an antenna into the discharge region through a ceramic window. The gas species are introduced through a needle valve. The required magnetic field is high (0.18 T as compared to 0.0875 T for ECR plasmas). One version of this source uses a slit-shaped beam to produce 10 mA of P$^+$ and As$^+$ for 100-200 hours, which is several times longer than for the Freeman and Bernas (hot-cathode type) sources [14]. A microwave ion source is used for oxygen implantation at a beam current of $\simeq$ 100 mA to create buried oxide layers [17]. For most doping materials, however, the source operating time is limited mainly by material deposition on the ceramic window and insulators.

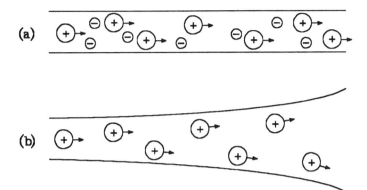

**Fig. 6.7** Effect of neutralization on beam divergence. (a) Space-charge neutralized; attraction and repulsion. (b) No space-charge neutralization; repulsion only.

## 6.1.2 Analyzer Magnet

In a preanalysis system, the beam passes through an analyzer magnet before it is accelerated. It is directed normal to the strong magnetic field that separates the ions according to their mass to charge ratio. The magnetic field deflects the species to a 60°- 110° angle (typically 90°) such that the beam of selected ions passes through an adjustable resolving aperture (typically 1.25 cm x 5 cm), while other species are screened out. For a magnetic field of intensity $B$, ions of mass $M_I$ and charge $Q$ move in a circular path of radius $R$. The plane of the circle remains normal to the magnetic field. The radius $R$ is found from

$$R = \sqrt{\frac{2M_I V}{QB^2}} \,, \qquad\qquad 6.2$$

where $V$ is the accelerating voltage which, in a preanalysis system, is approximately the source extraction voltage. The magnetic field intensity is chosen to make the radius of the selected ion-path coincide with the physical radius of the magnet. If other ions having a ratio $M_I/Q$ very close to that of the selected ions are present, they will be deflected by about the same angle, limiting the mass resolution. The quality of mass separation is defined by the resolving power $M_I/\Delta M_I$ where $\Delta M_I$ is the minimum resolvable mass difference. Mass resolutions of 60-100 are adequate for implantation of dopant species. The resolving power depends also upon the energy spread of the ion source, the beam diameter, and the aperture width. Two ion beams are resolvable if the separation of the beams $(D)$ is greater than the sum of the beam width $(W)$ and the aperture width $(S)$. For a mass difference of 1 amu, the resolving power is calculated as [2]

$$\frac{M_I}{\Delta M_I} = \frac{D\,M_I}{W+S}\,.\qquad\qquad 6.3$$

For magnets, this value is always given for $S = 0$, and $W$ is approximated as the full width half maximum (FWHM) of the ion beam.

### 6.1.3 Acceleration Column

Ion implantation machines can be categorized with respect to their accelerating voltage as low-energy ($\leq$ 20 keV), medium-energy (20-400 keV), and high-energy (400 keV - 3 MeV) implanters. The required energy depends upon the desired penetration depth of the species used. There is no single accelerator that satisfies all energy ranges. In the preanalysis arrangement shown in Fig. 6.1, most of the ion acceleration occurs after the beam leaves the magnet. The beam is then focused and directed into the field of an accelerating stack where the ions gain the desired kinetic energy. For example, a final maximum ion energy 200 keV can be achieved by using a beam with an extraction potential of 40 keV and post-acceleration of 160 keV. Another arrangement used in some machines is the postanalysis configuration where the beam is accelerated to its final velocity and then analyzed with the magnet. In this case, a larger and more expensive magnet is required because of the higher ion velocity. The minimum implantation energy that can be obtained without the use of molecular ions or deceleration is typically determined by the extraction potential. The extraction voltage, however, cannot be reduced below a certain limit without seriously affecting the beam current and ability to focus. To reduce the ion energy below this limit, some implanters operate in the deceleration mode where the extracted ions are slowed down by reversing the voltage polarity of the acceleration column. Therefore, for low-voltage machines, preanalysis combined with a decelerating voltage may be necessary, if energies below $\simeq 20$ keV are required. One problem associated with deceleration is the presence of neutrals created by charge exchange with residual gas molecules. Although the pressure in the accelerating column of the implanter is kept below $10^{-6}$ Torr to minimize ion scattering by gas molecules, some neutral species can be created by collisions of beam-ions with residual gas molecules, whereby ions acquire electrons from the molecules and are neutralized. Since neutral species cannot be decelerated nor deflected, they proceed at high energy on a straight-line path toward the target. Their energy depends on where in the accelerating column they are created.

The maximum energy is limited primarily by the breakdown field of the atmosphere around the high-voltage equipment. For example, an implanter which can operate at 200 keV at sea level is specified to operate at only 180 keV if installed at approximately 650 m above sea level because the breakdown field decreases as the atmospheric gas concentration drops. To prevent breakdown at accelerating voltages above $\simeq 400$ keV, some designs enclose the entire high-voltage section in an insulating pressurized gas, such as sulfur hexafluoride ($SF_6$) at about 7 atm, which increases the breakdown voltage by a factor of 20 [18]. Such an insulating gas is also used in conjunction with a tandem arrangement shown schematically in Fig. 6.8 [19]. In this system, most of the equipment remains at ground potential and the species are accelerated to their final kinetic energy in two stages. First, the extracted positive ions pass through a charge exchange chamber in which they are charged negatively by collisions with, e.g., a gas of magnesium atoms. The negatively charged ions are injected into the accelerator where they gain kinetic energy from ground to half of the desired voltage. The ions then pass through a "stripper canal" where collisions with, e.g., nitrogen gas molecules change a fraction of the negative ions to positive charge. The positive ions are accelerated again as they move from high voltage to ground in the second half of the accelerating column. The gas pressure in the exchange chamber and stripper canal directly affects the rate of charge exchange. For example, the gas pressure in the stripper canal determines the fraction of neutrals, negative ions, and singly and double charged positive ions that emerge from the stripper.

The beam current is limited by the charge exchange efficiency ($\simeq$ 12%, [20]).

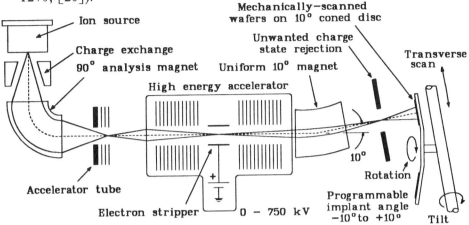

**Fig. 6.8** Schematic of a tandem high-energy ion-implantation machine [19].

In a linear accelerator, several accelerating sections are arranged along a straight path and alternately connected to the two poles of an ac voltage source. Acceleration occurs between sections, and the frequency of the ac voltage is chosen such that the ions are subjected to an accelerating phase at the end of each section. Therefore, the ion transit time through each section must be equal to half the period of the ac signal. This is why the sections must increase in length as the ions gain kinetic energy.

### 6.1.4 Scanning System

Since the ion beam is typically 1-5 cm in diameter, it is swept both vertically and horizontally across the wafer to distribute the dopants homogeneously over the surface of the wafer which is mounted in the target chamber. This is done either with electrostatic deflection plates (Fig. 6.9), or by keeping the beam stationary and scanning the wafer mechanically, or by a combination of electrostatic (or magnetic) and mechanical scanning. The electrostatic scan frequency is typically 1-2 kHz. Full mechanical scanning is done in both direction at a frequency of 0.5-15 Hz. In hybrid scan combinations, the beam is scanned electrostatically (1-2 kHz) or magnetically (1-10 Hz) in the x-direction while the wafer is moved slowly (0.5 Hz) in the y-direction (Fig. 6.10).

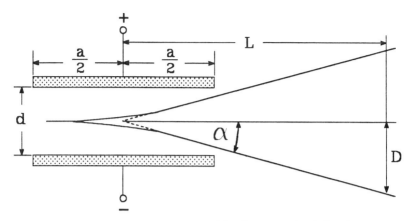

**Fig. 6.9** Beam scanning with electrostatic deflection plates.

Processes are sensitive to variations of the incident beam angle across the wafer and from wafer to wafer. With simple electrostatic scanning, however, the beam angle with respect to the wafer can change as the beam is scanned from the center of the wafer to its periphery. For a 200 mm or larger wafers, this can cause the angle to change by $> \pm 5°$. The total deflection is found as (Fig. 6.9)

$$D = L \tan \alpha \simeq L \, \frac{a \, V_{\text{def}}}{d \, V_{acc}} \, , \qquad 6.4$$

where $a$ is the length of the plates, $d$ the distance between the plates, $L$ the distance between the center of the plates and the target plane, $V_{\text{def}}$ the deflection voltage, and $V_{acc}$ the acceleration voltage. It is important that the wafer be scanned an adequate number of times so that the scan lines overlap. Parallel-beam scanning is required to reduce shadowing and channeling effects, whereby the beam is scanned at the same nominal incident angle over the entire wafer surface with control of incident angle variation to less than 0.5°. Channeling and shadowing effects are discussed later.

For high-current beams, electrostatic scanning becomes difficult because of space-charge effects. The electrons which initially neutralize the positive charge of the beam are stripped by the

deflection plates. causing a space-charge "blow-up" [21]. Therefore, for high beam current densities, the wafer is either scanned
mechanically in both directions past a stationary beam, or by a
hybrid combination of magnetic and mechanical scanning,
avoiding the disruption of space-charge neutrality. For a perfectly
flat wafer, mechanical scanning with a stationary beam has the
advantage of keeping the angle of incidence constant across the
wafer.

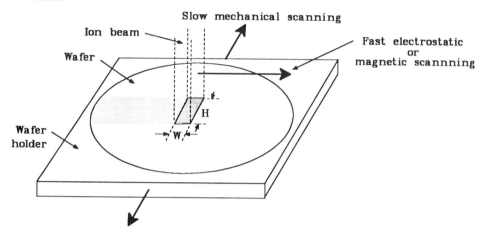

**Fig. 6.10** Hybrid beam scanning.

Beam purity is required to maintain the predicted device
characteristics. As mentioned above, some of the ions can be neutralized after mass separation and acceleration due to collisions
with residual gas atoms, and also by capturing trapped electrons.
For gas-phase charge exchange, the fraction of contamination with
neutrals is found as [22]:

$$\frac{\Delta N}{N} = 3.31 \times 10^{16} \, P \, L \, \sigma_{10} \, \frac{A}{a} \, , \qquad 6.5$$

where $N$ is the total number of atoms, $P$ is the pressure (Torr), $L$
the ion path length (cm), $\sigma_{10}$ the effective cross-section for
neutralization of single-charged ions ($\simeq 10^{-16} - 10^{-15} \, cm^2$), $A$ the
area over which the beam is scanned, and $a$ the cross-sectional
area of the focused beam. Depending on gas pressure, charge
exchange on walls and slit openings can become significant. Since
neutral species do not contribute to the beam current, the rate of
their deposition cannot be measured with the rest of the beam.

Also, since neutrals cannot be deflected by an electric field, they travel in a straight line without being scanned with the rest of the beam across the wafer and can create a spot on the wafer with a greatly increased flux. The fraction of neutrals reaching the surface can be reduced by improving the vacuum or by a "neutral-beam trap". The purpose of the bend in the beam immediately before impact on the wafer is to separate neutral species which can contaminate the beam. Neutral atoms are prevented from reaching the wafer by off-setting the target with respect to the beam. For this purpose, a 6.5-15° electrostatic deflector is placed typically in front of the target and a dc voltage is applied to the deflector to remove the neutrals and act as an energy filter. A controlled slit is also placed at the entrance of the target chamber to improve the energy resolution.

## 6.1.5 Target Chamber

Demands for high throughput, improved tool reliability, increased automation and yield, and the continuing increase in wafer size, has made the target chamber a key area of development. Target chambers are constructed for single wafer or batch processing, depending on the beam power density and ion flux. For batch processing, typical target chambers contain several wafers (up to nine 200-mm wafers) mounted on a spinning disk (or platen) which is also moved laterally across the beam. The batch size depends on wafer diameter and implantation system. Wafer cassettes and load-locks are used to load and unload wafers to minimize particulate contamination during handling and improve throughput.

### Wafer Orientation

The disc can be tilted, rotated, (10-1200 rpm) and translated across the beam for proper wafer orientation and uniform distribution of dopants. Wafer rotation and tilt determine the angle of incidence with respect to crystallographic orientation and the wafer surface (Fig. 6.11). Wafer orientation is defined as the angle of wafer rotation (azimuthal angle, or twist) with respect to a plane normal to the wafer contains the beam and a reference major crystal axis - typically (110). This angle can be varied during implantation, uniformly or in steps, from 0-300°. Wafer tilt is the angle between the beam and the normal to the wafer surface. In most machines, the wafer is tilted 7-10° away from the normal to avoid channeling. Modern machines allow an increase in tilt to 90° for implantation at large tilt-angles. Variable tilt and twist increase process flexibility and reduce channeling, as discussed later.

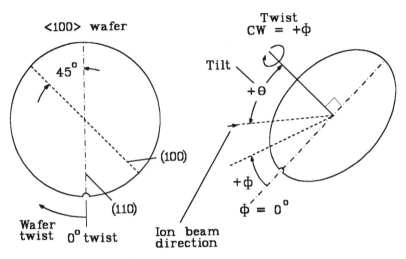

**Fig. 6.11** Definition of tilt and twist

## Dose

The total number of implanted ions per unit wafer area is called the flux, or **dose.** Doses range from $\simeq 10^{10}$ *atoms/cm²* for MOSFET threshold voltage adjustment to $\simeq 10^{18}$ *ions/cm²* for buried insulator layers. A dose variation of $< 0.7\%$ across 200 mm wafers can be achieved in modern implanters. The dose is determined by integrating the beam current in a completed circuit between target and ion source (Fig. 6.12), i.e., by measuring the total electron charge which is necessary to neutralize the implanted positive ions.

For rotating targets, the dose is defined by the integrated charge corrected by the fraction of time the ion beam hits the sample. For a beam of current *I*, swept over an aperture of area *A*, the dose $\phi$ is

$$\phi = \frac{1}{QA} = \int_0^{t_I} I dt \quad (atoms/cm^2), \qquad 6.6$$

where $t_I$ is the implantation time and $Q$ the ion charge. It follows that an increase in beam current improves the throughput by reducing implantation time. A Faraday cage that surrounds the target and has an opening facing the low-energy aperture, is maintained at a small positive voltage (a few 100 V). It collects secondary electrons, emitted from the target by incident ions, and returns them to the integrating circuit. If these secondary electrons were lost to the chamber walls, there would be an error in the dose measurement.

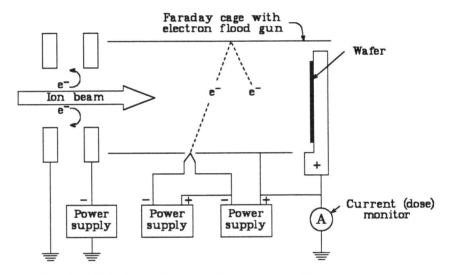

**Fig. 6.12** Principle of implant dose monitor. Wafer and electron flood part of Faraday cage (electron flood discussed below).

### Charging Effects

Implantation is frequently done over an insulating surface on the wafer. Although the ion beam is about 98% neutralized by trapped electrons [23], positive charge can accumulate on the wafer surface as a result of impinging ions and emitted secondary electrons. This can charge up the insulator surface high enough to give rise to substantial current through the insulator, causing its degradation or permanent damage. The charge accumulation can also cause a distortion of the ion beam by stripping the space-charge compensating electrons from the ion beam. The disturbance of space-charge neutrality of the beam often results in the so-called beam blow-up. To reduce charging, the beam is neutralized with excess electrons at the wafer surface [23]. The most commonly used method to provide electrons at low energy is by secondary emission (Fig. 6.13).

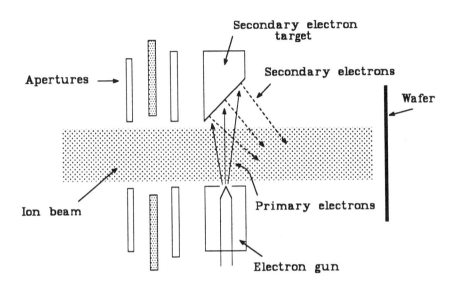

**Fig. 6.13** Schematic of an electron shower (Eaton NV10)

A low electron energy (5-10 eV) is necessary to avoid damaging the insulator by a build-up of negative ions. If the maximum electron energy is $E_0$, then the wafer will charge to only $-E_0$. Therefore, efficient neutralization can be achieved by immersing the wafer in a "shower" or "flood" of low-energy electrons [24]. When a wafer is implanted within a Faraday cage, it is important that the shower be mounted within the Faraday, and to make sure that electrons cannot escape and cause a dose error (Fig. 6.12). An electron shower is not required if charging is minimized, e.g., by reducing the beam current density, increasing the beam spot, increasing the implanted area, or increasing the rotation speed of the disk.

**Wafer Temperature Control**

Control of wafer temperature during implantation is of great importance, particularly at high beam power densities, to ensure pattern integrity and consistent properties of implanted films. In the presence of a photoresist mask, the wafer temperature may not exceed $\simeq 100°C$.

As the implanted ions lose their kinetic energy by collisions with target atoms, almost all the energy is converted into heat. Since the ions typically stop within a few tenths of a micron from the surface, the energy is absorbed by a very small volume of the target material immediately beneath the ion beam. The energy is then dissipated through conduction and radiation. Radiative cooling follows essentially Stefan-Boltzmann's law, and is rather ineffective in cooling wafers during implantation. It is only suited for special cases, such as batch high-energy implanters at low currents (Genus 1500), and high-current oxygen implanters at medium energy (Eaton NV200). In the first case, the beam power is low and distributed over a large target area by spinning the wafer holder (platen). In the second case, no photoresist is used and it is desired to achieve a high wafer temperature ($\simeq 500°$) during implantation of oxygen, as discussed later.

For silicon at low to moderate temperatures, conduction is the primary heat dissipation mechanism. To first approximation, the wafer is assumed to be isothermal and, if some type of conductive cooling is provided to the back of the wafer, such as elastometric or gas cooling, efficient heat conduction from the back of the wafer into the cooled metal platen can be achieved [24]. In elastometric cooling, the metal platen is covered with a thin layer of elastometer which yields under contact pressure and conforms to the microscopic backside surfaces, allowing a larger contact area to the wafer [25]. Because of its simplicity, elastometric cooling is used for batch processing. Also, wafer clamping is not necessary because pressure is provided by the centrifugal force of the spinning platen and by surface tension forces between the wafer and the elastometer. Gas cooling is used mostly in single-wafer implanters. In this case, the wafer is sealed by an O-ring against the disk and an inlet is used to introduce a suitable gas (helium or hydrogen) at an appropriate pressure behind the wafer. High cooling coefficients can be achieved with gas cooling [26, 27]. The use of an O-ring and clamps, however, has the disadvantages of reducing the "useful" wafer area, generating particulates, and creating a convex curvature on the wafer that causes variations in the implant angle.

The rise of wafer temperature can be estimated by making the simplifying assumptions that the wafer is isothermal, radiation and lateral heat conduction are negligible, and the beam scan period is short compared to the cooling decay time [24, 28]. In this case, the wafer temperature rise during implantation, $\Delta T$, can be approximated as

$$\Delta T(t) = \frac{1}{h} \frac{P_B}{A_S} (1 - e^{-t/\tau}),$$ 6.7

where $t$ is the time, $h$ the cooling coefficient, $P_B$ the beam power, $A_S$ the total area scanned by the beam, and $\tau$ the temperature decay time due to the conduction from the back of the wafer to the cooled wafer platen, given by

$$\tau = \frac{\rho L C_P}{h},$$ 6.8

where $\rho$ and $C_P$ and $L$ are the density, specific heat at constant pressure, and thickness of the wafer, respectively. For silicon, $\rho = 2.33 g/cm^3$, $C_P = 0.75 J/g/K$, and $L \simeq 600 \mu m$ (200 mm wafers). The cooling coefficient $h$ ranges typically from 10-25 $mW/cm^2/K$.

When the implantation time $t >> \tau$, Eq. 6.6 simplifies to

$$\Delta T_{max} \simeq \frac{1}{h} \frac{P_B}{A_S}.$$ 6.9

In high current machines, the total area is about 8000 $cm^2$ for 200 mm wafers. With a beam power of 5000 W, the power density is about 0.63 $W/cm^2$. Assuming a cooling coefficient of 10 $mW/cm^2/K$, the maximum temperature rise would be 63 K. Assuming the disk temperature is 293 K, the maximum wafer temperature would then be 356 K (83 °C). The temperature rise is higher when $t$ is comparable to $\tau$. The implant time is related to the total dose as

$$t = \frac{q \phi A_S}{I_B},$$ 6.10

where $\phi$ is the dose, and $I_B$ the beam current. From the above discussion, it becomes evident that wafer temperature rise can be reduced by increasing the cooling coefficient, reducing the beam power density, and reducing the dose.

## 6.2 Energy Loss and Range Distribution

The implanted ions travel some distance in the target material before they are brought to rest by a combination of energy losses. A Monte Carlo simulation of ion tracks for 50 keV boron implanted into silicon is shown in Fig. 6.14.

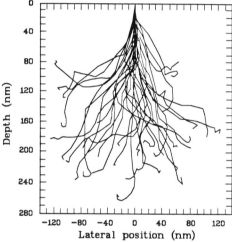

**Fig. 6.14** Two-dimensional projection of Monte Carlo calculation of 128 ion tracks for 50 keV boron implanted into silicon.

## 6.2.1 Energy Loss Mechanisms

An implanted ion is slowed down by several collisions, losing energy during each collision, until its energy drops below some pre-specified value where the ion is assumed to be at rest. The two major mechanisms for energy loss are inelastic electronic stopping and elastic nuclear stopping. The stopping power of the target is the loss of energy per unit distance -dE/ds, defined as

$$-\frac{dE}{ds} = N\left[S_e(E) + S_n(E)\right],$$  6.11

where $E$ is the ion energy, $s$ a coordinate along the ion path whose direction changes as a result of binary nuclear collisions, $N$ the density of atoms in the target material, $S_e$ the electronic stopping power, and $S_n$ the nuclear stopping power. The particles move in straight-line paths between nuclear collisions, and the stopping events are considered to be independent of each other. Therefore, ions lose energy in discrete amounts in nuclear collisions and lose energy continuously from electronic interactions. If expressions for $S_e$ and $S_n$ are known, the ion range can be calculated. The total distance an ion travels in an amorphous medium is called the ion **range** and is found from Eq. 6.11 as

$$R = \int ds = \frac{1}{N} \int_0^E \frac{dE}{S_e(E) + S_n(E)} \, . \qquad 6.12$$

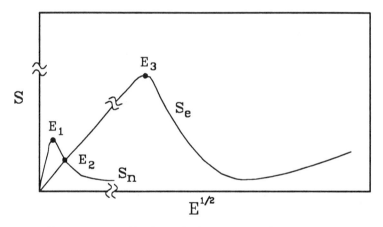

**Fig. 6.15** Qualitative of electronic and nuclear stopping powers as a function of energy Values for $E_1, E_2,$ and $E_3$ are given in Table 6.2 for typical dopants in silicon.

Fig. 6.15 shows schematically how the electronic and nuclear stopping powers vary with energy. Both stopping powers initially increase linearly, reach a maximum, and then decrease. Which of the two mechanisms predominates depends upon the energy and mass of the implanted ion, and the mass and atomic number of the target atom. While both mechanisms must be considered in ion implantation, electronic stopping dominates at high energy per unit ion mass, and nuclear stopping dominates at low energy per unit ion mass. There is a decrease of nuclear stopping efficiency at high energy because the ions move past target atoms at velocities too high to effectively transfer energy to them. Of particular interest are energies $E_1$, $E_2$, and $E_3$ in Fig. 6.15. These values are given in Table 6.2 for silicon as a target material [29]. The cross-over energy, $E_2$, at which electronic stopping becomes more efficient than nuclear stopping is higher for heavier implanted atoms. Since in ion implantation most of the energies are below $E_3$, the region beyond $E_3$ (called the Bethe-Bloch region) will not be considered.

**Table 6.2** Energies $E_1$, $E_2$, and $E_3$ of Fig. 6.15 for dopants in silicon [29]

| Ion | $E_1$ (keV) | $E_2$ (keV) | $E_3$ (keV) |
|-----|-------------|-------------|-------------|
| B | 3 | 17 | $3x10^3$ |
| P | 17 | 140 | $3x10^4$ |
| As | 73 | 800 | $2x10^5$ |
| Sb | 180 | 2000 | $6x10^5$ |

**Electronic Stopping**

Electronic stopping can be described as a result of a continuous drag-force exerted by a "sea" of bound and free electrons on the ion. These collisions are inelastic and result in small energy losses in which the electrons are excited or ejected from their shells and then dissipate their energy through thermal vibrations of the target. Because of the negligible electron mass, electronic collisions do not result in an appreciable change in the direction of the incident ion. In the range of ion implantation energies for dopants in silicon, the stopping contribution due to electronic collisions can be approximated by a velocity-proportional stopping power [30]. Since the ion velocity is proportional to the square root of energy, the relation can be expressed as:

$$-\frac{d\varepsilon}{d\rho} = k\sqrt{\varepsilon} \; . \qquad\qquad 6.13$$

The factor $k$ is defined as [30]

$$k \simeq \frac{0.0793 Z_1^{2/3} Z_2^{1/2} (M_1 + M_2)^{3/2}}{(Z_1^{2/3} + Z_2^{2/3})^{3/4} M_1^{3/2} M_2^{1/2}} \; , \qquad\qquad 6.14$$

where $Z$ and $M$ are the atomic number and mass, and the subscripts 1 and 2 refer to the incident ion and target atoms, respectively. For typical ion and target materials $k$ ranges from 0.1-0.3. The linear part of electronic stopping in Fig. 6.15 should therefore be a set of curves for different k-values. In Eq. 6.13 $\varepsilon$ and $\rho$ are dimensionless energy and length, respectively, given by [30]

$$\varepsilon = \frac{a M_2}{e^2 Z_1 Z_2 (M_1 + M_2)} E \; , \qquad\qquad 6.15$$

and

$$\rho = \frac{4\pi a^2 N M_1 M_2}{(M_1 + M_2)^2} s \; , \qquad\qquad 6.16$$

where

$$a = \frac{\frac{1}{2} \left( \frac{3}{4} \pi \right)^{2/3}}{(Z_1^{2/3} + Z_2^{2/3})^{1/2}} a_o \; , \qquad\qquad 6.17$$

$a_o$ is the Bohr radius ($0.529 x 10^{-8}$ cm), $N$ is the number of atoms per unit target volume, $E$ is the ion energy, $s$ the ion path length, and $e$ the electronic charge. A dimensionless unit for $\varepsilon$ in Eq. 6.15 is obtained by using $cm$ for $a$, $eV$ for energy, and converting the electronic charge $e$ to $1.602 x 10^{-19} x 9 x 10^9$ $V \, cm$. Combining Eqs. 6.17 and 6.15 gives a the practical relation

$$\varepsilon = \frac{32.53 M_2}{Z_1 Z_2 (M_1 + M_2)(Z_1^{2/3} + Z_2^{2/3})^{1/2}} E \; , \qquad\qquad 6.18$$

where $E$ is in $keV$. The theory which led to Eqs. 6.13-6.17 was developed by Lindhard, Scharff, and Schiott for homogeneous, isotropic material, and is subsequently referred to as the LSS theory [31].

From a slightly different model, the electronic stopping power can be reduced to [2]

$$S_e(v) = 2.34x10^{-23}(Z_1 + Z_2)\, v\ (eVcm^2),\qquad 6.19$$

where v is the speed of the projectile in cm/s. A more detailed analysis of electronic stopping shows that stopping goes as $S_e \propto v^{0.7}$ in silicon (and germanium) for low-velocity ions with atomic number less than 19 [32]. Another useful approximation for the electronic stopping power is

$$S_e(E) \simeq 2x10^{-16}\sqrt{E}\quad (eVcm^2).$$

Calculated values of electronic energy loss for boron, arsenic, and phosphorus in silicon are shown as a function of energy in Fig. 6.16 [31].

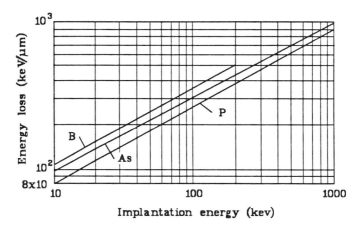

**Fig. 6.16** Calculated values for electronic energy loss for boron, arsenic, and phosphorus in silicon as a function of ion energy [31].

**Nuclear Stopping**

Nuclear collisions are Coulombic interactions between the ion, screened by its orbital electrons, and the target atom which is also a nucleus screened by its orbital electrons. As the incident ion approaches the target atom, it experiences an increasing electrostatic repulsion which deflects its path. When the energy transferred to an atom is larger than a threshold value of approximately 10-25 eV, the crystal atom will be dislodged from its lattice site and a vacancy is created. In ion implantation, the energy transferred to a crystal atom is far in excess of the displacement energy so that the recoiled atom will in turn cause the displacement of other lattice atoms, resulting in a cascade (Fig. 6.17). An ion can penetrate several monolayers of a target, constantly losing energy to electrons and following a linear path, before there is a collision with a target atom which is "hard" enough to displace that atom. Depending on its mass and energy, an implanted atom can displace as many as $10^4$ silicon atoms by nuclear collisions before it comes to rest (Sec. 6.2.6).

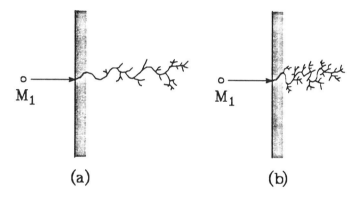

**(a)**                                           **(b)**

**Fig. 6.17** Schematic illustrating cascades of nuclear collisions causing end-of-range crystal damage. (a) Light ions, $M_1 < M_2$. (b) Heavy ions, $M_1 > M_2$.

If the incident ion and target atom were bare nuclei, then the Coulombic potential between them would be given by

$$V(r) = \frac{e^2 Z_1 Z_2}{4\pi\varepsilon_o r} \, , \qquad 6.20$$

where $r$ is the separation between the nuclei, and $\varepsilon_0$ the permittivity of free space ($\varepsilon_0 \simeq 8.86 x 10^{-14}$ F/cm). This potential would define the force of repulsion as a function of distance between the incident ion and target atom, the resulting angles of recoil and deflection, and the associated loss of incident ion energy. Since electrons screen the nuclear charge, the potential in Eq. 6.20 must be reduced by a screening function, such as the Thomas-Fermi screened potential described in [32]. Given the more realistic potential, the scattering angle and energy loss for any incident ion can be numerically calculated. In its simplest form, the collision event can be described by a two-body system, considering the conservation of energy and momentum (Fig. 6.18).

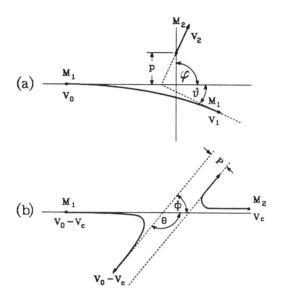

**Fig. 6.18** Scattering variables of a two-body collision. (a) Laboratory system. (b) Center-of-mass coordinates. Details in text [32].

The form of the equations involved is greatly simplified by a transformation to the center-of-mass (CM) coordinates [32, 33]. In the laboratory system, the incident ion with mass $M_1$ has an initial velocity $v_o$ and an impact parameter $p$ with the target atom. The ion is deflected by an angle $\theta$ with a final velocity $v_1$. The target atom with mass $M_2$ recoils at an angle $\phi$ with velocity $v_2$. In center-of-mass coordinates, the system has zero total momentum and moves with velocity $v_c$ relative to the laboratory coordinates. The new angles of scatter are $\Theta$ and $\Phi$ [32]. The scattering angle $\Theta$ can be found as a function of the impact parameter $p$ (Fig. 6.18).

From classical mechanics, the energy lost, $T_N(p, E)$, in an individual collision is

$$T_n(p,E) = \frac{4M_1M_2}{(M_1 + M_2)^2} E \sin^2\left[\frac{\Theta(p)}{2}\right]. \qquad 6.21$$

The energy lost by nuclear collisions is proportional to the atomic density and to the total energy transferred in all individual collisions:

$$S_n(E) = \frac{1}{N}\left(\frac{dE}{ds}\right)_n = \int_0^\infty T_n(E,p)\, d\sigma, \qquad 6.22$$

where $d\sigma = 2\pi p dp$ is the differential cross-section. For practical calculations, the universal nuclear stopping can be approximated as [32]

$$S_n(E) = \frac{8.462 \times 10^{-15} Z_1 Z_2 M_1 S_n(\varepsilon)}{(M_1 + M_2)(Z_1^{0.23} + Z_2^{0.23})} \quad eV/(atom/cm^2), \qquad 6.23$$

where, for $\varepsilon \leq 30$ (energy range of interest)

$$S_n(\varepsilon) \simeq \frac{\ln(1 + 1.1383\,\varepsilon)}{2[\varepsilon + 0.01321\varepsilon^{0.21226} + 0.19593\varepsilon^{0.5}]}, \qquad 6.24$$

and $\varepsilon$ is defined in Eqs. 6.15 and 6.18.

## 6.2.2 Range Distribution

The total distance an ion travels before it comes to rest is called the range, R, defined by Eq. 6.12. Of practical interest is the average depth below the surface an ion penetrates. This depth is typically shorter than the actual distance the ion travels. It is referred to as the **mean projected range, $R_p$,** and defined as

$$R_p = \langle x \rangle = \frac{1}{N}\sum_i x_i, \qquad 6.25$$

where $x_i$ is the projected range of ion "i" on the x-axis, i.e. the perpendicular distance from the surface to the end of the ion track. The relation between $R$ and $R_p$ is approximated as [29, 30]

$$R_p \simeq \frac{R}{1 + [\frac{M_2}{3M_1}]} \ . \qquad 6.26$$

Some ions stop at a depth smaller than $R_p$ and others at a depth larger than $R_p$. For amorphous material, the distribution about $R_p$ can be approximated by a Gaussian with a standard deviation, or **straggling, $\Delta R_p$**, defined as

$$\Delta R_p = \left[ \frac{1}{N} \left( \sum_i x_i^2 \right) - R_p^2 \right]^{1/2} . \qquad 6.27$$

For implantation at normal incidence to the wafer, through a mask with vertical edges, there is a lateral spread of ions which can also be approximated by a Gaussian with a lateral standard deviation, or **transverse straggle, $\Delta R_T$,** defined as

$$\Delta R_T = \left[ \frac{1}{N} \sum_i \left( \frac{y_i + z_i}{2} \right)^2 \right]^{1/2} , \qquad 6.28$$

where $y$ and $z$ are the coordinates of the plane normal to $x$, i.e. parallel to the wafer surface. Fig. 6.19 illustrates the relation between R, $R_p$ $\Delta R_p$, and $\Delta R_T$.

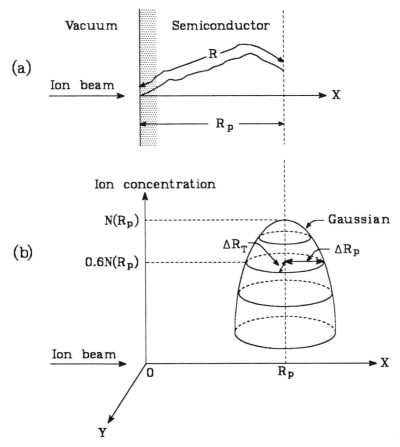

**Fig. 6.19** Definition of range parameters. (a) Ion range $R$ and mean projected range $R_p$. (b) Range straggling $\Delta R_p$ and $\Delta R_T$.

In the ideal case of a uniform, amorphous or preamorphized target in which the number of ion collisions and the energy transfer per collision are random, the final distribution of ions can be approximated by a simple Gaussian function, using only two moments, the projected range, $R_p$, and the range straggling, $\Delta R_p$:

$$N(x) = N_{max}e^{-(x-R_p)^2/2\Delta R_p^2}, \qquad 6.29$$

where $N(x)$ is the concentration of implanted ions as a function of depth, and $N_{max}$ the peak concentration of implanted ions. Eq. 6.29 can be rewritten as:

$$N(x) = \frac{\phi\, e^{-(x-R_p)^2/2\Delta R_p^2}}{\Delta R_p \sqrt{2\pi}} \qquad\qquad 6.30$$

where $N_{max}$ is substituted by

$$N_{\max} = \frac{\phi}{\Delta R_p\sqrt{2\pi}} \simeq \frac{0.4\phi}{\Delta R_p}\,, \qquad\qquad 6.31$$

and $\phi$ is the dose (or flux) defined by Eq. 6.6. A detailed analysis shows that $R_P$ depends mainly on the energy and mass of the incident ion while $\Delta R_P$ depends on the ratio of the mass of the incident ion to that of the host lattice atom.

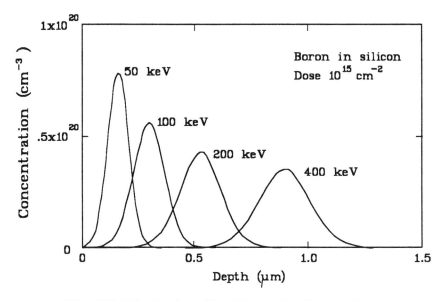

**Fig. 6.20** Calculated profile of boron in silicon using the range parameters in Fig. 6.21a (linear scale). [34].

Typical Gaussian distributions obtained from Eq. 6.29 for boron implanted in silicon at different energies and the same dose are shown on a linear scale in Fig. 6.20. As the implant energy increases, the average depth increases and the peak concentration decreases because the range straggling increases. In practice, however, the profiles are plotted on a semi-logarithmic scale for concentrations over several orders of magnitude. Several numerical methods have been used to calculate $R_p$ and $\Delta R_p$ in amorphous materials based on the LSS theory [30, 34 − 36]. While an appreciable variation in range statistics is observed in published data, the selection of values for profile calculations depends on how well the profile agrees with measurement. Typical $R_p$ and $\Delta R_p$ values for implanted dopants in silicon, silicon dioxide, silicon-nitride, and photoresist are plotted as a function of implanted energy on a log-log scale in Figs. 6.21. Range tables with more detailed information on implanted species are shown in Table 6.9 at the end of the chapter. An interesting result is that the ratio $\Delta R_p / R_p$ decreases as the implant energy increases and as the atomic number of the implanted ions increases.

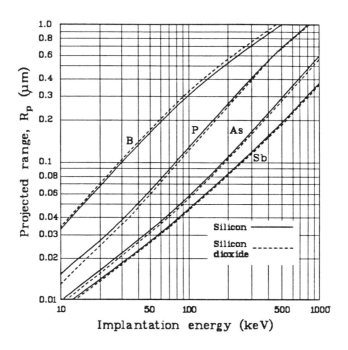

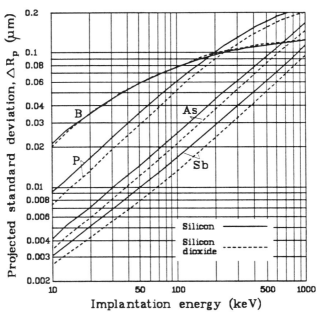

**Fig. 6.21a** Projected range $\Delta R_p$ and range straggling $\Delta R_p$ for antimony, arsenic, boron, and phosphorus in silicon and silicon dioxide [34].

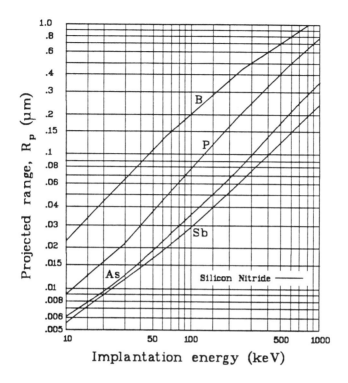

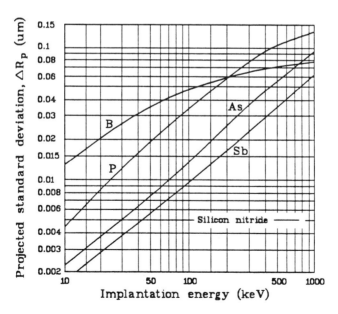

**Fig. 6.21b** Projected range $R_p$ and range straggling $\Delta R_p$ for antimony, arsenic, boron, and phosphorus in silicon nitride [34].

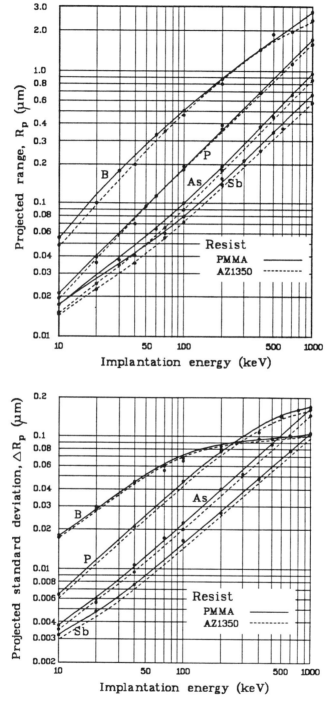

**Fig. 6.21c** Projected range $R_p$ and range straggling $\Delta R_p$ for antimony, arsenic, boron, and phosphorus in photoresist, AZ1350 and PMMA [36].

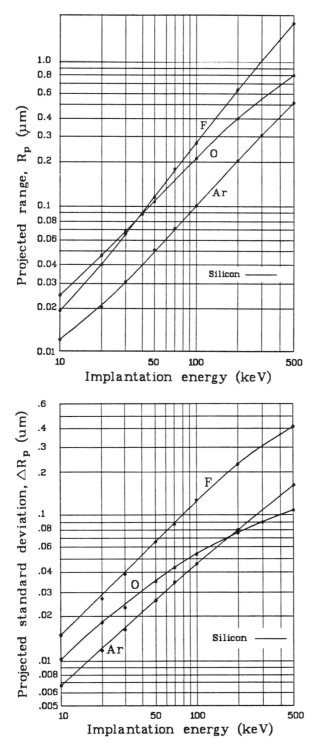

**Fig. 6.21d** Projected range $R_p$ and range straggling $\Delta R_p$ for argon, fluorine, and oxygen in silicon [36].

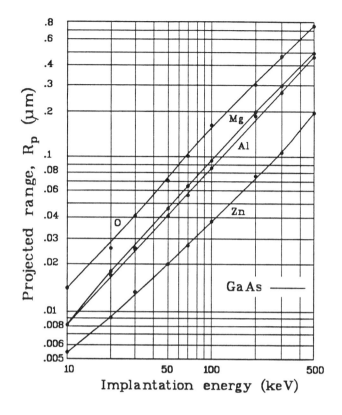

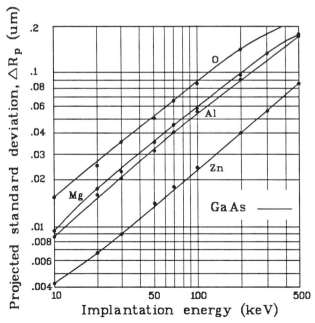

**Fig. 6.21e** Projected range $R_p$ and range straggling $\Delta R_p$ for aluminum, magnesium, oxygen, and zinc in gallium arsenide [36].

## 6.2.3 Departure from the Normal Distribution

Measured implanted profiles are rarely quite Gaussian in shape. Although a good fit to a Gaussian distribution is observed near the peak, there can be a pronounced skewness in the actual distribution (Fig. 6.22). The departure from the normal distribution is not a problem if the thermal cycles that follow implantation result in a substantial diffusion of dopants, so that non-ideal shapes are "smeared-out". When shallow and ultra-shallow implanted profiles are required, however, the thermal budget is considerably reduced to avoid substantial movement of impurities. A "tail" in the dopant profile, whether caused during implantation or by enhanced low-temperature diffusion due to implant damage, can seriously affect device parameters. Implant damage is discussed in more detail in Sec. 6.2.6 and solid-state diffusion in Chap. 7.

### Skewness

When boron, arsenic, or phosphorus is implanted into silicon, the dopant profiles will be typically non-symmetrical. For boron, there will be more ions for $x < R_P$ than for $x > R_P$, as shown in Fig. 6.22a. In the case of arsenic, more ions will be found for $x > R_P$ than for $x < R_P$ (Fig. 6.22b). The boron implant is said to be skewed toward the surface while the arsenic implant is skewed away from the surface. For phosphorus, the profile is skewed away from the surface at energies below $\simeq$ 130 keV and skewed toward the surface for higher energies. At the intermediate energy of $\simeq$ 130 keV, the phosphorus implant profile is almost exactly Gaussian (Fig. 6.22c). The different skewness can be visualized by thinking of forward momentum [37]. For implanted light atoms, such as boron, they will be a large amount of backward scattering resulting in filling-in of the distribution toward the surface. For heavy ions, such as arsenic, there is a disproportionate amount of forward scattering, resulting in filling-up of the distribution away from the mean projected range.

The curves can be approximated by a three-moment approach, using two Gaussians with their straggle $\Delta R_{p1}$ and $\Delta R_{p2}$, and joining their halves at their "modal range" $R_M = R_p - 0.8(\Delta R_{p1} - \Delta R_{p2})$, following the relations [38,39]

$$N(x) = \frac{2\phi}{\sqrt{2\pi}\,(\Delta R_{p1} + \Delta R_{p2})}\; e^{-(x - R_M)^2/2\Delta R_{p1}^2} \quad x \geq R_M, \qquad 6.32a$$

$$N(x) = \frac{2\phi}{\sqrt{2\pi}\,(\Delta R_{p1} + \Delta R_{p2})}\; e^{-(x - R_M)^2/2\Delta R_{p2}^2} \quad x \leq R_M. \qquad 6.32b$$

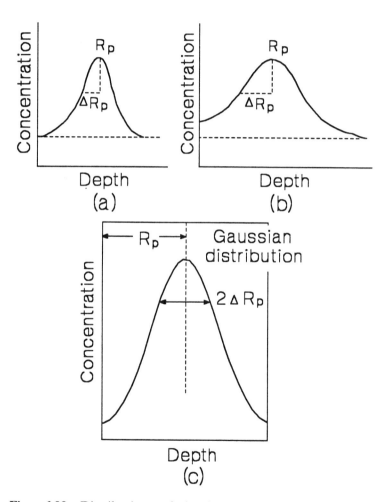

**Fig. 6.22** Distribution of implanted boron, arsenic and phosphorus to illustrate skewness. (a) Boron, skewed to the surface. (b) Arsenic, skewed away from the surface. (c) Phosphorus at an intermediate energy of 130 keV, Gaussian.

A more accurate distribution can be obtained by using two moments in addition to $R_p$ and $\Delta R_p$, namely the skewness defined as

$$\gamma = \frac{1}{N\Delta R_p^3} \sum_i (x_i - R_p)^3 \,, \qquad\qquad 6.33$$

and **kurtosis** defined as

$$\beta = \frac{1}{N\Delta R_p^4} \sum_i (x_i - R_p)^4 \,, \qquad\qquad 6.34$$

where $N$ is the total number of implanted ions. For $\gamma < 0$ the skewness is toward the surface, and $\gamma > 0$ places places the peak of the distribution closer to surface than $R_p$. Kurtosis describes whether the peak of the distribution is sharp or flat. For $\beta = 3$ and $\gamma = 0$, the distribution is normal (Gaussian). For $\beta < 3$ the distribution is flatter at the top, and $\beta > 3$ indicates a more pointed curve. Several relations lead to a distribution known as Pearson-IV [40], which in many cases is combined with an exponential function to improve the calculated profile of the distribution tail [41,42,43]. A comparison between measured profiles and fitted four-moment distributions for boron in silicon is shown for energies 30-800 keV in Fig. 6.23 [41]. Details on the features of Pearson-IV distributions and the coefficients used to calculate them can be found in [44,45].

**Channeling**
So far, the target was assumed to be amorphous and the scattering events randomly distributed and isotropic. Range distributions in single-crystal semiconductor targets, however, differ from those in amorphous targets because of the possibility of implanted ions to find their way into open directions, or "channels", in the crystal and penetrate deeper than projected for amorphous targets. Directional effects are best visualized by looking at a two-dimensional projection of a silicon crystal model in Fig. 6.24 [29]. When the crystal is oriented in the $<110>$ direction normal to the plane, it faces the ion with large open channels which extend deep into the crystal (Fig. 6.24a). With a tilt of 10° with respect to the normal, the crystal appears as an almost randomly distributed system of atoms (Fig. 6.24b).

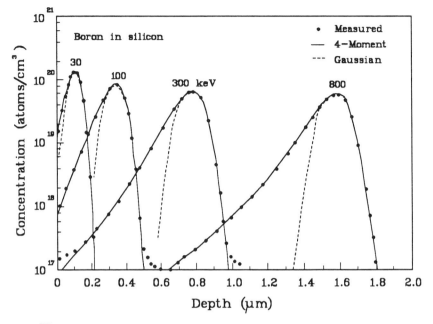

**Fig. 6.23** As-implanted boron profiles in amorphous silicon. Comparison with Gaussian approximation and Pearson-IV at four different energies. Ions were implanted into fine-grained silicon to avoid channeling [41].

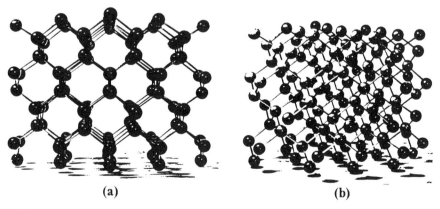

**Fig. 6.24** Two-dimensional projection of atomic configuration in a silicon crystal. (a) Viewed normal to the < 110 > direction. (b) Tilted ≃ 10° from the < 110 > direction [29].

When an ion enters a crystal within a certain angle of a major axis (such as <110>, <111>, or <100>) or a crystallographic plane, the rate of its energy loss to the crystal is reduced [2,29,42,43,44]. Within these "channels", there is practically no nuclear collisions. Each time the ion approaches the aligned rows (axial channeling) or planes (planar channeling) of lattice atoms, the repulsion force between the screened Coulomb fields of the ion and lattice atoms is sufficient to steer it away, preventing "violent" nuclear collisions from occurring over a considerable distance. Only electronic stopping slows the ion down and constrains it to move along a crystallographic axis (axial channeling), or between a set of planes (planar channeling). The relations for electronic stopping in Eqs. 6.13-6.19, however, cannot be used for channeling without modification because they are derived for uniform, amorphous materials.

The ion beam does not have to be exactly aligned to a major axis or plane for channeling to occur. There is an angle of incidence to a major axis or plane, called "critical angle", $\psi_c$, below which an ion can penetrate deep in the channel without appreciable nuclear collisions. The concept of a critical angle can be best understood by splitting the ion energy into its two components, one parallel and one normal to the axis of the channel (Fig. 6.25). As long as the normal component is smaller than the repelling potential of the atomic chain, the ions remains in the channel. In the range of typical ion implantation energies, the critical angle is defined as [29]

$$\psi_c = \left( \psi_1 \frac{a}{d} \right)^{1/2}, \qquad 6.35$$

where $a$ is defined by Eq. 6.17, $d$ is the atomic spacing along the aligned rows, and $\psi_1$ is given by

$$\psi_1 = \left[ \frac{2Z_1 Z_2 e^2}{Ed} \right]^{1/2}.$$

There is also a probability of the ion's being deflected out of the channel (de-channeled), particularly at elevated crystal temperatures where lattice atom vibrations increase in amplitude. Therefore, even if one orients the wafer precisely in a channel direction, there will be a large fraction of ions that are de-channeled and the profile will have an "amorphous component". Also, channeling can still occur even at incident angles larger than $\psi_c$ if subsequent nuclear collisions steer the ion in a channel direction. Critical angles for channeling of boron, phosphorus, and antimony in silicon are given in Table 6.3 [2, 29].

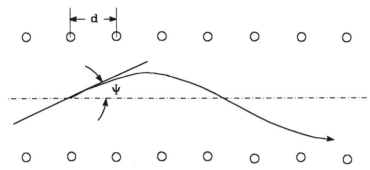

**Fig. 6.25** Schematic of an ion path in an axial channel for $\psi \leq \psi_c$.

**Table 6.3** Critical angle for channeling of boron, phosphorus, and antimony in silicon [2, 29]

| | | Critical angle (°) | | |
|---|---|---|---|---|
| Ion | Energy (keV) | < 100 > | < 110 > | < 111 > |
| Boron | 10 | 4.76 | 6.97 | 5.30 |
| | 100 | 2.67 | 3.47 | 2.98 |
| | 300 | 2.03 | 2.98 | 2.26 |
| Phosphorus | 10 | 5.79 | 7.51 | 6.45 |
| | 100 | 3.26 | 4.22 | 3.63 |
| | 300 | 2.47 | 3.21 | 2.76 |
| Antimony | 10 | 6.95 | 9.01 | 7.74 |
| | 100 | 3.91 | 5.07 | 4.35 |
| | 300 | 2.97 | 3.84 | 3.31 |

From Table 6.3 it can be seen that as the implant energy decreases, more tilt angle is required to reduce channeling. Experimentally, a large number of "microchannels" have been found within 20° of the < 100 > direction in silicon [47]. Calculations for boron implanted into <100> silicon show enhanced axial channeling in <100> and <110> [48]. Channeling is therefore extremely dependent on the crystal orientation with respect to the beam, as can be seen for phosphorus and boron in Figs. 6.26a and 6.26b [49,50]. The wafers were annealed at 850 °C to activate the

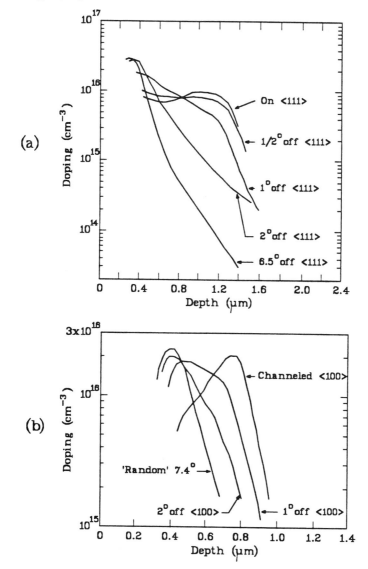

**Fig. 6.26** Dependence of channeling effect on tilting. (a) Phosphorus implanted at an energy of 300 keV and a dose of $10^{12}\,cm^{-2}$ into <111> silicon [49]. (b) Boron implanted at an energy of 150 keV and a dose of $7x10^{11}\,cm^{-2}$ into <100> silicon [50]. The dopants were activated at 850°C.

dopants without appreciable movement of impurities (Sec. 6.2.6), so that capacitance-voltage profiling of the free carrier concentrations could be made. Typical implantation conditions are chosen to minimize axial and planar channeling, and the wafer is tilted and twisted during implantation to maximize channeling. Channeling can be considerably reduced, however, not eliminated by tilting the major crystal axis 7° to 10° away from the incident direction [51], and twisting the wafer by 30° with respect to the <110> direction [52,53].

The number of channeling ions can be substantially reduced by placing an amorphous layer on the target surface which scatters the ions randomly before they impact on the crystal. This can be achieved by amorphizing the crystal before implanting dopants, e.g., by implanting a high dose of inert ions, silicon, germanium, or $BF_2$ into a shallow region of the surface [54], or by depositing or growing an amorphous layer (such as $SiO_2$, $Si_3N_4$) on the crystal and implanting through it. The effectiveness of the amorphous layer depends on its thickness compared to the mean projected range. This is illustrated in Fig. 6.27 for phosphorus implanted into silicon through an oxide film of varying thickness [2, 49].

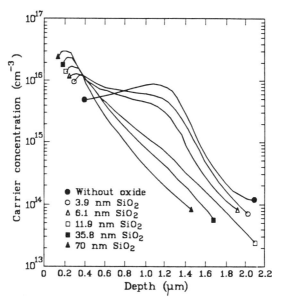

**Fig. 6.27** Channeling of phosphorus implanted at an energy of 300 keV and a dose of $10^{12}\, cm^{-2}$ into <111> silicon as a function of thickness of an oxide layer. The dopants were activated at 850°C [49].

Channeling also depends on dose and temperature. As the dose increases, more crystal damage is created. In the extreme case, the implanted layer becomes amorphous and channeling is eliminated (crystal damage is discussed Section 6.2.6). The effect of crystal damage is shown in Fig. 6.28 for phosphorus implanted at 40 keV and different doses into < 110 > silicon. In this particular case, there is a critical dose of $\simeq 5x10^{14}\ cm^{-2}$ above which negligible channeling is observed [29]. The tail at high doses is due to channeling before the amorphous layer is formed.

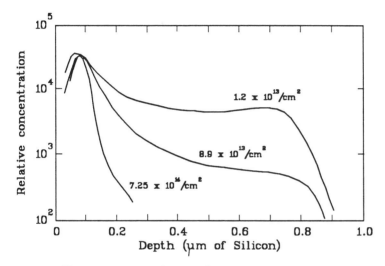

**Fig. 6.28** Dependence of channeling effect on dose for phosphorus implanted in silicon [51].

As the crystal temperature is increased, two competing effects occur. First, the amplitude of thermal vibrations increases, causing increased dechanneling of ions. Second, the crystal damage created at high doses is partially or completely annealed (Sec. 6.2.6), resulting in more lattice order and hence increased channeling probability. This is illustrated in Fig. 6.29 for phosphorus implanted at an energy of 40 keV and a dose of $1.2x10^{13}\ cm^{-2}$ into < 111 > silicon at two different temperatures, showing a net decrease in channeling at elevated temperature [51].

The three regions of a channeled profile are illustrated in Fig. 6.30. Region 1 corresponds to implanted ions into an amorphous region. The plateau in region 2 is created by de-channeled ions which are scattered out of the channel by irregularities in the

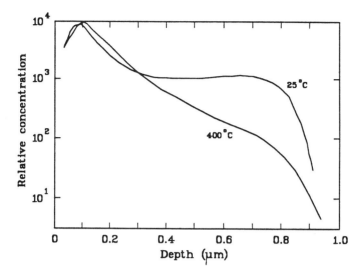

**Fig. 6.29** Channeling effect of phosphorus implanted at an energy of 40 keV and a dose of $1.2x10^{13}\ cm^{-2}$ into <111> silicon at 25°C and 400°C [51].

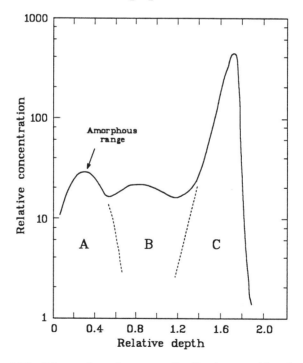

**Fig. 6.30** Schematic of range distributions with channeling showing the three important regions of the profile.

crystal lattice. A distinct separate channeling peak which is much deeper than in amorphous targets can be seen in region 3. The practical use of channeling to obtain deep buried layers corresponding to region 3, however, requires careful alignment of the crystal surface and a constant angle of incidence over the entire wafer surface. This may be very difficult and complicated by dechanneling effects caused by beam heating.

## Implantation into a Two-Layer Structure

Ion implantation is frequently done through a thin insulating amorphous layer (typically silicon dioxide) to reduce channeling, prevent contaminants from penetrating the crystal, and provide a means to place the peak of the distribution close to the semiconductor surface, for example, when adjusting the MOSFET threshold voltage. Sputtered metal atoms which originate from the mask or from the implantation system can have serious effects on device characteristics if they reach the semiconductor surface. This can be avoided by partially or completely etching the protective film after implantation. When the stopping powers of the insulator and silicon are approximately the same, the dual structure can be treated as one layer of a single material and the implanted profile predicted using the simple range statistics described earlier. This method can be applied, for example, to boron implanted into silicon (assumed amorphous) through an oxide film without creating an appreciable error. For other combinations of amorphous films, the range distribution can be predicted by numerically solving Boltzmann transport equations [45]. For crystalline and amorphous material, accurate distributions can be obtained by using the Monte Carlo approach. In this method, the history of an energetic ion is followed through successive collisions with target atoms assuming binary collision events [33]. The calculation of each trajectory begins with a given energy, position, and direction. A profile is generated by calculating a large number of trajectories ($\geq 10^3$), and plotting histograms of the number of ions stopped within each depth interval. The simplest method to estimate the profile in multi-layered targets, however, is to transform the films into an equivalent thickness of substrate material, $t_{EQ}$, where each layer is scaled in thickness according to the relative ranges of the ions in that layer and in the substrate. For example, the equivalent thickness of a silicon dioxide film on silicon, $t_{EQ}$, is

$$t_{EQ} = t_{Ox} \frac{R_p^{Si}}{R_p^{Ox}} , \qquad 6.36$$

where $Si$ and $Ox$ are for silicon and silicon dioxide, respectively. Eq. 6.36 makes the distribution in the scaled target the same as in an unlayered silicon target. The actual implantation profile is then obtained by reversing the scaling, i.e., by multiplying the concentration at any point in each layer by the ratio $R_p^{Si}/R_p^{Ox}$. A refinement of this approximation can be found in [55]. For an insulator film of thickness $t_i$ on a substrate of infinite thickness, one can estimate the profile by constructing two Gaussian profiles using the following approximation [56]:

$$0 \leq x \leq t_i \qquad N_1(x) \simeq \frac{\phi}{\sqrt{2\pi} \, \Delta R_{p1}} \, e^{-(R_{p1} - x)^2/2\Delta R_{p1}^2} , \qquad 6.37$$

$$x \geq t_i \qquad N_2(x) \simeq \frac{\phi}{\sqrt{2\pi} \, \Delta R_{p1}} \, e^{-B} , \qquad 6.38$$

where

$$B = \frac{\left[ t_i + \dfrac{(R_{p1} - t_i)\Delta R_{p2}}{\Delta R_{p1} - x} \right]^2}{2\Delta R_{p2}^2} ,$$

$R_p$ and $\Delta R_p$ are the projected range and range straggling, and the subscripts 1 and 2 are for the thin insulator film and substrate, respectively. The above approximation will result in a discontinuity at $x = t_i$. Table 6.4 summarizes the mathematical models used to predict impurity profiles and the limitations of each model.

## 6.2.4 Masking

The "ultimate" implantation system would be one which does not necessitate a mask. Such a system would allow direct implantation into selected regions by programming the movement of a finely focussed ion beam (FIB). This technique, also called direct beam writing, requires high-intensity beams with spot sizes $\simeq 0.1$ nm for realistic writing speeds and dimensions. While sources with high brightness have been developed for focused beams of dopants such as boron and arsenic [57, 58], the implantation times are still prohibitively long, particularly when large wafers are used.

**Table 6.4**

| Model | Limitations |
|---|---|
| Gaussian 2-moments | Amorphous material, $R_p$ , $\Delta R_p$ , $\Delta R_T$ |
| Joined-half Gaussian 3-moments | Amorphous material, skewness $R_p$ , $\Delta R_{p1}$ , $\Delta R_{p2}$ , $\Delta R_T$ |
| Pearson-IV 4-moments | Amorphous material, skewness, kurtosis, $R_p$ , $\Delta R_p$ , $\Delta R_T$ |
| Pearson-IV + exponential tail | Same as Pearson-IV with good approximation of channeling |
| Boltzmann transport equations | Multilayered structures of amorphous material. CPU intensive. |
| Monte-Carlo approach | Single or multilayered structures of crystalline or amorphous material. Very CPU intensive. |

In typical implantation systems, the whole wafer is implanted by scanning the beam and/or moving the wafer. When selective doping is required, a patterned film of suitable material is used, which is sufficiently thick to act as a mask that prevents penetration of dopants into substrate regions shielded by the film. The most commonly used masking materials for ion implantation are silicon dioxide, silicon nitride, polysilicon, and photoresist. Metals are used in some cases. When an insulator or polysilicon is used for masking, the temperature rise during implantation does not affect the mask. Also, the layers can be left on the surface during high-temperature treatment after implantation. In many designs, the layers remain permanently as part of the finished structure (Fig. 6.31).

The use of photoresist as a mask allows selective doping at essentially room temperature. In this case, the film temperature during implantation may not exceed ≃ 100 °C, otherwise distortion of the resist pattern can occur. One important advantage of photoresist is that the same film which is employed to pattern and etch layers to define device structures can be used as an implantation mask.

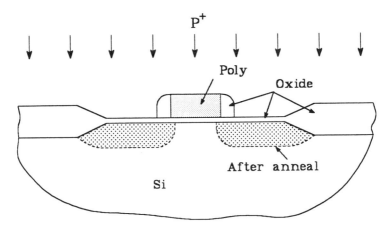

**Fig. 6.31** Patterned polysilicon gate with oxide sidewalls for "self-aligned" implantation of dopants to form the source and drain of a MOSFET. The gate and sidewalls remain part of the final structure.

The use of two photoresist layers for masking boron in selected local oxidation (LOCOS, Chap. 2) regions is illustrated in Fig. 6.32. The first mask is used to selectively etch the nitride, while a thin oxide film is left in the windows, which is practically transparent to the boron ions at the chosen energy. The first resist film is then "hardened" so that when a second "blocking" resist mask is defined, the first resist remains unaffected. This double patterning has the advantage of allowing the boron to reach selected regions which have been etched and defined by the first resist, while other etched windows are shielded by the second resist.

Problems can be encountered with resist masks when the implantation dose reaches levels above $\simeq 10^{14}\, cm^{-2}$. If no precautions are taken to control the wafer temperature by cooling the platen or by limiting the power density (e.g., platen rotation), pattern distortions can occur as a result of resist edge profile changes. Also, blistering and cracking in the resist can result from local increase in temperature during high-dose implantation. Negative working resists can tolerate temperatures up to $\simeq$ 150 °C, while positive working resists begin to degrade at $\simeq$ 100 °C [59]. Another problem which is not directly related to thermal effects is carbonization of resist at high doses. When energetic ions strike photoresist, they free hydrogen atoms from the hydrocarbon matrix. This causes outgassing of (primarily) hydrogen during implantation and affects the implanter's vacuum and dose monitoring system [60]. The resulting hydrogen-depleted "carbonized" layer of the resist film changes to disordered graphite and makes it difficult to remove after implantation (resist hardening) [61, 62]. If the resist thickness is larger than the ion range, the lower layers of the film will not harden and can be dissolved in the stripping solvent. Otherwise, an oxygen plasma becomes necessary to remove the resist.

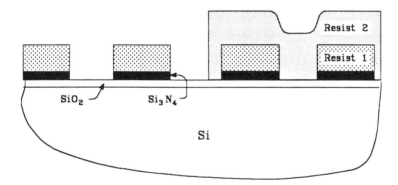

**Fig. 6.32** Dual photoresist masking to selectively implant boron in LOCOS regions. The first resist defines LOCOS windows and the second resist blocks implantation from selected LOCOS regions.

The ion ranges in photoresist (mostly carbon) are larger than in silicon, silicon-dioxide and silicon-nitride (Figs. 6.21), so that thicker layers of photoresist are necessary for masking. To define the minimum mask thickness of any material, one must know the range distribution in the masking material and the allowable concentration of ions penetrating into the semiconductor surface. The effectiveness of the mask can be defined in terms of the fraction of the dose it absorbs, or the fraction of the dose that penetrates beyond the mask (Fig. 6.33) [63]. For a mask thickness

$t_M$, the fraction of the dose that penetrates the mask is the area under the curve of concentration versus distance from $t_M$ to infinity divided by the total dose $\phi$. For a Gaussian profile, this fraction is found as

$$\frac{\int_{t_M}^{\infty} N(x)dx}{\phi} = \int_{t_M}^{\infty} \frac{1}{\sqrt{2\pi}\,\Delta R_p}\, e^{-(x-R_p)^2/2\Delta R_p^2}\, dx\,, \qquad 6.39$$

where $R_p$ and $\Delta R_p$ are the range and straggling in the mask material, respectively. Using the relation

$$erfc\,(z) = \frac{2}{\sqrt{\pi}} \int_{z}^{\infty} e^{-z^2} dz\,,$$

Eq. 6.39 reduces to

$$\frac{\int_{t_M}^{\infty} N(x)dx}{\phi} = \frac{1}{2}\, erfc\left(\frac{t_M - R_p}{\sqrt{2}\,\Delta R_p}\right). \qquad 6.40$$

Similarly, the dose absorbed by the mask, $\phi_M$ is found as

$$\phi_M = \frac{\phi}{2}\,\left(1 + erf\,\frac{t_M - R_p}{\sqrt{2}\,\Delta R_p}\right). \qquad 6.41$$

Values for the complementary error function are given in appendix A. at the end of the book. A plot of the absorbed and penetrated fractions of the dose is given for a Gaussian profile as a function of the mask parameter $(R_p - t_M)/\Delta R_p$ in Fig. 6.33. The lower horizontal scale shows the percentage of the dose beyond the mask and the upper horizontal scale shows the percentage of the dose stopped in the mask [63].

A minimum mask thickness is specified to ensure that the fraction of the dose that penetrates beyond the mask remains below a tolerable value. This thickness, however, should not be larger than necessary to allow the resolution of small features. Since the range parameters are not precisely known, it is advisable to make some allowance for error. The required mask thickness is typically defined in terms of the range and range straggling in the mask material:

$$t_M = R_p + k\Delta R_p,$$

6.42

where $k$ is a factor that depends on the required masking efficiency. For example, to ensure that no more that 0.001% of the implanted dose penetrates into the substrate, $k$ must be larger than 4.3.

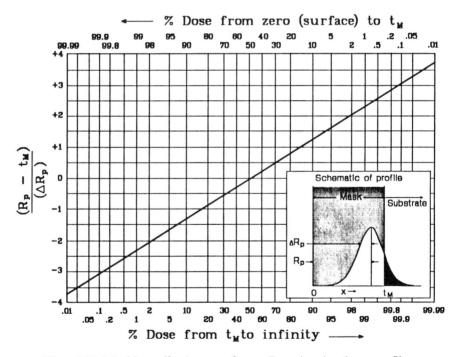

Fig. 6.33 Masking effectiveness for a Gaussian implant profile. $R_p$ and $\Delta R_p$ are the range and straggling in the mask material. The portion of the implanted profile that penetrates beyond the mask of thickness $t_M$ is shown as the solid black region in the inset [63].

**410**

While a fraction of 0.001% appears to be negligible, it can appreciably affect the semiconductor surface at high doses. For example, if the total implanted dose is $10^{16}\ cm^{-2}$, the dose that penetrates into the semiconductor becomes $10^{11}\ cm^{-2}$, which is comparable to doses used to adjust MOSFET threshold voltages. The factor $k$ can also be estimated from the specified concentration of ions that penetrate the mask:

$$\frac{N_{t_M}}{N_{Mmax}} \simeq e^{-k^2/2}, \qquad\qquad 6.43$$

where $N_{t_M}$ is the allowed concentration at interface between mask and shielded surface, and $N_{Mmax}$ the peak concentration in the masking material. The minimum thickness of different masking materials for a masking effectiveness of 99.999% can be estimated from Figs. 6.34.

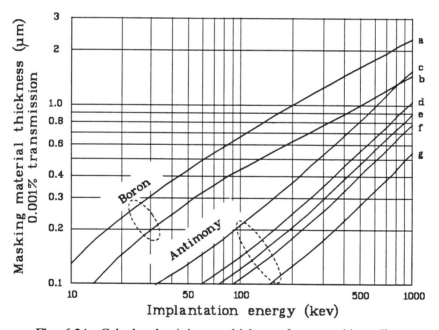

Fig. **6.34a** Calculated minimum thickness for a masking effectiveness of 99.999% for boron and antimony. (a) Oxide or polysilicon. (b) Positive or negative resist, or nitride. (c) Nitride. (d) Oxide or aluminum. (e) Polysilicon. (f) Negative resist. (g) Positive resist [64].

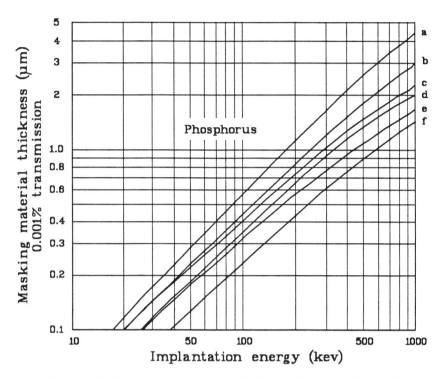

**Fig. 6.34b** Calculated minimum thickness for a masking effectiveness of 99.999% for phosphorus. (a) Positive resist. (b) Negative resist. (c) Polysilicon. (d) Oxide. (e) Aluminum. (f) Nitride [64].

## Mask Edge Effects

The mask edge profile can seriously affect the lateral distribution of implanted ions. Two typical mask edge slopes are shown schematically in Fig. 6.35. The tapered slope in Fig. 6.35a allows the penetration of ions in the thinner regions of the tapered profile, increasing the dopant concentration near the masked edge. Fig. 6.35b illustrates the effect of a tilt angle on the lateral impurity profile. The thick mask casts a shadow at the base which is $\simeq 12\%$ of the mask height for a typical tilt of 7° used to minimize channeling. This shadowing effect is particularly important, for example, when self-aligned source and drain junctions or junction extensions are implanted using a tall gate stack as a mask. In this case, shadowing can seriously affect the MOSFET threshold voltage, channel length and extrinsic resistances.

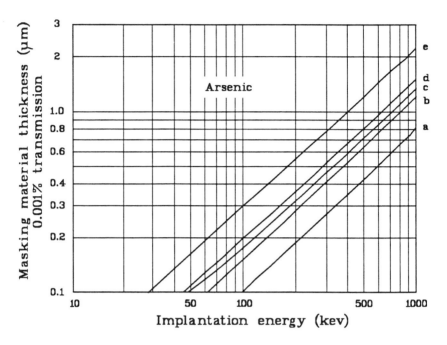

**Fig. 6.34c** Calculated minimum thickness for a masking effectiveness of 99.999% for arsenic. (a) Nitride. (b) Oxide or aluminum. (c) Polysilicon. (d) Negative resist. (e) Positive resist [64].

## 6.2.5 Lateral Spread of Implanted Ions

So far, we discussed the one-dimensional distribution of implanted ions, using the projected range and vertical straggling. When a mask is used to implant selected regions of the wafer, there is also lateral scattering perpendicular to the path of the incident beam that causes a transverse spread, $\Delta R_T$, of implanted ions from the mask edge. Referring to Fig. 6.19, the x-axis is chosen in the direction of incident ions normal to the surface passing through the mask-edge. The transverse straggle, $\Delta R_T$, is then the radial distance from the x-axis. As the dimensions of modern semiconductor devices shrink, $\Delta R_T$ becomes a very important parameter. Some calculated values for $R_p$, $\Delta R_p$ and $\Delta R_T$ are shown in Table 6.5 for boron, phosphorus, arsenic, and antimony in silicon [65] The lateral spread is of the same order as the vertical straggling, and both spreads increase as the ion mass and incident energy increase.

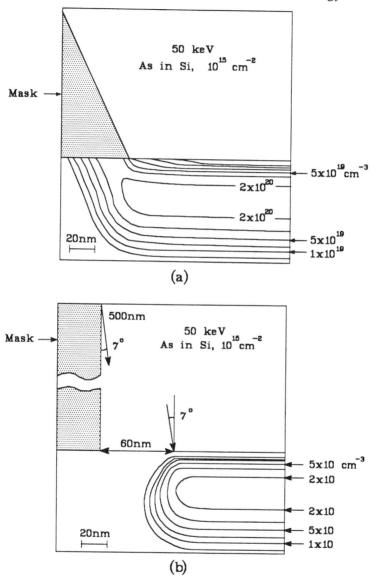

**Fig. 6.35** Schematic illustration of masking edge profiles. (a) Tapered profile, ions penetrate through the thinner regions of the mask. (b) Abrupt profile, shadowing effect.

**Table 6.5** Calculated values for $R_P$, $\Delta R_P$ and $\Delta R_T$, in nm, for boron, phosphorus, arsenic and antimony in silicon [65]

| E(keV)-> | | 20 | 40 | 60 | 80 | 100 | 120 | 140 | 160 | 180 | 200 |
|---|---|---|---|---|---|---|---|---|---|---|---|
| | $R_P$ | 77 | 157 | 234 | 308 | 381 | 450 | 516 | 580 | 642 | 701 |
| B | $\Delta R_P$ | 35 | 58 | 75 | 89 | 101 | 111 | 119 | 126 | 133 | 139 |
| | $\Delta R_T$ | 37 | 65 | 90 | 110 | 127 | 142 | 155 | 167 | 177 | 187 |
| | $R_P$ | 25 | 49 | 73 | 97 | 123 | 148 | 173 | 199 | 225 | 251 |
| P | $\Delta R_P$ | 11 | 20 | 29 | 37 | 45 | 52 | 58 | 64 | 70 | 76 |
| | $\Delta R_T$ | 9 | 18 | 25 | 32 | 39 | 46 | 53 | 60 | 66 | 72 |
| | $R_P$ | 15 | 26 | 37 | 47 | 58 | 68 | 79 | 89 | 100 | 111 |
| As | $\Delta R_P$ | 6 | 10 | 13 | 17 | 20 | 24 | 27 | 30 | 33 | 37 |
| | $\Delta R_T$ | 4 | 7 | 10 | 12 | 15 | 17 | 20 | 22 | 24 | 27 |
| | $R_P$ | 13 | 22 | 30 | 38 | 45 | 52 | 59 | 66 | 73 | 80 |
| Sb | $\Delta R_P$ | 4 | 7 | 9 | 12 | 14 | 16 | 18 | 20 | 22 | 24 |
| | $\Delta R_T$ | 3 | 5 | 7 | 8 | 10 | 11 | 13 | 14 | 15 | 17 |

When implantation is done through an infinitely long mask of width $2a$, the two-dimensional profile is given by

$$N(x,y) = \frac{N(x)}{2} \left\{ erfc\left[ \frac{y - a}{\sqrt{2}\,\Delta R_T} \right] - erfc\left[ \frac{y + a}{\sqrt{2}\,\Delta R_T} \right] \right\}, \qquad 6.44$$

where $N(x)$ is the one-dimensional impurity profile away from the mask edge given by Eq. 6.29. A calculated two-dimensional boron profile immediately after implantation through a window in a thick ($t_M >> R_p + \Delta R_p$) oxide mask is shown in Fig. 6.36 [65]. Assuming a peak boron concentration of $10^{20}$ $cm^{-3}$ and a background n-type doping of $10^{15}$ $cm^{-3}$, then the pn junction where the boron concentration drops to the background level is located laterally $\simeq$ 350 nm from the mask edge.

One important result of the above calculations is that the dopant concentration already decreases from center to mask edge before the edge of the mask is reached. For $a >> R_T$, $N(x,a) = N_{max}/2$, indicating that the concentration at the depth $R_p$ below the edge of the mask is half the peak concentration at the same depth along the center of the window. This is shown in Fig. 6.37, where the normalized boron concentration is plotted at a depth $R_p$ for the conditions given in Fig. 6.36. A comparison of the lateral spreads of boron, phosphorus, and antimony is shown for 70 keV, at a concentration of $10^{-3}$ relative to the maximum concentration in Fig. 6.38, illustrating how the lateral spread changes with the ion mass.

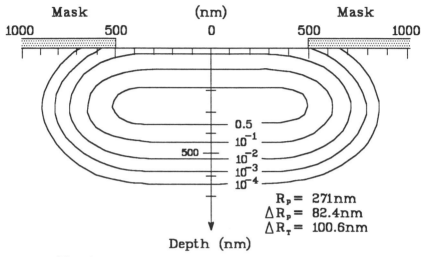

**Fig. 6.36** Equal-ion-concentration contours (normalized to 1) on an x-y plane, using a 1 $\mu$m wide, thick oxide mask with vertical edges, calculated for 70 keV boron [65].

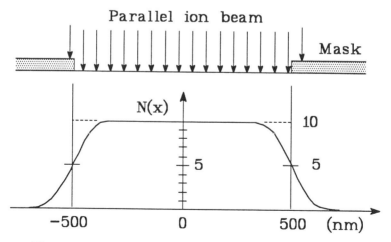

**Fig. 6.37** Normalized boron concentration at the depth $R_p$ below the surface after implantation under the conditions in Fig. 6.36 [65].

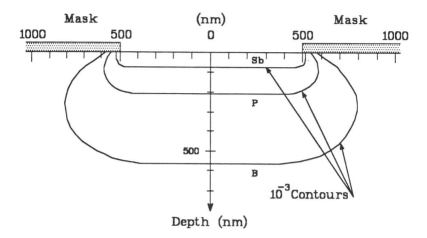

**Fig. 6.38** Contour lines for a relative concentration of $10^{-3}$ for boron, phosphorus, and antimony, implanted under the conditions in Fig. 6.35 [65].

The above discussion assumes that the mask edge is vertical and implantation is normal to the surface. Real situations, however, are more complex. There is an additional lateral spread which can be caused by a taper in the the mask edge and by tilting the wafer, as discussed above. An analysis of the lateral spread of arbitrarily shaped masks can be found in [66].

## 6.3 Crystal Damage and Dopant Activity

When a crystal is implanted with energetic (keV-MeV) ions, tracks of primary lattice disorder are produced. Also, immediately after implantation, only a small fraction of the species occupy substitutional sites where they are electrically active. Therefore, the semiconductor is typically subjected to a high-temperature treatment after implantation to restore the crystalline structure and activate the dopants. During this thermal treatment, however, secondary (or residual) defects can be formed depending on the degree of primary damage produced.

### 6.3.1 Primary Damage

Primary defects are formed by introducing energetic particles into the crystal [67]. An implanted ion makes a number of nuclear collisions with lattice atoms before it comes to rest. In each collision it can transfer sufficient energy to displace an atom from its lattice site (Fig. 6.39). The minimum energy required to break four covalent bonds and dislodge a lattice atom is called the threshold energy for displacement, $E_d$ (For silicon, $E_d \simeq 14$ eV [68]). The dislodged atom travels through the target as a second projectile and can possess enough energy to displace other atoms, and so on, creating a cascade of atomic collisions as shown in Fig. 6.17. Eventually, the energy per particle drops below $E_d$ and the cascade stops.

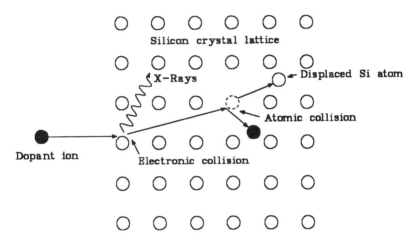

**Fig. 6.39** Displacement of lattice atom by incident high energy ion.

The simplest point defects are produced when atoms are displaced from their lattice positions to interstitial sites, resulting in vacancy-interstitial pairs called Frenkel-type defects [69]. Vacancies can have several charged states (neutral, positive, negative, double-negative). They are typically negative for n-type and neutral for p-type silicon [2]. The distribution of vacancies and interstitial atoms can cause other types of lattice disorder in the region around the ion track. For example, they can form complexes with impurity atoms and influence their diffusion (Chap. 7). Double vacancies can be formed from two single vacancies or when an incident ion displaces two neighboring lattice atoms.

As a measure of implant damage, one can use the fraction of energy that is transferred to the solid in form of nuclear collisions [70], or the number of vacancies produced by nuclear collisions. The total amount of disorder and its distribution depend on ions species, temperature, energy, total dose, and channeling effects. Light and heavy implanted ions have different damage configurations along their track. Light ions (such as B in Si) initially lose most of their energy by electronic collisions (which do not produce displacement damage). As ions penetrates deeper, they gradually lose energy until the cross-over energy between electronic and nuclear stopping is reached (about 10 keV for boron), beyond which nuclear stopping becomes predominant. Most of the damage is therefore created in that part of the light ion trajectory which lies beyond the cross-over point, so that the displaced atom profile has a buried peak concentration. In contrast, heavy ions (such as P, As, Sb, or Ar in Si) can transfer more energy per collision to lattice atoms and displace a larger volume of target atoms than lighter ones. Therefore, heavy ions lose most of their energy by nuclear stopping near the surface and the displaced atom profile is shallow.

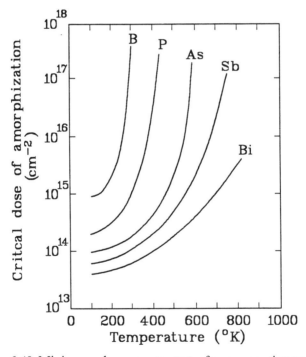

**Fig. 6.40** Minimum dose necessary to form a continuous amorphous layer as a function of temperature for different ion masses [29, 74 − 76].

If many lattice atoms are displaced by incident ions within small volumes of the crystal, local amorphous zones (clusters) about 10 nm in diameter can be produced (the term amorphous describes the absence of long-range crystallographic order). As the dose increases, the disordered regions begin to overlap and, at some point where there is enough overlap of damage, a "continuous" amorphous layer is formed to a certain depth. The transition from the crystalline to the amorphous state occurs at a threshold energy density of about $10^{24}$ eV/cm$^3$ [71]. It can be visually detected since it induces a change in color of the semiconductor. For example, silicon becomes lighter (milky appearance) when it amorphizes at the surface.

**Table 6.6** Minimum dose that amorphizes silicon at room temperature [74]

| Species | Mass of main isotope | Dose (cm$^2$) |
|---------|----------------------|---------------|
| B | 11 | $8x10^{16}$ |
| N | 14 | $2x10^{15}$ |
| Ne | 20 | $1x10^{14}$ |
| P | 31 | $6x10^{14}$ |
| Ar | 40 | $4x10^{14}$ |
| BF$_2$ | 49 | $3x10^{14}$ |
| Ga | 70 | $2x10^{14}$ |
| As | 75 | $2x10^{14}$ |
| Kr | 84 | $2x10^{14}$ |
| Sb | 122 | $1x10^{14}$ |
| In | 204 | $1x10^{14}$ |
| Bi | 209 | $5x10^{13}$ |

Table 6.6 gives the minimum ("threshold") dose necessary to amorphize silicon for different ions with the wafer held at room temperature [74]. As expected, this dose decreases as the mass of the implanted ion increases. Also, the lower the temperature

during ion implantation, the smaller the dose required for amorphizing (Fig. 6.40). Consequently, a high dose-rate has two competing effects on defect formation. If the crystal is heated by the ion beam, some of the radiation damage can be removed immediately, leaving a residue of defects that are difficult to annihilate, as discussed below [72]. If the wafer temperature does not increase, higher defect concentrations are typically obtained with increasing dose rate [73].

### Density and Distribution of Displaced Atoms

To estimate the number of displaced atoms, the partition of energy between electronic and nuclear collisions must be known for both the primary (incident) and recoil (knock-on) particles. The approximate energy available for nuclear collisions of energetic ions incident on silicon is given in Table 6.7 [29].

**Table 6.7** Approximate amount of energy (keV) available for nuclear collisions [29]

| Incident ion | Incident energy (keV) | | | | | | |
|:---:|:---:|:---:|:---:|:---:|:---:|:---:|:---:|
| | 1 | 3 | 10 | 30 | 100 | 300 | 1000 |
| B | 0.80 | 2.2 | 5.9 | 14 | 27 | 41 | 54 |
| P, Al | 0.83 | 2.4 | 7.3 | 19 | 51 | 100 | 170 |
| Ge,Ga,As | 0.84 | 2.5 | 7.7 | 21 | 63 | 160 | 370 |
| In,Sn,Sb | 0.85 | 2.5 | 7.9 | 22 | 68 | 180 | 460 |

An implanted ion can increase the number of recoil atoms only if it possesses and energy greater than $2E_d$. If the ion energy is less than $2E_d$ but larger than $E_d$, it can dislodge only one atom and come to rest. Therefore, by treating collisions as isolated two-body elastic scattering events and assuming that both the primary and recoil atoms are available to make further collisions, the number of displaced atoms can be approximated as [77]

$$N(E) = \frac{0.5E_n}{E_d} ,$$

6.45

where $E_n$ is the total energy loss of an ion in primary and recoil collisions. The essential feature of Eq. 6.45 is that the number of displaced atoms increases linearly with incident energy where electronic stopping can be neglected. In estimating the number of displaced atoms one is faced with several uncertainties. The displacement processes are not quite separate and the energy of displacement decreases in overlap regions. Also, there is a competing effect of vacancy-interstitial recombination at room temperature that reduces the number of stable displaced atoms. Another competing effect is self-annealing as the target temperature increases, repairing some or all of the damage as it is created. If the energy density along an ion track is sufficiently large, a thermal spike can occur causing the crystal to locally melt. As the superheated region cools down, it expands into an amorphous volume much larger than the collision cascade [78]. In this case, the cascade theory is no longer adequate to predict the number of displaced atoms. Finally, the displacement energy depends significantly on the recoil direction and the mean value of $E_d$ can be larger than the threshold energy used to calculate the number of displaced atoms. This can lead to a large variation in the projected density of displaced atoms. For example, for a heavy atom such as antimony in silicon at about 40 keV, the different theories suggest that between 1200 and 2400 silicon atoms would be displaced, depending on whether isolated defects or clusters are formed.

Fig. 6.41 shows the distribution of impurity ions (solid line) and of vacancies (dashed line) produced by boron ions at an energy of 60 keV. The data indicate that the maximum in the radiation damage distribution is closer to the surface than the maximum of ion distribution [79]. For boron in silicon, $X_d \simeq 0.8R_p$, and $\Delta X_d \simeq 0.75\Delta R_p$, where $X_d$ is the range of radiation damage, and $\Delta X_d$ its standard deviation [2]. This should be expected because a fraction of energy is lost to create the damage, while all energy is available for ion distribution.

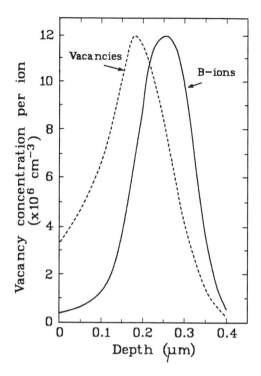

**Fig. 6.41** Distribution of implanted boron atoms (solid line) and vacancies (dashed line) for 60 keV boron in silicon. The maximum of crystal damage is closer to the surface than the maximum of boron distribution [29, 79].

## 6.3.2 Knock-On Ranges

Implantation is frequently done through films that are deposited or grown on the semiconductor surface. If the mass of implanted ions is comparable to that of the atoms in the film, the colliding ions can transfer sufficient energy to dislodge film-atoms and "implant" them into the substrate (knock-on or recoil implantation). In the presence of a silicon dioxide layer, for example, oxygen or silicon can be knocked into the underlying silicon, causing substantial disorder in the crystal [80]. When arsenic is implanted into bare silicon, no knock-on damage is observed in the crystal. This is also the case when the oxide layer is sufficiently thick to stop all the implanted and knock-on ions. Dislocations are observed, however, for oxide films of intermediate thickness that allow knock-on oxygen to enter the crystal [81]. Fig. 6.42 shows simulated profiles of knock-on oxygen and arsenic for implantation of arsenic into silicon at an energy of 150 keV and a dose of $10^{16}$ $cm^{-2}$ through 43 nm of $SiO_2$. For comparison, a simulated oxygen profile is shown for oxygen implanted into bare silicon at 10 keV and a dose of $10^{16}$ $cm^{-2}$ [82].

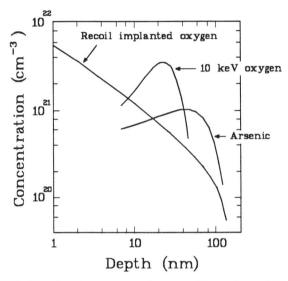

**Fig. 6.42** Simulated concentration profiles of recoil-implanted oxygen and arsenic for arsenic implanted through 42 nm $SiO_2$ into Si at an energy of 150 keV and a dose of $10^{16}$ $cm^{-2}$. Also shown is the simulated profile of oxygen implanted into bare silicon at an energy of 10 keV and a dose of $10^{16}$ $cm^{-2}$ [82].

### 6.3.3 Annealing and Secondary Defects

For low to medium implant energies and doses below $\simeq 10^{14}$ $cm^{-2}$, primary damage in silicon is in the form of isolated disordered regions that anneal at temperatures as low as 400 °C. Complete annealing of implant damage, however, is not easy to determine because even with a high number of dislocations and stacking faults, a semiconductor shows electrical parameters which are difficult to distinguish from those of a perfectly crystalline material. Sophisticated characterization techniques are required to determine the extent of any residual damage. Such techniques are discussed in a separate chapter.

For higher implant doses, more complex defects, known as extended, secondary, or residual defects, can result from aggregation of primary defects during annealing. This is also observed at high energies with doses above $\simeq 10^{13}$ cm$^{-2}$. In this case, complete annealing of defects becomes difficult [83]. For example, interstitials are very mobile and can diffuse over large distances even at anneal temperatures as low as 200°C. At high concentration, they can coalesce with other displaced atoms to form extended defects before moving into empty sites and repair damage [84 – 89]. Vacancies can also diffuse and combine to form extended defects.

Secondary defects have been classified into five categories [89, 90]. Defects belonging to the first category are formed when the implant dose is below the threshold for amorphization and the amount of primary implant damage is sufficiently high to produce extended defects after annealing, i.e., a large number of defects and radiation-damage clusters is created without producing a continuous amorphous layer. This residual damage is in the form of dislocations and stacking faults caused by agglomeration of silicon interstitials [89, 91]. In this case, damage removal can require significantly more energy than in the case of completely disordered, amorphous material, and complete restoration of the lattice is only possible at temperatures above $\simeq$ 1000 °C. The formation of pre-amorphization damage after high temperature annealing is found to depend critically on the total number of displaced silicon atoms [89]. The number of displaced Si atoms required to generate secondary defects, $N_{Crit}$, ranges from $1.5 \times 10^{16}\, cm^{-2}$ for [11]B to $2 \times 10^{17}$ cm$^{-2}$ for [121]Sb ions [89]. The increase in $N_{Crit}$ with the mass of implanted ion is attributed to the higher collision cascade density as the ion mass increases, whereby a large fraction of displaced silicon atoms constitutes small amorphous clusters that anneal at temperatures as low as 200 °C, leaving only a small fraction of mobile interstitials to form extended defects at higher anneal temperatures [89]. Therefore, when heavy ions (such as As and Sb) are implanted at low to medium energies, $N_{Crit}$ may not be reached before amorphization and no secondary defects are observed with doses up to the amorphization threshold. At higher implant energies, the density of collision cascades decreases and more mobile Si interstitials are available for residual defect formation, explaining why extended defects are observed for heavy ions at MeV energies and not at low to medium energies [89].

Residual defects that are generated just beyond an amorphous-crystalline interface belong to the second defect category. If the implant dose is above the amorphization threshold, a continuous amorphous layer is formed, extending to a certain depth. If the substrate temperature increases during implantation, the individual disordered regions can anneal before the next ion strikes in their vicinity, preventing formation of an amorphous layer and requiring significantly higher doses to produce the same amount of defects as at room temperature. When the energy is low, the amorphous layer extends to the surface, leaving only one interface with the single-crystal semiconductor. For high-energy (MeV) implantation, a buried amorphous layer with two interfaces with the crystal is produced and the region near the surface remains undamaged. There is an important difference between the anneal behavior of an amorphous layer and that of isolated disordered regions around the ion track. Continuous amorphous regions are easier to regrow into damage-free films by solid-phase epitaxy than are regions that have not been fully amorphized [92]. The recrystallization of an amorphous layer starts typically from the undamaged crystal surface. The amorphous/crystalline (a/C) interface moves toward the surface at a velocity that depends on temperature, doping, and crystal orientation (Fig. 6.43) [93]. At low to moderate dopant concentration, the activation energy for recrystallization is 2.3 eV, implying bond-breaking at the interface. The growth-rate increases at high dopant concentrations ( $\geq 10^{20} \, cm^{-3}$), probably because silicon bonds are weakened by the incorporation of dopants and bond-breaking requires less energy [94]. During epitaxial recrystallization, dislocation loops and stacking faults can be created from accumulation of excess silicon interstitials or grow during annealing, starting from an unannealed damaged region into the undamaged crystal [91]. Dislocation loops and stacking faults only anneal at temperatures above $\simeq$ 1000 °C or do not anneal at all. If annealing is accompanied by high-temperature ($\geq$ 1000 °C) oxidation, the growth of these defects can be enhanced by the injection of interstitials during oxidation. A low-temperature ($\simeq$ 800 °C) oxidation for a few minutes preceding the high-temperature oxidation appears to reduce the defect density by orders of magnitude [95].

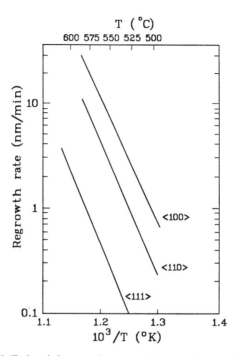

**Fig. 6.43** Epitaxial growth rates of amorphous silicon as a function of temperature for (100), (110), and (111) oriented silicon [93].

Defects, such as micro-twins, related to imperfect epitaxial growth belong to the third category. Solid-phase epitaxy of a buried amorphous layer results in regrowth from the two amorphous-crystalline interfaces, starting from the undamaged surface and the undamaged substrate. Recrystallization of the two regions is not correlated and does not result in a perfectly recrystallized crystal, forming defects of the fourth category where the two interfaces meet. Defects of the fifth category result from exceeding the solid solubility limit during annealing [90].

## 6.3.4 Annealing and Electrical Activity

The anneal temperature, duration and ambient must be optimized to remove primary damage and achieve the desired dopant profile with adequate dopant activation, without leaving residual defects that degrade the device's electrical parameters. Also, when ultra-shallow impurity profiles are formed, activation must achieved without significant movement of dopant atoms. The two heating regimes of interest are conventional furnace anneal over a period of minutes to hours, and rapid thermal anneal (RTA) lasting for 2-100s [96 − 98]. Pulsed laser or electron-beam annealing with time ranging from $\mu$s - ms is also used for special applications. In the first two regimes, the anneal time is typically longer than the thermal time constant, $\tau$, of the semiconductor. Considering a wafer of thickness $d$, the time constant for the heat flow from the top surface to the back-side of the wafer is approximated as [96].

$$\tau \simeq \frac{d^2}{\pi^2 D_T} , \qquad 6.46$$

where $D_T$ is the thermal diffusivity in the semiconductor, defined as

$$D_T = \frac{K}{\rho C} , \qquad 6.47$$

where $K$ is the semiconductor thermal conductivity, $\rho$ its density, and $C$ its specific heat. For example, if the temperature of both wafer faces is suddenly increased by $T$, the time-dependent temperature difference, $\Delta T$ (t), between either surface of the wafer and its body center is approximated as

$$\Delta T(t) \simeq e^{-t/\tau} . \qquad 6.48$$

In pulsed anneal, the heating time is typically comparable or less than $\tau$ and only a thin layer at the surface is allowed to heat and melt so that regrowth occurs by liquid-phase rather than solid-phase epitaxy.

As with damage anneal, dopant activation depends strongly on the implanted dose. With increasing dose below the amorphization threshold, high anneal temperatures are required to achieve a given level of electrical activity. Above the threshold for amorphization, a considerably lower temperature (500-650°C for Si) is adequate to achieve nearly complete electrical activation. This is because silicon recrystallizes in the same temperature range and ions are incorporated into lattice sites during epitaxial growth without requiring additional energy.

Incomplete annealing results in a reduction in the fraction of active dopants, mainly near the peak of the dopant distribution where damage is greatest [99]. Fig. 6.44 compares the boron and hole profiles in silicon obtained after annealing the wafer for 35 minutes at 800°C and 900°C for boron implantation at an energy of 70 keV and a dose of $10^{15}\, cm^{-2}$ which is below the amorphization threshold [100,101]. The boron profile is measured by secondary ion mass spectroscopy (SIMS), and the hole concentration is obtained from Hall measurements. Since the dose is below the threshold for amorphization, annealing at 800°C leaves a large fraction around the peak electrically inactive and annealing at or above 900°C becomes necessary to activate most of the implanted boron.

It is important to know whether residual defects will cause junction degradation. If after activation the damaged front does not penetrate as deep as the metallurgical junction and the depletion region does not spread into the damaged region, the damaged region causes no increase in junction leakage. This is experimentally shown on junctions as shallow as $\simeq 0.2\ \mu m$ [103], and 0.1 $\mu m$ [104]. In many cases the defective region even improves the junction quality by local gettering of impurities, particularly heavy metals, which can be present in the vicinity of the junction. Another example of gettering is the intentional damage of a buried region by high energy (MeV) implantation for the precipitation of heavy metals and other defects prior to device fabrication on the front side.

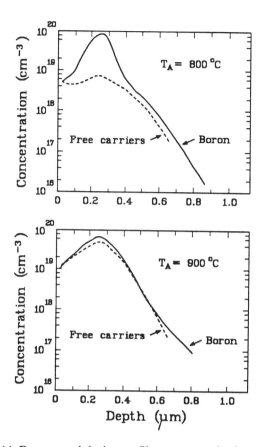

**Fig. 6.44** Boron and hole profiles measured after annealing at two different temperatures for 35 minutes. Activation of boron results in an increase in the hole concentration. Activation is incomplete at 800°C but essentially complete at 900°C [97,98].

### Isochronal and Isothermal Annealing

A measure of electrical activity is the sheet carrier concentration, i.e., the carrier concentration per unit area of the implanted film, $N_{square}$, which is determined by the anneal temperature and time. Isochronal annealing refers to a fixed anneal time and varying temperature, while in isothermal annealing the temperature is kept constant and the time is varied. An example of isochronal annealing is shown in Fig. 6.45. A slow, steady increase in electrical activity is observed for a low boron dose. If the implant dose is increased above $\simeq 10^{13}\,cm^{-2}$, but kept below the threshold for amorphization, three different regions can be distinguished in the anneal characteristic [92]. At temperatures below 500°C (region I), annealing of point defects reduces the carrier trap density and increases the free carrier concentration. In the range 500-600°C (region II), a reverse annealing behavior is seen for $N_{square}$. This is explained in terms of pair formation that reduces the number of substitutional boron atoms and causes a net decrease in $N_{square}$. These pairs dissociate again at higher temperature (region III) and the atoms are incorporated individually into the crystal lattice, increasing $N_{square}$. A similar annealing behavior is observed for phosphorus at low doses [105]. For doses above the amorphization threshold, the annealing temperature is essentially fixed at about 600°C, which is the temperature for solid-phase epitaxy silicon. Substitutional impurities are incorporated into the lattice as the crystal regrows, resulting in high electrical activity at the low anneal temperature.

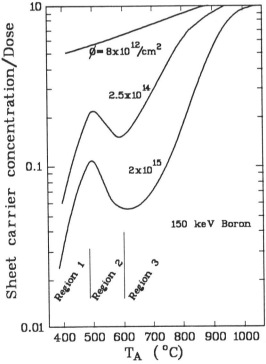

**Fig. 6.45** Sheet carrier concentration as a function of anneal temperature [92,102].

Additional information on annealing kinetics can be obtained by annealing at a fixed temperature for varying times (isothermal annealing), particularly when one is limited as to choice of temperature or when isochronal-annealing results in curves that are too steep for an activation energy to be extracted.

## Transient Annealing

Transient annealing falls into two groups, pulsed and rapid thermal anneals. For pulsed anneal, a laser or electron-beam irradiates the surface for the duration of $\simeq 1\mu s$ to ms, during which the silicon surface melts locally to a depth less than 0.1 $\mu m$, while the rest of the wafer remains essentially at ambient temperature. Pulsed annealing is adiabatic because the heating time is very short compared to the thermal time-constant of the material [106]. When radiation is removed, the molten region re-solidifies into a crystalline structure by liquid-phase epitaxial growth on the substrate. Pulsed annealing is used mainly to recrystallize polysilicon over insulator films to form single-crystal silicon on insulator (SOI) structures, or to locally dope the semiconductor by immersing the molten region in a gas of the dopant species, as discussed in the next chapter. Zone-melting recrystallization (ZMR) and other techniques to prepare SOI material are discussed in Process Integration.

Rapid thermal anneal (RTA) allows the removal of defects introduced by ion implantation, and activation of species with little movement of the dopants. Since the activation of damage removal is about 5 eV and diffusion has an activation energy of $\simeq$ 3 eV, the ratio of damage removal to dopant diffusion is greatest at high temperature. The diffusion can be short, of the order of nm, and the implanted profile can be activated without being significantly broadened during RTA. This is important when shallow and ultra-shallow impurity profiles are formed. While conventional furnaces are capable of supplying high temperature, the slow insertion and removal of the wafers and the heat capacity of the system require "ramp-up" and "ramp-down" times in the range of minutes and causing excessive movement of dopants. In contrast, during RTA the whole wafer is heated uniformly in seconds and, after annealing the wafer is cooled in seconds; annealing is typically isothermal.

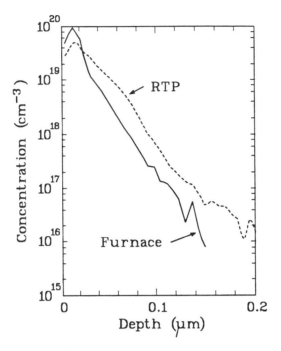

**Fig. 6.46** Comparison of RTA and furnace annealing of implanted boron showing less trapping at the damage peak with RTA than with furnace anneal and a slightly deeper distribution with RTA. Implant conditions: $BF_2$, 15 keV, $10^{15}$ $cm^{-2}$ into bare silicon. Furnace anneal conditions: 850°C, 21 min, in argon. RTA conditions: 1020°C, 5s, in argon [116].

RTA has emerged as a very attractive technique to overcome some limitations of conventional furnace anneals. Among these are: the control of boron penetration through the gate oxide while degenerately doping p-type polysilicon at its oxide interface in dual work-function CMOS designs [107, 108], silicide formation with improved uniformity over furnace anneal [109, 110], control of out-diffusion of ultra-shallow junctions from silicides and polysilicon films (Chap. 8) [111, 112], control of analog precision resistors and capacitors [113, 114], and inter-conductor dielectric densification and leveling [115]. As discussed in Process Integration, the optimum thermal budget is chosen to form the desired impurity profiles, while minimizing junction leakage and maintaining oxide integrity and stability of electrical parameters. While very short ramp times can be achieved with RTA ($\simeq 150°C/s$ as compared to $\simeq 1°C/s$ with furnace processing), an operating temperature above $\simeq 1000°C$ is required to maintain an acceptable throughput. Even with very short operating times (5-100 s), however, the repetition of high-temperature RTA cycles can cause substantial diffusion of impurities.

Fig. 6.46 compares the results of furnace annealing to RTA of high-concentration boron in silicon. Furnace annealing for 21 min at 850°C in argon leaves a fraction of dopants trapped by defects near the profile peak. Less trapping is seen with RTA at 1020°C for 5s in argon. The boron profile, however, has broadened [116]. Mechanisms for the diffusion of impurities in semiconductors are discussed in the next chapter.

# PROBLEMS

**6.1** Sketch the implanted ion profile for boron, phosphorus, arsenic, and antimony implanted into bare silicon at an energy of 180 keV and a dose of $5x10^{12}$ $cm^{-2}$. Assume normal incidence and a Gaussian distribution.

**6.2** Phosphorus is implanted at an energy of 200 keV and a dose of $10^{13}$ $cm^{-2}$ into bare silicon. Assume a Gaussian distribution and a uniform background boron concentration of $10^{16}$ $cm^{-3}$ in silicon, and find the junction depth (defined as the depth below the silicon surface where the phosphorus and boron concentrations are equal).

**6.3** a) Find the energy required to place the peak of an implanted boron profile at the center of a 400-nm polysilicon film. Assume a Gaussian distribution.
b) Calculate the dose necessary to produce a boron concentration of $10^{20}$ $cm^{-3}$ in polysilicon at the end of all anneal cycles. Assume that the implanted boron is distributed uniformly during annealing.

**6.4** A 400 nm layer of buried $SiO_2$ is to be formed inside a 200-mm silicon wafer. What is the minimum number of oxygen ions that must be implanted?

**6.5** Assume a beam current of 30 mA and find the time necessary to implant oxygen at a dose of $1.8x10^{18}$ $cm^{-2}$ into a silicon wafer having a diameter of 200 mm.

**6.6** The following table gives the ranges of 100 keV boron in titanium, tungsten, and their silicides [117]. Sketch the profiles of 100-keV boron implanted into silicon through 150-nm $TiSi_2$ and 75-nm $WSi_2$. Explain your assumptions.

**Table 6.8** Range parameters for 100-keV boron in Ti, W, TiSi$_2$ and WSi$_2$ [117].

| Material | Symbol | Density $(g/cm^3)$ | R$_p$ (nm) | $\Delta$R$_p$ (nm) |
|---|---|---|---|---|
| Titanium | Ti | 4.52 | 254.6 | 95.1 |
| Titanium silicide | TiSi$_2$ | 4.04 | 215.4 | 56.3 |
| Tungsten | W | 19.30 | 82.4 | 61.8 |
| Tungsten silicide | WSi$_2$ | 9.86 | 144.0 | 55.5 |

**6.7** Assume that a channeled ion experiences only electronic stopping and find the maximum range of a channeled 100 keV boron ion in silicon. What is the critical angle for [100] axial channeling at this energy?

**6.8** Estimate the number of silicon atoms displaced by an impinging 100-keV arsenic ion. Disregard channeling effects.

**6.9** The vacuum in the region between the mass-separation magnet and the target is typically $10^{-6}$ Torr. Assume a cross section of $10^{-16}$ $cm^2$/atom for the neutralizing reaction between positive ions (I$^+$) and neutral atom (N$_2$)

$$I^+ + N_2 \rightarrow I + N_2^+ .$$

Calculate the percentage of ions that are charge exchanged at a distance of 1 m.

**6.10** Can an analyzing magnet resolve AsF$_4^{++}$ and F$_4^+$ from As$^+$?

**6.11** An ion implanter has an extraction voltage of 30 keV, a 0.5-m analyzer magnet radius, and a phosphorus ion beam current of 0.1 mA. Assume a Gaussian distribution and calculate the magnetic field strength, the accelerating voltage, and the exposure time necessary to achieve a projected range of 100 nm and a peak concentration of $10^{20}\,cm^{-3}$ for phosphorus implanted into a 200-mm silicon wafer.

**6.12** Silicon-dioxide is used as a mask against phosphorus implantation. The increase in phosphorus concentration at the silicon surface covered by the mask may not exceed $10^{15}\,cm^{-3}$. For an implant energy of 50 keV and a dose of $5x10^{15}\,cm^{-2}$, find the minimum oxide thickness that is required for masking.

**6.13** To adjust the MOSFET threshold voltage, boron is implanted into silicon through a thin silicon dioxide film. Calculate the oxide thickness, implant energy, and total dose that result in an implanted peak boron concentration of $5x10^{17}\,cm^{-3}$ at the silicon surface, and an implanted boron dose of $2.5x10^{12}\,cm^{-2}$ in silicon.

**6.14** Plot the phosphorus profile of double ionized phosphorus implanted into bare silicon at a total energy of 200 keV and a dose of $2x10^{13}\,cm^{-2}$ for the following cases:
a) No contamination by charge exchange.
b) Contamination with single-ionized ions: 10% of the double-charged ions are charge exchanged to singly ionized phosphorus immediately before acceleration. Assume the extraction potential to be 25 keV.

**6.15** Using LOCOS as a mask, boron is implanted into silicon at an energy of 80 keV and a dose of $10^{14}\,cm^{-2}$.
a) How thick should the masking oxide be to ensure that only 0.001% of the boron dose penetrates into the silicon surface?
b) Assume that the silicon surface is covered with a 15 nm "screening" oxide film and determine the peak concentration of the implanted profile.
c) Approximate the LOCOS "bird's beak" with two planes intersecting at the silicon surface, each making an angle of 45° with the normal to the silicon surface, and trace the two-dimensional contour at peak concentration.

**6.16** a) Find the boron dose and energy required to form a buried p-type layer with a peak concentration of $10^{19}\,cm^{-3}$ located at a depth of 2.0 $\mu$m.

b) How thick should a photoresist (AZ1350) mask be to ensure that a maximum of 0.001% of the dose penetrates into the silicon surface covered by resist? Neglect the effect of a thin oxide film under the mask.

c) Assume a uniform arsenic background concentration of $10^{16}\ cm^{-3}$ and vertical mask edges. The junction is defined as the locus of points where the boron and arsenic concentrations are equal. How wide should the mask be to ensure that the junctions formed at both edges of the mask remain 1 $\mu$ apart?

# References

[1] W. Shockley, *"Forming Semiconductor Devices by Ionic Bombardment,"* U.S. Patent 2,787,564, filed Oct. 28, 1954.

[2] H. Ryssel and I. Ruge, Ion Implantation, John Wiley (1986).

[3] R. B. Simonton, *"Ion Source Operation and Maintenance in a Semiconductor Production Environment,"* in Ion Implantation Technology, J. F. Ziegler, Ed., North Holland (1992).

[4] D. Aitken, *"Ion Sources,"* in Ion Implantation Techniques, H. Ryssel and H. Glawischnig, Eds., Springer-Verlag, 23-71, New York (1982).

[5] A. Axmann, *"Ionizable Materials to Produce Ions for Implantation,"* Solid-State Technology, 17 (11), 36 (Nov. 1974)

[6] K. Brack, W. Euen and D. Hagmann, *"Discharge of $P^{2+}$ Ions in the Beamline of Medium Current Implanters,"* Nucl. Instr. and Methods in Phys. Res., B21, 405 (1987).

[7] P. Spinelli, J. Escaron, A. Soubie and M. Bruel, *"High Energy Ion Implantation for C-MOS Isolation n-Wells Technology: Problems Related to the Use of Multicharged Phosphorus Ions in an Industrial Context,"* Nucl. Instr. and Meth. in Phys. Res., B6, 283, North-Holland (1985).

[8] P. J. Wright, M. Wong and K.C. Saraswat, *"The Effect of Fluorine on Gate Dielectric Properties,"* IEDM Tech. Digest, p. 574 (1987).

[9] M. Y. Tsai and B. G. Streetman, *"Recrystallization of Implanted Amorphous Silicon Layers. I. Electrical Properties of Silicon Implanted with $BF_2^+$ or $Si^+ + B^+$,"* J. Appl. Phys., 50 (1), 183 (1979).

[10] G. Dearnaley, J. H. Freeman, R. S. Nelson, J. Stephen, Ion Implantation, North Holland, (1973).

[11] R. G. Wilson and G. R. Brewer, Ion Beams, John Wiley (1973).

[12] J. H. Freeman, Ion Implantation, G. Dearnaley, J. H. Freeman, J. Stephen, Eds., North Holland, p. 369 (1973).

[13] N. Sakudo, *"Microwave Ion Source for Ion Implantation,"* Nuc. Instr. and Meth. Phys. Res. B21, 168-177 (1987).

[14] K. G. Stephens, *"Ion Source Fundamentals,"* in Handbook of Ion Implantation Technology, J. F. Ziegler, Ed., p. 455, North Holland, Amsterdam (1992).

[15] J. H. Freeman, *"A New Ion Source for Electromagnetic Isotope Separators,"* Nucl. Instr. Meth. 22, 306 (1963).

[16] N. White, *"Ion Sources for Use in Ion Implantation,"* Nucl. Meth. Phys. Res. B37/38, 78 (1989).

[17] R. Simonton and F. Sinclair, *"Emerging Ion Implantation Processes for the 1990's,"* Advanced Ion Implantation Survey Report, Eaton Corp., Report No. 8500208 (1990).

[18] J. O'Connor, Genus Corp., private communication, June 1993.

[19] K. H. Purser, M. Cleland, H. Taylor, and T. H. Smick, *"The IONEX MeV Implantation System,"* SPIE, 530, 14 (1985).

[20] J. O'Connor. Genus Corp., private communication, June 1993.

[21] A. B. El-Kareh and J. C. El-Kareh Electron Beam Lenses and Optics, Vols. 1 and 2, Academic Press, New York (1970).

440

[22] G. Ryding and A. B. Wittkower, *"Industrial Ion Implantation Machines,"* IEEE Trans. Manuf. Tech., MTF-4 (1), 21-31 (1975).

[23] A. J. T Holmes, *"Theoretical and Experimental Study of Space Charge in Intense Ion Beams,"* Phys. Rev. A19 (1), 389-407 (1979).

[24] M. E. Mack, *"Wafer Cooling and Wafer Charging,"* in Handbook of Ion Implantation Technology, J. F. Ziegler, Ed., p. 599, North-Holland (1992).

[25] V. Benveniste, in Ion Implantation Technology, M. I. Current, N. W. Cheung, W. Weisenberg, and B. Kirby, Eds., p. 366, North Holland Physics, Amsterdam (1987).

[26] M. King and P. H. Rose, *"Experiments on Gas Cooling of Wafers,"* Nucl. Instr. Meth., 189, 169-173 (1981).

[27] D. C. Evans, in Ion Implantation Technology, M. I. Current, N. W. Cheung, W. Weisenberg, and B. Kirby, Eds., p. 385, North Holland Physics, Amsterdam (1987).

[28] P. D. Parry, *"Localized Substrate Heating During Ion Implantation,"* J. Vac. Sci. Tech. 15 (1), 111-115 (1978).

[29] J. W. Mayer, L. Erikson and J. A. Davies, Ion Implantation in Semiconductors, Academic Press, p. 24, New York (1970).

[30] J. Lindhard, M. Scharff, and H. Schiott, *"Range Concepts and Heavy Ion Ranges,"* Mat.-Fys.Med. Dan.Vid Selsk, 33 (14), 1 (1963).

[31] B. Smith, Ion Implantation Range Data for Silicon and Germanium Device Technologies, Research Studies, Forest Grove, Oregon, 1977.

[32] J. F. Ziegler *"Ion Implantation Physics,"* in Handbook of Ion Implantation Technologies, J. Z. Zielger, Ed., p. 1, North Holland, New York (1992).

[33] J. P. Biersack and L. G. Haggmark, *"A Monte Carlo Computer Program for the Transport of Energetic Ions in Amorphous Targets,* Nucl. Instr. and Meth, 174, 257 (1980).

[34] J. F. Gibbons, W. S. Johnson, and S. W. Mylroie, Projected Range Statistics, Dowden, Hutchison, and Ross, Eds., Academic Press, Vol. 2, New York (1975).

[35] J. F. Ziegler, *"The Stopping and Range of Ions in Solids,"* Ion Implantation Science and Technology, J. F. Ziegler Ed., Academic Press, (1984)

[36] J. F. Ziegler, J. P. Biersack, and U. Littmark, *"The Stopping and Range of Ions in Solids,"* in The Stopping and Ranges of Ions in Matter, J. F. Ziegler, Ed., Vol.1, Pergamon Press, New York (1985).

[37] T. E. Seidel, *"Ion Implantation,"* VLSI Technology, S. M. Sze, Ed., p. 219, McGraw-Hill, New York (1983).

[38] B. Mylroie and J.F. Gibbons, *"Computation of Third Central Moments for Projected Range Distributions of Common Ion-Implanted Dopants in Silicon,"* Ion Implantation in Semiconductors and other Materials, B. L. Crowder, Ed., Plenum Press, N.York 1973

[39] J. F. Gibbons and S. Mylroie, *"Estimation of Impurity Profiles in Ion-Implanted Amorphous Targets Using Joint Half-Gaussian Distributions,"* Appl. Phys. Lett., 22 (6), 568-569 (1973).

[40] M. G. Kendall and A. Stuart, The Advanced Theory of Statistics, Charles Griffin, London (1958).

[41] W. K. Hofker, *"Implantation of Boron in Silicon,"* Philips Res. Repts., Suppl #8, 1-121 (1975)

[42] H. Ryssel, *"Range Distributions,"* in Ion Implantation Techniques, H. Ryssel and H. Glawischnig, Eds., Springer Series in Electrophysics, New York, 10, 177 (1982).

[43] R. Simonton and A. F. Tasch, *"Channeling Effects in Ion Implantation,"* in Handbook of Ion Implantation Technology, J. B. Ziegler, Ed., p. 119, North Holland, new York (1992).

[44] H. Ryssel, H. Kranz, K. Muller, R. A. Henkelmannn, and J. Biersack, *"Comparison of Range and Range Straggling of implanted $^{10}B$ and $^{11}B$ in Silicon,"* Appl. Phys. Lett., 30 (4), 399-401 (1977).

[45] L. A. Christel, J. F. Gibbons, and S. Mylroie, *"Application of the Boltzmann Transport Equation to Ion Range and Damage Distributions in Multilayered Targets,"* J. Appl. Phys., 51 (12), 6176-6182 (1980).

[46] D. V. Morgan, Ed., Channeling, John Wiley and Sons, London (1973).

[47] J. F. Ziegler, *"The Channeling of Ions near the < 100 > Axis,"* in Ion Beam processing in Advanced Electronic Materials and Device Technology, B. R. Appleton et. al., Eds., Vol. 45, Mater. Res. Soc. (1985).

[48] K. V. Brannon and R. F. Lever, *"Computational Investigation of Channeling of Boron in Silicon,"* Electrochem. Soc. Ext. Abst. 86-2, p. 813 (1986).

[49] R. A. Moline and G. W. Reutlinger, *"Phosphorus Channeled in Silicon: Profiles and Electrical Activity,"* in Ion Implantation in Semiconductors, I. Ruge and J. Graul, Eds., p. 58, Springer Verlag, New York (1971).

[50] T. E. Seidel, *"Distribution of Boron Implanted Silicon,"* in Ion Implantation in Semiconductors, I. Ruge and J. Graul, Eds., p. 47, Springer Verlag, New York (1971).

[51] G. Dearnaley, J. H. Freeman, G. A. Gard, and M. A. Wilkins, *"Implantation Profiles of $^{32}P$ Channeled into Silicon Crystals,"* Can. J. Phys., 46 (6), 587-595 (1968).

[52] M. Miyake, M. Yoshizawa and H. Harada, *"Incidence Angle Dependence of Planar Channeling in Boron Ion Implantation into Silicon,"* J. Electrochem. Soc., 130, 716-719, (1973).

[53] M. I. Current, N. L. Turner, T. C. Smith and D. Crane, *"Planar Channeling Effects in Si(100),"* Proc. Fifth Intern. Conf. on Ion Implantation Equipment & Techn., p. 336 North-Holland (1984)

[54] I. Suni, et al., *"Effect of Preamorphization Depth on Channeling Tails in B $^+$ and As $^+$ Implanted Silicon,"* in Ion Beam Processes in Advanced Electronic Materials and Device Technology, Vol. 45, Mater. Res. Soc., 1985.

[55] H. Russel, J. Lorenz, and K. Hoffmann, *"Models of Implantation into Multilayered Targets,"* Appl. Phys. A, 41, 201 (1986).

[56] H. Ishihara, H. Furukawa, S. Yamada and J. Kawamura, *"Projected Range Distribution of Implanted Ions in a Double-Layer Substrate,"* in Ion Implantation in Semiconductors, S. Namba, Ed., p. 423-428, Plenum Press, New York (1975).

[57] K. Gamo, T. Ukegawa, and S. Namba, *"Field Ion Sources Using Eutectic Alloys,"* Jpn. J. Appl. Phys., 19 (7), L379-L382 (1980).

[58] R. L. Kubena, C. L. Anderson, R. L. Seliger, R. A. Jullens, and E. H. Stevens, *"FET Fabrication Using Maskless Ion Implantation,"* J. Vac. Sci. Technol., 19 (4), 916-920 (1981).

[59] F. H. Dill and J. M. Shaw, *"Thermal Effects on the Photoresist AZ1350J,"* IBM J. Res. Develop., p. 210, May 1977.

[60] P. Burggraaf, *"Resist Implant Problems: Some Solved, Others Not,"* Semiconductor International, p. 66, June 1992.

[61] Y. Okuyama, T. Hashimoto, and T. Koguchi, *"High Dose Implantation into Photoresist,"* J. Electrochem. Soc., 125 (8), 1293 (1978).

## 442

[62] T. C. Smith, *"Wafer Cooling and Photoresist Masking Problems in Ion Implantation,"* in Ion Implantation Equipment and Techniques, H. Ryssel and H. Glawischnig, Eds., p. 196, Springer-Verlag, New York (1983).

[63] T. C. Smith, *"Photoresist Problems and Particle Contamination,"* in Ion Implantation Science and Technology, 2nd Edition, J. F. Ziegler, Ed., p. 345, Academic Press, San Diego, CA (1988).

[64] *"Ion Implantation,"* Practical VLSI Fabrication for the 90s, R. Bowman, G. Fry, J. Griffin, R. Potter, and R. Skinner, Eds., Integrated Circuit Engineering Corp., Arizona (1990).

[65] S. Furukawa, H. Matsumura and H. Ishiwara, *"Theoretical Consideration on Lateral Spread of Implanted Ions,"* Jpn. J. Appl. Phys. 11, 134, (1972).

[66] H. Runge, *"Distribution of Implanted Ions under Arbitrary Shaped Mask Edges,"* Phys. Status Solidi (a) 39, 595-599 (1977).

[67] Y. Akasaka, K. Horie, K. Yoneda, T. Sakurai, H. Nishi, S. Kawabe, and E. Tohi, *"Depth Distributions of Defects and Impurities on 100-keV B $^+$ Ion Implanted Silicon,"* J. Appl. Phys., 44 (1), 220-224 (1973).

− 68 − R. Bauerlein, in Radiation Damage in Solids, D.S. Bellington, Ed., p. 358, Academic Press New York (1969).

[69] J. F. Gibbons, *"Ion Implantation in Semiconductors - Part II, Damage Production and Annealing,"* Proc. IEEE, 60, 1062 (1972).

[70] D. K. Brice, *"Ion Implantation Depth Distributions: Energy Deposition into Atomic Processes and Ion Locations,"* Appl. Phys. Lett., 16 (3), 103-106 (1970)

[71] W. P. Maszara and G. A. Rozgonyi, *"Kinetics of Damage Production in Silicon during Self-Implantation,"* J. Appl. Phys. 60, 2310 (1986).

[72] B. L. Crowder, *"The Role of Damage in the Annealing Characteristics of Ion Implanted Si,"* J. Electrochem. Soc. 117 (5), 671-674 (1970).

[73] F. H. Eisen and B. Welch, *"Flux and Fluence Dependence of Disorder Produced during Implantation of $^{11}B$ in Silicon,"* in First International Conference on Ion Implantation, F. Eisen and L. Chadderton, Eds., Gordon and Breach, 459-464, New York (1971).

[74] F.F. Morehead and B. L. Crowder, *"A Model for the formation of Amorphous Silicon by Ion Implantation,"* in First International Conference on Ion Implantation, F. Eisen and L. Chadderton, Eds., Gordon and Breach, p. 25, New York (1971).

[75] B. L. Crowder, R. S. Title, M. H. Brodsky, and G. D. Pettit, *"ESR and Optical Absorption Studies of Ion-Implanted Silicon,"* Appl. Phys. Lett. 16 (5), 205-208 (1970).

[76] J. E. Westmoreland, J. W. Mayer, F. H. Eisen, and B. Welch, *"Production and Annealing of Lattice Disorder in Silicon by 200-keV Boron Ions,"* Appl. Phys. Lett. 15 (9), 308-310 (1969).

[77] G. H. Kinchin and R. S. Pease, *"The Displacement of Atoms in Solids by Radiation,"* Rep. Prog. Phys. 18, 1 (1955):

[78] W. L. Brown, *"Collision Cascades, Ionization, Spikes and Energy Transfer,"* Mat. Sci. Res. Symp. 51, 53 (1985).

[79] P. V. Pavlov, D. I. Tetel'baum, E. I. Zorin, and V. I. Alekseev, *"Distribution of Implanted Atoms and Radiation Defects in the Ion Bombardment of Silicon (Monte Carlo Calculation Method).* Sov. Phys.-Solid State 8 (9), 2141-2146 (1967).

[80] T. Hirao, G. Fuse, K. Inoue, S. Takayanagi, Y. Yaegashi, and S. Ichikawa, *"Electrical Properties of Si Implanted with As through $SiO_2$ Films,"* J. Appl. Phys., 51, 262 (1980).

[81] T. C. Cass and V. G. K. Reddi, *"Anomalous Residual Damage in Si after Annealing of "Through-Oxide" Arsenic Implantation,"* Appl. Phys. Lett., 23 (5), 268-270 (1973).

[82] R. Moline and A. G. Cullis, *"Residual Defects in Si Produced by Recoil Implantation of Oxygen,"* Appl. Phys. Lett., 26 (10), 551-553 (1975).

[83] T. Y. Tan, *"Dislocation Nucleation Models from Point Defect Condensations in Silicon and Germanium,"* Mat. Res. Soc. Symp. Proc. 2, 163-178 (1981).

[84] S. Mader, *"Ion Implantation Damage in Silicon,"* Ion Implantation Science and Technology, J.F. Ziegler, Ed., Academic Press, (1984)

[85] W. X. Lu, Y. H. Qian, R. H. Tian, Z. L. Wang, R. J. Schreutelkamp, J. R. Liefting, and F. W. Saris, *"Reduction of Secondary Defect Formation in MeV B $^+$ Ion-Implanted Si (100),"* Appl. Phys. Lett., 55 (18), 1838-1840 (1989).

[86] V. Raineri, R. J. Schreutelkamp, F. W. Saris, K. T. F. Janssen, and R. A. Kaim, *"Reduction of Boron Diffusion in Silicon by 1 MeV $^{29}$Si $^+$ Irradiation,"* Appl. Phys. Lett., 58 (9), 922-924 (1991).

[87] P. B. Griffin, P. M. Fahey, J. D. Plummer, and R. W. Dutton, *"Measurement of Silicon Interstitial Diffusivity,"* Appl. Phys. Lett., 47 (3), 319-321 (1985).

[88] P. B. Griffin, S. T. Ahn, W. A. Tiller, and J. D. Plummer, *"Model for Bulk Effects on Si Interstitial Diffusivity in Silicon,"* Appl. Phys. Lett., 51 (2), 115-117 (1987).

[89] R. J. Schreutelkamp, J. S. Custer, J. R. Liefting, W. X. Lu, and F. W. Saris, *"Pre-Amorphization Damage in Ion-Implanted Silicon,"* Mater. Sci. Rep., 6 (7/8), 275-366 (1991)

[90] K. S. Jones, S. Prussin, and E. R. Weber, *"A Systematic Analysis of Defects in Ion-Implanted Silicon,"* Appl. Phys. A, 45, 1-34 (1988).

[91] T.E. Seidel and L. A Larson, *"Status of Low-Dose Implantation for VLSI,"* Material Res. Soc. 17 (6), 34 (1992).

[92] T. E. Seidel and A. U. MacRae, *"The Isothermal Annealing of Boron Implanted Silicon,"* in First International Conference on Ion Implantation, F. Eisen and L. Chadderton, Eds., Gordon and Beach, New York (1971).

[93] L. Csepregi, E. F. Kennedy, J. W. Mayer, and T. W., Sigmon, *"Substrate-Orientation Dependence of the Epitaxial Regrowth Rate from Si-Implanted Amorphous Si,"* J. Appl. Phys.,49 (7), 3906-3911 (1978).

[94] L. Csepregi, E. F. Kennedy, T. J. Gallangher, J. W. Mayer, and T. W. Sigmon, *"Reorder of Amorphous Layers of Si Implanted with P, As, and B Ions,"* J. Appl. Phys., 48 (10), 4234-4240 (1977).

[95] D. Hagmann, D. Steiner, and T. Schellinger, *"A Method to Impede the Formation of Crystal Defects after High Dose Arsenic Implants,"* J. Electrochem. Soc. 133 (12), 2597-2600 (1986).

[96] T. O. Sedgewick, *"Short Term Annealing,"* J. Electrochem. Soc. 130, 484 (1983).

[97] R. Kwor, D. L. Kwong, C. C. Ho, B. Y. Tsaur, and S. Baumann, *"Effect of Furnace Pre-Anneal and Rapid Thermal Anneal on Arsenic-Implanted Silicon,"* J. Electrochem. Soc. 132 (5), 1201-1206 (1985).

[98] S. K. Tiku and W. M. Duncan, *"Self-Compensation in Rapid Thermal Annealed Silicon-Implanted Gallium Arsenide,"* J. Electrochem. Soc. 132, 2237 (1985).

[99] M. D. Giles, *"Ion Implantation,"* in VLSI Technology, S. M. Sze, Ed., p. 327, John Wiley, New York (1983).

**444**

[100] W. K. Hofker, D. P. Oosthoek, N. J. Koenman, and H. A. M. De Grefte, *"Concentration Profiles of Boron Implantations in Amorphous and Polycrystalline Silicon,"* Rad. Effects, 24, 223 (1975).

[101] W. K. Hofker, H. W. Werner, D. P. Oosthoek, and N. J. Koenman, *"Boron Implantation in Silicon: A Comparison of Charge Carriers and Boron Concentration Profiles,"* Appl. Phys., 4, 125 (1974).

[102] R. F. Webber, R. S. Thorn, L. N. Large, *"The Measurement of Electrical Activity and Hall Mobility of Boron and Phosphorus Ion-Implanted Layers in Silicon,"* Int. J. Electronics, 26 (2), 163-172 (1969).

[103] H. J. Geipel and W. K. Tice, *"Critical Microstructure for Ion-Implantation Gettering Effects in Silicon,"* Appl. Phys. Lett., 30 (7), 325-327 (1977).

[104] B. El-Kareh, *"Ultra-Shallow Doped Film Requirements for Future Technologies,"* J. Vac. Sci. Technol., Jan./Feb. 1994.

[105] M. Tamura, T. Ikeda, and T. Takuyama, *"Crystal Defects and Electrical Properties of Implanted Silicon,"* in Second International Conference on Ion Implantation, I. Ruge and J. Graul, Eds., Plenum Press, New York (1972).

[106] C. Hill, *"Beam Processing in Silicon Device Technology,"* in Laser and Electron Beam Interaction and Material Processing, J. F. Gibbons, L. D. Hess, and T. W. Sigmon, Eds., North-Holland, New York (1981).

[107] J. Y. Sun, C. Wong, Y. Taur, and C. H. Hsu, *"Study of Boron Penetration Through Thin Oxide with $P^+$ Polygate,"* VLSI Tech. Digest., p. 17 (1989).

[108] B. El-Kareh, W. W. Abadeer, and W. Tonti, *"Design of Submicron PMOSFETs for DRAM Array Applications,"* IEDM Tech. Digest, p. 379 (1991).

[109] C. M. Osburn, *"Silicides,"* Rapid Thermal Processing, Science and Technology, R. B. Fair, Ed., Chap. 6, Academic Press, San Diego, CA (1993). See also references at end of chapter.

[110] L. Van den Hove and R. F. DeKeersmaecker, *"Silicide by Rapid Thermal Processing,"* Reduced Thermal Processing for VLSI, R. A. Levy, Ed., Plenum Press, New York (1989).

[111] H. Jiang, C. M. Osburn, P. Smith, Z.-G. Xiao, D. Griffis, G. McGuire, and G. A. Rozgonyi, *"Ultra Shallow Junction Formation Using Diffusion from Silicides,"* J. Electrochem. Soc., 139 (10, 196 (1992).

[112] M. H. Juang and H. C. Cheng, *"Effect of Rapid Thermal Annealing on the Formation of Shallow $P^+ - N Junction by Implanting :f. BF_2$ Ions into Thin Metal Films on Si Substrate,"* J. Appl. Phys., 71 (3), 1271 (1992).

[113] J. Suarez, B. E. Johnson, and B. El-Kareh, *"Thermal Stability of Polysilicon Resistors,"* IEEE Trans. Comp. Manuf. Tech., 15 (3), 392 (1992).

[114] S. A. StOnge, S. G. Franz, A. F. Puttlitz, A. Kalinoski, B. E. Johnson, and B. El-Kareh, *"Design of Precision Capacitors for Analog Applications,"* IEEE Trans. Comp. Manuf. Tech., 15 (6), 1064 (1992).

[115] J. S. Mercier, *"Rapid Flow of Doped Glasses for VLSI Fabrication,"* Solid-State Technology, 30 (7), 85 (1987).

[116] B. El-Kareh, *"Comparison of RTP and Furnace Processing for Quarter-Micron CMOS,"* SPIE Proceedings, Vol. 2091 (1994).

[117] L. J. Parkes and J. P. Lavine, *"Calculated Moments for the Implantation of Boron into Silicides,"* J. Appl. Phys., 60, 14 (1986).

**Table 6.9** Range statistics for boron in silicon and polysilicon
Courtesy: Technology Modeling Associates (TMA).

| Energy | $R_P$ | $\Delta R_P$ | $\gamma$ | $\beta$ | $\Delta R_T$ |
|--------|-------|--------------|----------|---------|--------------|
| 5 | 0.0231 | 0.0119 | 0.321 | 2.6020 | 0.0142 |
| 10 | 0.0439 | 0.0257 | 0.924 | 4.9986 | 0.0250 |
| 15 | 0.0618 | 0.0309 | 0.941 | 4.8528 | 0.0337 |
| 20 | 0.0815 | 0.0387 | 0.698 | 4.0924 | 0.0424 |
| 25 | 0.0957 | 0.0422 | 0.463 | 3.8330 | 0.0497 |
| 30 | 0.1164 | 0.0486 | 0.395 | 3.5937 | 0.0569 |
| 40 | 0.1326 | 0.0455 | -0.611 | 3.5045 | 0.0694 |
| 50 | 0.1629 | 0.0515 | -0.718 | 3.7393 | 0.0803 |
| 60 | 0.1921 | 0.0566 | -0.811 | 3.9744 | 0.0899 |
| 70 | 0.2202 | 0.0610 | -0.892 | 4.2078 | 0.0986 |
| 80 | 0.2473 | 0.0649 | -0.964 | 4.4390 | 0.1064 |
| 90 | 0.2736 | 0.0683 | -1.030 | 4.6675 | 0.1135 |
| 100 | 0.2990 | 0.0714 | -1.091 | 4.8935 | 0.1200 |
| 110 | 0.3237 | 0.0742 | -1.147 | 5.1169 | 0.1260 |
| 120 | 0.3476 | 0.0767 | -1.199 | 5.3380 | 0.1316 |
| 130 | 0.3710 | 0.0790 | -1.248 | 5.5569 | 0.1368 |
| 140 | 0.3937 | 0.0811 | -1.294 | 5.7737 | 0.1416 |
| 150 | 0.4158 | 0.0830 | -1.337 | 5.9885 | 0.1462 |
| 160 | 0.4375 | 0.0848 | -1.379 | 6.2014 | 0.1504 |
| 170 | 0.4586 | 0.0865 | -1.418 | 6.4127 | 0.1545 |
| 180 | 0.4793 | 0.0881 | -1.456 | 6.6223 | 0.1583 |
| 190 | 0.4995 | 0.0896 | -1.492 | 6.8304 | 0.1619 |
| 200 | 0.5194 | 0.0909 | -1.527 | 7.0371 | 0.1653 |
| 220 | 0.5580 | 0.0935 | -1.592 | 7.4449 | 0.1716 |
| 240 | 0.5952 | 0.0957 | -1.654 | 7.8482 | 0.1774 |
| 260 | 0.6311 | 0.0978 | -1.711 | 8.2473 | 0.1827 |
| 280 | 0.6660 | 0.0996 | -1.765 | 8.6426 | 0.1876 |
| 300 | 0.6998 | 0.1013 | -1.816 | 9.0344 | 0.1922 |
| 320 | 0.7326 | 0.1028 | -1.865 | 9.4230 | 0.1964 |
| 340 | 0.7646 | 0.1043 | -1.911 | 9.8086 | 0.2003 |
| 360 | 0.7958 | 0.1056 | -1.955 | 10.1913 | 0.2040 |
| 380 | 0.8262 | 0.1068 | -1.998 | 10.5715 | 0.2074 |
| 400 | 0.8560 | 0.1080 | -2.039 | 10.9493 | 0.2107 |
| 420 | 0.8851 | 0.1090 | -2.078 | 11.3247 | 0.2138 |
| 440 | 0.9135 | 0.1100 | -2.116 | 11.6980 | 0.2167 |
| 460 | 0.9414 | 0.1110 | -2.152 | 12.0692 | 0.2194 |
| 480 | 0.9688 | 0.1119 | -2.188 | 12.4385 | 0.2221 |
| 500 | 0.9956 | 0.1127 | -2.222 | 12.8060 | 0.2246 |
| 550 | 1.0607 | 0.1146 | -2.302 | 13.7088 | 0.2303 |
| 600 | 1.1231 | 0.1163 | -2.377 | 14.6027 | 0.2354 |
| 650 | 1.1831 | 0.1178 | -2.448 | 15.4883 | 0.2401 |
| 700 | 1.2410 | 0.1192 | -2.514 | 16.3661 | 0.2443 |
| 750 | 1.2969 | 0.1204 | -2.577 | 17.2368 | 0.2482 |
| 800 | 1.3512 | 0.1215 | -2.636 | 18.1008 | 0.2517 |
| 850 | 1.4038 | 0.1225 | -2.693 | 18.9585 | 0.2550 |
| 900 | 1.4550 | 0.1235 | -2.747 | 19.8102 | 0.2580 |
| 1000 | 1.5535 | 0.1252 | -2.849 | 21.4971 | 0.2635 |
| 2000 | 2.972 | 0.2060 | -1.940 | 15.78 | 0.2060 |
| 3000 | 4.023 | 0.2290 | -2.460 | 19.38 | 0.2290 |
| 4000 | 5.336 | 0.2520 | -2.980 | 21.42 | 0.2520 |

**Table 6.9 (Cont'd)** Range statistics for boron in silicon dioxide
Courtesy: Technology Modeling Associates (TMA).

| Energy | $R_P$ | $\Delta R_P$ | $\gamma$ | $\beta$ | $\Delta R_T$ |
|---|---|---|---|---|---|
| 5 | 0.0194 | 0.0094 | -0.0054 | 0. | 0.0112 |
| 10 | 0.0355 | 0.0167 | -0.033 | 2.9117 | 0.0224 |
| 20 | 0.0713 | 0.0280 | -0.286 | 3.0378 | 0.0395 |
| 30 | 0.1068 | 0.0370 | -0.463 | 3.2485 | 0.0542 |
| 40 | 0.1414 | 0.0443 | -0.600 | 3.4836 | 0.0670 |
| 50 | 0.1748 | 0.0505 | -0.713 | 3.7265 | 0.0781 |
| 60 | 0.2070 | 0.0557 | -0.810 | 3.9711 | 0.0881 |
| 70 | 0.2381 | 0.0602 | -0.894 | 4.2151 | 0.0970 |
| 80 | 0.2681 | 0.0642 | -0.970 | 4.4574 | 0.1051 |
| 90 | 0.2972 | 0.0678 | -1.039 | 4.6978 | 0.1124 |
| 100 | 0.3253 | 0.0709 | -1.102 | 4.9359 | 0.1191 |
| 110 | 0.3526 | 0.0738 | -1.160 | 5.1719 | 0.1253 |
| 120 | 0.3792 | 0.0763 | -1.214 | 5.4058 | 0.1311 |
| 130 | 0.4050 | 0.0787 | -1.265 | 5.6377 | 0.1364 |
| 140 | 0.4302 | 0.0809 | -1.313 | 5.8677 | 0.1414 |
| 150 | 0.4547 | 0.0829 | -1.359 | 6.0959 | 0.1461 |
| 160 | 0.4787 | 0.0847 | -1.402 | 6.3223 | 0.1505 |
| 170 | 0.5021 | 0.0864 | -1.443 | 6.5472 | 0.1546 |
| 180 | 0.5250 | 0.0880 | -1.482 | 6.7706 | 0.1586 |
| 190 | 0.5474 | 0.0895 | -1.520 | 6.9926 | 0.1623 |
| 200 | 0.5694 | 0.0909 | -1.556 | 7.2132 | 0.1658 |
| 220 | 0.6121 | 0.0935 | -1.624 | 7.6486 | 0.1723 |
| 240 | 0.6533 | 0.0958 | -1.687 | 8.0798 | 0.1783 |
| 260 | 0.6931 | 0.0979 | -1.747 | 8.5071 | 0.1837 |
| 280 | 0.7316 | 0.0997 | -1.803 | 8.9308 | 0.1887 |
| 300 | 0.7690 | 0.1015 | -1.856 | 9.3511 | 0.1933 |
| 320 | 0.8054 | 0.1030 | -1.906 | 9.7682 | 0.1976 |
| 340 | 0.8408 | 0.1045 | -1.954 | 10.1824 | 0.2017 |
| 360 | 0.8753 | 0.1058 | -2.000 | 10.5938 | 0.2054 |
| 380 | 0.9089 | 0.1071 | -2.044 | 11.0027 | 0.2089 |
| 400 | 0.9418 | 0.1082 | -2.087 | 11.4092 | 0.2123 |
| 420 | 0.9740 | 0.1093 | -2.127 | 11.8134 | 0.2154 |
| 440 | 1.0054 | 0.1103 | -2.167 | 12.2154 | 0.2183 |
| 460 | 1.0363 | 0.1113 | -2.204 | 12.6154 | 0.2211 |
| 480 | 1.0665 | 0.1122 | -2.241 | 13.0134 | 0.2238 |
| 500 | 1.0962 | 0.1130 | -2.276 | 13.4095 | 0.2263 |
| 550 | 1.1681 | 0.1149 | -2.359 | 14.3808 | 0.2322 |
| 600 | 1.2370 | 0.1166 | -2.437 | 15.3435 | 0.2374 |
| 650 | 1.3033 | 0.1182 | -2.509 | 16.2979 | 0.2421 |
| 700 | 1.3672 | 0.1195 | -2.577 | 17.2444 | 0.2463 |
| 750 | 1.4290 | 0.1208 | -2.642 | 18.1837 | 0.2502 |
| 800 | 1.4889 | 0.1219 | -2.703 | 19.1160 | 0.2538 |
| 850 | 1.5470 | 0.1229 | -2.762 | 20.0419 | 0.2571 |
| 900 | 1.6036 | 0.1238 | -2.817 | 20.9617 | 0.2602 |
| 950 | 1.6586 | 0.1247 | -2.871 | 21.8756 | 0.2630 |
| 1000 | 1.7123 | 0.1255 | -2.922 | 22.7839 | 0.2657 |

**Table 6.9 (Cont'd)** Range statistics for boron in silicon nitride
Courtesy: Technology Modeling Associates (TMA).

| Energy | $R_P$ | $\Delta R_P$ | $\gamma$ | $\beta$ | $\Delta R_T$ |
|---|---|---|---|---|---|
| 5 | 0.0120 | 0.0070 | 0.120 | 2.9300 | 0.0100 |
| 10 | 0.0243 | 0.0119 | 0.014 | 2.9103 | 0.0159 |
| 20 | 0.0499 | 0.0205 | -0.226 | 2.9899 | 0.0289 |
| 30 | 0.0758 | 0.0277 | -0.398 | 3.1591 | 0.0404 |
| 40 | 0.1013 | 0.0337 | -0.531 | 3.3575 | 0.0506 |
| 50 | 0.1263 | 0.0387 | -0.641 | 3.5669 | 0.0597 |
| 60 | 0.1506 | 0.0432 | -0.736 | 3.7804 | 0.0679 |
| 70 | 0.1743 | 0.0470 | -0.818 | 3.9950 | 0.0753 |
| 80 | 0.1972 | 0.0504 | -0.892 | 4.2092 | 0.0820 |
| 90 | 0.2195 | 0.0535 | -0.960 | 4.4225 | 0.0882 |
| 100 | 0.2412 | 0.0562 | -1.021 | 4.6345 | 0.0940 |
| 110 | 0.2623 | 0.0587 | -1.078 | 4.8450 | 0.0993 |
| 120 | 0.2828 | 0.0610 | -1.131 | 5.0541 | 0.1042 |
| 130 | 0.3029 | 0.0631 | -1.181 | 5.2617 | 0.1088 |
| 140 | 0.3224 | 0.0650 | -1.228 | 5.4679 | 0.1131 |
| 150 | 0.3416 | 0.0668 | -1.273 | 5.6727 | 0.1172 |
| 160 | 0.3603 | 0.0685 | -1.315 | 5.8761 | 0.1210 |
| 170 | 0.3786 | 0.0700 | -1.355 | 6.0784 | 0.1246 |
| 180 | 0.3966 | 0.0714 | -1.394 | 6.2794 | 0.1281 |
| 190 | 0.4141 | 0.0728 | -1.431 | 6.4793 | 0.1313 |
| 200 | 0.4314 | 0.0741 | -1.466 | 6.6782 | 0.1344 |
| 220 | 0.4650 | 0.0764 | -1.533 | 7.0713 | 0.1402 |
| 240 | 0.4974 | 0.0785 | -1.595 | 7.4611 | 0.1454 |
| 260 | 0.5288 | 0.0804 | -1.653 | 7.8478 | 0.1503 |
| 280 | 0.5592 | 0.0821 | -1.709 | 8.2315 | 0.1547 |
| 300 | 0.5888 | 0.0836 | -1.761 | 8.6124 | 0.1588 |
| 320 | 0.6176 | 0.0851 | -1.811 | 8.9908 | 0.1627 |
| 340 | 0.6456 | 0.0864 | -1.858 | 9.3669 | 0.1663 |
| 360 | 0.6730 | 0.0876 | -1.903 | 9.7407 | 0.1697 |
| 380 | 0.6997 | 0.0888 | -1.946 | 10.1125 | 0.1728 |
| 400 | 0.7257 | 0.0898 | -1.988 | 10.4824 | 0.1758 |
| 420 | 0.7513 | 0.0908 | -2.028 | 10.8504 | 0.1786 |
| 440 | 0.7763 | 0.0918 | -2.067 | 11.2166 | 0.1813 |
| 460 | 0.8008 | 0.0927 | -2.104 | 11.5812 | 0.1838 |
| 480 | 0.8249 | 0.0935 | -2.140 | 11.9443 | 0.1862 |
| 500 | 0.8485 | 0.0943 | -2.175 | 12.3058 | 0.1885 |
| 550 | 0.9057 | 0.0961 | -2.257 | 13.1944 | 0.1938 |
| 600 | 0.9606 | 0.0977 | -2.334 | 14.0763 | 0.1985 |
| 650 | 1.0134 | 0.0991 | -2.406 | 14.9516 | 0.2028 |
| 700 | 1.0644 | 0.1003 | -2.473 | 15.8207 | 0.2067 |
| 750 | 1.1137 | 0.1015 | -2.537 | 16.6841 | 0.2103 |
| 800 | 1.1615 | 0.1025 | -2.598 | 17.5421 | 0.2135 |
| 850 | 1.2079 | 0.1035 | -2.656 | 18.3949 | 0.2166 |
| 900 | 1.2531 | 0.1044 | -2.711 | 19.2428 | 0.2194 |
| 950 | 1.2970 | 0.1052 | -2.764 | 20.0862 | 0.2220 |
| 1000 | 1.3399 | 0.1059 | -2.815 | 20.9251 | 0.2244 |

**Table 6.9 (Cont'd)** Range statistics for boron in aluminum
Courtesy: Technology Modeling Associates (TMA).

| Energy | $R_P$ | $\Delta R_P$ | $\gamma$ | $\beta$ | $\Delta R_T$ |
|---|---|---|---|---|---|
| 10 | 0.0344 | 0.0186 | 0.036 | 2.9120 | 0.0255 |
| 20 | 0.0702 | 0.0321 | -0.188 | 2.9655 | 0.0459 |
| 30 | 0.1069 | 0.0434 | -0.350 | 3.1028 | 0.0640 |
| 40 | 0.1434 | 0.0529 | -0.477 | 3.2692 | 0.0803 |
| 50 | 0.1793 | 0.0611 | -0.581 | 3.4471 | 0.0949 |
| 60 | 0.2145 | 0.0683 | -0.671 | 3.6296 | 0.1082 |
| 70 | 0.2487 | 0.0747 | -0.749 | 3.8138 | 0.1202 |
| 80 | 0.2821 | 0.0803 | -0.819 | 3.9980 | 0.1313 |
| 90 | 0.3147 | 0.0854 | -0.883 | 4.1816 | 0.1415 |
| 100 | 0.3465 | 0.0900 | -0.942 | 4.3643 | 0.1510 |
| 110 | 0.3775 | 0.0942 | -0.996 | 4.5458 | 0.1598 |
| 120 | 0.4078 | 0.0980 | -1.046 | 4.7260 | 0.1680 |
| 130 | 0.4374 | 0.1016 | -1.094 | 4.9050 | 0.1757 |
| 140 | 0.4663 | 0.1048 | -1.138 | 5.0828 | 0.1829 |
| 150 | 0.4946 | 0.1079 | -1.181 | 5.2594 | 0.1897 |
| 160 | 0.5224 | 0.1107 | -1.221 | 5.4349 | 0.1962 |
| 170 | 0.5496 | 0.1133 | -1.259 | 5.6093 | 0.2023 |
| 180 | 0.5763 | 0.1158 | -1.296 | 5.7827 | 0.2081 |
| 190 | 0.6024 | 0.1181 | -1.331 | 5.9551 | 0.2136 |
| 200 | 0.6281 | 0.1203 | -1.365 | 6.1266 | 0.2188 |
| 220 | 0.6782 | 0.1243 | -1.428 | 6.4658 | 0.2286 |
| 240 | 0.7267 | 0.1279 | -1.487 | 6.8021 | 0.2375 |
| 260 | 0.7736 | 0.1312 | -1.543 | 7.1356 | 0.2458 |
| 280 | 0.8192 | 0.1342 | -1.596 | 7.4667 | 0.2534 |
| 300 | 0.8636 | 0.1369 | -1.646 | 7.7955 | 0.2605 |
| 320 | 0.9067 | 0.1394 | -1.693 | 8.1221 | 0.2671 |
| 340 | 0.9488 | 0.1418 | -1.738 | 8.4468 | 0.2733 |
| 360 | 0.9899 | 0.1439 | -1.782 | 8.7697 | 0.2791 |
| 380 | 1.0301 | 0.1459 | -1.823 | 9.0908 | 0.2846 |
| 400 | 1.0694 | 0.1478 | -1.863 | 9.4103 | 0.2898 |
| 420 | 1.1078 | 0.1496 | -1.902 | 9.7284 | 0.2947 |
| 440 | 1.1455 | 0.1512 | -1.939 | 10.0450 | 0.2993 |
| 460 | 1.1824 | 0.1528 | -1.975 | 10.3603 | 0.3037 |
| 480 | 1.2187 | 0.1542 | -2.009 | 10.6743 | 0.3079 |
| 500 | 1.2543 | 0.1556 | -2.043 | 10.9871 | 0.3119 |
| 550 | 1.3407 | 0.1588 | -2.122 | 11.7578 | 0.3211 |
| 600 | 1.4236 | 0.1616 | -2.196 | 12.5231 | 0.3294 |
| 650 | 1.5035 | 0.1641 | -2.265 | 13.2831 | 0.3369 |
| 700 | 1.5806 | 0.1664 | -2.331 | 14.0383 | 0.3437 |
| 750 | 1.6552 | 0.1684 | -2.393 | 14.7889 | 0.3499 |
| 800 | 1.7276 | 0.1703 | -2.451 | 15.5353 | 0.3557 |
| 850 | 1.7979 | 0.1720 | -2.508 | 16.2777 | 0.3610 |
| 900 | 1.8663 | 0.1736 | -2.561 | 17.0163 | 0.3660 |
| 950 | 1.9329 | 0.1750 | -2.613 | 17.7514 | 0.3706 |
| 1000 | 1.9980 | 0.1764 | -2.662 | 18.4831 | 0.3749 |

**Table 6.9 (cont'd)** Range statistics for boron in photoresist (type AZ-111)
Courtesy: Technology Modeling Associates (TMA).

| Energy | $R_P$ | $\Delta R_P$ | $\gamma$ | $\beta$ | $\Delta R_T$ |
|---|---|---|---|---|---|
| 10 | 0.1086 | 0.0270 | -0.384 | 3.1420 | 0.0312 |
| 20 | 0.2267 | 0.0475 | -0.634 | 3.5516 | 0.0591 |
| 30 | 0.3446 | 0.0635 | -0.836 | 4.0442 | 0.0834 |
| 40 | 0.4587 | 0.0763 | -0.997 | 4.5496 | 0.1046 |
| 50 | 0.5683 | 0.0868 | -1.133 | 5.0609 | 0.1231 |
| 60 | 0.6736 | 0.0955 | -1.250 | 5.5672 | 0.1395 |
| 70 | 0.7747 | 0.1030 | -1.356 | 6.0827 | 0.1541 |
| 80 | 0.8721 | 0.1095 | -1.451 | 6.5934 | 0.1673 |
| 90 | 0.9661 | 0.1151 | -1.538 | 7.1037 | 0.1792 |
| 100 | 1.0569 | 0.1202 | -1.620 | 7.6239 | 0.1901 |
| 110 | 1.1450 | 0.1247 | -1.695 | 8.1348 | 0.2002 |
| 120 | 1.2305 | 0.1288 | -1.766 | 8.6507 | 0.2094 |
| 130 | 1.3137 | 0.1325 | -1.834 | 9.1754 | 0.2180 |
| 140 | 1.3947 | 0.1359 | -1.898 | 9.6977 | 0.2260 |
| 150 | 1.4738 | 0.1391 | -1.958 | 10.213 | 0.2336 |
| 160 | 1.5511 | 0.1420 | -2.017 | 10.746 | 0.2406 |
| 170 | 1.6267 | 0.1447 | -2.072 | 11.266 | 0.2472 |
| 180 | 1.7007 | 0.1472 | -2.126 | 11.800 | 0.2535 |
| 190 | 1.7732 | 0.1495 | -2.177 | 12.325 | 0.2594 |
| 200 | 1.8445 | 0.1518 | -2.227 | 12.861 | 0.2651 |
| 220 | 1.9831 | 0.1558 | -2.320 | 13.914 | 0.2756 |
| 240 | 2.1173 | 0.1595 | -2.407 | 14.969 | 0.2851 |
| 260 | 2.2476 | 0.1628 | -2.490 | 16.042 | 0.2939 |
| 280 | 2.3742 | 0.1658 | -2.569 | 17.126 | 0.3020 |
| 300 | 2.4976 | 0.1685 | -2.645 | 18.229 | 0.3095 |
| 320 | 2.6182 | 0.1711 | -2.718 | 19.346 | 0.3166 |
| 340 | 2.7360 | 0.1734 | -2.787 | 20.457 | 0.3232 |
| 360 | 2.8515 | 0.1757 | -2.854 | 21.588 | 0.3293 |
| 380 | 2.9647 | 0.1777 | -2.919 | 22.736 | 0.3352 |
| 400 | 3.0759 | 0.1797 | -2.922 | 22.790 | 0.3407 |
| 420 | 3.1852 | 0.1815 | -3.044 | 25.092 | 0.3459 |
| 440 | 3.2927 | 0.1832 | -3.103 | 26.275 | 0.3509 |
| 460 | 3.3986 | 0.1849 | -3.161 | 27.483 | 0.3557 |
| 480 | 3.5030 | 0.1865 | -3.217 | 28.694 | 0.3602 |
| 500 | 3.6059 | 0.1880 | -3.272 | 29.927 | 0.3646 |
| 550 | 3.8576 | 0.1914 | -3.395 | 32.847 | 0.3747 |
| 600 | 4.1022 | 0.1946 | -3.514 | 35.896 | 0.3839 |
| 650 | 4.3406 | 0.1974 | -3.627 | 39.007 | 0.3923 |
| 700 | 4.5736 | 0.2001 | -3.737 | 42.248 | 0.4001 |
| 750 | 4.8019 | 0.2025 | -3.843 | 45.579 | 0.4072 |
| 800 | 5.0260 | 0.2048 | -3.946 | 49.021 | 0.4141 |
| 850 | 5.2464 | 0.2069 | -4.047 | 52.602 | 0.4205 |
| 900 | 5.4634 | 0.2089 | -4.145 | 56.279 | 0.4265 |
| 950 | 5.6775 | 0.2109 | -4.240 | 60.043 | 0.4322 |
| 1000 | 5.8889 | 0.2127 | -4.334 | 63.966 | 0.4376 |

**Table 6.9 (Cont'd)** Range statistics for boron from $BF_2$ in silicon
Courtesy: Technology Modeling Associates (TMA).

| Energy | $R_P$ | $\Delta R_P$ | $\gamma$ | $\beta$ | $\Delta R_T$ |
|---|---|---|---|---|---|
| 5 | $6.380x10^{-3}$ | $4.344x10^{-3}$ | $8.708x10^{-3}$ | 0.0 | $5.21x10^{-3}$ |
| 10 | $1.081x10^{-}$ | $6.770x10^{-3}$ | $9.034x10^{-3}$ | 0.0 | $8.12x10^{-3}$ |
| 20 | $1.956x10^{-2}$ | $1.092x10^{-2}$ | $2.051x10^{-2}$ | 0.0 | $1.31x10^{-2}$ |
| 25 | 0.02583 | 0.01446 | -0.65139 | 11.96689 | 0.01736 |
| 50 | 0.04610 | 0.02453 | -0.11660 | 6.26944 | 0.02713 |
| 75 | 0.05901 | 0.03129 | -0.23726 | 15.47629 | 0.03690 |
| 100 | 0.07556 | 0.03703 | -0.19004 | 7.74301 | 0.04595 |
| 120 | 0.08207 | 0.03490 | -0.36106 | 7.15967 | 0.05246 |

**Table 6.9 (Cont'd)** Range statistics for boron from $BF_2$ in polysilicon

| Energy | $R_P$ | $\Delta R_P$ | $\gamma$ | $\beta$ | $\Delta R_T$ |
|---|---|---|---|---|---|
| 5 | $6.380x10^{-3}$ | $4.344x10^{-3}$ | $8.708x10^{-3}$ | 0.0 | $5.21x10^{-3}$ |
| 10 | $1.081x10^{-3}$ | $6.770x10^{-3}$ | $9.034x10^{-3}$ | 0.0 | $8.12x10^{-3}$ |
| 20 | $1.956x10^{-2}$ | $1.092x10^{-2}$ | $2.051x10^{-2}$ | 0.0 | $1.31x10^{-2}$ |
| 45 | 0.0351 | 0.0182 | -0.078 | 2.9195 | 0.0250 |
| 89 | 0.0685 | 0.0296 | -0.313 | 3.0633 | 0.0424 |
| 134 | 0.1011 | 0.0384 | -0.480 | 3.2746 | 0.0569 |

**Table 6.9 (Cont'd)** Range statistics for boron from $BF_2$ in silicon dioxide
Courtesy: Technology Modeling Associates (TMA).

| Energy | $R_P$ | $\Delta R_P$ | $\gamma$ | $\beta$ | $\Delta R_T$ |
|---|---|---|---|---|---|
| 5 | $5.377x10^{-3}$ | $3.642x10^{-3}$ | $2.805x10^{-2}$ | 0.0 | $4.37x10^{-3}$ |
| 10 | $9.769x10^{-3}$ | $5.266x10^{-3}$ | $-4.637x10^{-3}$ | 0.0 | $6.31x10^{-3}$ |
| 20 | $1.767x10^{-2}$ | $8.783x10^{-3}$ | $3.837x10^{-3}$ | 0.0 | $1.05x10^{-2}$ |
| 45 | 0.0355 | 0.0167 | -0.033 | 2.9117 | 0.0224 |
| 89 | 0.0713 | 0.0280 | -0.286 | 3.0378 | 0.0395 |
| 134 | 0.1068 | 0.0370 | -0.463 | 3.2485 | 0.0542 |

**Table 6.9 (Cont'd)** Range statistics for boron from $BF_2$ in silicon nitride
Courtesy: Technology Modeling Associates (TMA).

| Energy | $R_P$ | $\Delta R_P$ | $\gamma$ | $\beta$ | $\Delta R_T$ |
|--------|-------|--------------|----------|---------|--------------|
| 5 | 0.0040 | 0.0020 | 0.525 | 3.3500 | 0.0020 |
| 10 | 0.0070 | 0.0033 | 0.500 | 3.3000 | 0.0036 |
| 20 | 0.0130 | 0.0056 | 0.450 | 3.2300 | 0.0063 |
| 25 | 0.0243 | 0.0119 | 0.014 | 2.9103 | 0.0159 |
| 45 | 0.0243 | 0.0119 | 0.014 | 2.9103 | 0.0159 |
| 89 | 0.0499 | 0.0205 | -0.226 | 2.9899 | 0.0289 |
| 134 | 0.0758 | 0.0277 | -0.398 | 3.1591 | 0.0404 |

**Table 6.9 (Cont'd)** Range statistics for boron from $BF_2$ in aluminum

| Energy | $R_P$ | $\Delta R_P$ | $\gamma$ | $\beta$ | $\Delta R_T$ |
|--------|-------|--------------|----------|---------|--------------|
| 25 | 0.0344 | 0.0186 | 0.036 | 2.9120 | 0.0255 |
| 45 | 0.0344 | 0.0186 | 0.036 | 2.9120 | 0.0255 |
| 89 | 0.0702 | 0.0321 | -0.188 | 2.9655 | 0.0459 |
| 134 | 0.1069 | 0.0434 | -0.350 | 3.1028 | 0.0640 |

**Table 6.9 (Cont'd)** Range statistics for boron from $BF_2$ in photoresist (type AZ-111)

| Energy | $R_P$ | $\Delta R_P$ | $\gamma$ | $\beta$ | $\Delta R_T$ |
|--------|-------|--------------|----------|---------|--------------|
| 25 | 0.1086 | 0.0270 | -0.384 | 3.1420 | 0.0312 |
| 45 | 0.1086 | 0.0270 | -0.384 | 3.1420 | 0.0312 |
| 89 | 0.2267 | 0.0475 | -0.634 | 3.5516 | 0.0591 |
| 134 | 0.3446 | 0.0635 | -0.836 | 4.0442 | 0.0834 |

**Table 6.9 (Cont'd)** Range statistics for phosphorus in silicon and polysilicon
Courtesy: Technology Modeling Associates (TMA).

| Energy | $R_P$ | $\Delta R_P$ | $\gamma$ | $\beta$ | $\Delta R_T$ |
|---|---|---|---|---|---|
| 5 | 0.0105 | 0.0054 | 0.624 | 3.214 | 0.0064 |
| 10 | 0.0160 | 0.0080 | 0.631 | 3.5455 | 0.0095 |
| 20 | 0.0283 | 0.0135 | 0.544 | 3.3797 | 0.0159 |
| 30 | 0.0407 | 0.0186 | 0.467 | 3.2550 | 0.0220 |
| 40 | 0.0534 | 0.0236 | 0.401 | 3.1630 | 0.0279 |
| 50 | 0.0663 | 0.0284 | 0.342 | 3.0932 | 0.0337 |
| 60 | 0.0794 | 0.0331 | 0.288 | 3.0398 | 0.0395 |
| 70 | 0.0928 | 0.0376 | 0.239 | 2.9994 | 0.0452 |
| 80 | 0.1064 | 0.0421 | 0.194 | 2.9689 | 0.0508 |
| 90 | 0.1201 | 0.0464 | 0.153 | 2.9465 | 0.0563 |
| 100 | 0.1339 | 0.0506 | 0.115 | 2.9306 | 0.0618 |
| 110 | 0.1479 | 0.0547 | 0.080 | 2.9199 | 0.0672 |
| 120 | 0.1619 | 0.0587 | 0.047 | 2.9135 | 0.0725 |
| 130 | 0.1760 | 0.0626 | 0.018 | 2.9105 | 0.0777 |
| 140 | 0.1902 | 0.0664 | -0.010 | 2.9102 | 0.0829 |
| 150 | 0.2044 | 0.0701 | -0.036 | 2.9120 | 0.0880 |
| 160 | 0.2187 | 0.0737 | -0.061 | 2.9159 | 0.0930 |
| 170 | 0.2329 | 0.0772 | -0.085 | 2.9213 | 0.0979 |
| 180 | 0.2472 | 0.0807 | -0.107 | 2.9280 | 0.1028 |
| 190 | 0.2615 | 0.0840 | -0.128 | 2.9356 | 0.1076 |
| 200 | 0.2758 | 0.0873 | -0.147 | 2.9436 | 0.1123 |
| 220 | 0.3044 | 0.0936 | -0.185 | 2.9633 | 0.1215 |
| 240 | 0.3328 | 0.0996 | -0.221 | 2.9862 | 0.1305 |
| 260 | 0.3612 | 0.1053 | -0.254 | 3.0114 | 0.1392 |
| 280 | 0.3895 | 0.1108 | -0.286 | 3.0384 | 0.1476 |
| 300 | 0.4176 | 0.1161 | -0.316 | 3.0666 | 0.1558 |
| 320 | 0.4455 | 0.1211 | -0.344 | 3.0962 | 0.1638 |
| 340 | 0.4733 | 0.1259 | -0.371 | 3.1269 | 0.1716 |
| 360 | 0.5009 | 0.1306 | -0.397 | 3.1586 | 0.1791 |
| 380 | 0.5283 | 0.1350 | -0.422 | 3.1905 | 0.1864 |
| 400 | 0.5555 | 0.1393 | -0.445 | 3.2229 | 0.1935 |
| 420 | 0.5825 | 0.1434 | -0.468 | 3.2557 | 0.2005 |
| 440 | 0.6093 | 0.1473 | -0.490 | 3.2890 | 0.2072 |
| 460 | 0.6358 | 0.1511 | -0.511 | 3.3227 | 0.2138 |
| 480 | 0.6622 | 0.1548 | -0.531 | 3.3568 | 0.2202 |
| 500 | 0.6884 | 0.1584 | -0.551 | 3.3915 | 0.2265 |
| 550 | 0.7530 | 0.1667 | -0.603 | 3.4895 | 0.2415 |
| 600 | 0.8164 | 0.1744 | -0.652 | 3.5893 | 0.2556 |
| 650 | 0.8785 | 0.1816 | -0.697 | 3.6891 | 0.2690 |
| 700 | 0.9395 | 0.1882 | -0.739 | 3.7889 | 0.2816 |
| 750 | 0.9994 | 0.1943 | -0.778 | 3.8883 | 0.2936 |
| 800 | 1.0582 | 0.2001 | -0.816 | 3.9873 | 0.3050 |
| 850 | 1.1160 | 0.2055 | -0.851 | 4.0859 | 0.3158 |
| 900 | 1.1727 | 0.2105 | -0.884 | 4.1839 | 0.3262 |
| 950 | 1.2286 | 0.2153 | -0.916 | 4.2814 | 0.3361 |
| 1000 | 1.2835 | 0.2198 | -0.946 | 4.3782 | 0.3455 |
| 2000 | 1.848 | 0.2840 | -0.160 | 6.260 | 0.2840 |
| 3000 | 2.457 | 0.2960 | -0.590 | 7.220 | 0.2960 |
| 4000 | 2.944 | 0.3080 | -0.9600 | 8.500 | 0.3080 |
| 5000 | 3.375 | 0.3200 | -1.270 | 10.10 | 0.3200 |
| 6000 | 3.816 | 0.3320 | -1.520 | 12.02 | 0.3320 |
| 7000 | 4.333 | 0.3440 | -1.710 | 14.26 | 0.3440 |

**Table 6.9 (Cont'd)** Range statistics for phosphorus in silicon dioxide
Courtesy: Technology Modeling Associates (TMA).

| Energy | $R_P$ | $\Delta R_P$ | $\gamma$ | $\beta$ | $\Delta R_T$ |
|--------|-------|--------------|----------|---------|--------------|
| 5 | 0.0092 | 0.0042 | 0.663 | 3.4790 | 0.0050 |
| 10 | 0.0147 | 0.0065 | 0.557 | 3.4025 | 0.0074 |
| 20 | 0.0263 | 0.0112 | 0.488 | 3.2872 | 0.0126 |
| 30 | 0.0381 | 0.0155 | 0.430 | 3.2020 | 0.0176 |
| 40 | 0.0500 | 0.0197 | 0.379 | 3.1356 | 0.0224 |
| 50 | 0.0623 | 0.0238 | 0.332 | 3.0825 | 0.0271 |
| 60 | 0.0747 | 0.0278 | 0.288 | 3.0397 | 0.0318 |
| 70 | 0.0874 | 0.0316 | 0.247 | 3.0057 | 0.0364 |
| 80 | 0.1002 | 0.0353 | 0.210 | 2.9787 | 0.0409 |
| 90 | 0.1131 | 0.0390 | 0.175 | 2.9577 | 0.0453 |
| 100 | 0.1261 | 0.0425 | 0.143 | 2.9418 | 0.0497 |
| 110 | 0.1392 | 0.0459 | 0.113 | 2.9301 | 0.0540 |
| 120 | 0.1524 | 0.0491 | 0.086 | 2.9216 | 0.0583 |
| 130 | 0.1656 | 0.0523 | 0.061 | 2.9159 | 0.0624 |
| 140 | 0.1788 | 0.0554 | 0.037 | 2.9121 | 0.0665 |
| 150 | 0.1920 | 0.0585 | 0.015 | 2.9104 | 0.0706 |
| 160 | 0.2053 | 0.0614 | -0.006 | 2.9101 | 0.0745 |
| 170 | 0.2185 | 0.0642 | -0.025 | 2.9110 | 0.0784 |
| 180 | 0.2318 | 0.0670 | -0.044 | 2.9131 | 0.0822 |
| 190 | 0.2450 | 0.0697 | -0.063 | 2.9162 | 0.0860 |
| 200 | 0.2581 | 0.0723 | -0.079 | 2.9197 | 0.0896 |
| 220 | 0.2844 | 0.0772 | -0.122 | 2.9331 | 0.0968 |
| 240 | 0.3106 | 0.0820 | -0.160 | 2.9502 | 0.1038 |
| 260 | 0.3365 | 0.0865 | -0.196 | 2.9698 | 0.1105 |
| 280 | 0.3623 | 0.0907 | -0.228 | 2.9913 | 0.1169 |
| 300 | 0.3878 | 0.0948 | -0.258 | 3.0140 | 0.1232 |
| 320 | 0.4132 | 0.0986 | -0.286 | 3.0379 | 0.1293 |
| 340 | 0.4383 | 0.1023 | -0.312 | 3.0628 | 0.1351 |
| 360 | 0.4632 | 0.1058 | -0.337 | 3.0885 | 0.1408 |
| 380 | 0.4878 | 0.1092 | -0.361 | 3.1150 | 0.1463 |
| 400 | 0.5123 | 0.1124 | -0.383 | 3.1412 | 0.1517 |
| 420 | 0.5365 | 0.1155 | -0.405 | 3.1682 | 0.1569 |
| 440 | 0.5605 | 0.1184 | -0.426 | 3.1955 | 0.1619 |
| 460 | 0.5843 | 0.1213 | -0.446 | 3.2233 | 0.1668 |
| 480 | 0.6078 | 0.1240 | -0.465 | 3.2514 | 0.1716 |
| 500 | 0.6311 | 0.1266 | -0.484 | 3.2797 | 0.1762 |
| 550 | 0.6885 | 0.1327 | -0.536 | 3.3659 | 0.1872 |
| 600 | 0.7446 | 0.1383 | -0.585 | 3.4536 | 0.1976 |
| 650 | 0.7995 | 0.1434 | -0.629 | 3.5408 | 0.2073 |
| 700 | 0.8531 | 0.1482 | -0.669 | 3.6273 | 0.2165 |
| 750 | 0.9057 | 0.1525 | -0.707 | 3.7130 | 0.2252 |
| 800 | 0.9572 | 0.1566 | -0.743 | 3.7979 | 0.2334 |
| 850 | 1.0077 | 0.1604 | -0.776 | 3.8820 | 0.2412 |
| 900 | 1.0573 | 0.1639 | -0.807 | 3.9652 | 0.2486 |
| 950 | 1.1059 | 0.1673 | -0.837 | 4.0477 | 0.2556 |
| 1000 | 1.1536 | 0.1704 | -0.866 | 4.1292 | 0.2624 |

**Table 6.9 (Cont'd)** Range statistics for phosphorus in silicon nitride
Courtesy: Technology Modeling Associates (TMA).

| Energy | $R_P$ | $\Delta R_P$ | $\gamma$ | $\beta$ | $\Delta R_T$ |
|--------|-------|-------|-------|-------|-------|
| 10 | 0.0096 | 0.0043 | 0.528 | 3.3520 | 0.0049 |
| 20 | 0.0172 | 0.0073 | 0.458 | 3.2419 | 0.0083 |
| 30 | 0.0248 | 0.0101 | 0.398 | 3.1591 | 0.0115 |
| 40 | 0.0325 | 0.0128 | 0.344 | 3.0958 | 0.0146 |
| 50 | 0.0404 | 0.0155 | 0.295 | 3.0462 | 0.0177 |
| 60 | 0.0484 | 0.0180 | 0.249 | 3.0074 | 0.0207 |
| 70 | 0.0566 | 0.0204 | 0.208 | 2.9775 | 0.0237 |
| 80 | 0.0648 | 0.0228 | 0.169 | 2.9546 | 0.0266 |
| 90 | 0.0731 | 0.0251 | 0.133 | 2.9377 | 0.0294 |
| 100 | 0.0814 | 0.0274 | 0.100 | 2.9258 | 0.0323 |
| 110 | 0.0898 | 0.0295 | 0.070 | 2.9176 | 0.0350 |
| 120 | 0.0982 | 0.0316 | 0.042 | 2.9127 | 0.0377 |
| 130 | 0.1066 | 0.0336 | 0.016 | 2.9104 | 0.0404 |
| 140 | 0.1151 | 0.0356 | -0.009 | 2.9101 | 0.0430 |
| 150 | 0.1235 | 0.0375 | -0.033 | 2.9117 | 0.0456 |
| 160 | 0.1320 | 0.0393 | -0.054 | 2.9146 | 0.0481 |
| 170 | 0.1404 | 0.0411 | -0.075 | 2.9187 | 0.0506 |
| 180 | 0.1488 | 0.0429 | -0.094 | 2.9239 | 0.0530 |
| 190 | 0.1572 | 0.0446 | -0.113 | 2.9299 | 0.0554 |
| 200 | 0.1656 | 0.0462 | -0.130 | 2.9364 | 0.0578 |
| 220 | 0.1823 | 0.0493 | -0.173 | 2.9565 | 0.0624 |
| 240 | 0.1989 | 0.0523 | -0.212 | 2.9801 | 0.0668 |
| 260 | 0.2154 | 0.0551 | -0.247 | 3.0059 | 0.0710 |
| 280 | 0.2317 | 0.0578 | -0.280 | 3.0333 | 0.0751 |
| 300 | 0.2479 | 0.0603 | -0.311 | 3.0618 | 0.0791 |
| 320 | 0.2639 | 0.0627 | -0.340 | 3.0913 | 0.0829 |
| 340 | 0.2798 | 0.0650 | -0.367 | 3.1217 | 0.0866 |
| 360 | 0.2955 | 0.0672 | -0.393 | 3.1531 | 0.0902 |
| 380 | 0.3111 | 0.0693 | -0.417 | 3.1843 | 0.0937 |
| 400 | 0.3266 | 0.0713 | -0.441 | 3.2162 | 0.0971 |
| 420 | 0.3419 | 0.0733 | -0.463 | 3.2485 | 0.1003 |
| 440 | 0.3570 | 0.0751 | -0.484 | 3.2811 | 0.1035 |
| 460 | 0.3720 | 0.0769 | -0.505 | 3.3140 | 0.1066 |
| 480 | 0.3868 | 0.0785 | -0.525 | 3.3472 | 0.1096 |
| 500 | 0.4015 | 0.0802 | -0.545 | 3.3807 | 0.1125 |
| 550 | 0.4377 | 0.0840 | -0.598 | 3.4785 | 0.1195 |
| 600 | 0.4730 | 0.0875 | -0.646 | 3.5775 | 0.1260 |
| 650 | 0.5075 | 0.0907 | -0.691 | 3.6758 | 0.1321 |
| 700 | 0.5413 | 0.0936 | -0.733 | 3.7732 | 0.1379 |
| 750 | 0.5744 | 0.0963 | -0.771 | 3.8697 | 0.1433 |
| 800 | 0.6067 | 0.0988 | -0.807 | 3.9653 | 0.1484 |
| 850 | 0.6385 | 0.1012 | -0.842 | 4.0600 | 0.1533 |
| 900 | 0.6696 | 0.1034 | -0.874 | 4.1536 | 0.1580 |
| 950 | 0.7001 | 0.1054 | -0.904 | 4.2464 | 0.1624 |
| 1000 | 0.7301 | 0.1074 | -0.934 | 4.3382 | 0.1666 |

**Table 6.9 (Cont'd)** Range statistics for phosphorus in aluminum
Courtesy: Technology Modeling Associates (TMA).

| Energy | $R_P$ | $\Delta R_P$ | $\gamma$ | $\beta$ | $\Delta R_T$ |
|---|---|---|---|---|---|
| 10 | 0.0137 | 0.0067 | 0.598 | 3.4791 | 0.0079 |
| 20 | 0.0243 | 0.0113 | 0.516 | 3.3310 | 0.0132 |
| 30 | 0.0350 | 0.0155 | 0.437 | 3.2109 | 0.0183 |
| 40 | 0.0459 | 0.0196 | 0.369 | 3.1243 | 0.0232 |
| 50 | 0.0569 | 0.0236 | 0.309 | 3.0601 | 0.0280 |
| 60 | 0.0681 | 0.0274 | 0.255 | 3.0121 | 0.0327 |
| 70 | 0.0795 | 0.0311 | 0.207 | 2.9768 | 0.0373 |
| 80 | 0.0910 | 0.0346 | 0.162 | 2.9511 | 0.0418 |
| 90 | 0.1026 | 0.0381 | 0.121 | 2.9330 | 0.0462 |
| 100 | 0.1143 | 0.0415 | 0.084 | 2.9209 | 0.0506 |
| 110 | 0.1260 | 0.0447 | 0.049 | 2.9137 | 0.0549 |
| 120 | 0.1378 | 0.0479 | 0.017 | 2.9104 | 0.0591 |
| 130 | 0.1496 | 0.0509 | -0.014 | 2.9103 | 0.0633 |
| 140 | 0.1614 | 0.0539 | -0.041 | 2.9126 | 0.0674 |
| 150 | 0.1733 | 0.0568 | -0.067 | 2.9171 | 0.0714 |
| 160 | 0.1851 | 0.0595 | -0.090 | 2.9227 | 0.0753 |
| 170 | 0.1970 | 0.0623 | -0.114 | 2.9303 | 0.0792 |
| 180 | 0.2088 | 0.0649 | -0.135 | 2.9385 | 0.0829 |
| 190 | 0.2206 | 0.0675 | -0.155 | 2.9477 | 0.0867 |
| 200 | 0.2324 | 0.0700 | -0.174 | 2.9571 | 0.0903 |
| 220 | 0.2559 | 0.0748 | -0.214 | 2.9817 | 0.0974 |
| 240 | 0.2792 | 0.0793 | -0.252 | 3.0094 | 0.1043 |
| 260 | 0.3024 | 0.0836 | -0.287 | 3.0394 | 0.1110 |
| 280 | 0.3255 | 0.0877 | -0.320 | 3.0711 | 0.1174 |
| 300 | 0.3483 | 0.0916 | -0.351 | 3.1040 | 0.1236 |
| 320 | 0.3710 | 0.0953 | -0.381 | 3.1380 | 0.1296 |
| 340 | 0.3934 | 0.0989 | -0.409 | 3.1728 | 0.1354 |
| 360 | 0.4157 | 0.1023 | -0.435 | 3.2082 | 0.1411 |
| 380 | 0.4378 | 0.1055 | -0.460 | 3.2444 | 0.1465 |
| 400 | 0.4597 | 0.1086 | -0.484 | 3.2808 | 0.1519 |
| 420 | 0.4813 | 0.1116 | -0.507 | 3.3176 | 0.1570 |
| 440 | 0.5028 | 0.1145 | -0.530 | 3.3552 | 0.1620 |
| 460 | 0.5241 | 0.1173 | -0.552 | 3.3930 | 0.1669 |
| 480 | 0.5452 | 0.1199 | -0.573 | 3.4312 | 0.1716 |
| 500 | 0.5661 | 0.1225 | -0.593 | 3.4692 | 0.1762 |
| 550 | 0.6175 | 0.1284 | -0.646 | 3.5770 | 0.1872 |
| 600 | 0.6678 | 0.1339 | -0.696 | 3.6859 | 0.1975 |
| 650 | 0.7170 | 0.1389 | -0.741 | 3.7943 | 0.2072 |
| 700 | 0.7652 | 0.1436 | -0.784 | 3.9020 | 0.2164 |
| 750 | 0.8124 | 0.1479 | -0.823 | 4.0089 | 0.2250 |
| 800 | 0.8587 | 0.1519 | -0.861 | 4.1149 | 0.2332 |
| 850 | 0.9041 | 0.1557 | -0.896 | 4.2203 | 0.2410 |
| 900 | 0.9486 | 0.1592 | -0.929 | 4.3247 | 0.2484 |
| 950 | 0.9924 | 0.1625 | -0.961 | 4.4282 | 0.2554 |
| 1000 | 1.0353 | 0.1656 | -0.992 | 4.5307 | 0.2622 |

**Table 6.9 (Cont'd)** Range statistics for phosphorus in photoresist (type AZ-111)
Courtesy: Technology Modeling Associates (TMA).

| Energy | $R_P$ | $\Delta R_P$ | $\gamma$ | $\beta$ | $\Delta R_T$ |
|---|---|---|---|---|---|
| 10 | 0.0481 | 0.0114 | -0.026 | 2.9111 | 0.0117 |
| 20 | 0.0866 | 0.0198 | -0.076 | 2.9190 | 0.0200 |
| 30 | 0.1256 | 0.0277 | -0.134 | 2.9380 | 0.0290 |
| 40 | 0.1654 | 0.0353 | -0.180 | 2.9606 | 0.0357 |
| 50 | 0.2060 | 0.0427 | -0.220 | 2.9857 | 0.0433 |
| 60 | 0.2474 | 0.0499 | -0.256 | 3.0126 | 0.0503 |
| 70 | 0.2895 | 0.0569 | -0.288 | 3.0400 | 0.0582 |
| 80 | 0.3320 | 0.0636 | -0.318 | 3.0687 | 0.0655 |
| 90 | 0.3750 | 0.0702 | -0.346 | 3.0980 | 0.0727 |
| 100 | 0.4182 | 0.0765 | -0.371 | 3.1264 | 0.0798 |
| 110 | 0.4617 | 0.0826 | -0.395 | 3.1556 | 0.0868 |
| 120 | 0.5053 | 0.0886 | -0.417 | 3.1840 | 0.0936 |
| 130 | 0.5490 | 0.0943 | -0.437 | 3.2112 | 0.1004 |
| 140 | 0.5927 | 0.0999 | -0.457 | 3.2397 | 0.1070 |
| 150 | 0.6365 | 0.1052 | -0.475 | 3.2666 | 0.1135 |
| 160 | 0.6803 | 0.1104 | -0.492 | 3.2929 | 0.1199 |
| 170 | 0.7239 | 0.1155 | -0.509 | 3.3202 | 0.1262 |
| 180 | 0.7675 | 0.1203 | -0.525 | 3.3468 | 0.1323 |
| 190 | 0.8110 | 0.1250 | -0.539 | 3.3708 | 0.1384 |
| 200 | 0.8544 | 0.1296 | -0.553 | 3.3955 | 0.1443 |
| 220 | 0.9406 | 0.1383 | -0.583 | 3.4506 | 0.1559 |
| 240 | 1.0262 | 0.1465 | -0.613 | 3.5089 | 0.1670 |
| 260 | 1.1110 | 0.1543 | -0.640 | 3.5641 | 0.1778 |
| 280 | 1.1951 | 0.1616 | -0.665 | 3.6175 | 0.1882 |
| 300 | 1.2783 | 0.1685 | -0.688 | 3.6686 | 0.1982 |
| 320 | 1.3606 | 0.1750 | -0.711 | 3.7216 | 0.2078 |
| 340 | 1.4421 | 0.1812 | -0.732 | 3.7717 | 0.2172 |
| 360 | 1.5227 | 0.1871 | -0.753 | 3.8235 | 0.2262 |
| 380 | 1.6024 | 0.1927 | -0.773 | 3.8743 | 0.2349 |
| 400 | 1.6813 | 0.1980 | -0.791 | 3.9213 | 0.2434 |
| 420 | 1.7593 | 0.2031 | -0.809 | 3.9695 | 0.2516 |
| 440 | 1.8364 | 0.2079 | -0.825 | 4.0135 | 0.2595 |
| 460 | 1.9128 | 0.2125 | -0.842 | 4.0612 | 0.2672 |
| 480 | 1.9883 | 0.2169 | -0.857 | 4.1043 | 0.2747 |
| 500 | 2.0631 | 0.2212 | -0.873 | 4.1512 | 0.2820 |
| 550 | 2.2465 | 0.2310 | -0.912 | 4.2698 | 0.2992 |
| 600 | 2.4253 | 0.2399 | -0.949 | 4.3879 | 0.3154 |
| 650 | 2.5998 | 0.2480 | -0.984 | 4.5049 | 0.3304 |
| 700 | 2.7701 | 0.2554 | -1.016 | 4.6162 | 0.3446 |
| 750 | 2.9365 | 0.2623 | -1.047 | 4.7282 | 0.3579 |
| 800 | 3.0993 | 0.2685 | -1.076 | 4.8368 | 0.3705 |
| 850 | 3.2586 | 0.2744 | -1.103 | 4.9411 | 0.3824 |
| 900 | 3.4146 | 0.2798 | -1.129 | 5.0447 | 0.3937 |
| 950 | 3.5675 | 0.2848 | -1.154 | 5.1471 | 0.4044 |
| 1000 | 3.7175 | 0.2895 | -1.177 | 5.2438 | 0.4146 |

457

**Table 6.9 (Cont'd)** Range statistics for arsenic in silicon and polysilicon
Courtesy: Technology Modeling Associates (TMA).

| Energy | $R_P$ | $\Delta R_P$ | $\gamma$ | $\beta$ | $\Delta R_T$ |
|---|---|---|---|---|---|
| 5 | 0.0076 | 0.0024 | 0.275 | 2.5780 | 0.0029 |
| 10 | 0.0121 | 0.0045 | 0.639 | 3.5623 | 0.0050 |
| 20 | 0.0191 | 0.0070 | 0.625 | 3.5327 | 0.0078 |
| 30 | 0.0254 | 0.0093 | 0.617 | 3.5167 | 0.0101 |
| 40 | 0.0314 | 0.0114 | 0.610 | 3.5038 | 0.0123 |
| 50 | 0.0371 | 0.0134 | 0.604 | 3.4918 | 0.0144 |
| 60 | 0.0428 | 0.0154 | 0.599 | 3.4804 | 0.0164 |
| 70 | 0.0484 | 0.0173 | 0.593 | 3.4695 | 0.0183 |
| 80 | 0.0539 | 0.0192 | 0.587 | 3.4590 | 0.0202 |
| 90 | 0.0595 | 0.0210 | 0.582 | 3.4488 | 0.0221 |
| 100 | 0.0650 | 0.0228 | 0.577 | 3.4390 | 0.0239 |
| 110 | 0.0706 | 0.0247 | 0.572 | 3.4293 | 0.0258 |
| 120 | 0.0761 | 0.0264 | 0.566 | 3.4198 | 0.0276 |
| 130 | 0.0817 | 0.0282 | 0.561 | 3.4103 | 0.0293 |
| 140 | 0.0872 | 0.0300 | 0.556 | 3.4011 | 0.0311 |
| 150 | 0.0928 | 0.0317 | 0.551 | 3.3919 | 0.0329 |
| 160 | 0.0984 | 0.0335 | 0.546 | 3.3824 | 0.0346 |
| 170 | 0.1040 | 0.0352 | 0.541 | 3.3737 | 0.0364 |
| 180 | 0.1097 | 0.0369 | 0.535 | 3.3644 | 0.0381 |
| 190 | 0.1153 | 0.0386 | 0.532 | 3.3580 | 0.0399 |
| 200 | 0.1210 | 0.0403 | 0.527 | 3.3499 | 0.0416 |
| 220 | 0.1324 | 0.0437 | 0.505 | 3.3129 | 0.0450 |
| 240 | 0.1439 | 0.0470 | 0.485 | 3.2817 | 0.0484 |
| 260 | 0.1554 | 0.0503 | 0.468 | 3.2566 | 0.0518 |
| 280 | 0.1670 | 0.0535 | 0.453 | 3.2345 | 0.0551 |
| 300 | 0.1787 | 0.0567 | 0.437 | 3.2116 | 0.0584 |
| 320 | 0.1904 | 0.0598 | 0.426 | 3.1961 | 0.0617 |
| 340 | 0.2022 | 0.0630 | 0.413 | 3.1782 | 0.0650 |
| 360 | 0.2141 | 0.0660 | 0.401 | 3.1637 | 0.0683 |
| 380 | 0.2260 | 0.0691 | 0.394 | 3.1547 | 0.0715 |
| 400 | 0.2379 | 0.0721 | 0.380 | 3.1373 | 0.0747 |
| 420 | 0.2499 | 0.0751 | 0.374 | 3.1300 | 0.0779 |
| 440 | 0.2619 | 0.0780 | 0.366 | 3.1204 | 0.0811 |
| 460 | 0.2739 | 0.0809 | 0.363 | 3.1169 | 0.0843 |
| 480 | 0.2860 | 0.0837 | 0.361 | 3.1149 | 0.0874 |
| 500 | 0.2981 | 0.0866 | 0.352 | 3.1042 | 0.0905 |
| 550 | 0.3285 | 0.0935 | 0.316 | 3.0666 | 0.0982 |
| 600 | 0.3589 | 0.1002 | 0.285 | 3.0376 | 0.1058 |
| 650 | 0.3895 | 0.1067 | 0.260 | 3.0157 | 0.1133 |
| 700 | 0.4201 | 0.1130 | 0.238 | 2.9983 | 0.1206 |
| 750 | 0.4508 | 0.1192 | 0.219 | 2.9851 | 0.1278 |
| 800 | 0.4814 | 0.1251 | 0.202 | 2.9741 | 0.1349 |
| 850 | 0.5121 | 0.1309 | 0.188 | 2.9651 | 0.1419 |
| 900 | 0.5427 | 0.1365 | 0.174 | 2.9571 | 0.1487 |
| 950 | 0.5733 | 0.1419 | 0.162 | 2.9509 | 0.1554 |
| 1000 | 0.6038 | 0.1472 | 0.151 | 2.9456 | 0.1620 |
| 3000 | 1.890 | 0.3820 | -0.400 | 3.350 | 0.3820 |
| 4000 | 2.352 | 0.3960 | -0.510 | 3.910 | 0.3960 |
| 5000 | 2.750 | 0.4100 | -0.620 | 4.470 | 0.4100 |
| 6000 | 3.096 | 0.4240 | -0.730 | 5.030 | 0.4240 |
| 7000 | 3.402 | 0.4380 | -0.840 | 5.590 | 0.4380 |
| 8000 | 3.680 | 0.4520 | -0.950 | 6.150 | 0.4520 |
| 9000 | 3.942 | 0.4660 | -1.060 | 6.710 | 0.4660 |
| 10000 | 4.200 | 0.4800 | -1.170 | 7.270 | 0.4800 |

**Table 6.9 (Cont'd)** Range statistics for arsenic in silicon dioxide
Courtesy: Technology Modeling Associates (TMA).

| Energy | $R_P$ | $\Delta R_P$ | $\gamma$ | $\beta$ | $\Delta R_T$ |
|---|---|---|---|---|---|
| 5 | 0.0073 | 0.0020 | 0.512 | 2.9750 | 0.0024 |
| 10 | 0.0115 | 0.0038 | 0.558 | 3.4036 | 0.0041 |
| 20 | 0.0183 | 0.0060 | 0.547 | 3.3842 | 0.0064 |
| 30 | 0.0244 | 0.0079 | 0.540 | 3.3732 | 0.0084 |
| 40 | 0.0301 | 0.0097 | 0.535 | 3.3635 | 0.0102 |
| 50 | 0.0357 | 0.0115 | 0.529 | 3.3535 | 0.0119 |
| 60 | 0.0412 | 0.0132 | 0.523 | 3.3433 | 0.0136 |
| 70 | 0.0467 | 0.0149 | 0.517 | 3.3329 | 0.0152 |
| 80 | 0.0521 | 0.0165 | 0.510 | 3.3225 | 0.0168 |
| 90 | 0.0575 | 0.0181 | 0.504 | 3.3120 | 0.0184 |
| 100 | 0.0629 | 0.0197 | 0.497 | 3.3016 | 0.0200 |
| 110 | 0.0683 | 0.0213 | 0.491 | 3.2912 | 0.0215 |
| 120 | 0.0737 | 0.0229 | 0.484 | 3.2809 | 0.0230 |
| 130 | 0.0792 | 0.0245 | 0.478 | 3.2708 | 0.0245 |
| 140 | 0.0846 | 0.0260 | 0.471 | 3.2607 | 0.0260 |
| 150 | 0.0901 | 0.0276 | 0.464 | 3.2507 | 0.0275 |
| 160 | 0.0956 | 0.0291 | 0.457 | 3.2403 | 0.0290 |
| 170 | 0.1011 | 0.0306 | 0.450 | 3.2295 | 0.0305 |
| 180 | 0.1066 | 0.0321 | 0.445 | 3.2219 | 0.0320 |
| 190 | 0.1121 | 0.0337 | 0.438 | 3.2132 | 0.0334 |
| 200 | 0.1177 | 0.0351 | 0.432 | 3.2048 | 0.0349 |
| 220 | 0.1289 | 0.0381 | 0.414 | 3.1802 | 0.0378 |
| 240 | 0.1402 | 0.0410 | 0.397 | 3.1584 | 0.0407 |
| 260 | 0.1515 | 0.0439 | 0.382 | 3.1391 | 0.0435 |
| 280 | 0.1629 | 0.0468 | 0.368 | 3.1223 | 0.0464 |
| 300 | 0.1744 | 0.0496 | 0.354 | 3.1067 | 0.0492 |
| 320 | 0.1859 | 0.0524 | 0.342 | 3.0933 | 0.0520 |
| 340 | 0.1975 | 0.0552 | 0.329 | 3.0797 | 0.0548 |
| 360 | 0.2092 | 0.0579 | 0.318 | 3.0687 | 0.0576 |
| 380 | 0.2209 | 0.0606 | 0.308 | 3.0586 | 0.0604 |
| 400 | 0.2326 | 0.0632 | 0.296 | 3.0478 | 0.0631 |
| 420 | 0.2444 | 0.0658 | 0.288 | 3.0398 | 0.0658 |
| 440 | 0.2562 | 0.0684 | 0.277 | 3.0303 | 0.0685 |
| 460 | 0.2680 | 0.0710 | 0.271 | 3.0247 | 0.0712 |
| 480 | 0.2799 | 0.0735 | 0.264 | 3.0193 | 0.0739 |
| 500 | 0.2918 | 0.0760 | 0.253 | 3.0105 | 0.0765 |
| 550 | 0.3216 | 0.0820 | 0.227 | 2.9906 | 0.0831 |
| 600 | 0.3516 | 0.0879 | 0.203 | 2.9744 | 0.0896 |
| 650 | 0.3816 | 0.0936 | 0.182 | 2.9617 | 0.0959 |
| 700 | 0.4116 | 0.0991 | 0.163 | 2.9515 | 0.1021 |
| 750 | 0.4417 | 0.1044 | 0.146 | 2.9434 | 0.1082 |
| 800 | 0.4717 | 0.1096 | 0.131 | 2.9369 | 0.1142 |
| 850 | 0.5018 | 0.1146 | 0.117 | 2.9314 | 0.1201 |
| 900 | 0.5317 | 0.1194 | 0.105 | 2.9272 | 0.1259 |
| 950 | 0.5616 | 0.1241 | 0.094 | 2.9237 | 0.1316 |
| 1000 | 0.5915 | 0.1287 | 0.083 | 2.9208 | 0.1372 |

**Table 6.9 (Cont'd)** Range statistics for arsenic in silicon nitride
Courtesy: Technology Modeling Associates (TMA).

| Energy | $R_P$ | $\Delta R_P$ | $\gamma$ | $\beta$ | $\Delta R_T$ |
|---|---|---|---|---|---|
| 10 | 0.0076 | 0.0025 | 0.553 | 3.3947 | 0.0027 |
| 20 | 0.0120 | 0.0039 | 0.541 | 3.3751 | 0.0042 |
| 30 | 0.0160 | 0.0052 | 0.535 | 3.3641 | 0.0055 |
| 40 | 0.0197 | 0.0064 | 0.530 | 3.3545 | 0.0068 |
| 50 | 0.0234 | 0.0076 | 0.524 | 3.3448 | 0.0079 |
| 60 | 0.0270 | 0.0087 | 0.518 | 3.3349 | 0.0090 |
| 70 | 0.0305 | 0.0098 | 0.512 | 3.3249 | 0.0101 |
| 80 | 0.0341 | 0.0109 | 0.506 | 3.3148 | 0.0112 |
| 90 | 0.0376 | 0.0120 | 0.499 | 3.3046 | 0.0122 |
| 100 | 0.0411 | 0.0130 | 0.493 | 3.2945 | 0.0132 |
| 110 | 0.0447 | 0.0141 | 0.487 | 3.2845 | 0.0142 |
| 120 | 0.0482 | 0.0151 | 0.480 | 3.2745 | 0.0153 |
| 130 | 0.0518 | 0.0161 | 0.474 | 3.2646 | 0.0163 |
| 140 | 0.0553 | 0.0172 | 0.467 | 3.2548 | 0.0173 |
| 150 | 0.0589 | 0.0182 | 0.461 | 3.2451 | 0.0182 |
| 160 | 0.0625 | 0.0192 | 0.454 | 3.2351 | 0.0192 |
| 170 | 0.0661 | 0.0202 | 0.448 | 3.2263 | 0.0202 |
| 180 | 0.0697 | 0.0212 | 0.440 | 3.2160 | 0.0212 |
| 190 | 0.0733 | 0.0222 | 0.434 | 3.2063 | 0.0222 |
| 200 | 0.0770 | 0.0232 | 0.426 | 3.1966 | 0.0231 |
| 220 | 0.0843 | 0.0251 | 0.409 | 3.1730 | 0.0251 |
| 240 | 0.0917 | 0.0271 | 0.392 | 3.1514 | 0.0270 |
| 260 | 0.0991 | 0.0290 | 0.376 | 3.1326 | 0.0289 |
| 280 | 0.1065 | 0.0308 | 0.362 | 3.1163 | 0.0308 |
| 300 | 0.1140 | 0.0327 | 0.348 | 3.1006 | 0.0326 |
| 320 | 0.1216 | 0.0345 | 0.336 | 3.0872 | 0.0345 |
| 340 | 0.1291 | 0.0364 | 0.323 | 3.0741 | 0.0363 |
| 360 | 0.1367 | 0.0382 | 0.311 | 3.0620 | 0.0382 |
| 380 | 0.1444 | 0.0399 | 0.302 | 3.0535 | 0.0400 |
| 400 | 0.1520 | 0.0417 | 0.290 | 3.0421 | 0.0418 |
| 420 | 0.1597 | 0.0434 | 0.282 | 3.0345 | 0.0436 |
| 440 | 0.1675 | 0.0451 | 0.273 | 3.0268 | 0.0454 |
| 460 | 0.1752 | 0.0468 | 0.266 | 3.0210 | 0.0472 |
| 480 | 0.1829 | 0.0484 | 0.259 | 3.0147 | 0.0490 |
| 500 | 0.1907 | 0.0501 | 0.251 | 3.0082 | 0.0508 |
| 550 | 0.2102 | 0.0541 | 0.223 | 2.9879 | 0.0551 |
| 600 | 0.2298 | 0.0580 | 0.198 | 2.9712 | 0.0594 |
| 650 | 0.2494 | 0.0617 | 0.176 | 2.9585 | 0.0636 |
| 700 | 0.2690 | 0.0653 | 0.157 | 2.9483 | 0.0677 |
| 750 | 0.2886 | 0.0689 | 0.139 | 2.9404 | 0.0718 |
| 800 | 0.3082 | 0.0723 | 0.124 | 2.9340 | 0.0758 |
| 850 | 0.3279 | 0.0756 | 0.110 | 2.9289 | 0.0797 |
| 900 | 0.3474 | 0.0788 | 0.097 | 2.9248 | 0.0835 |
| 950 | 0.3670 | 0.0819 | 0.086 | 2.9215 | 0.0873 |
| 1000 | 0.3864 | 0.0849 | 0.075 | 2.9188 | 0.0910 |

**Table 6.9 (Cont'd)** Range statistics for arsenic in aluminum
Courtesy: Technology Modeling Associates (TMA).

| Energy | $R_P$ | $\Delta R_P$ | $\gamma$ | $\beta$ | $\Delta R_T$ |
|---|---|---|---|---|---|
| 10 | 0.0105 | 0.0038 | 0.633 | 3.5490 | 0.0042 |
| 20 | 0.0166 | 0.0060 | 0.619 | 3.5202 | 0.0066 |
| 30 | 0.0221 | 0.0080 | 0.611 | 3.5042 | 0.0086 |
| 40 | 0.0273 | 0.0098 | 0.604 | 3.4908 | 0.0105 |
| 50 | 0.0324 | 0.0115 | 0.597 | 3.4783 | 0.0123 |
| 60 | 0.0373 | 0.0132 | 0.591 | 3.4662 | 0.0140 |
| 70 | 0.0423 | 0.0149 | 0.585 | 3.4545 | 0.0156 |
| 80 | 0.0471 | 0.0165 | 0.579 | 3.4431 | 0.0173 |
| 90 | 0.0520 | 0.0181 | 0.573 | 3.4320 | 0.0189 |
| 100 | 0.0569 | 0.0197 | 0.567 | 3.4212 | 0.0205 |
| 110 | 0.0618 | 0.0213 | 0.561 | 3.4106 | 0.0220 |
| 120 | 0.0667 | 0.0228 | 0.556 | 3.4002 | 0.0236 |
| 130 | 0.0716 | 0.0244 | 0.550 | 3.3899 | 0.0251 |
| 140 | 0.0765 | 0.0259 | 0.544 | 3.3797 | 0.0267 |
| 150 | 0.0815 | 0.0274 | 0.538 | 3.3697 | 0.0282 |
| 160 | 0.0864 | 0.0289 | 0.532 | 3.3595 | 0.0297 |
| 170 | 0.0914 | 0.0304 | 0.527 | 3.3502 | 0.0312 |
| 180 | 0.0964 | 0.0319 | 0.519 | 3.3363 | 0.0327 |
| 190 | 0.1014 | 0.0334 | 0.510 | 3.3221 | 0.0342 |
| 200 | 0.1064 | 0.0349 | 0.502 | 3.3090 | 0.0357 |
| 220 | 0.1165 | 0.0378 | 0.481 | 3.2757 | 0.0387 |
| 240 | 0.1267 | 0.0407 | 0.462 | 3.2477 | 0.0416 |
| 260 | 0.1370 | 0.0435 | 0.447 | 3.2252 | 0.0445 |
| 280 | 0.1473 | 0.0463 | 0.432 | 3.2039 | 0.0474 |
| 300 | 0.1576 | 0.0491 | 0.419 | 3.1867 | 0.0503 |
| 320 | 0.1680 | 0.0519 | 0.407 | 3.1713 | 0.0531 |
| 340 | 0.1785 | 0.0546 | 0.395 | 3.1558 | 0.0560 |
| 360 | 0.1890 | 0.0572 | 0.382 | 3.1398 | 0.0588 |
| 380 | 0.1995 | 0.0599 | 0.370 | 3.1248 | 0.0616 |
| 400 | 0.2101 | 0.0625 | 0.358 | 3.1112 | 0.0644 |
| 420 | 0.2208 | 0.0651 | 0.341 | 3.0921 | 0.0672 |
| 440 | 0.2314 | 0.0676 | 0.332 | 3.0835 | 0.0699 |
| 460 | 0.2421 | 0.0701 | 0.326 | 3.0773 | 0.0726 |
| 480 | 0.2528 | 0.0726 | 0.318 | 3.0692 | 0.0754 |
| 500 | 0.2636 | 0.0750 | 0.319 | 3.0694 | 0.0781 |
| 550 | 0.2905 | 0.0810 | 0.289 | 3.0406 | 0.0847 |
| 600 | 0.3176 | 0.0868 | 0.261 | 3.0171 | 0.0913 |
| 650 | 0.3447 | 0.0924 | 0.237 | 2.9982 | 0.0977 |
| 700 | 0.3718 | 0.0979 | 0.217 | 2.9840 | 0.1040 |
| 750 | 0.3990 | 0.1031 | 0.200 | 2.9724 | 0.1102 |
| 800 | 0.4262 | 0.1083 | 0.185 | 2.9636 | 0.1163 |
| 850 | 0.4533 | 0.1132 | 0.172 | 2.9563 | 0.1223 |
| 900 | 0.4804 | 0.1180 | 0.160 | 2.9498 | 0.1282 |
| 950 | 0.5075 | 0.1227 | 0.148 | 2.9440 | 0.1340 |
| 1000 | 0.5345 | 0.1272 | 0.137 | 2.9395 | 0.1397 |

**Table 6.9 (Cont'd)** Range statistics for arsenic in photoresist (type AZ-111)
Courtesy: Technology Modeling Associates (TMA).

| Energy | $R_P$ | $\Delta R_P$ | $\gamma$ | $\beta$ | $\Delta R_T$ |
|---|---|---|---|---|---|
| 10 | 0.0412 | 0.0078 | 0.166 | 2.9531 | 0.0076 |
| 20 | 0.0673 | 0.0126 | 0.152 | 2.9461 | 0.0123 |
| 30 | 0.0908 | 0.0168 | 0.130 | 2.9364 | 0.0163 |
| 40 | 0.1129 | 0.0207 | 0.117 | 2.9314 | 0.0199 |
| 50 | 0.1343 | 0.0244 | 0.106 | 2.9275 | 0.0234 |
| 60 | 0.1553 | 0.0280 | 0.097 | 2.9247 | 0.0268 |
| 70 | 0.1760 | 0.0315 | 0.088 | 2.9221 | 0.0300 |
| 80 | 0.1966 | 0.0349 | 0.080 | 2.9200 | 0.0332 |
| 90 | 0.2171 | 0.0382 | 0.072 | 2.9181 | 0.0363 |
| 100 | 0.2375 | 0.0415 | 0.064 | 2.9164 | 0.0393 |
| 110 | 0.2579 | 0.0448 | 0.056 | 2.9149 | 0.0423 |
| 120 | 0.2783 | 0.0480 | 0.048 | 2.9136 | 0.0453 |
| 130 | 0.2987 | 0.0512 | 0.040 | 2.9125 | 0.0482 |
| 140 | 0.3192 | 0.0543 | 0.032 | 2.9116 | 0.0511 |
| 150 | 0.3397 | 0.0575 | 0.025 | 2.9110 | 0.0540 |
| 160 | 0.3602 | 0.0606 | 0.017 | 2.9105 | 0.0569 |
| 170 | 0.3808 | 0.0636 | 0.010 | 2.9102 | 0.0597 |
| 180 | 0.4015 | 0.0667 | 0.003 | 2.9100 | 0.0626 |
| 190 | 0.4222 | 0.0697 | -0.004 | 2.9100 | 0.0654 |
| 200 | 0.4430 | 0.0727 | -0.012 | 2.9102 | 0.0682 |
| 220 | 0.4847 | 0.0787 | -0.025 | 2.9110 | 0.0737 |
| 240 | 0.5266 | 0.0846 | -0.038 | 2.9123 | 0.0792 |
| 260 | 0.5688 | 0.0904 | -0.051 | 2.9141 | 0.0847 |
| 280 | 0.6113 | 0.0961 | -0.062 | 2.9160 | 0.0901 |
| 300 | 0.6539 | 0.1018 | -0.073 | 2.9183 | 0.0955 |
| 320 | 0.6967 | 0.1073 | -0.084 | 2.9210 | 0.1008 |
| 340 | 0.7397 | 0.1128 | -0.095 | 2.9241 | 0.1061 |
| 360 | 0.7829 | 0.1183 | -0.105 | 2.9272 | 0.1114 |
| 380 | 0.8263 | 0.1237 | -0.115 | 2.9306 | 0.1166 |
| 400 | 0.8698 | 0.1290 | -0.124 | 2.9340 | 0.1218 |
| 420 | 0.9135 | 0.1342 | -0.131 | 2.9368 | 0.1270 |
| 440 | 0.9572 | 0.1394 | -0.139 | 2.9402 | 0.1321 |
| 460 | 1.0011 | 0.1445 | -0.146 | 2.9433 | 0.1372 |
| 480 | 1.0451 | 0.1495 | -0.152 | 2.9461 | 0.1422 |
| 500 | 1.0892 | 0.1545 | -0.159 | 2.9495 | 0.1473 |
| 550 | 1.1998 | 0.1666 | -0.174 | 2.9573 | 0.1597 |
| 600 | 1.3107 | 0.1784 | -0.190 | 2.9664 | 0.1719 |
| 650 | 1.4219 | 0.1898 | -0.205 | 2.9757 | 0.1839 |
| 700 | 1.5333 | 0.2008 | -0.219 | 2.9850 | 0.1957 |
| 750 | 1.6447 | 0.2115 | -0.232 | 2.9942 | 0.2073 |
| 800 | 1.7560 | 0.2219 | -0.243 | 3.0024 | 0.2187 |
| 850 | 1.8673 | 0.2319 | -0.254 | 3.0110 | 0.2299 |
| 900 | 1.9784 | 0.2416 | -0.265 | 3.0200 | 0.2409 |
| 950 | 2.0893 | 0.2510 | -0.274 | 3.0276 | 0.2517 |
| 1000 | 2.2000 | 0.2602 | -0.283 | 3.0355 | 0.2623 |

**Table 6.9 (Cont'd)** Range statistics for antimony in silicon and polysilicon
Courtesy: Technology Modeling Associates (TMA).

| Energy | $R_P$ | $\Delta R_P$ | $\gamma$ | $\beta$ | $\Delta R_T$ |
|--------|-------|-------|-------|--------|-------|
| 5 | 0.0081 | 0.0021 | 0.273 | 2.6220 | 0.0025 |
| 10 | 0.0121 | 0.0037 | 0.546 | 3.3824 | 0.0040 |
| 20 | 0.0184 | 0.0056 | 0.542 | 3.3752 | 0.0060 |
| 30 | 0.0237 | 0.0071 | 0.542 | 3.3757 | 0.0077 |
| 40 | 0.0286 | 0.0086 | 0.542 | 3.3765 | 0.0092 |
| 50 | 0.0331 | 0.0100 | 0.542 | 3.3766 | 0.0106 |
| 60 | 0.0375 | 0.0112 | 0.542 | 3.3761 | 0.0119 |
| 70 | 0.0417 | 0.0125 | 0.541 | 3.3750 | 0.0132 |
| 80 | 0.0458 | 0.0137 | 0.541 | 3.3736 | 0.0144 |
| 90 | 0.0499 | 0.0149 | 0.540 | 3.3720 | 0.0156 |
| 100 | 0.0539 | 0.0161 | 0.539 | 3.3701 | 0.0167 |
| 110 | 0.0578 | 0.0172 | 0.537 | 3.3680 | 0.0178 |
| 120 | 0.0617 | 0.0184 | 0.536 | 3.3659 | 0.0189 |
| 130 | 0.0655 | 0.0195 | 0.535 | 3.3636 | 0.0200 |
| 140 | 0.0693 | 0.0206 | 0.534 | 3.3613 | 0.0211 |
| 150 | 0.0731 | 0.0217 | 0.532 | 3.3589 | 0.0221 |
| 160 | 0.0769 | 0.0228 | 0.531 | 3.3565 | 0.0232 |
| 170 | 0.0807 | 0.0239 | 0.529 | 3.3540 | 0.0242 |
| 180 | 0.0844 | 0.0249 | 0.528 | 3.3514 | 0.0252 |
| 190 | 0.0882 | 0.0260 | 0.526 | 3.3488 | 0.0262 |
| 200 | 0.0919 | 0.0271 | 0.525 | 3.3462 | 0.0272 |
| 220 | 0.0993 | 0.0292 | 0.517 | 3.3338 | 0.0292 |
| 240 | 0.1068 | 0.0312 | 0.510 | 3.3222 | 0.0312 |
| 260 | 0.1142 | 0.0333 | 0.504 | 3.3119 | 0.0331 |
| 280 | 0.1216 | 0.0353 | 0.498 | 3.3023 | 0.0351 |
| 300 | 0.1290 | 0.0374 | 0.492 | 3.2933 | 0.0370 |
| 320 | 0.1364 | 0.0394 | 0.487 | 3.2848 | 0.0389 |
| 340 | 0.1439 | 0.0414 | 0.482 | 3.2766 | 0.0407 |
| 360 | 0.1513 | 0.0434 | 0.476 | 3.2687 | 0.0426 |
| 380 | 0.1588 | 0.0454 | 0.471 | 3.2610 | 0.0445 |
| 400 | 0.1663 | 0.0474 | 0.466 | 3.2534 | 0.0463 |
| 420 | 0.1738 | 0.0493 | 0.461 | 3.2461 | 0.0482 |
| 440 | 0.1813 | 0.0513 | 0.457 | 3.2394 | 0.0500 |
| 460 | 0.1888 | 0.0532 | 0.451 | 3.2312 | 0.0518 |
| 480 | 0.1963 | 0.0551 | 0.447 | 3.2255 | 0.0537 |
| 500 | 0.2039 | 0.0571 | 0.442 | 3.2188 | 0.0555 |
| 550 | 0.2229 | 0.0618 | 0.425 | 3.1952 | 0.0600 |
| 600 | 0.2420 | 0.0665 | 0.410 | 3.1747 | 0.0644 |
| 650 | 0.2612 | 0.0712 | 0.396 | 3.1566 | 0.0689 |
| 700 | 0.2804 | 0.0758 | 0.383 | 3.1404 | 0.0733 |
| 750 | 0.2998 | 0.0803 | 0.371 | 3.1259 | 0.0777 |
| 800 | 0.3193 | 0.0848 | 0.359 | 3.1125 | 0.0820 |
| 850 | 0.3389 | 0.0892 | 0.348 | 3.1007 | 0.0864 |
| 900 | 0.3585 | 0.0936 | 0.338 | 3.0893 | 0.0907 |
| 950 | 0.3783 | 0.0980 | 0.328 | 3.0787 | 0.0949 |
| 1000 | 0.3980 | 0.1022 | 0.319 | 3.0698 | 0.0992 |

**Table 6.9 (Cont'd)** Range statistics for antimony in silicon dioxide
Courtesy: Technology Modeling Associates (TMA).

| Energy | $R_P$ | $\Delta R_P$ | $\gamma$ | $\beta$ | $\Delta R_T$ |
|---|---|---|---|---|---|
| 5 | 0.0079 | 0.0017 | 0.489 | 0.0 | 0.0020 |
| 10 | 0.0117 | 0.0031 | 0.464 | 3.2504 | 0.0033 |
| 20 | 0.0179 | 0.0048 | 0.461 | 3.2461 | 0.0051 |
| 30 | 0.0231 | 0.0061 | 0.462 | 3.2472 | 0.0065 |
| 40 | 0.0279 | 0.0074 | 0.463 | 3.2488 | 0.0078 |
| 50 | 0.0324 | 0.0086 | 0.464 | 3.2497 | 0.0090 |
| 60 | 0.0367 | 0.0097 | 0.464 | 3.2500 | 0.0101 |
| 70 | 0.0408 | 0.0108 | 0.464 | 3.2497 | 0.0112 |
| 80 | 0.0449 | 0.0119 | 0.463 | 3.2489 | 0.0122 |
| 90 | 0.0489 | 0.0129 | 0.463 | 3.2478 | 0.0132 |
| 100 | 0.0528 | 0.0139 | 0.462 | 3.2465 | 0.0142 |
| 110 | 0.0567 | 0.0149 | 0.461 | 3.2449 | 0.0151 |
| 120 | 0.0605 | 0.0159 | 0.459 | 3.2431 | 0.0161 |
| 130 | 0.0643 | 0.0169 | 0.458 | 3.2411 | 0.0170 |
| 140 | 0.0680 | 0.0179 | 0.457 | 3.2391 | 0.0179 |
| 150 | 0.0718 | 0.0188 | 0.455 | 3.2369 | 0.0188 |
| 160 | 0.0755 | 0.0198 | 0.453 | 3.2346 | 0.0197 |
| 170 | 0.0792 | 0.0207 | 0.452 | 3.2322 | 0.0206 |
| 180 | 0.0829 | 0.0217 | 0.450 | 3.2298 | 0.0214 |
| 190 | 0.0866 | 0.0226 | 0.448 | 3.2273 | 0.0223 |
| 200 | 0.0903 | 0.0235 | 0.447 | 3.2247 | 0.0232 |
| 220 | 0.0977 | 0.0254 | 0.440 | 3.2155 | 0.0249 |
| 240 | 0.1050 | 0.0272 | 0.434 | 3.2067 | 0.0265 |
| 260 | 0.1124 | 0.0291 | 0.428 | 3.1987 | 0.0282 |
| 280 | 0.1197 | 0.0309 | 0.422 | 3.1910 | 0.0299 |
| 300 | 0.1271 | 0.0327 | 0.417 | 3.1838 | 0.0315 |
| 320 | 0.1344 | 0.0344 | 0.412 | 3.1768 | 0.0331 |
| 340 | 0.1418 | 0.0362 | 0.406 | 3.1701 | 0.0347 |
| 360 | 0.1492 | 0.0380 | 0.401 | 3.1635 | 0.0364 |
| 380 | 0.1566 | 0.0397 | 0.396 | 3.1571 | 0.0380 |
| 400 | 0.1640 | 0.0415 | 0.391 | 3.1510 | 0.0395 |
| 420 | 0.1714 | 0.0432 | 0.386 | 3.1447 | 0.0411 |
| 440 | 0.1789 | 0.0449 | 0.381 | 3.1384 | 0.0427 |
| 460 | 0.1863 | 0.0466 | 0.376 | 3.1327 | 0.0443 |
| 480 | 0.1938 | 0.0484 | 0.372 | 3.1273 | 0.0459 |
| 500 | 0.2013 | 0.0501 | 0.367 | 3.1217 | 0.0474 |
| 550 | 0.2202 | 0.0543 | 0.353 | 3.1059 | 0.0513 |
| 600 | 0.2391 | 0.0584 | 0.339 | 3.0909 | 0.0552 |
| 650 | 0.2582 | 0.0626 | 0.327 | 3.0775 | 0.0590 |
| 700 | 0.2774 | 0.0666 | 0.315 | 3.0654 | 0.0628 |
| 750 | 0.2967 | 0.0706 | 0.303 | 3.0543 | 0.0666 |
| 800 | 0.3160 | 0.0746 | 0.293 | 3.0442 | 0.0703 |
| 850 | 0.3355 | 0.0785 | 0.282 | 3.0350 | 0.0740 |
| 900 | 0.3550 | 0.0824 | 0.272 | 3.0262 | 0.0777 |
| 950 | 0.3746 | 0.0862 | 0.263 | 3.0183 | 0.0814 |
| 1000 | 0.3943 | 0.0900 | 0.254 | 3.0111 | 0.0851 |

**Table 6.9 (Cont'd)** Range statistics for antimony in silicon nitride
Courtesy: Technology Modeling Associates (TMA).

| Energy | $R_P$ | $\Delta R_P$ | $\gamma$ | $\beta$ | $\Delta R_T$ |
|---|---|---|---|---|---|
| 10 | 0.0077 | 0.0021 | 0.465 | 3.2511 | 0.0022 |
| 20 | 0.0117 | 0.0032 | 0.462 | 3.2467 | 0.0034 |
| 30 | 0.0151 | 0.0041 | 0.462 | 3.2476 | 0.0043 |
| 40 | 0.0183 | 0.0049 | 0.463 | 3.2492 | 0.0052 |
| 50 | 0.0212 | 0.0057 | 0.464 | 3.2502 | 0.0059 |
| 60 | 0.0240 | 0.0064 | 0.464 | 3.2505 | 0.0067 |
| 70 | 0.0267 | 0.0071 | 0.464 | 3.2503 | 0.0074 |
| 80 | 0.0294 | 0.0078 | 0.464 | 3.2496 | 0.0081 |
| 90 | 0.0320 | 0.0085 | 0.463 | 3.2486 | 0.0087 |
| 100 | 0.0345 | 0.0092 | 0.462 | 3.2474 | 0.0094 |
| 110 | 0.0370 | 0.0099 | 0.461 | 3.2459 | 0.0100 |
| 120 | 0.0395 | 0.0105 | 0.460 | 3.2442 | 0.0107 |
| 130 | 0.0420 | 0.0112 | 0.459 | 3.2423 | 0.0113 |
| 140 | 0.0445 | 0.0118 | 0.457 | 3.2404 | 0.0119 |
| 150 | 0.0469 | 0.0124 | 0.456 | 3.2383 | 0.0125 |
| 160 | 0.0494 | 0.0131 | 0.455 | 3.2361 | 0.0131 |
| 170 | 0.0518 | 0.0137 | 0.453 | 3.2338 | 0.0136 |
| 180 | 0.0542 | 0.0143 | 0.451 | 3.2315 | 0.0142 |
| 190 | 0.0566 | 0.0149 | 0.450 | 3.2290 | 0.0148 |
| 200 | 0.0590 | 0.0156 | 0.448 | 3.2265 | 0.0154 |
| 220 | 0.0638 | 0.0168 | 0.441 | 3.2170 | 0.0165 |
| 240 | 0.0686 | 0.0180 | 0.435 | 3.2079 | 0.0176 |
| 260 | 0.0734 | 0.0192 | 0.429 | 3.1995 | 0.0187 |
| 280 | 0.0782 | 0.0204 | 0.423 | 3.1918 | 0.0198 |
| 300 | 0.0830 | 0.0216 | 0.417 | 3.1844 | 0.0209 |
| 320 | 0.0878 | 0.0227 | 0.412 | 3.1773 | 0.0220 |
| 340 | 0.0926 | 0.0239 | 0.407 | 3.1704 | 0.0230 |
| 360 | 0.0975 | 0.0251 | 0.401 | 3.1637 | 0.0241 |
| 380 | 0.1023 | 0.0262 | 0.396 | 3.1572 | 0.0252 |
| 400 | 0.1071 | 0.0274 | 0.391 | 3.1508 | 0.0262 |
| 420 | 0.1120 | 0.0285 | 0.386 | 3.1447 | 0.0273 |
| 440 | 0.1169 | 0.0297 | 0.381 | 3.1389 | 0.0283 |
| 460 | 0.1217 | 0.0308 | 0.376 | 3.1326 | 0.0294 |
| 480 | 0.1266 | 0.0319 | 0.371 | 3.1267 | 0.0304 |
| 500 | 0.1315 | 0.0331 | 0.367 | 3.1215 | 0.0314 |
| 550 | 0.1438 | 0.0358 | 0.352 | 3.1050 | 0.0340 |
| 600 | 0.1562 | 0.0386 | 0.338 | 3.0897 | 0.0366 |
| 650 | 0.1687 | 0.0413 | 0.325 | 3.0759 | 0.0391 |
| 700 | 0.1812 | 0.0440 | 0.313 | 3.0637 | 0.0416 |
| 750 | 0.1938 | 0.0467 | 0.302 | 3.0526 | 0.0441 |
| 800 | 0.2064 | 0.0493 | 0.290 | 3.0422 | 0.0466 |
| 850 | 0.2191 | 0.0519 | 0.280 | 3.0331 | 0.0491 |
| 900 | 0.2318 | 0.0544 | 0.270 | 3.0242 | 0.0516 |
| 950 | 0.2446 | 0.0569 | 0.260 | 3.0162 | 0.0540 |
| 1000 | 0.2575 | 0.0594 | 0.22 | 3.0092 | 0.0564 |

**Table 6.9 (Cont'd)** Range statistics for antimony in aluminum
Courtesy: Technology Modeling Associates (TMA).

| Energy | $R_P$ | $\Delta R_P$ | $\gamma$ | $\beta$ | $\Delta R_T$ |
|---|---|---|---|---|---|
| 10 | 0.0105 | 0.0031 | 0.537 | 3.3671 | 0.0034 |
| 20 | 0.0160 | 0.0047 | 0.533 | 3.3604 | 0.0051 |
| 30 | 0.0206 | 0.0061 | 0.533 | 3.3610 | 0.0066 |
| 40 | 0.0249 | 0.0074 | 0.534 | 3.3619 | 0.0079 |
| 50 | 0.0289 | 0.0085 | 0.534 | 3.3621 | 0.0090 |
| 60 | 0.0327 | 0.0096 | 0.534 | 3.3615 | 0.0102 |
| 70 | 0.0364 | 0.0107 | 0.533 | 3.3604 | 0.0113 |
| 80 | 0.0400 | 0.0118 | 0.532 | 3.3589 | 0.0123 |
| 90 | 0.0436 | 0.0128 | 0.531 | 3.3570 | 0.0133 |
| 100 | 0.0471 | 0.0138 | 0.530 | 3.3550 | 0.0143 |
| 110 | 0.0505 | 0.0148 | 0.529 | 3.3528 | 0.0152 |
| 120 | 0.0539 | 0.0158 | 0.527 | 3.3504 | 0.0162 |
| 130 | 0.0573 | 0.0168 | 0.526 | 3.3480 | 0.0171 |
| 140 | 0.0607 | 0.0177 | 0.524 | 3.3454 | 0.0180 |
| 150 | 0.0640 | 0.0187 | 0.523 | 3.3428 | 0.0189 |
| 160 | 0.0673 | 0.0196 | 0.521 | 3.3401 | 0.0198 |
| 170 | 0.0707 | 0.0206 | 0.519 | 3.3373 | 0.0207 |
| 180 | 0.0740 | 0.0215 | 0.518 | 3.3345 | 0.0216 |
| 190 | 0.0773 | 0.0224 | 0.516 | 3.3317 | 0.0225 |
| 200 | 0.0806 | 0.0233 | 0.514 | 3.3288 | 0.0234 |
| 220 | 0.0871 | 0.0252 | 0.507 | 3.3166 | 0.0251 |
| 240 | 0.0937 | 0.0270 | 0.500 | 3.3052 | 0.0268 |
| 260 | 0.1003 | 0.0288 | 0.493 | 3.2949 | 0.0284 |
| 280 | 0.1068 | 0.0305 | 0.487 | 3.2853 | 0.0301 |
| 300 | 0.1134 | 0.0323 | 0.481 | 3.2763 | 0.0317 |
| 320 | 0.1200 | 0.0341 | 0.476 | 3.2677 | 0.0334 |
| 340 | 0.1266 | 0.0358 | 0.470 | 3.2594 | 0.0350 |
| 360 | 0.1332 | 0.0376 | 0.465 | 3.2513 | 0.0366 |
| 380 | 0.1398 | 0.0393 | 0.460 | 3.2435 | 0.0382 |
| 400 | 0.1464 | 0.0410 | 0.454 | 3.2359 | 0.0398 |
| 420 | 0.1531 | 0.0427 | 0.449 | 3.2285 | 0.0414 |
| 440 | 0.1597 | 0.0444 | 0.444 | 3.2210 | 0.0430 |
| 460 | 0.1664 | 0.0461 | 0.438 | 3.2127 | 0.0446 |
| 480 | 0.1731 | 0.0478 | 0.432 | 3.2039 | 0.0462 |
| 500 | 0.1798 | 0.0495 | 0.425 | 3.1949 | 0.0478 |
| 550 | 0.1967 | 0.0536 | 0.409 | 3.1732 | 0.0517 |
| 600 | 0.2136 | 0.0577 | 0.394 | 3.1549 | 0.0556 |
| 650 | 0.2307 | 0.0617 | 0.381 | 3.1383 | 0.0594 |
| 700 | 0.2479 | 0.0657 | 0.368 | 3.1234 | 0.0632 |
| 750 | 0.2651 | 0.0697 | 0.356 | 3.1093 | 0.0670 |
| 800 | 0.2824 | 0.0736 | 0.346 | 3.0976 | 0.0708 |
| 850 | 0.2998 | 0.0774 | 0.334 | 3.0855 | 0.0746 |
| 900 | 0.3173 | 0.0812 | 0.324 | 3.0749 | 0.0783 |
| 950 | 0.3348 | 0.0850 | 0.314 | 3.0648 | 0.0820 |
| 1000 | 0.3524 | 0.0887 | 0.305 | 3.0561 | 0.0857 |

**Table 6.9 (Cont'd)** Range statistics for antimony in photoresist (type AZ-111)
Courtesy: Technology Modeling Associates (TMA).

| Energy | $R_P$ | $\Delta R_P$ | $\gamma$ | $\beta$ | $\Delta R_T$ |
|---|---|---|---|---|---|
| 10 | 0.0410 | 0.0065 | 0.194 | 2.9688 | 0.0063 |
| 20 | 0.0657 | 0.0104 | 0.179 | 2.9601 | 0.0099 |
| 30 | 0.0870 | 0.0137 | 0.156 | 2.9480 | 0.0131 |
| 40 | 0.1065 | 0.0166 | 0.145 | 2.9428 | 0.0159 |
| 50 | 0.1249 | 0.0194 | 0.139 | 2.9402 | 0.0185 |
| 60 | 0.1425 | 0.0221 | 0.135 | 2.9385 | 0.0209 |
| 70 | 0.1595 | 0.0246 | 0.132 | 2.9372 | 0.0233 |
| 80 | 0.1761 | 0.0270 | 0.129 | 2.9360 | 0.0256 |
| 90 | 0.1923 | 0.0294 | 0.126 | 2.9348 | 0.0278 |
| 100 | 0.2082 | 0.0317 | 0.124 | 2.9340 | 0.0299 |
| 110 | 0.2239 | 0.0340 | 0.121 | 2.9329 | 0.0320 |
| 120 | 0.2394 | 0.0363 | 0.119 | 2.9321 | 0.0341 |
| 130 | 0.2542 | 0.0385 | 0.116 | 2.9310 | 0.0361 |
| 140 | 0.2700 | 0.0406 | 0.114 | 2.9303 | 0.0381 |
| 150 | 0.2852 | 0.0428 | 0.111 | 2.9292 | 0.0400 |
| 160 | 0.3002 | 0.0449 | 0.109 | 2.9286 | 0.0420 |
| 170 | 0.3151 | 0.0470 | 0.106 | 2.9275 | 0.0439 |
| 180 | 0.3300 | 0.0491 | 0.104 | 2.9269 | 0.0457 |
| 190 | 0.3448 | 0.0511 | 0.102 | 2.9262 | 0.0476 |
| 200 | 0.3596 | 0.0531 | 0.099 | 2.9253 | 0.0494 |
| 220 | 0.3890 | 0.0572 | 0.092 | 2.9232 | 0.0531 |
| 240 | 0.4184 | 0.0611 | 0.084 | 2.9210 | 0.0567 |
| 260 | 0.4476 | 0.0651 | 0.078 | 2.9195 | 0.0602 |
| 280 | 0.4767 | 0.0689 | 0.071 | 2.9179 | 0.0637 |
| 300 | 0.5059 | 0.0728 | 0.066 | 2.9168 | 0.0671 |
| 320 | 0.5350 | 0.0766 | 0.060 | 2.9156 | 0.0706 |
| 340 | 0.5641 | 0.0804 | 0.054 | 2.9146 | 0.0739 |
| 360 | 0.5933 | 0.0841 | 0.049 | 2.9138 | 0.0773 |
| 380 | 0.6224 | 0.0878 | 0.044 | 2.9130 | 0.0806 |
| 400 | 0.6516 | 0.0915 | 0.039 | 2.9124 | 0.0840 |
| 420 | 0.6808 | 0.0951 | 0.035 | 2.9119 | 0.0872 |
| 440 | 0.7107 | 0.0988 | 0.030 | 2.9114 | 0.0905 |
| 460 | 0.7394 | 0.1024 | 0.025 | 2.9110 | 0.0938 |
| 480 | 0.7687 | 0.1059 | 0.021 | 2.9107 | 0.0970 |
| 500 | 0.7981 | 0.1095 | 0.016 | 2.9104 | 0.1002 |
| 550 | 0.8718 | 0.1183 | 0.006 | 2.9101 | 0.1082 |
| 600 | 0.9485 | 0.1270 | -0.005 | 2.9100 | 0.1161 |
| 650 | 1.0201 | 0.1355 | -0.016 | 2.9104 | 0.1239 |
| 700 | 1.0947 | 0.1440 | -0.025 | 2.9110 | 0.1317 |
| 750 | 1.1696 | 0.1523 | -0.034 | 2.9118 | 0.1393 |
| 800 | 1.2448 | 0.1605 | -0.042 | 2.9128 | 0.1470 |
| 850 | 1.3203 | 0.1686 | -0.050 | 2.9139 | 0.1545 |
| 900 | 1.3960 | 0.1766 | -0.058 | 2.9153 | 0.1620 |
| 950 | 1.4720 | 0.1845 | -0.065 | 2.9166 | 0.1695 |
| 1000 | 1.5482 | 0.1923 | -0.072 | 2.9181 | 0.1768 |

# Chapter 7

# Diffusion

## 7.0 Introduction

Diffusion is the result of random motion of particles in three dimensions in a gaseous, liquid, or solid medium. When the concentration of particles is uniform and their motion isotropic, the particle flux in one direction is, on the average, balanced by an equal flux in the opposite direction and the net transport of material is zero. In the presence of a concentration gradient, however, the random motion results in a net transport of particles in the direction of decreasing concentration. While the probability for a particle to move in one direction is the same as in any other direction, more particles are available to move from regions of higher concentration to regions of lower concentration than vice versa. The "driving force" of diffusion is the concentration gradient. Another factor that affects the motion of particles is their diffusivity, a property that describes the "ease" with which they move through the medium. Transport of material by diffusion was introduced in the preceding chapters. For thermal oxidation to proceed, for example, a gradient in the oxygen concentration is established between the top surface of the oxide and the oxide-silicon interface, causing oxygen to travel through the growing oxide to the silicon surface by diffusion. Similar transport processes occur during deposition or etch of materials.

Typical dopant impurities in silicon are the group-III acceptor elements boron (B), gallium (Ga), Indium (In), and aluminum (Al), and the group-V donor elements phosphorus (P), arsenic (As), and antimony (Sb). In earlier stages of manufacturing, the most practical method of introducing these impurities into the semiconductor was to expose a batch of wafers to a controlled gaseous source of dopants in a furnace at a controlled elevated temperature. In this process, impurities enter the semiconductor at the surface, establishing a concentration gradient between surface and bulk, and causing a net transport of dopants toward the bulk of the crystal by solid-state diffusion. Diffusion time and temperature determine the depth of dopant penetration. Increasing the temperature enhances the motion of dopants and reduces the time required to obtain a certain depth of dopant concentration. Selective doping is achieved by patterning a masking material, typically silicon dioxide of adequate thickness, allowing the impurities to diffuse vertically and laterally into the semiconductor only in those "windows" not covered by the mask.

As the wafer size increases, however, it becomes more difficult to achieve the required profile uniformity by solid-state diffusion in a furnace, batch process. Also, for some dopant sources this technique becomes impractical and less economical. More important, a concomitant decrease in the vertical and horizontal dimensions of doped regions, and the need for more complex impurity profiles to optimize the electrical characteristics of small devices necessitates the development of other methods to dope the semiconductor.

Ion implantation was introduced in Chap. 6 as a very efficient technique to introduce dopants into semiconductor crystals. Ion implantation with preamorphization is a promising technique to form shallow junctions [1]. Implanted layers are, however, typically inactive and accompanied by radiation damage that adversely affects junction characteristics [2,3]. This is why implantation is always followed by some high-temperature cycle to anneal the damage and activate the dopants. The anneal conditions vary, depending on the desired impurity profile and on where in the process sequence implantation is performed. When ultra-shallow layers are formed, annealing must be done with as little movement of impurities as possible. Defects induced by implantation, however, transiently enhance the diffusivity of some species even at very low anneal temperatures, making it difficult to achieve ultra-shallow profiles. Also, residual defects prohibitively increase junction leakage currents and require very high temperatures to anneal, a condition that is not compatible with ultra-shallow profiles.

Another challenge with ultra-shallow junctions is the formation of good contacts, i.e., contacts that allow passage of current in both directions with negligible resistance without increasing junction leakage. The most commonly used metallurgy to define contacts to silicon is a silicide formed by reacting a refractory metal (typically titanium) with silicon in the areas to be contacted (Chap. 8). Two major problems can arise when silicide is formed after implantation: metal penetration and dopant segregation. The former can cause an increase in junction leakage while the latter can increase contact resistance.

Problems associated with crystal damage and contact metallurgy can be avoided by implanting into a polysilicon or silicide film rather than directly into the crystal, and then using the implanted film as a low-temperature solid diffusion source to dope the crystal. In both cases, the contact metallurgy is removed from the doped region in the crystal, alleviating problems with metal penetration. For example, emitters of modern bipolar transistors are typically formed by implanting dopants into a polysilicon film and diffusing the species from the polysilicon into the crystal; contact is made to polysilicon without damaging the diffused emitter region. Silicide as a diffusion source (SADS) to form shallow junctions is also a subject of intense study [4]. In this technique, impurities are first implanted into a previously formed silicide and then diffused into the crystal, avoiding crystal damage, dopant segregation, and metal penetration.

An alternative method to form shallow impurity profiles is to diffuse dopants from doped-oxide or boron-nitride sources [5 − 7]. For example, shallow junctions have been diffused into silicon from borosilicate glass (BSG) and arsenosilicate glass (ASG) using rapid thermal annealing (RTA) [8].

Another technique to form ultra-shallow doped regions by diffusion from a gaseous source uses a laser to locally melt the semiconductor immersed in the gas. In this method, known as gas-immersion laser doping (GILD), dopants are introduced by liquid-phase diffusion (which is very rapid), and the dimensions of the doped region are essentially determined by the extent of the melt. Also, upon recrystallization, the dopants occupy substitutional sites and no activation anneal is necessary [9 − 11].

The foregoing discussion emphasizes the importance of predicting and controlling the diffusion of impurities in semiconductors. This chapter focuses on diffusion of materials used for the manufacture of semiconductor integrated circuits, in particular on diffusion of impurities in silicon.

## 7.1 Point Defects

Diffusion in a semiconductor crystal is fundamentally related to the interaction between impurities and point defects [12 − 15]. Point defects were introduced in Chap. 1. Native point defects are vacancies, interstitials, and **interstitialcy.** A vacancy ($V$) is an empty lattice site. An interstitial is an atom that resides in one of the interstices of the crystal lattice. A self-interstitial is an interstitial host atom. An interstitialcy defect consists of two atoms in non-substitutional positions configured about a single substitutional site. Throughout this chapter, no distinction is made between interstitialcies and self-interstitials and both defects are denoted by $I$. When activated, dopants almost completely occupy substitutional sites, but a finite fraction will always exist in a coupled dopant-defect state. Only the fraction of dopants that is paired with point defects is available to take part in the diffusion process [12 − 15]. The two coupled mechanisms that govern the motion of atoms in a silicon crystal are vacancy-assisted and interstitial(cy)-assisted diffusion, as illustrated in a simplified two-dimensional model in Fig. 7.1. The open circles represent host atoms and the solid circles represent impurity atoms. Associated with each diffusion mechanism is an activation energy, $E_A$. Vacancies and self-interstitials are present in a pure semiconductor at any temperature above 0 K, and their densities increase exponentially with temperature (Chap. 1). For silicon, the equilibrium concentration of neutral, single vacancies has been approximated as [16]

$$N_V \simeq 9x10^{23}e^{-2.6(eV)/kT} \quad cm^{-3} , \qquad 7.1$$

and that of neutral interstitials as [17 − 19]

$$N_I \simeq 1x10^{27}e^{-3.8(eV)/kT} \quad cm^{-3} , \qquad 7.2$$

where $k$ is the Boltzmann constant ($k = 8.62x10^{-5}eV/K$), and $T$ the absolute temperature in K. The definition of the actual coefficients and activation energies in Eqs. 7.1 and 7.2 is, however, still being researched.

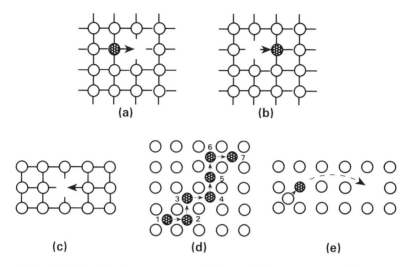

**Fig. 7.1** Models for vacancy and interstitial movement of atoms. (a) Jump of an impurity atom into a nearby vacancy. (b) Vacancy created after the move in (a). (c) Jump of a host atom into a nearby vacancy (self-diffusion). (d) Movement of an impurity atom by interstitial jumps. (e) Replacement mechanism [13,14,19].

When an impurity moves by jumping into a vacancy, diffusion is termed vacancy-assisted (Figs. 7.1a,b). The activation energy for this motion is essentially equal to the energy required to create the vacancy. Since bonds must be broken to remove a host atom from its lattice site and then be re-established to place an atom in a vacancy, this energy must be related to the energy-gap of the semiconductor ($E_A \simeq 3$ eV for Ge, $\simeq 4$ eV for Si). If the migrating atom is a host atom, the motion is called **self-diffusion** by vacancies (Fig. 7.1c). When an atom migrates by jumping from one interstitial site to another without occupying a regular lattice site, diffusion is called direct-interstitial (Fig. 7.1d). The activation energy for direct interstitial motion is only $\simeq 1$ eV. Interstitial diffusion can also occur through a dissociative mechanism by which atoms occupy both interstitial and substitutional sites. In this process, a substitutional atom becomes interstitial, moves interstitially at a significant rate, and then occupies another substitutional site. For example, copper, cobalt, gold, iron, platinum, and nickel are fast diffusants that move in silicon by this mechanism. Some of the heavy metals, such as Au, Cu and Fe, still diffuse significantly in silicon⁻ at room temperature. Heavy metals are important contaminants since they create one or more energetically deep-lying levels in silicon which reduce the minority-carrier lifetime. In typical process technologies, procedures are taken to avoid contamination with heavy metals, since a reduction in the minority-carrier lifetime adversely affects junction and transistor characteristics. In some cases, however, heavy metals (such as Au) are intentionally introduced to shorten the diffusion length of minority carriers. In either case, it is important to understand the diffusion mechanisms of these contaminants.

Figure 7.1e describes a replacement mechanism in which a self-interstitial atom displaces a substitutional impurity which subsequently travels interstitially to a nearby vacancy [13, 14]. There is a great deal of discussion about the relative importance of the vacancy-assisted and interstitial-assisted mechanisms, but it appears that boron, phosphorus, aluminum, and gallium diffuse in silicon predominantly by the interstitialcy mechanism, while the dominating mechanism for the motion of arsenic and antimony is diffusion by vacancies [13,14]. Direct-interstitial motion is responsible for the fast diffusion of small-size group I and group VIII elements in silicon crystals. These include hydrogen, helium, argon, and alkali metals, such as lithium, potassium, and sodium.

Point defects can exist in different multiple charged states which obey the laws of mass action under equilibrium conditions. Vacancies and interstitials can be neutral ($V^0$, $I^0$), positively charged by donating one electron ($V^+$, $I^+$), double positively charged by donating two electrons ($V^{++}$, $I^{++}$), negatively charged by accepting one electron, ($V^-$, $I^-$), and double-negatively charged by accepting two electrons ($V^=$, $I^=$). The probability for higher level charged states is very low. Dopants that have opposite charge sign to a defect experience a Coulombic attraction to that defect, while those with a similar charge sign experience a repulsion. The effective diffusivity is a combination of diffusivities due to neutral and electrically charged vacancies and interstitials.

## 7.2 Fick's Laws

While diffusion is a non-equilibrium process, quasi-equilibrium can be assumed at dilute (less than $\simeq 10^{18}\, cm^{-3}$) impurity concentrations. Fick's laws of diffusion are based on a simple quasi-equilibrium continuum theory which does not necessitate knowing the details of point defect reactions involved to model diffusion adequately. These simple laws are applicable wherever diffusion exhibits simple behavior where the diffusivity is isotropic and independent of concentration.

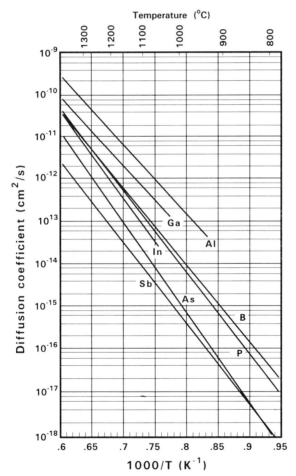

**Fig. 7.2** Average diffusion coefficients of acceptors and donors in silicon at concentrations below $\simeq 10^{18}\, cm^{-3}$ [12].

The first law relates the number of particles passing through a unit area per unit time, called flux, to the concentration gradient. In one dimension, the flux is expressed as

$$j_N = -D \frac{\partial N}{\partial x} , \qquad 7.3$$

where $N$ is the impurity concentration, $j_N$ the flux (or particle current density), $D$ the diffusion coefficient (or diffusivity), $\partial N/\partial x$ the concentration gradient, and $x$ the distance. The negative sign accounts for the fact that diffusion flux proceeds in the direction of decreasing concentration while the concentration gradient, by definition, points the other way. The diffusion coefficient, $D$, is expressed in $cm^2/s$ or in the more practical units of $\mu m^2/h$. Experimentally measured diffusion coefficients of various impurities at low concentration in silicon are shown in Figs. 7.2 and 7.3 [12,20,21]. The diffusion coefficients of Si (self-diffusion) and Ge in silicon are shown separately for low dopant concentrations as a function of temperature in Fig. 7.4 [12,22,23].

The temperature dependence of $D$ can be represented by

$$D = D_0 \, e^{-E_A/kT}, \qquad 7.4$$

where $D_0$ is a measure of the frequency with which an atom "attempts" to make a jump over the energy barrier ($10^{13} - 10^{14} \, Hz$). The exponential term is referred to as the Boltzmann factor which represents the probability than an atom will have an energy equal to or in excess of the activation energy $E_A$. Both $D_0$ and the exponential term are temperature dependent. Since the exponential term varies more rapidly with temperature, $D_0$ can be assumed constant within a small temperature range of approximately 200 K. Experimental values of diffusivities depend on the method and temperature range of measurements. Even with accurate control of diffusion temperature and time, calculated impurity profiles are only reasonable approximations of measurements. Also, diffusion already begins as the furnace is heated (ramp-up) and continues during cooling to room temperature (ramp-down), so that the effective diffusion temperature and duration are a function of ramp-rate and furnace geometry. It is therefore not surprising to find a large variation in the diffusivities obtained from different authors.

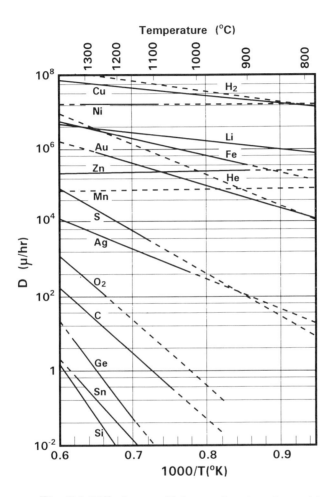

**Fig. 7.3** Diffusion coefficients of various impurities in silicon [12].

Approximate empirical low-concentration (less than $\simeq 10^{18}\,cm^{-3}$) values for $D_0$ and $E_A$ of important materials in silicon are given in Table 7.1. The values for self-diffusion by interstitials, Si(I), and by vacancies, Si(V) were obtained from the analysis of the diffusion of gold and platinum in silicon [19,25].

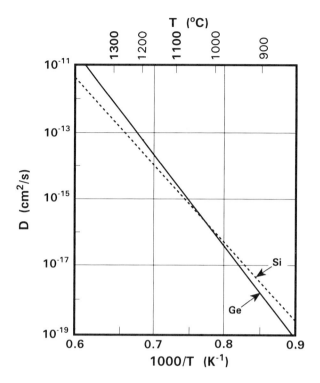

**Fig. 7.4** Diffusion coefficients of silicon and germanium in silicon as a function of temperature at low concentration [22,23].

**Table 7.1** Approximate low-concentration $D_0$ and $E_A$ of Si, group III, and group V elements in silicon.

| Impurity | $D_0$ $(cm^2/s)$ | $E_A$ $(eV)$ | Reference |
|----------|------------------|--------------|-----------|
| Si (I)   | 4000             | 5.0          | [19]      |
| Si (V)   | 40               | 4.6          | [19]      |
| Al       | 0.14             | 3.41         | [12]      |
| B        | 0.76             | 3.46         | [12]      |
| Ga       | 0.37             | 3.39         | [12]      |
| In       | 0.79             | 3.63         | [12]      |
| Bi       | 1.08             | 3.85         | [12]      |
| As       | 22.9             | 4.10         | [24]      |
| P        | 3.85             | 3.66         | [12]      |
| Sb       | 0.214            | 3.65         | [12]      |

To derive Fick's second law of diffusion, we note that conservation of matter requires that the rate of change of particle concentration in a volume element be equal to the change in the net flux in or out of the volume element, provided no material is created or annihilated within the volume element. This condition is summarized in the continuity relation

$$\frac{\partial N(x,t)}{\partial t} = - \frac{\partial J(x,t)}{\partial x} \, . \tag{7.5}$$

It is best visualized by considering a slab of unit cross-sectional area and thickness $\Delta x$ with the axis normal to the slab. A flux of particles $J_1$ enters the slab at one side and a flux $J_2$ leaves the slab at the other. For a very small slab thickness, the difference between the two fluxes can be approximated as

$$J_1 - J_2 = - \Delta x \, \frac{\partial J}{\partial x} \, .$$

Since the two fluxes are different, the particle concentration within the slab must change with time. For a slab of unit area, the slab volume is $\Delta x$ and the net change in the number of particles within the slab is

$$J_1 - J_2 = \Delta x \, \frac{\partial N}{\partial t} = - \Delta x \, \frac{\partial J}{\partial x} \, ,$$

which leads to Eq. 7.5. Substituting Eq. 7.3 in 7.5 and assuming constant $D_0$ gives Fick's second law of diffusion in one dimension:

$$\frac{\partial N(x,t)}{\partial t} = D \, \frac{\partial^2 N(x,t)}{\partial x^2} \tag{7.6}$$

An identical equation describes the conduction of heat in solids, except that the diffusivity $D$ is replaced by the thermal diffusivity $\kappa$. When $D$ is independent of concentration and position, Fick's second law can be solved with simple initial and boundary conditions. Two typical profiles of interest are described in the following sections.

### 7.2.1 Diffusion from a Constant Source

In practice, diffused layers are formed in two steps. In the first step, called **predeposition,** impurities are introduced into a very thin layer of the semiconductor while maintaining a constant surface concentration, $N_S$. They are then diffused to the desired depth into the semiconductor without adding more impurities to the surface. This second step is called **drive-in.** Predeposition is performed by placing the wafer in a furnace through which an inert carrier gas containing the desired impurity flows at a constant rate. Provided $N_S$ stays below the solid solubility limit in the semiconductor at the diffusion temperature (Chap. 1), a linear equilibrium relationship is established between the partial pressure $p$ of the impurity in the gas and the impurity concentration at the surface. Above solid solubility, $N_S$ saturates and becomes independent of partial pressure. Dopants in excess of this limit precipitate without contributing to the free-carrier concentration. Predeposition is typically done above the solid solubility limit to ensure a reproducible surface concentration even when the partial pressure fluctuates. In this case, the initial condition at time $t = 0$ is $N(x,0) = 0$, and the boundary conditions are $N(0,t) = N_S$ and $N(\infty,t) = 0$. Solving Fick's diffusion equation 7.6 subject to these initial and boundary conditions results in the following complementary error function

$$N(x,t) = N_S \, erfc\left[ \frac{x}{2\sqrt{Dt}} \right],\qquad\qquad 7.7$$

where $\sqrt{Dt}$ is the diffusion length. With increasing time, impurities penetrate deeper into the crystal. The total amount of impurities $Q(t)$ introduced into the crystal during the diffusion time $t$ is obtained by integrating Eq. 7.7 as

$$Q(t) = \frac{2}{\sqrt{\pi}} \, N_S \sqrt{Dt} \, .\qquad\qquad 7.8$$

A normalized complementary error profile is shown in Fig. 7.5. Some properties of the error function are given in Appendix A at the end of the book.

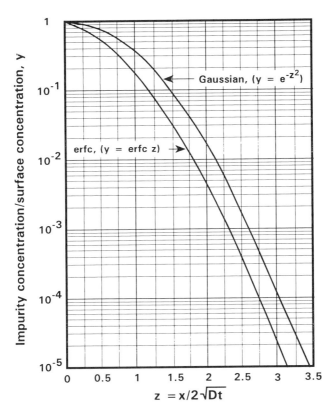

**Fig. 7.5** Normalized erfc and Gaussian profiles

## 7.2.2 Diffusion from an Instantaneous Source

An instantaneous source is a fixed amount of impurity atoms per unit area, $Q$, which is predeposited onto the silicon surface. It is typically driven into the bulk without addition or loss of impurities. The initial condition is $N(x,0) = 0$, and the boundary conditions are $N(\infty,t) = 0$ and $\int_0^\infty N(x,t)dx = Q$. Solving Fick's law subject to the above initial and boundary condition results in the following Gaussian distribution function

$$N(x, t) = \frac{Q}{\sqrt{\pi Dt}} \, e^{-x^2/4Dt}. \qquad 7.9$$

By setting $x = 0$ in Eq. 7.9, the surface concentration $N_S$ is found as

$$N_S = N(0, t) = \frac{Q}{\sqrt{\pi D t}} \; . \tag{7.10}$$

A normalized Gaussian profile is shown in Fig. 7.5.

### 7.2.3 Two-Dimensional Diffusion

The discussion so far has focused on one-dimensional motion in a direction normal to the semiconductor surface. To locally dope the semiconductor, however, impurities are introduced through openings in a mask of appropriate composition and thickness. The dopants are then annealed or diffused, resulting in a two-dimensional profile around the mask edge. The shape of this profile can strongly affect the electrical and geometrical character-istics of junctions and transistors. This lateral distribution is inherent with masked diffusion and, to a lesser extent, with ion implantation (Chap. 6). Techniques to simulate and measure the two-dimensional profile are described in a separate chapter. Simu-lated two-dimensional profiles are shown in Figs. 7.6, where isoconcentration contours are drawn rather than concentration versus depth [26]. Figure 7.6a shows the two-dimensional profile for a constant surface concentration. As mentioned earlier, a pn junction is formed when dopants are introduced into a semicon-ductor of opposite dopant polarity. The junction is a three-dimensional surface where dopants of opposite polarity exactly balance each other. The two most important junction dimensions are its depth below the semiconductor surface, $X_J$, and its lateral distance from the mask edge, $X_L$. Typically, $X_L < X_J$, and the dif-ference between both dimensions increases when the diffusivity is concentration dependent (Sec. 7.3.4) [27].

A simulated two-dimensional diffusion profile for an instan-taneous source, such as an ion-implanted layer of dose $\phi$, is shown in Fig. 7.6b. In this case, diffusion under the mask causes depletion of impurities within the opening, so that both the lateral extent and junction depth are affected by the window size [28]. This window-size dependence of the lateral impurity profile plays an important role in MOSFET and bipolar transistor characteristics.

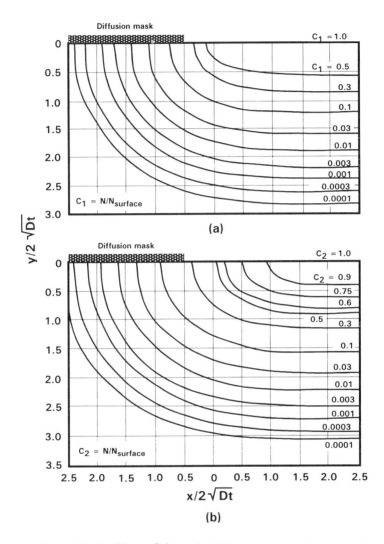

**Fig. 7.6** Profiles of lateral diffusion through an oxide window (a) Constant surface concentration. (b) Instantaneous source showing the source depletion inside the window [27].

One of the challenges in process design is to shape and characterize this portion of the junction. For example, in many junction designs, such as MOSFET source and drain junctions, a gradual rather than abrupt profile is typically required to reduce the two-dimensional electric field near the surface. For n⁺p junctions, this is best achieved by using phosphorus rather than arsenic at the edge of heavily doped n-type junctions.

## 7.3 Non-Constant Diffusivity

When the impurity concentration is low (smaller than the intrinsic carrier concentration, $n_i$), and the concentration gradient not larger than $\simeq 10^{22}\ cm^{-4}$, dopant impurities exhibit a simple behavior and the concentration-independent diffusion coefficients in Fig. 7.2 give good agreement between measured impurity profiles and Fick's diffusion laws. Diffusion under these conditions is commonly referred to as **intrinsic diffusion,** and the diffusion coefficient is termed intrinsic, $D_i$. At concentrations higher than $n_i$, complex interactions between impurities and point defects and between different impurities can cause the diffusivity to vary along the impurity concentration gradient. In this case, diffusion is termed **extrinsic** [29]. To properly model non-constant diffusivities, Fick's diffusion laws must be modified. The diffusivity becomes, however, difficult to interpret and the profile cannot be described with a simple closed form relation. Computer simulations are typically used to represent impurity profiles with non-constant diffusivities. The one-dimensional continuity relation which takes into account a non-constant diffusivity is [12 − 14]

$$\frac{\partial N}{\partial t} = \frac{\partial}{\partial x}\left(D\ \frac{\partial N}{\partial x}\right) + \frac{q}{kT}\ \frac{\partial}{\partial x}\cdot\left(DN'\ \frac{\partial \phi}{\partial x}\right),\qquad 7.11$$

where $D$ is the non-constant diffusion coefficient, $N$ the total impurity concentration, $N'$ the electrically active impurity concentration, and $\phi$ the electrostatic potential given by

$$\phi = \frac{kT}{q}\ \ln\ \frac{n}{n_i} = -\frac{kT}{q}\ \ln\ \frac{p}{n_i}\ .\qquad 7.12$$

In Eq. 7.12, $n$ and $p$ are the electron and hole concentrations, respectively, and $n_i$ is the intrinsic carrier concentration found for silicon as [30]

$$n_i \simeq 3.87x10^{16}\ T^{1.5}\ e^{-(0.605+\Delta E_g)/kT}\quad (cm^{-3}),\qquad 7.13$$

where $\Delta E_g$ is the energy-gap lowering approximated by

$$\Delta E_g \simeq 7.1 x 10^{-10} (n_i/T)^{1/2} \quad (eV)$$

The first term in Eq. 7.11 represents Fick's second law including non-constant diffusivity. The second term incorporates the flux induced by the **built-in electric field** (defined by $\partial\phi/\partial x$) including the non-constant diffusivity, as described next.

## 7.3.1 Effect of Electric Field

The motion of impurities can be accelerated or retarded in the presence of an electric field, depending on the direction of the electric field and the sign of the ionized impurities. When the net impurity concentration is not uniform, the Fermi level varies over the spatial extent of the diffused profile, creating an internal electric field in the crystal. When only one type of dopant is present, this built-in field is always in a direction that accelerates the motion of impurities. Consider, for example, a non-uniformly doped n-type semiconductor. The free electrons created by the donors tend to diffuse in the direction of decreasing concentration with a flux defined by a relation similar to Eq. 7.3. Since electrons move much faster than ions, the ionized impurities appear immobile to the diffusing electrons so that local charge separation occurs between positive donors and negative electrons. This separation is self-limiting since it creates an electric field that gives rise to a drift current component of electrons in a direction opposite to that of the diffusion current component. Eventually, when the two current components exactly balance each other, no further charge separation occurs. The semiconductor is now left with a built-in field that is proportional to the density of "separated" charge. This field exerts an accelerating force on donors, creating an additional component to diffusion at elevated temperature. A similar situation occurs in a non-uniformly doped p-type semiconductor where the field points toward the positively charged acceptors, accelerating their motion at elevated temperature. In both cases, the field is defined in one dimension by the term $\partial\phi/\partial x$ in Eq. 7.11. When the approximation $n \simeq N_D$ or $p \simeq N_A$ can be made, Eq. 7.12 gives

$$E_x = -\frac{\partial \phi}{\partial x} = \pm kT/q \frac{1}{N} \frac{dN}{dx} , \qquad 7.14$$

where the "+" sign is for a p-type ($N \simeq N_A$) and the "-" sign for an n-type ($N \simeq N_D$) semiconductor. The field-enhancement of particle motion is found by adding a drift component to the flux in Eq. 7.3. Using Einstein's relation $D = \mu kT/q$ we get

$$J = -D \frac{\partial N}{\partial x} \pm D \frac{q}{kT} NE_x .\qquad 7.15$$

For an n-type semiconductor, Eq. 7.15 can be written as

$$J = -Dh \frac{\partial N_D}{\partial x} ,$$

where $h$ is the "electric-field enhancement factor" given by

$$h = 1 + N_D \frac{\partial}{\partial N_D} \ln\left(\frac{N_D}{n_i}\right),$$

or [12,13]

$$h = 1 + \frac{N_D}{2n_i} \cdot \frac{1}{\left[\left(\frac{N_D}{2n_i}\right)^2 + 1\right]^{1/2}} \cdot \qquad 7.16$$

The field-enhancement factor $h$ ranges from 1 for $N_D << n_i$ to a maximum value of 2 for $N_D >> n_i$.

## 7.3.2 Dependence of Diffusion on Surface Reactions

Diffusion of most impurities in silicon require point defects as the diffusion "vehicle". Therefore, processes that affect the concentration of point defects have a direct impact on the diffusivity of impurities in the crystal. Such processes are oxidation, nitridation, ion implantation, and doping at high concentration. In silicon, both self-interstitials and vacancies contribute to diffusion, so that enhanced or retarded diffusion can be induced by deviations of the fractions of vacancies or interstitials from their equilibrium values.

Oxidation- and nitridation-enhanced or retarded diffusion of groups III and group V elements and the so-called anomalous diffusion phenomena in silicon (discussed later) are typical examples of these effects.

### Oxidation- and Nitridation-Dependent Diffusion

The generation of point defects during oxidation modifies their relative equilibrium concentrations and leads to enhanced or retarded diffusion of group III and group V elements in Si, depending on the mechanism by which the dopants diffuse. Oxidation of silicon has long been observed to enhance the diffusivity of boron and, to a lesser extent, phosphorus and arsenic [31 − 34]. Figure 7.7 compares the oxidation-enhanced diffusivities for boron and phosphorus with those for inert ambient conditions [32]. The effect of oxidation-enhanced diffusion (OED) is illustrated for boron in Fig. 7.8 which compares the profile obtained for an oxidized silicon surface to that for a non-oxidized surface under otherwise the same thermal and ambient conditions. The figure also shows the decrease in the boron concentration near the surface caused by boron segregation into the oxide, as discussed later. This observation is of great importance because thermal oxidation of silicon is one of the most common steps in the manufacture of integrated circuits and is typically performed in the presence of implanted or diffused impurity profiles within the affected range.

The growth or shrinkage behavior of extrinsic stacking faults is accepted as a test of which type of defect, interstitial or vacancy, is injected during a surface reaction. An extrinsic stacking fault is an extra half plane of silicon atoms (Chap. 1). It can grow by absorbing excess interstitials or shrink by absorbing excess vacancies [13,32]. It is now known that oxidation of silicon causes the nucleation and growth of interstitial-type dislocation loops on {111} planes which contain extrinsic stacking faults, commonly termed oxidation-induced stacking faults (OSF, Chap. 2). The growth of interstitial-type OSF implies that during oxidation, excess silicon is injected from the silicon-oxide interface into the bulk where it occupies interstitial sites (self-interstitials), enhancing the interstitial component of impurity diffusion. The supersaturation of self-interstitials can affect a region as deep as $\simeq 10\ \mu m$ in silicon [29].

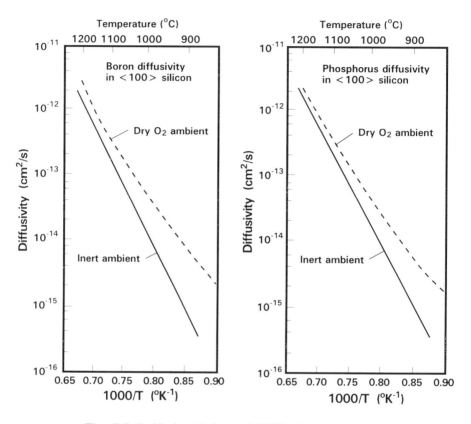

**Fig. 7.7** Oxidation-Enhanced Diffusivities for boron and phosphorus (dashed lines) compared with those for inert ambient conditions (solid lines) [32].

As pointed out in [13,35], oxidation-enhanced diffusion is closely related to the growth of OSF. This can be deduced from the observation that under the same oxidation conditions, both the diffusion enhancement of boron and the growth rate of OSF increase when the orientation of the oxidized surface is varied from <111> to <110> to <100>. Both phenomena are related to a change in the concentration of point defects induced by oxidation.

Under the same conditions that lead to enhanced diffusion of boron, arsenic and phosphorus, oxidation-retarded diffusion (ORD) of antimony is observed [13,31,36,37]. Following a similar argument as in [32], it is concluded that a supersaturation of interstitials during oxidation reduces the vacancy population from its equilibrium level by interstitial-vacancy (I-V) recombination [13], and that antimony diffuses predominantly by the vacancy mechanism [13,38]. The simultaneous enhancement of one dopant while another is retarded is evidence of two complementary diffusion mechanisms.

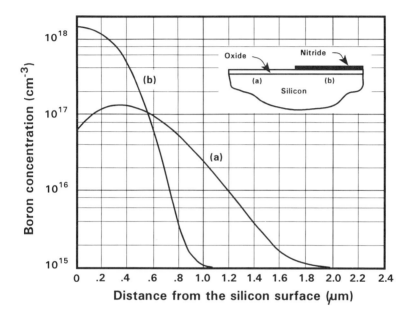

**Fig. 7.8** Simulated boron profiles after a 2 Hr wet oxidation cycle at 1000 °C illustrating oxidation enhanced diffusion. (a) Surface covered with 20 nm silicon-dioxide. (b) Surface covered with 20 nm silicon dioxide and 20 nm silicon nitride.

Thermally grown silicon nitride or silicon dioxide layers nitrided by exposure to an ammonia ambient (typically used as alternative dielectrics in DRAM node capacitors) also affect diffusion in the underlying layer by injecting point defects from the surface [13]. Direct nitridation by exposing the silicon surface to $NH_3$ ambient results in the injection of vacancies while thermal nitridation of a pre-grown silicon dioxide film causes interstitial injection. Direct nitridation of silicon causes enhanced diffusion of antimony and retarded diffusion of boron and phosphorus.

Enhanced diffusivity has been measured for phosphorus in an oxidizing ambient, and for antimony in a nitridizing ambient on test structures, as illustrated schematically in Fig. 7.9 [13].

**Oxidation**

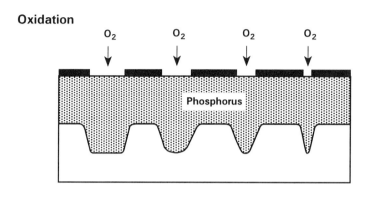

**Nitridation**

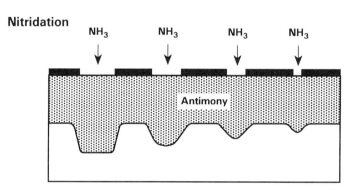

**Fig. 7.9** Schematic illustration of enhanced diffusion of phosphorus in an oxidizing ambient and of antimony in a nitridizing ambient [13].

### Redistribution of Impurities during Oxidation

During thermal oxidation, dopants redistribute between the growing oxide and silicon as the Si-SiO$_2$ boundary moves into the silicon bulk. This can cause considerable changes in the impurity in silicon near its surface. An abrupt change in concentration typically occurs across the interface of both materials (Fig. 7.10) [39].

The physical "preference" of dopants for either oxide or silicon is quantified by the **segregation coefficient,** m, defined as the ratio of the thermal-equilibrium dopant concentration in silicon to that in the $SiO_2$ near the interface. When m < 1, the growing oxide depletes the silicon from impurities as the dopants segregate into the oxide. Even for m = 1, dopant depletion is observed because of the larger volume for redistribution offered by the growing oxide. For m > 1, the oxide ejects excess impurities back into the silicon, causing a "pile-up" of impurities near the silicon surface. Redistribution of dopants is also affected by their diffusivity in the oxide and by the oxidation rate, as illustrated in Fig. 7.10. The relative rate at which the oxide-silicon boundary moves into silicon compared to the diffusion rate of the impurity in the oxide is an important factor that determines the redistribution of the impurity concentration in silicon.

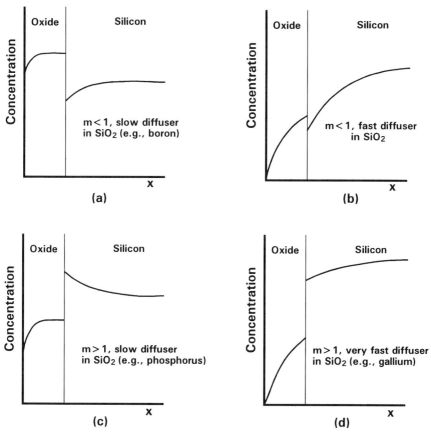

**Fig. 7.10** Impurity concentrations at the oxide-silicon interface for different values of segregation coefficient *m*.

**490**

Most of the experimental measured segregation coefficients are reported for boron, with $m$ typically ranging from 0.1-0.3, resulting in boron depletion during oxidation [40,41]. To illustrate this effect, an "ideal" half-Gaussian distribution having its peak at the interface is used as the initial profile prior to oxidation (Fig. 7.11) [40]. As oxidation proceeds, the boron concentration near the surface drops well below the value that would be obtained from diffusion alone (solid curves).

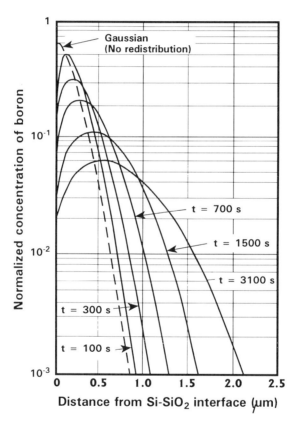

Fig. 7.11 Illustration of boron segregation during oxidation of silicon. Dashed curve: "ideal" half-Gaussian profile without redistribution. Solid curves: redistribution with segregation. Assumptions: $m = 0.1$, D in Si $= 0.6 \times 10^{-12}\ cm^2/s$, D in $SiO_2 = 1.02 \times 10^{-12}\ cm^2/s$, T $= 1200\ °C$ [40].

For very thin oxides, a large fraction of boron can diffuse through the oxide and escape into the ambient atmosphere, enhancing the depletion of boron at the silicon surface.

For boron, the segregation coefficient increases with increasing temperature and is crystal-orientation dependent, with values for <100> being greater than for <111> orientation [41, 42]. For <100> oriented silicon, the coefficient is approximated as [41]

$$\text{Dry } O_2: \quad m_{eff} \simeq 13.4e^{-0.33eV/kT}, \qquad 7.17$$

and for wet oxidation

$$\text{Wet } O_2: \quad m_{eff} \simeq 104.0e^{-0.663eV/kT}. \qquad 7.18$$

For phosphorus, arsenic and antimony, m > 1 and the dopants segregate into silicon, causing a "pile-up" of dopants at the silicon surface during oxidation. The segregation coefficient can be as high as 10 for phosphorus and 800 for arsenic [39].

Redistribution of impurities during thermal oxidation can seriously affect device parameters, particularly MOSFET threshold voltages, and must be taken into account when designing a process technology.

## 7.3.3 Diffusion of Implanted Profiles

It is necessary, by a suitable heat treatment, to place the implanted ions on electrically-active sites and anneal the implant damage. Annealing, however, is typically accompanied by a redistribution of impurities and methods must be developed to predict the profile that results after annealing. The simplest situation occurs when a damage-free low-dose buried region is implanted away from the semiconductor surface and can be approximated by a symmetrical Gaussian profile. In this case, the profile remains Gaussian after annealing. The effect of annealing is to lower the peak concentration and widen the distribution (Fig. 7.12). The total dose remains unchanged after annealing. To obtain the annealed profile, the standard deviation, $\Delta R_p$ in Eqs. 6.11-6.13 is simply replaced by $\sqrt{(\Delta R_p^2 + 2Dt)}$ [43].

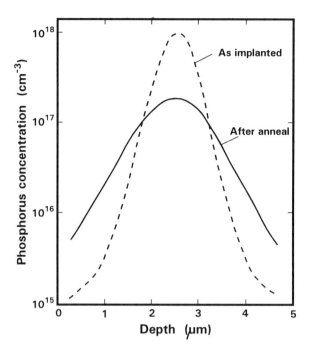

**Fig 7.12** Redistribution of a symmetrical Gaussian phosphorus profile after anneal.

When the implanted region is close to the semiconductor surface, the profile can no longer be described by a simple Gaussian expression due to the reflection of impurities at the semiconductor surface and to segregation effects at the semiconductor-insulator interface [43].

As discussed in Chap. 6, ion implantation displaces silicon atoms from their lattice sites. Each displaced atom becomes an interstitial or interstitialcy and leaves a vacancy behind, creating a Frenkel pair. Implantation at high doses results in a supersaturation of these point defects. Upon heating to activate the dopants and anneal the damage, point defects created by implantation become mobile and can diffuse and agglomerate to form extended defects which can absorb or release interstitials and whose concentration changes with time and position during annealing. In extreme cases, extended defects appear as dislocation loops or stacking faults. Interactions between dopants and point defects induced by implantation can cause anomalous enhanced diffusion of dopants during transient heat treatment [44 – 46]. Even when a rapid thermal cycle is used, more diffusion than predicted by diffusion theory equations is typically observed [47]. Enhanced diffusion during rapid-thermal annealing was observed for ion-implanted boron, and phosphorus at high doses [48 – 51]. For phosphorus, a transient enhancement in diffusivity by a factor of 1000 or higher has been observed during the initial stages of annealing [49,50]. Enhanced diffusion was also observed during RTA for phosphorus implanted at a dose below the amorphization threshold to eliminate the effect of high concentration and extended defects on diffusivity [52,53]. No appreciable enhancement has been observed for arsenic or antimony [53 – 55]. Extended defects act as sinks for point defects and cause a large spatial variation in the transient enhancement or retardation of diffusion [56,57]. They are believed to trap interstitials and immobilize boron, causing a peak in the heavily damaged zone (Fig. 7.13) [58]. Release of interstitials deeper than the heavily damaged zone enhances the diffusivity of boron, creating a tail in the profile [59]. Residual damage and anomalous diffusivity of dopants during annealing limits the capability of forming ultrashallow doped films by implanting directly into the silicon crystal.

### 7.3.4 Concentration-Dependent Diffusivity

At dopant concentrations above $n_i$, detailed interactions between impurities and point defects must be taken into account to properly model dopant diffusion. Doping the crystal increases the coupling between impurities and point defects, the relative density of charged point defects, and the total density of point defects [13,14]. The neutral vacancy concentration remains the same throughout the semiconductor because it is not affected by the Fermi level or concentration gradient. It follows that the extrinsic concentration of neutral defects will be equal to the concentration in intrinsic material, which is only a function of temperature [60]. Since the probability for a defect to donate or accept one or two electrons and become charged depends on the Fermi level with respect to the energy level of the point defect, the relative population in the different charged states, and hence the diffusivity of impurities, can be altered by changing the Fermi level in the semiconductor. This is accomplished by doping because a variation in the dopant concentration causes a change in the Fermi level.

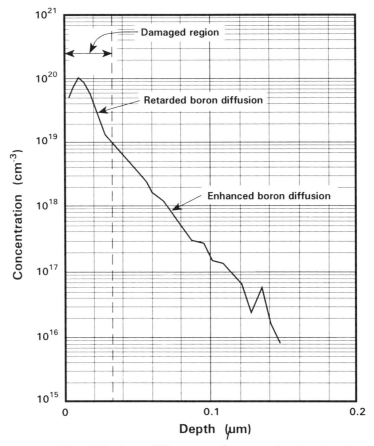

**Fig 7.13** Immobilization of boron in the heavily-damaged implanted region and enhanced diffusion deeper in silicon.

By assuming Boltzmann statistics, one can predict the changes in point-defect concentrations as a function of dopant concentrations using the following simple relations [12 − 14, 61,62]:

$$\frac{N_X{}^-}{N_X^{i\,-}} = \frac{n}{n_i}\,, \quad \frac{N_X{}^=}{N_X^{i\,=}} = \left(\frac{n}{n_i}\right)^2, \qquad 7.19a$$

$$\frac{N_{X^+}}{N_{X^+}^i} = \frac{p}{n_i} \, , \quad \frac{N_{X^{++}}}{N_{X^{++}}^i} = \left(\frac{p}{n_i}\right)^2 \, , \qquad 7.19b$$

where $X$ can be a vacancy ($V$) or an interstitial ($I$) and $i$ denotes intrinsic. The above relations show that relative to the intrinsic condition, the concentrations of single ($X^-$) and double ($X^=$) negatively charged point defects are higher in n-type semiconductors than in p-type. Similarly, the concentrations of single ($X^+$) and double ($X^{++}$) positively charged defects are higher in p-type than in n-type semiconductors.

If the dopant concentration is sufficiently high, the number of available sites at which a point defect can be considered unperturbed by the presence of dopant atoms will diminish significantly. At a concentration of about $3x10^{20} \, cm^{-3}$ all point defects can be considered coupled with dopant atoms [63]. Above a certain concentration (solid solubility limit), excess dopant atoms no longer dissolve completely on substitutional sites but precipitate in a second phase. Precipitation is responsible for the electrical inactivity of boron, phosphorus, and antimony [56,64,65]. For example, when the boron concentration is above $\simeq 10^{20} \, cm^{-3}$, the concentration of electrically active boron is less than the total boron concentration [66].

For arsenic, two models have been proposed to account for its electrical inactivity at high concentrations. In the first model, multiple As atoms form some new configuration, called a cluster, which is inactive at room temperature [67,68]. One clustering model consists of three arsenic atoms and one electron that are electrically neutral at room temperature but active at the anneal temperature [69]; the clusters dissociate at elevated temperature into separate arsenic atoms. The second model attributes the inactivity to precipitation, similar to that found for phosphorus [70]. Both models explain the experimental observation that above a certain arsenic concentration, the free electron concentration saturates to a maximum that depends on temperature but not on the arsenic concentration above this limit (Fig. 7.14) [68,71].

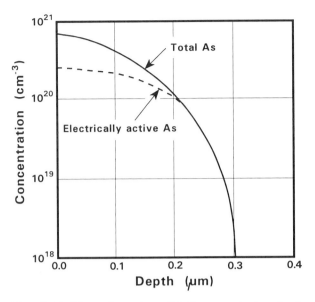

**Fig. 7.14** Plot of total (solid line) and electrically active (dashed line) arsenic concentration versus depth [68].

### Single-Species Diffusion

For single-species diffusion, the continuity relation for a concentration-dependent diffusivity can be written as

$$\frac{\partial N}{\partial t} = \frac{\partial}{\partial x}\left(D^*\frac{\partial N}{\partial x}\right),$$

where $D^*$ is the effective diffusivity approximated by combining the intrinsic diffusivities due to the different charged states of point defects. For example, from Fermi-Dirac statistics with the Boltzmann approximation, the effective diffusivity in n-type silicon is found as as [12 − 14,61,62]

$$D^* = h\left[D_{X^0}^i + D_{X^-}^i\left(\frac{n}{n_i}\right) + D_{X^=}^i\left(\frac{n}{n_i}\right)^2\right], \qquad 7.20$$

where the $i$ superscripts denote intrinsic conditions, and $D_{X^0}^i$, $D_{X^-}^i$, $D_{X^=}^i$ represent the intrinsic diffusivity due to impurity coupling with neutral, single negative and double negative point defects, respectively. The multiplier $h$ is a correction factor due to a spatially varying Fermi level, defined in Eq. 7.16 [12,13]. For p-type silicon $n$ is replaced by $p$ and only positively-charged point defects are considered. For each type of impurity, one of the terms in Eq. 7.20 has been found dominant in determining the effective diffusivity. The dominant mechanism cannot be obtained from just fitting Eq. 7.20 to the measured impurity profile; it must be inferred from separate measurements.

**Table 7.2** Ratio of extrinsic to intrinsic diffusivities of Si and Ge in Si $(D_e/D_i)$ for different dopant concentrations, N.

| System | Dopant | N ($cm^{-3}$) | T (°C) | $D_e/D_i$ | Ref. |
|---|---|---|---|---|---|
| $^{31}Si$ in Si | P, As | $8 \times 10^{19} - 2 \times 10^{20}$ | 1090 - 1197 | 1.35-3.15 | [72] |
| $^{31}Si$ in Si | B | $8 \times 10^{19} - 2 \times 10^{20}$ | 1090 - 1197 | 1-1.75 | [72] |
| $^{71}Ge$ in Si | P | $1.1 \times 10^{20}$ | 1200 - 1370 | $10 - 2.5$ | [73] |
| $^{71}Ge$ in Si | As | $4 \times 10^{19}$ | 890- - 1220 | $1.8 - 2.5$ | [71] |
| $^{71}Ge$ in Si | B | $6 \times 10^{18} - 1.2 \times 10^{19}$ | 855 - 910 | 0.7-0.9 | [74] |

Doping with n-type impurities raises the Fermi level toward the conduction band and increases the equilibrium density of negatively charged (acceptor type) vacancies. This enhances the diffusion of Si and Ge in Si via the vacancy mechanism, whereas doping with p-type impurities have the opposite effect [38]. Table 7.2 shows the effect of dopant concentration, $N$, on self-diffusion and the diffusion of germanium in silicon. The ratio of extrinsic to intrinsic diffusivities of Si and Ge in Si $(D_e/D_i)$ is also given for different dopant concentration and temperature ranges.

The above mechanisms are also applicable to the dopant concentration dependence of the diffusion of group III and group V elements in silicon. Boron and phosphorus show a dominant component of interstitial-assisted diffusion. The diffusivity of phosphorus is, however, not well understood. Phosphorus has long shown "anomalous" diffusion behavior, manifested by a "kink" in the profile at moderate concentrations and a "tail" at low concentrations (Fig. 7.15). Boron shows some of these effects, but to a much lesser extent [75,76]. Several quantitative models based on the coupling between dopants and point defects have been developed to explain this anomaly [12,13, 77,78]. There is, however, a lot of discussion on whether interactions with vacancies or interstitials are the predominant coupling mechanisms for phosphorus.

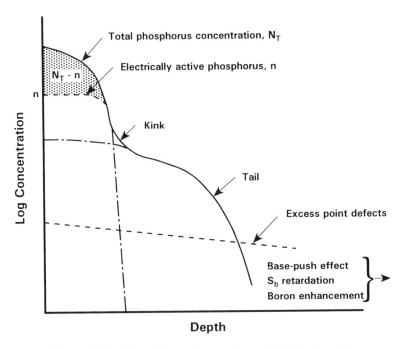

**Fig. 7.15** The formation of a "kink" in the phosphorus profile and a suggested model to explain the enhanced boron diffusion in the presence of phosphorus.

Boron as an acceptor is negatively charged when active in silicon. Its diffusivity is primarily determined by the concentration of positively charged and neutral point defects. The dependence of the boron diffusion coefficient on boron concentration is represented by [14]

$$D_B(e) \simeq D_B(i)^0 + D_B(i)^+ \frac{p}{n_i} \, ,$$

or [79]

$$D_B(e) = D_B(i) \left( \frac{N_A}{2n_i} \right) \left( 1 + \sqrt{1 + \frac{4n_i^2}{N_A^2}} \right) , \qquad 7.21$$

where $N_A$ is the active boron concentration, $D_B(e)$ the extrinsic boron diffusivity, and $D_B(i)$ the intrinsic boron diffusivity given by [12]

$$D_B(i) = 0.77e^{-3.47eV/kT}.$$  7.22

Arsenic is a donor and hence positively charged in the silicon crystal. Its diffusivity is primarily dependent on the concentration of negatively charged and neutral point defects and is be approximated as [14]

$$D_{As}(e) \simeq D_{As}(i)^0 + D_{As}(i)^- \frac{n}{n_i} ..$$

A useful approximation is [80]

$$D_{As} \simeq 0.066e^{-3.44/kT} + 12.0\left(\frac{n}{n_i}\right)e^{-4.05/kT}.$$  7.23

When the arsenic surface concentration is above $\simeq 10^{20}\,cm^{-3}$, the measured profile has a nearly flat portion near the silicon surface and then drops abruptly, creating a very high electric field at the boundary with a p-type region (junction). The dependence of arsenic diffusivity on arsenic concentration is illustrated in Fig. 7.16 for different ambient conditions [24].

Because of its very small diffusivity relative to arsenic and phosphorus, antimony is an excellent candidate for ultra-shallow n-type regions [81,82]. While the solubility limit of antimony in silicon is only $\simeq 2x10^{19}\,cm^{-3}$ when incorporated in silicon at 1000 °C, a rapid epitaxial regrowth of an amorphous layer produced by implanting antimony at high dose into silicon can place Sb atoms into substitutional sites up to a concentration threshold of $\simeq 3.5x10^{20}\,cm^{-3}$ [83]. The diffusion coefficient of antimony is based on interactions between Sb atoms with neutral and double-negatively charged point defects [84]:

$$D_{eff} \simeq 17.5e^{-4.05eV/kT} + 0.01\left(\frac{n}{n_i}\right)^2 e^{-3.75eV/kT}.$$  7.24

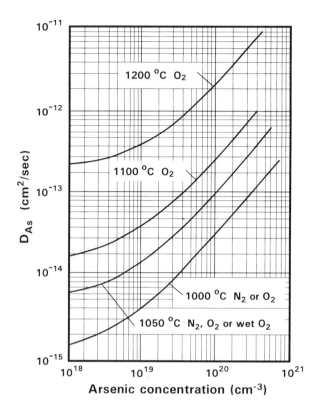

**Fig. 7.16** Dependence of arsenic diffusivity on concentration in silicon [24].

## Multiple-Species Diffusion

The interaction between dopant species at high concentrations results in a dramatic change in the diffusion coefficient. When acceptors and donors are diffused through the same window, the diffusion of both types of impurities are not independent of each other and exhibit a complex behavior. Fig. 7.17 shows the impurity profile for arsenic and gallium as a result of their interaction during diffusion [85]. A similar behavior is observed when boron and arsenic are diffused in the same region. The dip in the Ga or B profile is very characteristic of this type of diffusion and was predicted by simulations [86]. Both the amount of gallium that lies outside the arsenic-doped region and the depth of gallium penetration are substantially reduced by the presence of the As. It is well known that these same effects occur for boron in place of gallium, and must be taken into account when defining bipolar transistors with As-doped emitters and B-doped bases. The width of the base and its total dose are critical transistor parameters that are affected by the presence of arsenic. The retardation of boron or gallium in the arsenic region is also an important consideration when defining MOSFET source and drain junctions.

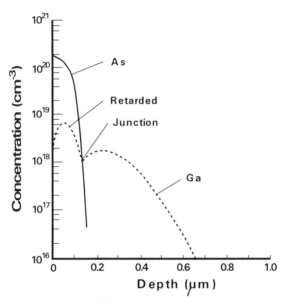

**Fig. 7.17** Retarded diffusion of gallium in the presence of arsenic at high concentration. A similar behavior is observed for boron in an arsenic-doped region [13].

Three factors contribute to inhibiting gallium and boron diffusion in arsenic-doped regions [13]: (i) The change in the Fermi level reduces the concentration of $X^+$ and $X^{++}$ defects and thus reduces the diffusivity of the negatively charged acceptor ions due to coupling with these defects. (ii) The electric field factor is strong due to the sharp concentration gradient of the As and is in a direction opposing the diffusion of acceptors out of the As-doped region. (iii) Pairing between the acceptor ions and As$^+$ immobilizes some of the acceptor ions [87].

Phosphorus at high concentration enhances the diffusivity of boron and retards the diffusion of antimony. In earlier bipolar npn transistors, the emitter was formed by phosphorus diffusion into a boron-doped base. The diffusion of phosphorus caused boron to diffuse deeper in the base immediately beneath the emitter (intrinsic base) than in the base region outside the emitter (extrinsic base). The enhanced boron diffusion in the intrinsic base is referred to as the base-push or emitter-dip effect. It is now known that phosphorus diffusion injects excess point defects as part of the diffusion process, greatly increasing the diffusivity of boron, as indicated in Fig. 7.15 [12,13]. The effect of phosphorus at high concentration on lightly doped buried phosphorus and antimony is schematically illustrated in Fig. 7.18 [13]. Similarly to boron, the diffusivity of the buried phosphorus is enhanced under the heavily doped phosphorus region. The motion of antimony is, however, retarded.

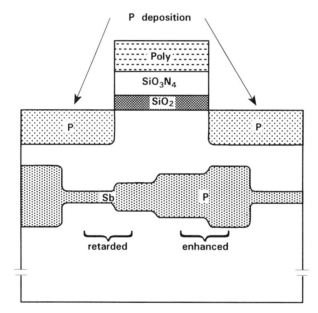

**Fig. 7.18** Schematic illustration of the effect of high-concentration phosphorus diffusion on lightly doped buried phosphorus and antimony layers [13].

## 7.4 Diffusion in Polysilicon

The deposition and properties of polycrystalline silicon (polysilicon) films were discussed in Chap. 3. Polysilicon is an important material which has found many applications in modern technologies. It is used as a self-aligned gate conductor in MOSFETs, as a capacitor plate in stacked-capacitor or trench-capacitor DRAM cells, or as a conductor for local interconnections. In three-dimensional integrated circuits, polysilicon constitutes the body of thin-film transistors (TFT). It is also used to define high-impedance load resistors for small-size memory cells, for electrically trimmable resistors, and for precision resistors in analog designs.

Polysilicon films deposited directly on silicon substrates have found important applications as diffusion sources to produce shallow junctions in single-crystal silicon without damaging the crystal [88 − 97]. The films are used to define self-aligned emitter and base contacts in high-performance bipolar structures [88,91,92, 98 − 102], and to form elevated silicon regions over the source and drain areas of MOSFETs [89,90,103 − 106]. The films are typically doped by ion implantation but can also be doped as deposited (in situ). The dopants are then diffused into the crystal to form the shallow junctions.

Polysilicon consists of an aggregate of single-crystal silicon grains of different orientation and size, separated by grain boundaries of amorphous material (Chap. 3). Typically, the grain size is in the submicrometer range and the grain boundaries are less than 10-nm thick. Thick polysilicon films deposited at temperatures above ≃ 800 °C exhibit a columnar grain-growth with a predominant < 110> orientation normal to the surface. If deposited below ≃ 600 °C, the films are initially amorphous, but grow into large crystallites when subjected to a high-temperature heat treatment [107]. Diffusion in polysilicon is strongly affected by the motion of dopants along grain boundaries (also labeled as "diffusion pipelines"), where the diffusivity can be orders of magnitude higher than within crystallites [108 − 112]. This is important in several process-integration situations where a rapid and uniform distribution of dopants in the polysilicon film is desired without a large increase in the thermal budget. The net diffusivity of impurities in polysilicon depends on the texture of the film. It is therefore affected by the method and temperature of deposition, dopant type and concentration, film thickness and composition of layer on which polysilicon is deposited. Because of this complexity, the diffusion mechanisms of dopants in polysilicon are rather difficult to predict. For boron, the enhanced diffusivity along grain boundaries ($D_{PolyGB}$) is estimated as [111]

$$D_{PolyGB} \simeq 0.040\, e^{-1.87\, eV/kT}\ (cm^2/s)\,.\qquad\qquad 7.25$$

The net diffusivity in polysilicon is found to be intermediate between the diffusivity along its grain boundaries and that within the crystallites. It can be approximated for boron as [113]

$$D_{Poly} \simeq 0.277\, e^{-2.51\, eV/kT}\ (cm^2/s)\,.\qquad\qquad 7.26$$

The diffusivity of boron, as estimated from Eq. 7.26, is a factor 200-100 larger than in single-crystal silicon. The factor is even greater for arsenic and and phosphorus because of their larger segregation into grain boundaries [114].

The effect of thermal annealing on the conductivity of doped polysilicon is different than for single-crystal silicon [115 – 118]. For arsenic- or phosphorus-doped films, the resistivity repeatedly increases and decreases by successively annealing the same sample at 800 °C and 1000 °C [115]. It increases substantially when the film is subjected to a final low-temperature anneal, and decreases rapidly when this anneal is followed by RTA above $\simeq$ 800 °C [118]. This effect is attributed to dopant segregation into grain boundaries during the lower temperature anneal, where they become inactive, and then de-segregation (and activation) at the higher temperature anneal [115,119 – 121]. It is not an easy task, however, to take segregation and de-segregation effects into account when modeling the diffusivity in polysilicon.

Another important effect on diffusivity in polysilicon is the change in morphology during annealing. An enhancement in grain-growth rate is observed, for example, at high arsenic concentrations after RTA above 1100°C [122,123]. The growth of grains reduces the density of grain boundaries ("diffusion pipelines"), resulting in a decrease in diffusivity.

Dopants in polysilicon behave differently when they are co-diffused with other impurities than when they are diffused alone, but the mechanism for this "interaction" is different than in single-crystal silicon. For example, arsenic at low to moderate concentration diffuses much faster alone than in the presence of boron at high concentration. It is believed that impurities at high concentration saturate grain-boundary "traps" and inhibit the motion along boundaries of other dopants at lower concentration [124].

## 7.5 Diffusion in Insulators

Insulators are deposited or grown at different stages of the process to fulfill various tasks. With very few exceptions, an important property of insulators is to block the migration of impurities, particularly when the film is used as a mask during diffusion, implantation, or oxidation.

The three most commonly used insulators in the manufacture of integrated circuits are pure or doped silica films, silicon nitride, and polyimides. Silicon dioxide and silicon nitride are the most frequently used masking materials, while phosphosilicate glass (PSG), borosilicate glass (BSG), phospho-borosilicate glass (BPSG), and polyimide are typically used for interlevel dielectrics, leveling ("planarization"), and passivation.

For a material to be effective as a mask, the diffusivity of impurities in the material must be considerably smaller than in silicon. Approximate diffusivities of some elements in $SiO_2$ are given in Table 7.3. As can be seen, silicon dioxide is an effective barrier against most elements commonly used to dope silicon (boron, arsenic, phosphorus, and antimony), but not against gallium. The diffusion mechanisms in silicon dioxide are, however, not as well understood as for silicon, and the diffusion coefficients and activation energies reported in the literature vary widely. The diffusivities are only approximations which strongly depend on the impurity concentration and defect structure of the insulator and hence on sample preparation and measurement conditions.

Figures 7.19a & b show the thickness of $SiO_2$, grown in dry oxygen, that is required to mask against $P_2O_5$ and $B_2O_3$ vapors as a function of time with diffusion temperature as parameter [130,131]. Initially, only the top oxide reacts with $P_2O_5$ or $B_2O_3$ to form a phospho- or borosilicate glass, and the film is effective in masking against $P_2O_5$ and $B_2O_3$ in the gas phase. As the reaction proceeds, the glass thickness increases until the entire $SiO_2$ film is converted to glass and the film loses its effectiveness as a mask [130, 132, 133].

**Table 7.3** Diffusivities of some elements in $SiO_2$.

| Element | $D_0$ $(cm^2/s)$ | $E_A$ (eV) | $C_S$ $(cm^{-3})$ | Source, ambient | Ref. |
|---------|------------------|------------|-------------------|-----------------|------|
| P | $1.2x10^{-2}$ | 5.3 | $< 10^{19}$ | Doped poly | [125] |
| P | $5.7x10^{-5}$ | 2.3 | $\simeq 10^{21}$ | $P_2O_5$ Vap, $N_2$ | [126] |
| P | $1.86x10^{-1}$ | 4.0 | $< 10^{19}$ | PSG, $N_2$ | [127] |
| P | 7.23 | 4.44 | --- | Model | TMA* |
| B | 0.31 | 4.2 | $< 10^{19}$ | Poly | [125] |
| B | $7.23x10^{-6}$ | 2.4 | $\simeq 10^{20}$ | $B_2O_5$ Vap, $N_2$ | [126] |
| B | $1.23x10^{-4}$ | 3.4 | $\simeq 6x10^{18}$ | $B_2O_5$ Vap | [126] |
| B | $3.16x10^{-4}$ | 3.53 | --- | Model | TMA* |
| As | $2.3x10^3$ | 4.1 | $< 10^{19}$ | Poly | [125] |
| As | 67.25 | 4.7 | $< 5x10^{20}$ | Implant, $N_2$ | [128] |
| As | $3.7x10^{-2}$ | 3.7 | $< 5x10^{20}$ | Implant, $O_2$ | [128] |
| As | 1.75 | 4.89 | --- | Model | TMA*. |
| Ga | $1.04x10^5$ | 4.17 | --- | $Ga_2O_2$ Vap | [126] |
| Sb | $1.31x10^{16}$ | 8.75 | $\simeq 5x10^{19}$ | $Sb_2O_5$ Vap, $O_2 + N_2$ | [126] |
| $H_2$ | $5.65x10^{-4}$ | 0.45 | --- | --- | [107] |
| $H_2O$ | $10^{-6}$ | 0.79 | --- | --- | [107] |
| $O_2$ | $2.7x10^{-4}$ | 1.16 | --- | --- | [107] |
| Na | 0.05 | 0.66 | --- | --- | [129] |
| K | $8x10^{-4}$ | 1.09 | --- | --- | [129] |

* Technology Modeling Associates.

The minimum oxide thickness $(x_M)$ necessary to block diffusion is approximated for phosphorus as [132]

$$\frac{x_M^2}{t} \simeq 1.7x10^{-7}e^{-1.46eV/kT} \ (cm^2/s),$$ 7.27

and for boron as [133]

$$\frac{x_M^2}{t} \simeq 4.9x10^{-5}e^{-2.80eV/kT} \ (cm^2/s),$$ 7.28

where $t$ is the diffusion time. The values obtained from Eqs. 7.27 and 7.28 are conservative estimates since they are inferred from measurements made under high $P_2O_5$ and $B_2O_3$ pressures.

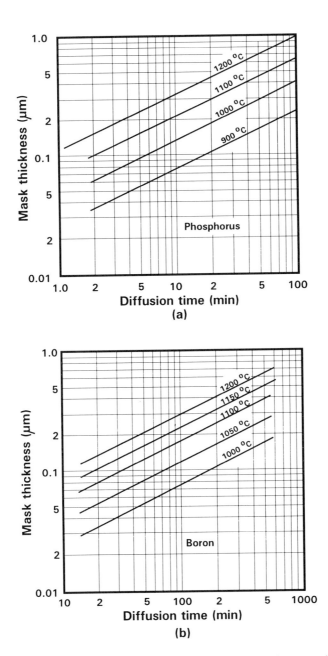

**Fig. 7.19** Minimum masking thickness of dry SiO$_2$ required to mask against impurities. (a) For phosphorus. (b) For boron [130, 131].

For boron at high concentration, an anomalous enhancement of diffusion attributed to a change in the $SiO_2$ bond characteristics is observed for long diffusion times and very thin $SiO_2$ [125]. Also, the diffusivity of boron in $SiO_2$ is found to increase by three orders of magnitude in a hydrogen ambient [134]. The diffusivity is enhanced during wet oxidation or nitride deposition because hydrogen is generated during these processes. The mechanism responsible for the hydrogen enhancement of boron diffusion is not well understood. It is, however, important to avoid a hydrogen ambient to obtain optimum masking by the $SiO_2$ film against boron. For example, after defining and etching polysilicon gates, an oxidation step is typically performed to grow a thin oxide film on the sidewalls (and top, when exposed) of polysilicon. When polysilicon is heavily doped with boron, it is important to avoid wet oxidation to prevent boron from diffusing through the thin gate oxide into the substrate [135, 136].

Compared to boron and phosphorus, the diffusivity of arsenic in $SiO_2$ increases only slightly with concentration (from $\simeq 2.6x10^{-16}$ to $\simeq 3.5x10^{-15} \, cm^2/s$ [137]). An oxygen ambient enhances the arsenic diffusivity in $SiO_2$ as can be seen from Table 7.3. The diffusivity of gallium in $SiO_2$ is $\simeq 5.2x10^{-11} \, cm^2/s$ at 1100 °C, orders of magnitude larger than the diffusivity in Si. Therefore, $SiO_2$ is not a suitable mask against gallium.

Amorphous silicon nitride ($Si_3N_4$) films are practically impermeable to most elements. The blocking efficiency of thin nitride films allows the use of this material as a barrier layer against Na migration (Fig. 7.20) [138, 139], and as a mask during local oxidation (LOCOS, Chap. 2). $Si_3N_4$ has a denser structure than $SiO_2$ so that the penetration depth of sodium in the nitride film is less than 50 nm. The smaller the crystallite size in the nitride film, the more effective the film becomes against sodium migration [139]. A thin layer of the nitride film will oxidize, however, at a considerable slower rate than silicon. The oxidation of $Si_3N_4$ is responsible for the enhanced diffusivity of some elements, such as Sb, in the film [140]. $Si_3N_4$ provides an effective barrier against gallium and is also used to prevent Ga from escaping GaAs.

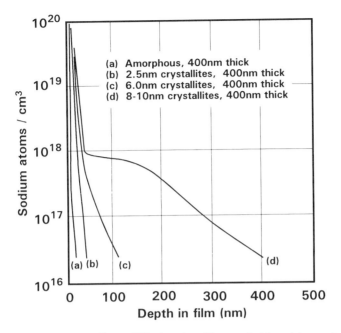

**Fig. 7.20** Sodium diffusion in silicon nitride with crystallite size in the nitride as parameter (After [139]).

## 7.6 Diffusion Sources

Impurities can be diffused into silicon from a gaseous, liquid, or solid source. Typical diffusion furnaces are shown schematically in Fig. 7.21.

The actual source for boron diffusion from a gaseous, liquid or solid source is a layer of boron silicate glass (BSG) which contains boron trioxide ($B_2O_3$) in contact with the silicon surface. The reaction of boron trioxide with silicon results in boron that diffuses into silicon and a simultaneous growth of oxide:

$$2B_2O_3 + 3Si \rightarrow 4B + 3SiO_2$$

Similarly, the source for arsenic or antimony diffusion is an arsenic trioxide (arsenic silicate glass, ASG) or antimony trioxide. Arsenic, for example, reacts with silicon as:

$$2As_2O_3 + 3Si \rightarrow 4As + 3SiO_2$$

For phosphorus, the final source is a layer of phosphosilicate glass which contains phosphorus pentoxide in contact with silicon. Phosphorus pentoxide reacts with silicon as:

$$2P_2O_5 + 5Si \rightarrow 4P + 5SiO_2$$

In all cases, the reaction of the source with silicon results in a grown oxide layer.

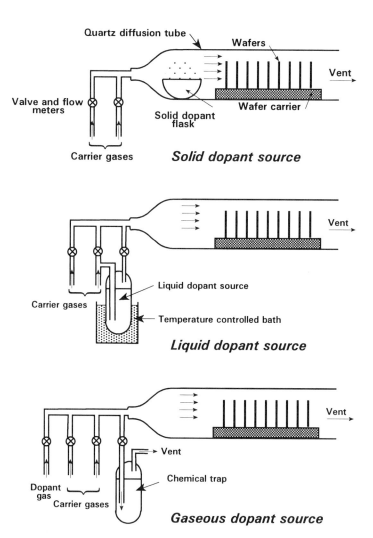

Fig. 7.21 Schematics of different diffusion systems.

## 7.7 Gettering in Silicon

Gettering was introduced in Chap. 1. It is discussed here again, in conjunction with diffusion processes. Gettering is the ability to "trap" elements, such as heavy metals, sodium, excess silicon, or oxygen, by providing regions of enhanced solubility of these elements, or "sinks" into which the elements precipitate, so that they are removed from other regions of the crystal. While precautions are taken to maintain ultra-clean processes, gettering steps are still required to remove traces of harmful elements from active regions of the crystal. Most metallic impurities are electrically active in silicon and traces of these impurities deposited on the wafer surface can be incorporated into the crystal during subsequent high-temperature steps. Their presence in the vicinity of the circuit reduces the minority-carrier lifetime and increases junction leakage [141 − 145], degrades the MOSFET gate oxide integrity [146], reduces the surface mobility in MOSFETs [147,148], and can even cause direct shorts between otherwise isolated transistor elements. When the concentration of heavy metals approaches that of dopants in silicon, the resistivity of both n- and p-regions begins to increase. Also, the presence of heavy metals can lead to the formation of stacking faults during oxidation. Stacking faults act as precipitation sites for metallic impurities which cluster around them to relieve the stress in the lattice. If precipitation occurs near junctions, the junction characteristics can be degraded.

**Table 7.4** Commonly used back-side gettering methods [150].

| Method | Comment | Ref. |
|---|---|---|
| Phosphorus diffusion | High concentration | [142,151] |
| Abrasion | Grinding, sandblasting. | |
| Ion implantation | Typically argon ions | [152 − 157] |
| Laser irradiation | Thermal shock Dislocation network is getter | [158, 159] |
| Nitride deposition | Typically 400 nm, stress-induced dislocation network is getter. | [160,161] |
| Polysilicon film, oxidation | Damage cause by stress, OSF | [161] |

The most common heavy metal elements in silicon are nickel, copper, iron, zinc, cobalt, titanium, and gold [147,149]. They can be removed from the top surface by damaging the backside, creating sinks that act as getters. Table 7.4 summarizes the most widely used methods of back-side gettering [150]. Heavily-doped phosphorus diffusion or implantation on the back of the wafer is a commonly used gettering technique. The highly-doped regions increase the solubility of acceptor-type heavy metals in silicon [162].

Although heavy metals are typically fast diffusers in silicon, backside gettering becomes increasingly inefficient as both diffusion time and temperature are reduced in low thermal-budget processes. In particular, rapid-thermal processing (RTP) is frequently considered inappropriate for gettering because of the short time involved which would not allow the impurities to migrate to gettering sites. Therefore, other techniques that place the gettering regions closer to the top surface of the crystal are receiving more attention. For example, high energy implantation is used to place a gettering region several microns below the active region, with implant doses ranging from $10^{14} - 10^{16} \, cm^{-2}$ [163]. The reliability of thin oxide films covered by polysilicon is increased by heavily doping the polysilicon with phosphorus, because the gettering effect of the heavily doped film induces the transport of alkaline contaminants and boron from the oxide into the heavily doped film [164]. Heavily-doped phosphorus regions created on the top of the wafer by conventional or rapid-thermal processing are found to be efficient in gettering heavy metals such as copper and gold [165].

**Intrinsic Gettering**

As mentioned in Chap. 1, oxygen and, to a lesser extent, carbon are present as impurities in silicon wafers. Crystals grown by the Czochralski technique can contain $1 - 2x10^{18} \, cm^{-3}$ of dissolved interstitial oxygen. These impurities form defect complexes with silicon and other impurities, pinning dislocations at these sites. Figure 7.22 shows the equilibrium oxygen interstitial solubility in silicon as a function of temperature [166]. Oxygen is initially electrically inactive and, when cooled to room temperature, is immobile and remains dispersed so that the crystal is supersaturated with oxygen [150]. During high-temperature processing, however, oxygen tends to precipitate in regions of stress in the form of large clumps 100-1000 nm in diameter and about 1 $\mu$ apart. These precipitates are known to getter heavy metals and it is important to place them away from the active devices. Intrinsic gettering consists of first forming an "oxygen-free" region below the top silicon surface that is thick enough to contain the devices. It is also important to keep this region "denuded" of oxygen during further thermal processing (Chap. 1). A high-temperature (950 − 1150°C) treatment for 2-4 hours dissolves any oxygen aggregates and allows the oxygen to out-diffuse, forming a "denuded" zone 5-6 $\mu$m below the silicon surface [167,168]. When this is followed by a low-temperature ($\simeq$600°C) heat cycle, oxygen precipitates form beneath the denuded zone and act as getters.

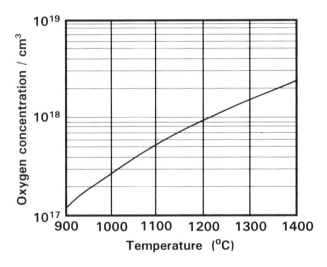

**Fig. 7.22** Oxygen interstitial solubility in silicon as a function of temperature ([166]).

**Halogenic Dopant Sources**

The use of halogenic sources during diffusion or oxidation serves to convert metallic impurities into their volatile halides, which then leave the system by incorporation into the gas stream (Chap. 2). Also, the presence of chlorine during oxidation is highly effective in preventing the formation of oxidation-induced stacking faults [169,170].

**Phosphosilicate and Borosilicate Glasses**

Both PSG and BSG in contact with silicon at elevated temperatures are very effective in reducing metallic impurities, stacking faults, and dislocations in silicon wafers [171,172]. They also provide viscous layers in which some of the impurities can be trapped. These layers are typically deposited from a halogenic source, and so have some chlorine or bromine incorporated in them. This further provides a gettering role by converting the impurity into its halide. PSG grown or deposited on top of $SiO_2$ chemically binds sodium in an electrically neutral form and inhibits its migration into oxide films near the silicon surface where it can cause severe device degradation.

# PROBLEMS

**7.1** Estimate the time required to form a junction 2 $\mu$m deep by performing a constant source, solid-solubility limited boron diffusion in silicon at 900 °C, 1000 °C, and 1100 °C, assuming a background arsenic concentration of $10^{16}\,cm^{-3}$.

**7.2** Phosphorus is implanted into a p-type silicon wafer having a uniform boron concentration of $5x10^{15}\,cm^{-3}$. The implant energy is 500 keV and the phosphorus dose is $1x10^{14}\,cm^{-2}$. The wafer is then subjected to an anneal temperature of 900 °C for 20 minutes. Assuming a symmetrical profile, plot the net donor concentration versus depth immediately after implantation and then after anneal.

**7.3** The boron concentration in the substrate region between the n-type source and drain of a MOSFET is $5x10^{17}\,cm^{-3}$ and the distance between source and drain is 0.25 $\mu$m. Assume that a carrier moves along a straight-line path between source and drain without deflection. How many boron atoms would the carrier encounter on its path?

**7.4** An instantaneous dopant source having a sheet concentration of $10^{14}\,cm^{-2}$ is diffused for a time $t$ at a temperature that gives a diffusivity of $8x10^{-14}\,cm^2/s$. Assuming the diffusivity to be constant, plot the dopant concentration for t = 30 minutes and t = 60 minutes.

**7.5** Compare the built-in fields in arsenic an phosphorus doped junctions. Both junctions have a concentration $N_D = 10^{19}\,cm^{-3}$ at a depth x = 0.1 $\mu$m. The concentration drops to $N_D = 10^{15}\,cm^{-3}$ at x = 0.15 $\mu$m for arsenic, and at x = 0.25 $\mu$ for phosphorus. Assume an exponential profile for arsenic and a Gaussian distribution for phosphorus and find the built-in fields at x = 0.12 $\mu$m, and x = 0.14 $\mu$m.

**7.6** Use Einstein's relation to find the mobilities of sodium and potassium in silicon dioxide. Assume the drift velocities of Na$^+$ and K$^+$ ions in $SiO_2$ to be proportional to the electric field. Find the time it takes sodium and potassium ions to traverse a 250-nm thick oxide under a bias of 5 V.

**7.7** In a metal-oxide-silicon (MOS) structure, the concentration of sodium ions in the oxide is $10^{18}\, cm^{-3}$ at one interface, and $10^{14}\, cm^{-3}$ at the other. Assuming that the structure is heated instantaneously to 400 °C, calculate the current density due to sodium diffusion. Disregard barrier effects at the interfaces.

**7.8** A silicon wafer is uniformly doped with boron at a concentration $N_A = 5x10^{15}\, cm^{-3}$. At what temperature will the intrinsic carrier concentration $n_i = 0.1\, N_D$?

**7.9** Phosphorus is selectively diffused into silicon at 1000 °C for 1 hr using an oxide pattern as a mask. Estimate the minimum oxide thickness required to block phosphorus diffusion under the oxide. How would this thickness change if the diffused species were boron?

**7.10** Phosphorus is diffused into silicon at 1000 °C for 60 minutes. Assume a complementary error function profile and a phosphorus surface concentration of $10^{20}\, cm^{-3}$. Find the total phosphorus dose in silicon and the concentration gradients at x = 0 and at a depth where the phosphorus concentration reaches $10^{16}\, cm^{-3}$.

# References

[1] T. E. Seidel, R. Knoell, G. Poli, and B. Schwartz, *"Rapid Thermal Annealing of Dopants Implanted into Preamorphized Silicon,"* J. Appl. Phys., 58 (2), 683-687 (1985).

[2] S. D. Brotherton, J. P. Gowers, N. D. Young, J. B. Clegg, and J. R. Ayers, *"Defects and Leakage Currents in $BF_2$ Implanted Preamorphized Silicon,"* J. Appl. Phys., 60 (10), 3567-3575 (1986).

[3] M. Miyake, S. Aoyama, S. Hiroto, and T. Kobayashi, *"Electrical Properties of Preamorphized and Rapid Thermal Annealed Shallow $p^+n$ Junctions,"* J. Electrochem. Soc., 135 (11), 2872-2876 (1988).

[4] C. M. Osburn, *"Formation of Silicided, Ultra-Shallow Junctions Using Low Thermal Budget Processing,"* J. Electron. Mat.., 19 (1), 67-88 (1990).

[5] H. Takemura, S. Ohi, M. Sugiyama, T. Tashiro, and M. Nakamae, *"BSA Technology for sub-100 nm Deep Base Bipolar Transistors,"* IEDM 1987 Tech. Digest, p. 375 (1987).

[6] E. Ling, P. D. Maguire, H. S. Gamble, and B. M. Armstrong, *"Very-Shallow Low-Resistivity $p^+$-n Junctions for CMOS Technology,"* Electron Dev. Lett., EDL-8 (3), 96 (1987).

[7] K.-T. Kim and C.-K. Kim, *"Formation of Shallow $p^+$-n Junctions Using Boron-Nitride Solid Diffusion Source,"* Electron Dev. Lett., EDL-8 (12), 569 (1987).

[8] M. Miyake, *"Diffusion of Boron into Silicon from Borosilicate Glass Using Rapid Thermal Processing,"* J. Electrochem. Soc., 138 (10), 3031-3039 (1991).

[9] P. G. Carey, K. Bezzian, T. W. Sigmon, P. Gildea, and T. S. Magee, *"Fabrication of Submicrometer MOSFET's Using Gas Immersion Laser Doping,"* Electron Dev. Lett., EDL-7 (7), 440 (1986).

[10] P. G. Carey, T. W. Sigmon, R. L. Press, and T. S. Fahlen, *"Ultra-Shallow High-Concentration Boron Profiles for CMOS Processing,"* Electron Dev. Lett., EDL-6 (6), 291 (1985).

[11] E. Ishida, K. T. Kramer, T. Talwar, T. W. Sigmon, K. H. Weiner, and W. T. Lynch, *"Shallow Junction Formation in Silicon: Dopant Incorporation and Diffusion Through Tungsten Silicide Films Using Gas Immersion Laser Doping (GILD),"* Mat. Res. Soc., Pittsburgh, Pennsylvania, pp. 673-678 (1992).

[12] R. B. Fair, *"Concentration Profiles of Diffused Dopants in Silicon,"* in Impurity Doping Processes in Silicon, F. F. Y. Wang, Ed., Chapter 7, North-Holland, Amsterdam (1981).

[13] P. M. Fahey, P. B. Griffin, and J. D. Plummer, *"Point Defects and Dopant Diffusion in Silicon,"* Reviews of Modern Physics, 61 (2), 289-384 (1989).

[14] R. W. Dutton and Z. Yu, Technology CAD, Computer Simulation of IC Processes and Devices, Kluwer Academic Publishers, Boston (1993).

[15] D. Mathiot and J. C. Pfister, *"Dopant Diffusion in Silicon: A Consistent View Involving Nonequilibrium Defects,"* J. Appl. Phys., 55 (10), 3518-3530 (1984).

[16] J. A. Van Vechten, March Meeting of the Am. Phys. Soc., New York, Paper BI 1 (1987).

[17] S. F. Mantovani, F. Nava, C. Nobili, and G. Ottaviani, *"In-Diffusion of Pt in Si from the PtSi/Si Interface,"* Phys. Rev. B33 (8), 5536-5544 (1986).

[18] A. Seeger, *"On the Theory of Diffusion of Gold into Silicon,"* Phys. Stat. Solidi, A61, 521-529 (1980).

**518**

[19] F. F. Morehead, *"The Diffusivity of Self-Interstitials in Silicon,"* in Defects in Electronic Materials, Mat. Res. Soc. Symp., M. Stavola, S. J. Pearton, and G. Davies, Eds., Mat. Res. Soc., Pittsburg, 104, 99-104 (1988).

[20] Atomic Diffusion in Semiconductors, D. Shaw, Ed., Plenum Press, New York (1973).

[21] Integrated Silicon Device Technology, Research Triangle Institute, Research Report ASD-TDR-63-316, Vol. 4 (1964).

[22] F. J. Demond, S. Kalbitzer, H. Mannsperger, and H. Damjantschitsch, *"Study of Si Seld-Diffusion by Nuclear Techniques,"* Phys. Lett., 93A (9), 503-506 (1983).

[23] P. Dorner, W. Gust, A. Lodding, H. Odelius, B. Predel, and U. Roll, in DIMETAL 82 - Diffusion in Metals and Alloys, F. J. Kedves and D. L. Beke, Eds., Monograph Series 7, Trans. Tech. Publications, 488 (1983).

[24] R. B. Fair and J. C. C. Tsai, *"The Diffusion of Ion-Implanted Arsenic in Silicon,"* J. Electrochem. Soc., 122 (12), 1689-1696 (1975).

[25] A. Seeger and K. P. Chik, *"Diffusion Mechnisms and Point Defects in Silicon and Germanium,"* Phys. Status Solidi A 29, 455-542 (1968).

[26] M. Miyake and H. Harada, *"Diffusion and Segregation of Low-Dose Implanted Boron in Silicon under Dry $O_2$ Ambient,"* J. Electrochem. Soc., 129 (5), 1097-1103 (1982).

[27] D. P. Kennedy and R. R. O'Brien, *"Analysis of the Impurity Atom Distribution near the Diffusion Mask for a Planar P-N Junction,"* IBM J. Res. Dev., 9, 179, May 1965.

[28] D. D. Warner and C. L. Wilson, , *"Two-Dimensional Concentration Dependent Diffusion,"* Bell Syst. Tech. J., 59, 1-41 (1980).

[29] U. Goesele, *"Current Understanding of Diffusion Mechanisms in Silicon,"* Semiconductor Silicon 1986, H. R. Huff, T. Abe, and B. Kolbesen, Eds., Electrochem. Soc., Pennington, NJ, 541-555 (1986).

[30] F. J. Morin and J. P. Maita, *"Electrical properties of Silicon Containing Arsenic and Boron,"* Phys. Rev., 96, 28 (1954).

[31] D. A. Antoniadis and I. Moskowitz, *"Diffusion of Substitutional Impurities in Silicon at Short Oxidation Times: An Insight into Point Defect Kinetics,"* J. Appl. Phys. 53 (10), 6788-6796 (1982).

[32] A. M. Lin, D. A. Antoniadis, and R. W. Dutton, *"The Oxidation Rate Dependence of Oxidation-Enhanced Diffusion of Boron and Phosphorus in Silicon,"* J. Electrochem. Soc., 128 (5), 1131-1137 (1981).

[33] S. T. Ahn, P. B. Griffin, J. D. Shott, J. D. Plummer, and W. A. Tiller, *"A Study of Silicon Interstitial Kinetics Using Silicon Membrane: Applications to 2D Dopant Diffusion,"* J. Appl. Phys. 62 (12), 4745-4755 (1987).

[34] Y. Shibata, S. Hashimoto, K. Taniguchi, and C. Hamaguchi, *"Oxidation Enhanced Diffusion of Phosphorus over a Wide Range of Oxidation Rates,"* J. Electrochem. Soc., 139 (1), 231-237 (1992).

[35] S. M. Hu, *"Formation of Stacking Faults and Enhanced Diffusion in the Oxidation of Silicon,"* J. Appl. Phys. 45 (4), 1567-1573 (1974).

[36] S. Mizuo and H. Higuchi, *" Retardation of Sb Diffusion in Si During Thermal Oxidation,"* Jap. J. Appl. Phys., 20 (4), 739-744 (1981).

[37] D. A. Antoniadis, *"Oxidation-Induced Point Defects in Silicon,"* J. Electrochem. Soc., 129 (5), 1093-1097 (1982).

[38] W. Frank, U. Goesele, H. Mehrer, and A. Seeger, *"Diffusion in Silicon and Germanium,"* G. E. Murch and A.S. Nowich, Eds., Diffusion in Crystalline Solids, Chapter 2, Academic Press, New York (1984).

[39] A. S. Grove, O. Leisteko, and C. T. Sah, *"Redistribution of Acceptor and Donor Impurities During Thermal Oxidation of Silicon,"* J. Appl. Phys. 35 (7), 2695-2397 (1964).

[40] T. Kato and W. Nishi, *"Redistribution of Diffused Boron in Silicon by Thermal Oxidation,"* Jpn. J. Appl. Phys., 3, 377 (1964).

[41] R. B. Fair and J. C. C. Tsai, *"Theory and Direct Measurement of Boron Segregation in $SiO_2$ during Dry, Near Dry, and Wet $O_2$ Oxidation,"* J. Electrochem. Soc. 125, 2050 (1978).

[42] S. P. Murarka, *"Diffusion and Segregation of Ion-Implanted Boron in Silicon in Dry Oxygen Ambient,"* Phys. Rev. B. 12, 2502 (1975).

[43] T. E. Seidel and A. U. MacRae, *"Some Properties of Ion Implanted Boron in Silicon,"* Trans. Met. Soc. AIME, 245, 491, (1969).

[44] A. E. Michel, W. Rausch, P. A. Ronsheim, and R. H. Kastl, *"Rapid Annealing and the Anomalous Diffusion of Ion Implanted Boron in Silicon,"* Appl. Phys. Lett. 50 (7), 416 (1987).

[45] M. Servidori, Z. Sonrek, and S. Solmi, *"Some Aspects of Damage Annealing in Ion-Implanted Silicon: Discussion in Terms of Dopant Anomalous Diffusion,"* J. Appl. Phys., 62 (5), 1723-1728 (1987).

[46] R. B. Fair, *"Point Defect Charge-State Effects on Transient Diffusion of Dopants in Si,"* J. Electrochem. Soc., 137 (2), 667-671 (1990).

[47] K. Cho, M. Numan, T. G. Finstad, and W. K. Chu, *"Transient Enhanced Diffusion during Rapid Thermal Annealing of Boron Implanted Silicon,"* Appl. Phys. Lett., 47 (12), 1321 (1985).

[48] R. Kalish, T. O. Segewick, S. Mader, and S. Shatas, *"Transient Enhanced Diffusion in Arsenic-Implanted Short Time Annealed Silicon,"* Appl. Phys. Lett. 44 (1), 107 (1983)}.

[49] N. E. B. Cowern, D. J. Godfrey, and D. E. Sykes, *"Transient Enhanced Diffusion of Phosphorus in Silicon,"* Appl. Phys. Lett., 49 (25), 1711 (1986).

[50] S. Solmi, F. Cembali, R. Fabbri, M. Servidori, and R. Canteri, *"Dependence of Anomalous Phosphorus Diffusion in Silicon on Depth Position of Defects Created by Ion Implantation,"* Appl. Phys. A 48, 255-260 (1989).

[51] M. D. Giles, *"Transient Phosphorus Diffusion Below the Amorphization Threshold,"* J. Electrochem. Soc., 138 (4), 1160-1165 (1991).

[52] M. D. Giles, *"Transient Phosphorus Diffusion from Silicon and Argon Implantation Damage,"* Appl. Phys. Lett. 62 (16), 1940 (1993).

[53] H. Park and M. E. Law, *"Effects of Low-Dose Si Implanted Damage on Diffusion of Phosphorus and Arsenic in Si,"* Appl. Phys. Lett., 58 (7), 732 (1991).

[54] R. Angelucci, F. Cembali, P. Negrini, M. Servidori, and S. Solmi, *"Temperature and Time Dependence of Dopant Enhanced Diffusion in Self-Ion Implanted Silicon,"* J. Electrochem. Soc., 134 (12), 3130-3134 (1987).

[55] T. O. Sedgwick, A. E. Michel, S. A. Cohen, V. R. Deline, and G. S. Oehrlein, *"Investigation of Transient Diffusion Effects in Rapid Thermally Processed Ion Implanted Arsenic in Silicon,"* Appl. Phys. Lett., 47 (8), 848 (1985).

[56] Y. Kim, H. Z. Massoud, and R. B. Fair, *"The Effect of Ion-Implantation Damage on Dopant Diffusion in Silicon During Shallow-Junction Formation,"* J. Electron Mat., 18 (2), 143-150 (1989).

# 520

[57] M. Servidori, S. Solmi, O. Zaumseil, U. Winter, and M. Anderle, *"Interaction between Point Defects and Dislocation Lopps as the Phenomenon Able to Reduce Anomalous Diffusion of Phosphorus Implanted in Silicon,"* J. Appl. Phys., 65 (1), 98-104 (1989).

[58] E. Guerrero, H. Poetzl, G. Stingeder, M. Grassbauer, and K. Piplitz, *"Annealing of High Dose Sb-Implanted Single-Crystal Silicon,"* J. Electrochem. Soc., 132 (12), 3048-3052 (1985).

[59] H. Kinoshita, C. Q Lo, D. L. Kwong, and S. Novak, *"Diffusion Modeling of Ion Implanted Boron in Si during RTA: Correlation of Extended Defect Formation and Annealing with the Enhanced Diffusion of Boron,"* J. Electrochem. Soc., 140 (1), 248-252 (1993).

[60] S. K. Ghandhi, VLSI Fabrication Principles, John Wliey, New York (1983).

[61] D. Shaw, *"Self- and Impurity Diffusion in Ge and Si,"* Phys. Stat. Solidi, B72, 11 (1975).

[62] A. F. W. Willoughby, *"Double-Diffusion Process in Silicon,"* in Impurity Doping Processes in Silicon, F. F. Y. Wang, Ed., Chapter 1, North-Holland, Amsterdam (1981).

[63] D. Mathiot, and J. C. Pfister, *"High Concentration Diffusion of P in Si: a Percolation Problem,"* J. Phys. Lett. (Paris), 43 (12), L453-L459 (1982)..

[64] R. Armigliato, D. Nobili, P. Ostoja, M. Servidori, and S. Solmi, *" Solubility and Precipitation of Boron in Silicon and Supersaturation Resulting by Thermal Predeposition,"* in Semiconductor Silicon, H. R. Huff and E. Sirtl, Eds., Electrochem. Soc., Pennington, New Jersey, pp. 638-647 (1977).

[65] D. Nobili, R. Armigliato, M. Finetti, and S. Solmi, *"Precipitation as the Phenomenon Responsible for the Electrically Inactive Phosphorus in Silicon,"* J. Appl. Phys. 53 (3), 1484-1491 (1982).

[66] H. Ryssel, K. Muller, K. Haberger, R. Henkelmann, and F. Jahael, *"High Concentration Effects of Ion Implanted Boron in Silicon,"* Appl. Phys., 22, 35 (1980).

[67] S. M. Hu, *"Diffusion in Silicon and Germanium,"* in Atomic Diffusion in Semiconductors, D. Shaw, Ed., Plenum Press, London, pp. 217-350 (1973).

[68] T. L. Chui and H. N. Ghosh, *"Diffusion Models for Arsenic in Silicon,"* IBM J. Res. Dev., 11, 472 (1971).

[69] M. Y. Y. Tsai, F. F. Morehead, J. E. E. Baglin, and A. E. Michel, *"Shallow Junctions by High Dose As Implanted in Si: Experiments and Modeling,"* J. Appl. Phys., 51 (6), 3230-3235 (1980).

[70] D. Nobili, A. Carabelas, G. Celotti, and S. Solmi, *"Precipitation as the Phenomenon Responsible for the Electrically Inactive Arsenic in Silicon,"* J. Electrochem. Soc. 130 (4), 922-928, (1983).

[71] E. Guerrero, H. Poetzl, R. Tielert, M, Grassbauer, and G. Stingeder, *" Generalized Model for the Clustering of As Dopants in Si,"* J. Electrochem. Soc. 129 (8), 1826-1831, (1982).

[72] J. M. Fairfield and B. J. Master, *"Self-Diffusion in Intrinsic and Extrinsic Silicon,"* J. Appl. Phys., 38 (10), 3148-3154 (1967).

[73] G. L. McVay and A. R. Du Charme, *"The Diffusion of $^{71}Ge$" in Si and Si-Ge Alloys: A Study of Self-Diffusion Mechanisms,"* Inst. Phys. Conf. Ser. 23, 91-102 (1975).

[74] G. Hettich, H. Mehrer, and K. Maier, *"Tracer Diffusion of $^{71}Ge$ and $_{31}Si$ in Intrinsic and Doped Silicon,"* Inst. Phys. Conf. Ser., 46, 500-507 (1979).

[75] S. Matsumoto, Y. Ishikawa, Y. Shirai, S. Sekine, and T. Nimii, *"Concentration Dependence of the Diffusion Coefficient of Boron in Silicon,"* Jap. J. Appl. Phys., 19 (1), 217-218 (1980).

[76] R. F. Lever, B. Garben, C. M. Hisieh, and W. A. Orr Arienzo *"Diffusion of Boron from Polysilicon at High Concentration,"* in Impurity Diffusion and Gettering in Silicon, R. B. Fair, C. W. Pearse, and J. Washburn, Eds., Mat. Res. Soc. Symp. Proc., Vol. 36, pp. 95-100, Pittsburgh, PA (1985).

[77] S. T. Dunham, *"A Quantitative Model for the Coupled Diffusion of Phosphorus and Point Defects in Silicon,"* J. Electrochem. Soc., 139 (9), 2628-2636 (1992).

[78] K. Nishi and D. A. Antoniadis, *"Observation of Silicon Self-Interstitial Supersaturation during Phosphorus Diffusion from Growth and Shrinkage of Oxidation-Induced Stacking Faults,"* J. Appl. Phys., 59 (4), 1117-1124 (1986).

[79] R. B. Fair, *"Boron Diffusion in Silicon - Concentration and Orientation Dependence, Background Effects, and Profile Estimation,"* J. Electrochem. Soc., 122 (6), 800-805 (1975).

[80] C. P. Ho and S. E. Hansen, Technical Report N0 SEL 83-001, Stanford University, Stanford, California (1983)

[81] G. A. Sai-Halasz, K. T. Short, and J. S. Williams, *"Antimony and Arsenic Segregation at Si-SiO₂ Interfaces,"* Electron Dev. Lett., EDL-6 (6), 285 (1985).

[82] G. A. Sai-Halasz and H. B. Harrison, *"Device-Grade Ultra-Shallow Junctions Fabricated with Antimony,"* Electron Dev. Lett., EDL-7 (9), 534 (1986).

[83] S. Solmi, F. Baruffaldi, and M. Derdour, *"Experimental Investigation and Simulation of Sb Diffusion in Si,"* J. Appl. Phys., 71 (2), 697-703 (1992).

[84] R. B. Fair, M. L. Manda, and J. J. Wortman, *"The Diffusion of Antimony in Heavily Doped N- and P-Type Silicon,"* J. Mater. Res. 1 (5), 705-711 (1986).

[85] N. Mallam, C. L. Jones, and A. F. W. Willoughby, *"Modeling of Double-Diffusion Processing,"* in Semiconductor Silicon, H. R. Huff, R. Kriegler, and Y. Takeishi, Eds., Electrochem. Soc., Pennington, New Jersey, 979-987 (1981).

[86] S. M. Hu and S. Schmidt, *"Interactions in Sequential Diffusion Processes in Semiconductors,"* J. Appl. Phys., 39, 4272-4283 (1968).

[87] A.F.W. Willoughby, A. G. R. Evans, P. Champ, K. J. Yallup, D. J. Godfrey, and M. D. Dowsett, *"Diffusion of Boron in Heavily Doped N- and P-Type Silicon,"* J. Appl. Phys. 59 (7), 2392-2397 (1986).

[88] T. Chen, C. T. Chuang, G. L. Li, S. Basavaiah, D. D. Tang, M. B. Ketchen, T. H. Ning, *"An Advanced Bipolar Transistor with Self-Aligned Ion-Implanted Base and W/Poly Emitter,"* IEEE Trans. Electron Devices, ED-35 (8), 1322-1327 (1988).

[89] M. K. Moravvej-Farshi and M. A. Green, *"Novel Self-Aligned Polysilicon-Gate MOSFETs with Polysilicon Source and Drain,"* Solid-State Electronics, 30 (10), 1053-1062 (1987).

[90] C. S. Oh and C. K. Kim, *"A New MOSFET Structure with Self-Aligned Polysilicon Source and Drain Electrodes,"* IEEE Electron Device Lett., EDL-5, 400 (1984).

[91] G. P. Burbuscia, G. Chin, R. W. Dutton, T. Alvarez, and L. Arledge, *"Modeling of Polysilicon Dopant Diffusion for Shallow Junction Bipolar Technology,"* IEDM Tech. Dig., 757-760 (1984).

[92] Y. Yamaguchi, Y.-C. S. Yu, E. E. Lane, J. S. Lee, E. E. Patton, R. D. Herman, D. R. Ahrendt, V. F. Drobny, T. H. Yuzuriha, and V. E. Garuts, *"Process and Device Performance of a High-Speed Double Poly-Si Bipolar Technology Using Borosenic-Poly Process with Coupling-Base Implant,"* IEEE Trans. Electron Dev., ED-35 (8), 1247-1255 (1988).

[93] W. A. Rausch, R. F. Lever, and R. H. Kastl, *"Diffusion of Boron in Polycrystalline Silicon from a Single-Crystal Source,"* J. Appl. Phys., 54 (8), 4405-4407 (1983).

# 522

[94] W. J. M. J. Josquin, P. R. Boudewijn, and Y. Tamminga, *"Effectiveness of Polycrystalline Silicon Source Diffusion,"* Appl. Phys. Lett., 43 (10), 960 (1983).

[95] H. Schaber, R. v. Criegern, and I. Weitzel, *"Analysis of Polycrystalline Silicon Diffusion Sources by Secondary Ion Mass Spectroscopy,"* J. Appl. Phys., 58 (11), 4036-4042 (1985).

[96] B. Garben, W. A. Orr-Arienzo, and R. F. Lever, *"Investigation of Boron Diffusion from Polycrystalline Silicon,"* J. Electrochem. Soc., 120 (10), 2152-2156 (1986).

[97] V. Probst, H. J. Boehm, H. Schaber, H. Oppolzer, and I. Weitzel, *"Analysis of Polycrystalline Silicon Sources,"* J. Electrochem. Soc., 135 (3), 671-676 (1988).

[98] T. H. Ning and R. D. Isaac, *"Effect of Emitter Contact on Current Gain of Silicon Bipolar Devices,"* IEEE Trans. Electron Dev., ED-27, 2051-2055 (1980).

[99] T. H. Ning, R. D. Isaac, P. M. Solomon, D. D. Tang, H. Yu, G. C. Feth, and S. K. Wiedmann, *"Self-Aligned Bipolar Transistors for High Performance and Low Power-Delay VLSI,"* IEEE Trans. Electron Dev., ED-28, 1010-1013 (1981).

[100] A. Cuthberson and P. Ashburn, *"Self-Aligned Transistors with Polysilicon Emitters for Bipolar VLSI,"* IEEE Trans. Electron Dev., ED-32, 242-247 (1985).

[101] G. L. Patton, J. C. Bravman, and J. D. Plummer, *"Physics, technology, and Modeling of Polysilicon Emitter Contacts for VLSI Bipolar Transistors,"* IEEE Trans. Electron Dev., ED-33 (11), 1754-1768 (1986).

[102] J. N. Burghartz, J. H. Comfort, G. L. Patton, J. D. Cressler, B. S. Meyerson, J. M. C. Stork, J. Y.-C. Sun, G. Scilla, J. Warnock, B. J. Ginsberg, K. Jenkins, K.-Y. Toh, D. L. Harame, and S. R. Mader, *"Sub-30ps ECL Circuits Using High-$f_T$ Si and SiGe Epitaxial Base SEEW Transistors,"* IEDM Tech. Dig., 297-300 (1990).

[103] S. Kimura and E. Takeda, *"Elevated Source and Drain 0.1 μm-MOSFET,"* Oyo Butsuri, Jap. Appl. Phys., 61 (11), 1143-1145 (1992).

[104] S. Kimura, H. Noda, D. Hisamoto, and E. Takeda, *"A 0.1 μm-Gate Elevated Source and Drain MOSFET Fabricated by Phase-Shifted Lithography,"* IEDM Tech. Dig., 950 (1991).

[105] M. Rodder and D. Yeakley, *"Raised Source/Drain MOSFET with Dual Sidewall Spacers,"* IEEE Electron Device Lett., EDL-12 (3), 89 (1991).

[106] T. Yamada, S. Samata, H. Takato, Y. Matsushita, K. Hieda, A. Natayama, F. Horiguchi and F. Masuoka, *"Spread Source/Drain (SSD) MOSFET Using Selective Silicon Growth for 64Mbit DRAMs,"* IEDM Tech. Dig., 35 (1989).

[107] J. C. C. Tsai, VLSI Technology, Chapter 5, S. M. Sze, Ed., McGraw-Hill, 1983.

[108] B. Swaminathan, K. C. Saraswat, R. W. Dutton, and T. I. Kamins, *"Diffusion of Arsenic in Polycrystalline Silicon,"* Appl. Phys. Lett., 40 (9), 795 (1982).

[109] S. Horiuchi and R. Blanchard, *"Boron Diffusion in Polycrystalline Silicon Layer,"* Solid-State Electronics, 18, 529-532 (1975).

[110] L. Mei and R. L. Dutton, *"A Process Simulation Model for Multilayer Structures Involving Polycrystalline Silicon,"* IEEE Trans. Electron Devices, ED-29 (11), 1726-1734 (1982).

[111] A. D. Buonaquisti, W. Carter, and P. H. Holloway, *"Diffusion Characteristics of Boron and Phosphorus in Polycrystalline Silicon,"* Thin Solid Films, 100, 235-248 (1983).

[112] T. I. Kamins, J. Manolin, and R. N. Tucker, *"Diffusion of Impurities in Polycrystalline Silicon,"* J. Appl. Phys., 43, 83 (1972).

[113] K. Suzuki, T. Fukano, and Y. Kataoka, *"Constant Boron Concentration Diffusion Source of Ion-Implanted Polycrystalline Silicon for Forming Shallow Junctions in Silicon,"* J. Electrochem. Soc., 138 (7), 2201-2205 (1991).

[114] I. R. C. Post and P. Ashburn, *"Investigation of Boron Diffusion in Polysilicon and its Application to the Design of P-N-P Polysilicon Emitter Bipolar Transistors with Shallow Emitter Junctions,"* IEEE Trans. Electron Devices, ED-38 (11), 2442-2451 (1991).

[115] M. M. Mandurah, K. C. Saraswat, C. R. Helms, and T. I. Kamins, *"Dopant Segregation in Polycrystalline Silicon,"* J. Appl. Phys., 51, 5755 (1980).

[116] W. K. Schubert, *"Properties of Furnace-Annealed, High-Resistivity, Arsenic-Implanted Polycrystalline Silicon Films,"* J. Mater. Res., 1, 311 (1986).

[117] A. Almagoussi, J. Sicart, J. L. Robert, G. Chaussemy, and A. Laugier," *"Electrical properties of Highly Boron-Implanted Polycrystalline Silicon after Rapid or Conventional Thermal Annealing,"* J. Appl. Phys. 66, 4301 (1989).

[118] J. E. Suarez, B. E. Johnson, and B. El-Kareh, *"Thermal Stability of Polysilicon Resistors,"* IEEE Trans. Comp. Manuf. Tech., 15 (3), 386-392 (1992).

[119] M. E. Cowher and T. O. Sedgwick, *"Chemical Vapor Deposited Polycrystalline Silicon,"* J. Electrochem. Soc., 119, 1565 (1972).

[120] A. L. Fripp and L. H. Stack, *"Resistivity of Doped Polycrystalline Films,"* J. Electrochem. Soc., 120, 145 (1973).

[121] C. R. M. Governor, P. E. Batson, D. A. Smith, and C. Y. Wong, *"Effect of Arsenic Segregation on the Electrical Properties of Grain Boundaries in Polycrystalline Silicon,"* J. Appl. Phys., 57 (2), 438-442 (1985).

[122] S. J. Krause, S. R. Wilson, W. M. Paulson, and R. B. Gregory, *"Grain Growth During Transient Annealing of As-Implanted Polycrystalline Silicon Films,"* Appl. Phys. Lett., 45, 778 (1984).

[123] A. G. O'Neill, C. Hill, J. King, and C. Please, *"A New Model for the Diffusion of Arsenic in Polycrystalline Silicon,"* J. Appl. Phys., 64 (1), 167-174 (1988).

[124] G. Gontrand, A. Merabet, S. Krieger-Kaddour, C. Dubois, and J. Vallard, *"Analysis of Rapid Thermal Annealing of Boron and Arsenic in Polysilicon Emitter Structures,"* J. Electron. Mat., 22 (1), 135-141 (1993).

[125] T. Matsuura, J. Murota, N. Mikoshiba, I. Kawashima, and T. Sawai *"Diffusion of As, P, and B from Doped Polysilicon through Thin $SiO_2$ Films into Si Substrates,"* J. Electrochem. Soc., 138 (11) 3474-3480 (1991).

[126] M. Ghezzo and D. M. Brown, *"Diffusivity Summary of B, Ga, P, As, and Sb in $SiO_2$,"* Films into Si Substrates," J. Electrochem. Soc., 120 146 (1973).

[127] R. N. Ghostagore, , *"Silicon Dioxide Masking of Phosphorus Diffusion in Silicon,"* Solid-State Electronics, 18, 399 (1975).

[128] Y. Wada and D. A. Antoniadis, *"Anomalous Arsenic Diffusion in Silicon Dioxide,"* J. Electrochem. Soc., 128, 1317 (1981)..

[129] J. P. Stagg, *"Drift Mobilities of Na$^+$ and K$^+$ Ions in $SiO_2$ Films,"* Appl. Phys. Lett., 31 (8), 532 (1977).

[130] E. H. Nicollian and J. R. Brews, MOS Physics and Technology, John Wiley, New York (1982).

[131] H. F. Wolfe, Silicon Semiconductor Data, p. 601, Pergamon Press, New York, 1969.

[132] C. T. Sah, H. Sello, and D. A. Tremere, *"Diffusion of Phosphorus in Silicon Oxide Films,"* J. Phys. Chem. Solids, 11, 288-298 (1959).

# 524

[133] S. Horiuchi and J. Yamaguchi, *"Diffusion of Boron in Silicon Through Oxide Layer,"* Jap. J. Appl. Phys. 1 (6), 314-323 (1962).

[134] A. S. Grove, O. Leistiko, and C. T. Sah, *"Diffusion of Gallium Through a Silicon Dioxide Layer,"* J. Phys. Chem. Solids, 25, 985-992 (1964).

[135] C. W. Wong and F. S. Lai, *"Ambient and Dopant Effects on Boron Diffusion in Oxides,"* Appl. Phys. Lett., 48 (24), 1658 (1986).

[136] C. W. Wong, J. W.-C. Sun, Y. Taur, C. S. Oh, R. Angelucci, and B. Davari, *"Doping of N + and P + Polysilicon in Dual-Gate CMOS Process,"* IEDM Tech. Digest, 238 (1988).

[137] M. Ghezo and D. M. Brown, *"Arsenic Glass Source Diffusion in Si and SiO$_2$,"* J. Electrochem. Soc. 120 (1), 110-116 (1973).

[138] T. E. Burgess, J. C. Baum, F. M. Fowkes, R. Holmstrom, an G. A. Shim, *"Thermal Diffusion of Sodium in Silicon Nitride Shielded Silicon Oxide Films,"* J. Electrochem. Soc. 116, 1005-1008 (1969).

[139] J. V. Dalton and J. Drobek, *"Structure and Sodium Migration in Silicon Nitride Films,"* J. Electrochem. Soc. 115, 865-868 (1968).

[140] M. Dydyk and K. Evans, *"Silicon Nitride Masking Integrity Characterization,"* J. Electrochem. Soc. 137 (12), 3882-3884 (1990).

[141] A Goetzberger and W. Shockley, *"Metal Precipitation in Silicon P-N Junctions,"* J. Appl. Phys., 31 (10), 1821-1824 (1960).

[142] A. Ourmazd and W. Schroeter, *"Phosphorus Gettering and Intrinsic Gettering of Nickel in Silicon,"* Appl. Phys. Lett., 45, 781 (1984).

[143] L. E. Katz, P. F. Schmidt, and C. W. Pearce, *"Neutron Activation Study of a Gettering Treatment for Czochralski Silicon Substrates,"* J. Electrochem. Soc., 128 (3), 620-624 (1981).

[144] T. E. Seidel, R. L. Meek and A. G. Cullis, *"Direct Comparison of Ion Damage Gettering and Phosphorus-Diffusion Gettering of Au in Si,"* J. Appl. Phys., 46, 600-609 (1975).

[145] L. Jastrebski, *"Origin and Control of Material Defects in Silicon VLSI Technologies: An Overview,"* IEEE Trans. Electron Dev., ED-29, 475 (1982).

[146] K. Hondo et al,, *"Breakdown in Silicon Oxides - Correlation with Cu Precipitates,"* Appl. Phys. Lett., 45, 270 (1984).

[147] E. R. Weber, *"Transition Metals in Silicon,"* Appl. Phys. A, 1-22 (1983).

[148] P. C. Parekh, *"Gettering of Gold and its Influence on Some Transistor Parameters,"* Solid-State Electronics, 13, 1401-1406 (1970).

[149] H. J. Rath, P. Stallhofer, D. Huber, P. Eichinger, and I. Ruge, *"Assessment of Metallic Trace Contaminants on Silicon Wafer Surfaces,"* in Impurity Diffuson and Gettering in Silicon, R. B. Fair, C. W. Pearse, and J. Washburn, Eds., Materials Res. Soc., 36, 13 (1985).

[150] W. R. Runyan and K. E. Bean, Semiconductor Integrated Circuit Processing Technology, Addison Wesley, New York (1990).

[151] L. Balsi et al., *"Heavy Metal Gettering in Silicon-Device Processing,"* J. Electrochem. Soc., 127, 164-169 (1980).

[152] H. J. Geipel and W. K. Tice, *"Reduction of Leakage by Implantation Gettering in VLSI Circuits,"* IBM J. Res. Dev., 24, 310 (1980).

[153] T. M. Buck et al., *"Gettering rates of Various Fast-Diffusing Metal Impurities at Ion-Damaged Layers on Silicon,"* Appl. Phys. Lett., 21, 485 (1972).

[154] T. W. Sigmon et al., *"Ion Implant Gettering of Gold in Silicon,"* J. Electrochem. Soc., 123, 1116-1117 (1976).

[155] M. R. Poponiak et al., *"Argon Implantation Gettering of Bipolar Devices,"* J. Electrochem. Soc., 124, 1802-1805 (1977).

[156] J. A. Topich, *"Reduction of Defects in Ion Implanted Bipolar Transistors by Argon Backside Damage,"* J. Electrochem. Soc., 128, 866-870 (1981).

[157] K. D. Beyer and T. H. Yeh, "Impurity Gettering of Silicon Damage by Ion Implantation Through $SiO_2$ Layers," J. Electrochem. Soc., 129, 2527-2530 (1982).

[158] Y. Hayafuji, T. Yanada, and Y. Aoki, *"Laser Damage Gettering and its Application to Lifetime Improvement in Silicon,"* J. Electrochem. Soc., 128, 1975 (1981).

[159] C. W. Pearce and V. J. Zalackas, *"A New Approach to Lattice Damage Gettering,"* J. Electrochem. Soc., 126, 1436-1437 (1979).

[160] P. M. Petroff, G. A. Rozgonyi, and T. T. Sheng, *"Elimination of Process-Induced Stacking Faults by Pre-Oxidation Gettering of Si Wafers,"* J. Electrochem. Soc., 123, 565 (1976).

[161] M. C. Chen and V. J. Silvestri, *"Post-Epitaxial Polysilicon and $Si_3N_4$ Gettering in Silicon,"* J. Electrochem. Soc. 129, 1294-1299 (1982).

[162] R. L. Meek, T. E. Seidel, and A. G. Cullis, *"Diffusion Gettering of Au and Cu in Silicon,"* J. Electrochem. Soc., 122 (6), 786-796 (1975).

[163] T. Kuroi, S. Komori, K. Fukumoto, Y. Mishiko, K. Tsukamoto, and Y. Akasaka, *"Proximity Gettering of Micro-Defects by High Energy Ion Implantation,"* Extended Abstracts of the 1991 International Conference on Solid State Devices and Material, Yokohama, 56-58 (1991).

[164] C. W. Pearce, J. L. Morre, and F. A. Stevie, *"Removal of Alkaline Impurities in a Polysilicon Gate Structure by Phosphorus Diffusion,"* J. Electrochem. Soc., 140 (5), 1409-1413 (1993).

[165] B. Hartiti, M. Hage-Ali, J. C. Muller, and P. Siffert, *"Rutherford Backscattering Analysis of Phosphorus Gettering of Gold and Copper,"* Appl. Phys. Lett., 62 (26), 3476 (1993).

[166] R. A. Craven, *"Oxygen Precipitation in Czochralski Silicon,"* in Semiconductor Silicon 1981, H. R. Huff and R. J. Krieger, Eds., pp. 254-271, Electrochem. Soc., Pennington, New Jersey (1981).

[167] R. A. Craven and H. W. Korb, *"Internal Gettering in Silicon,"* Solid-State Technology, p. 55, (July 1981).

[168] C. Y. Tan et al., *"Intrinsic Gettering by Oxide Precipitation Induced Dislocation in Czochralski Si,"* Appl. Phys. Lett., 30, 175 (1977).

[169] D. R. Young and C. M. Osburn, *"Minority Carrier Generation Studies in MOS Capacitors on n-type Silicon,"* J. Electrochem. Soc., 120, 1578-1581 (1973).

[170] P. D. Esqueda and M. B. Das, *"Dependence of Minority Carrier Bulk Generation in Silicon MOS Structures on HCl Concentration in an Oxidizing Ambient,"* Solid-State Electronics, 23, 741-746 (1980).

[171] M. Nakamura and N. Oi, *"A Study of Gettering Effect of Metallic Impurities in Silicon,"* Jap. J. Appl. Phys., 7, 512 (1968).

[172] S. P. Murarka, *"A Study of the Phosphorus Gettering of Gold in Silicon by the Use of Neutron Activation Analysis,"* J. Electrochem. Soc., 123, 765 (1976).

# Chapter 8

# Contact and Interconnect Technology

## With J. G. Ryan, IBM

## 8.0 Introduction

The preceding chapters described how to define p- and n-type doped regions in single crystal and polycrystalline material. Those regions are the basic elements of semiconductor devices that must be connected in a specific configuration to form the desired circuit. The circuit must also be accessible to the "outside world" through conducting pads for testing with metal probes and for bonding to metal pins to complete the packaged chip. While doped silicon and polysilicon conduct electricity, they are of limited use for interconnections, mainly because of their prohibitively large resistance and lack of interconnecting flexibility. Therefore, at least one low-resistance conductor film must be deposited and patterned to contact and interconnect the different regions on the chip. Several single- and multi-metal systems are available for this purpose. Because of its high conductivity, compatibility with a silicon-base technology, and low processing cost, aluminum is the most widely used interconnect material.

One simple metallization process sequence consists of covering the wafer with an insulator, patterning and etching contact openings in the insulator, and then depositing and defining an aluminum film to form both the contacts and interconnecting leads (Fig. 8.1). Aluminum, however, penetrates silicon to a certain depth that depends primarily on the metal composition and thermal cycle following metal deposition. As the minimum feature size is reduced and the device dimensions shrink, so does the junction depth, increasing the probability for the penetrating aluminum to reach the junction and cause excessive leakage. To avoid metal penetration, it is necessary to modify the contact metallurgy by, e.g., adding silicon to the metal, as discussed below.

Another concern with reducing the depth of the junction is the inevitable increase in its sheet and contact resistances. Metal silicides are frequently used to reduce these resistances. Barrier metals are deposited on silicides to protect the films and interfaces with silicon from diffusing species during subsequent processing.

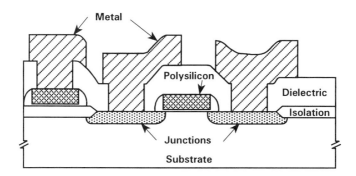

**Fig. 8.1** Simplified metallization process illustrating the definition of contacts and interconnections.

Reducing the minimum feature size is typically coupled with decreasing the depth of focus of the lithographic tools (Chap. 4). This sets a limit on the allowable wafer topography, requiring techniques to planarize (level) the wafer surface prior to contact and metal definition. Also, a concomitant increase in the achievable number of circuits per unit area necessitates the use of multilevel metallization (MLM) systems to increase the interconnecting flexibility, particularly in logic designs (Fig. 8.2).

The definition and performance of metallization systems are now dominant factors in determining circuit density and speed. Key features of a metallization system are the contacting scheme, the interlevel dielectrics, the interconnecting metals, and the reliability of the metallization system. The first part of this chapter deals with contact metallurgies and discusses the different materials and techniques used to form barriers and silicides. The second part describes materials used to insulate conductors and techniques to deposit and planarize the dielectrics. The deposition, definition and reliability of interconnecting metals are discussed in the third part.

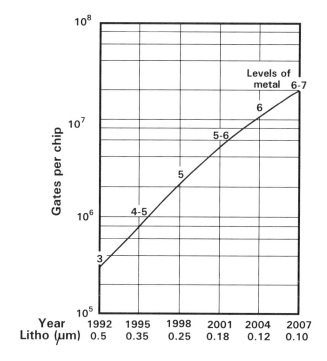

**Fig. 8.2** Projected circuit density and metal levels as a function of minimum feature size [1]

## 8.1 Contact Metallurgy

When contact is made between two dissimilar materials, a contact potential (barrier) typically develops at the interface between the materials. Contacts between metals and semiconductors can be categorized as non-rectifying (ohmic) and rectifying (Schottky-barrier type diodes). In an "ideal" ohmic contact, the barrier is "transparent" to current and a linear and symmetrical current-voltage characteristic with negligible resistance in both directions is displayed across the contact (Fig. 8.3). For this to occur, the semi-conductor surface concentration must be above $\simeq 10^{19}\,cm^{-3}$. Ohmic contacts are characterized by a specific contact resistance $R_c$ (in Ohm-cm²) that defines the average resistance across the interface between contacting materials. For concentrations $N$ above $\simeq 10^{19}\,cm^{-3}$, $R_c$ decreases roughly as $1/\sqrt{N}$. Surface dopant concentrations greater than $\simeq 10^{20}\,cm^{-3}$ are required to achieve sub-micron contacts of sufficiently low resistance. An "ideal" recti-fying contact allows the passage of current with little resistance in one direction but presents a barrier in the other direction (Fig. 8.4). For a rectifying contact to form, the semiconductor surface concentration must be less than $\simeq 5x10^{17}\ cm^{-3}$. A Schottky-barrier type contact is formed, for example, between aluminum or platinum silicide and n-type silicon having a surface concentration $N \leq 5x10^{17}\ cm^{-3}$.

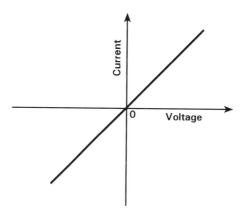

**Fig. 8.3** Linear and symmetrical current-voltage behavior of an ideal ohmic contact.

The contacting metal must exhibit good adhesion to the semiconductor, withstand temperature cycling without failure, allow high currents without changes in electrical properties, and not penetrate deep into the semiconductor, so that the current-voltage characteristics of contacted junctions are not affected by the metal. When defining a rectifying contact, the metal must also form a stable barrier that does not change with time. This section focuses on the formation of ohmic contacts to single-crystal and polysilicon. These contacts are required to interconnect semiconductor circuits and also make connections to the "outside world", using conducting lines and pads. One of the challenges in modern technologies is to make low-resistance and reliable ohmic contacts to semiconductor regions of sub-0.5 $\mu$m sizes without degrading contacted junctions. Considering that the number of such contacts approaches 100-500 million per chip, the task of making fault-free contacts in high-density product designs has become increasingly difficult as the contact size decreases, its aspect ratio increases, and the junction depth decreases.

## 8.1.1 The Aluminum-Silicon Contact

For junction depths larger than $\simeq 1$ $\mu m$ and surface concentration above $\simeq 10^{20}$ $cm^{-3}$, aluminum can be used to simultaneously contact junctions and provide low-resistance interconnects. This allows the deposition of both interconnect and contact metallurgies in one step. As the junction depth and contact size decrease, however, it becomes more difficult to form low-resistance aluminum contacts without causing excessive leakage due to metal spiking [2]. This section describes two limitations of an aluminum-base contact metallurgy.

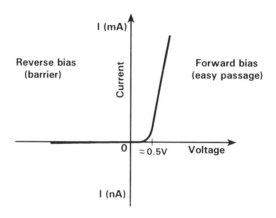

**Fig. 8.4** Current-voltage behavior of an ideal rectifying contact. Current passes only in one direction.

### Aluminum Spiking

After metal deposition and definition, a post-deposition heat treatment (sintering/annealing) is performed, typically at 400-450 °C for 20-30 min in forming gas (FG: 10% $H_2$, 90% $N_2$). The sintering step improves grain-growth in the metal, reduces any native or residual oxide between the two materials, and allows some interdiffusion and alloying to achieve good metal-to-semiconductor contacts. Another purpose of annealing is to reduce the density of surface states created, particularly during plasma processing steps (surface states are electronic "traps" that can degrade circuit performance and reliability). The aluminum-silicon phase diagram is shown in Fig. 8.5 [3]. Since the anneal temperature is below the eutectic point, aluminum forms a uniform alloy with silicon.

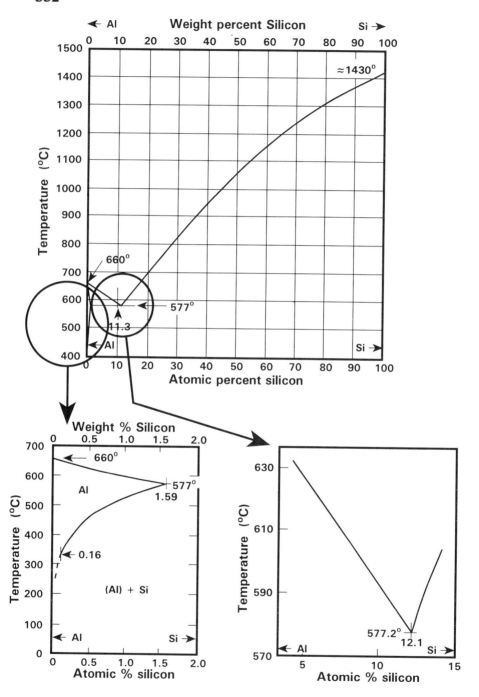

**Fig. 8.5** Aluminum-silicon phase diagram [3].

Because silicon has a finite solubility in aluminum, e.g., $\simeq 0.3\%$ at 400 °C [4], heat treatments result in dissolution of silicon, which tends to proceed more slowly along (111) than (110) or (100) planes. This creates voids in the crystal into which aluminum precipitates and forms conductive spikes that can cause high leakage or shorts in shallow junctions (Fig. 8.6). The net transport of material across such an interface is known as the **Kirkendall effect** (Sec. 8.6). Void formation is enhanced by the high diffusivity of silicon in aluminum (Fig. 8.7), allowing silicon to be be transported to a considerable distance in the metal during heat treatment [4]. This distance can be as large as 40 $\mu$m for a one-hour anneal at 400 °C. Void formation therefore depends on the amount of metal surrounding the junction.

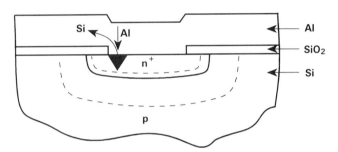

**Fig. 8.6** Illustration of aluminum spiking in a junction contact. Excessive junction leakage occurs when the spike approaches the space-charge region (dashed line).

A commonly used method to avoid aluminum spiking is to use an Al-Si alloy where the amount of silicon is in excess of the solubility limit, typically 1.0-1.5 wt%. This can be achieved by either covering the metal with a thin silicon film before sintering or by co-depositing the Al-Si alloy. When a top layer of Si is deposited, silicon diffuses into aluminum before a significant amount of silicon diffuses from the substrate into the metal during following heat treatment. Adding silicon to aluminum, however, can cause a substantial increase in contact resistance. Upon cooling, a solid solution of aluminum in silicon freezes out epitaxially on the silicon surface. The volume of silicon available for epitaxial growth is determined by the excess Si in the interconnect line within a diffusion length from the contact [2]. Since aluminum is an acceptor, a thin p-layer is formed on the silicon surface to which an alloy of aluminum-silicon is contacted. The recrystallized p-type film can cause non-ohmic behavior and a sharp increase in the effective resistance of contacts to n-type regions, particularly when the junction surface is not heavily doped and the contact size is reduced.

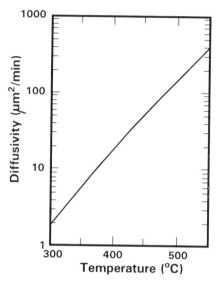

**Fig. 8.7** Diffusivity of silicon in deposited aluminum films [4].

## 8.1.2 Contact Materials and Barriers

In modern technologies, the junction depth is decreased considerably below 500 nm to increase circuit density and speed, and reduce power dissipation. This precludes the use of "pure" aluminum for contacts. Also, to achieve shallow junctions, the dopant concentration at the silicon surface is dropped below $\simeq 10^{20}\ cm^{-3}$. The combination of reduced surface concentration, junction depth and contact size results in prohibitively high contact resistances with aluminum-silicon metallurgies. Contact resistance and spiking problems can be solved by using appropriate contacting materials and barriers to inhibit silicon diffusion across the contact boundary. One attractive approach is to deposit a layered structure, such as Ti/Al-Si or Ti-W/Al-Si, as illustrated in Fig. 8.8 [2]. The presence of a titanium or tungsten barrier inhibits the diffusion of silicon from the Al-Si metal to the substrate. It also inhibits diffusion of silicon from the substrate into the metal, however, only efficiently when the barrier film is pinhole-free.

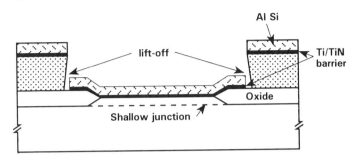

**Fig. 8.8** Barrier metals to inhibit silicon diffusion across the contact boundary [2]. As the resist is dissolved, it "lifts-off" the metal outside the contact (Sec. 8.3.2).

Another method for making contacts is to deposit and define an intermediate in-situ or ex-situ doped polysilicon film over contacts to junctions, forming "landing pads" to which the metal can be contacted and avoiding direct contact between metal and single-crystal silicon. While heavily-doped polysilicon can also be used for contacts and interconnects, its high resistivity compared to aluminum limits it to short interconnects and moderate current densities. Landing pads can also be formed by selectively growing silicon over exposed single-crystal silicon surfaces (Chap. 3), with the advantage of self-alignment but without the capability of using the film for short interconnections. In both cases, the polysilicon film can be used as a diffusion source to form underlying shallow junctions, and as a pad to form silicide (Fig. 8.9).

**Silicides**

Silicides for microelectronic applications are formed by reacting refractory or near noble metals with silicon. Among them are titanium silicide ($TiSi_2$), cobalt silicide ($CoSi_2$), tungsten silicide ( $WSi_2$), platinum silicide (PtSi), molybdenum silicide ($MoSi_2$), palladium silicide ($Pd_2Si$), and tantalum silicide ($TaSi_2$) [5 − 10]. Tables 8.1 and 8.2 compares some properties of important metals and silicides.

Silicide films in parallel with the higher-resistivity junction or polysilicon layers reduce series resistances in the different transistor regions. Another advantage of these contact metallurgies is

**536**

that they can withstand temperatures above 550 °C, such as PSG or BPSG reflow at 750-850 °C for planarization. Such temperatures are not compatible with aluminum. Because of their low resistivities and stability in contacts with polysilicon gates and junctions, $WSi_2$, $TiSi_2$ and $CoSi_2$ are commonly used silicides. Their properties have been extensively studied and reported, as detailed in a separate section.

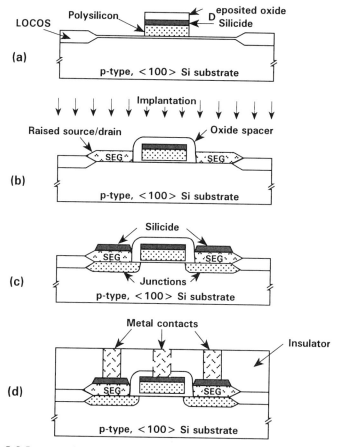

**Fig. 8.9** Intermediate polysilicon film to form contact to shallow and ultra-shallow junctions. (a) Defined polycide stack (b) Spacer and selective epitaxial growth (SEG) (c) Silicide formation. (d) Metal contacts to polysilicon and diffusions.

**Table 8.1** Some properties of metal films [11]

| Metal | Melting-Point (°C) | Resistivity ($\mu\Omega C$) |
|-------|-------------------|---------------------------|
| Al | 659 | 2.8 |
| Cu | 1083 | 1.7 |
| Au | 1063 | 2.4 |
| Pd | 1554 | 11 |
| Ti | 1670 | 55 |
| Co | 1497 | 12 |
| Cr | 1878 | 12 |
| Mo | 2615 | 5.7 |
| Pt | 1772 | 10 |
| W | 3417 | 5.5 |

**Table 8.2** Some properties of silicide films [11]

| Silicide | Typical forming temperature (°C) | Resistivity ($\mu\Omega C$) |
|----------|----------------------------------|----------------------------|
| $TiSi_2$ | 600-800 | 13-17 |
| $CoSi_2$ | 550-700 | 13-19 |
| $WSi_2$ | 900-1100 | 31 |
| $PtSi$ | 700-800 | 28-35 |
| $HfSi_2$ | 850-950 | 50 |
| $TaSi_2$ | 900-1100 | 35-55 |
| $MoSi_2$ | 900-1100 | 40-70 |
| $Pd_2Si$ | 400-500 | 30-35 |

To form the silicide selectively over exposed silicon regions, a refractory metal (such as Ti or Co) is first deposited over the entire wafer by sputtering, evaporation, or CVD (Chap. 3). For titanium, this is followed by a two-step furnace or rapid-thermal anneal. The first step consists of annealing at a temperature near 650 °C for about 20 minutes in a nitrogen atmosphere, during which nitrogen reacts with titanium to form titanium nitride (TiN) at the surface of the metal, while titanium reacts with silicon and forms silicide in those regions where it comes in direct contact with silicon [12]. At temperatures between 650-750 °C, a high-resistivity phase of $TiSi_2$ (called C49) forms more rapidly than TiN in the exposed silicon regions, while negligible reaction with silicon occurs where the metal is in contact with silicon dioxide or silicon nitride [13]. The metal composition in contact with insulators consists predominantly of TiN and unreacted Ti. It is removed with a peroxide solution, leaving silicided regions intact (Fig. 8.10). This results in self-alignment between silicide and exposed silicon regions. Typical etchants of "unreacted" metals are given in Table 8.3 [14]. If the polysilicon and diffusion patterns are both exposed to the metal, silicide forms simultaneously over both regions and the method is sometimes labeled as "salicide" to emphasize that silicides formed over polysilicon and single-crystal silicon are self-aligned to each other. The second annealing step is done typically at 800 °C for 15 minutes in forming gas to transform the silicide to its lower-resistivity phase (called C54) [13,14], as discussed later in this section.

**Table 8.3** Typical etchants of unreacted metals [14]

| Silicide | Metal etchant |
|---|---|
| $TiSi_2$ | $NH_4OH:H_2O_2$ ; piranha |
| $CoSi_2$ | $3HCl:H_2O_2$ ; piranha |
| PtSi | Aqua regia |
| $Pd_2Si$ | $KI:I_2$ |

For self-aligned silicides, the reaction between metal and silicon requires that silicon be supplied by the exposed substrate or polysilicon regions. With $TiSi_2$, silicon is the predominant diffusion species that moves through the growing silicide to react with the metal [15]. With $CoSi_2$, it is the metal that diffuses through the film toward the silicon-silicide interface [16]. Also, platinum and palladium form metal-rich or mono-silicides at temperatures near 600 °C and the metal rather than silicon diffuses through the growing silicide. In all cases silicon is consumed and the silicide penetrates into the regions both vertically and laterally. The theoretical ratio of silicide to metal thickness is 3.6 for $CoSi_2$ and 2.3 for $TiSi_2$. In practice, however, more silicon is consumed than theoretically predicted for $TiSi_2$ because of the enhanced diffusivity of silicon along grain boundaries of the silicide or metal and its subsequent reaction with the metal.

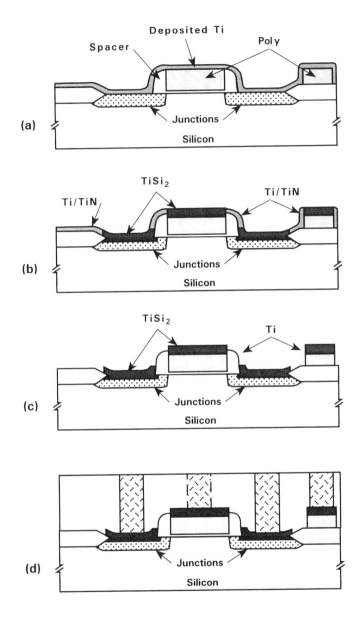

**Fig. 8.10** Schematic cross-section showing the self-aligned formation of silicides over exposed junctions and over polysilicon in a MOSFET, using TiSi₂ as an example. (a) Titanium deposition. (b) Silicide formation. (c) Removal of unreacted Ti. (d) Metal contacts to silicide.

In patterned geometries, lateral TiSi$_2$ overgrowth occurs as a result of the enhanced diffusion of silicon along grain boundaries. Therefore, when defining thick silicides, care must be taken to avoid excessive lateral overgrowth and shorting ("bridging") between polysilicon and single-crystal silicon along insulator spacers on polysilicon sidewalls. The enhanced diffusivity of silicon along grain boundaries also causes the silicide to be nonplanar and have a rough texture. Where lateral growth is confined, as in contact windows, expansion can create stresses high enough to cause the silicide to break away from the contact [17]. Lateral overgrowth and the associated "bridging" are much less severe with CoSi$_2$ because Co rather than Si is the predominant diffuser.

Several process integration issues must be considered when defining the silicide composition and thickness for a self-aligned contact metallurgy. Titanium forms compounds with boron, arsenic and phosphorus, reducing the dopant concentration at the silicide-silicon interface [13, 14]. Also, as the silicide penetrates deeper, its boundary moves into regions of lower concentration. The combination of dopant reaction with titanium and moving boundary results in an increase in contact and series resistances [18, 19]. If the silicide boundary approaches the metallurgical junction, laterally or vertically, high leakage currents can result, degrading the junction characteristics and increasing circuit power dissipation [18, 20]. Deep silicide asperities in polysilicon can severely reduce the breakdown field of underlying MOSFET gate dielectrics.

Another important factor is dopant segregation during silicide formation. For TiSi$_2$ and CoSi$_2$, the rapid diffusion of dopants along silicide grain boundaries causes segregation of dopants into the silicide and their depletion from the junction surface, further increasing contact and sheet resistances. In contrast, when PtSi or Pd$_2$Si is formed over As-doped regions, arsenic is driven ahead of Pt or Pd, resulting in arsenic pile-up at the silicide-silicon interface [21]. The rapid diffusion of dopants along grain boundaries in titanium and cobalt silicides can create problems when silicided n- and p-type regions are in close proximity to each other, because the diffusion of dopants of opposite polarities and their cancellation can cause severe changes in the net dopant concentration in each region [22]. This "inter-diffusion" depends strongly on the thermal budget that follows silicide formation. In practice, silicided regions of opposite dopant polarity are kept typically 1-3 $\mu$m apart to minimize the effect of inter-diffusion. Lowering the thermal budget by rapid thermal annealing minimizes dopant segregation and inter-diffusion.

The diffusion of impurities in silicon is also affected by the silicide process. The formation of titanium or cobalt silicide is accompanied by generation of excess point defects that alter the diffusivity of dopants in silicon (Chap. 7) [14, 23 − 27]. For example, the diffusion of silicon into $TiSi_2$ creates an excess in vacancies, enhancing the diffusivity of antimony and retarding the motion of boron in silicon [23,24]. Injection of vacancies during $TiSi_2$ silicidation also results in the dissolution of end-of-range defects (interstitial dislocation loops) caused by ion implantation, improving junction leakage [28].

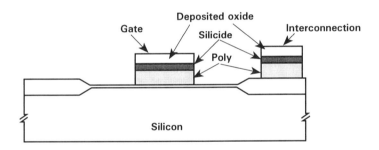

**Fig. 8.11** Silicide formed on polysilicon by co-depositing a refractory metal with silicon. The film is covered by an insulator and then patterned.

When used only for contacts, silicide films are formed very thin, in the range 30-100 nm [29, 30]. A rule-of-thumb is to keep the thickness of silicon consumed smaller than half the junction depth to avoid junction leakage. As the silicide thickness is reduced, however, the film's thermal stability greatly degrades and its sheet resistance increases [23]. Thin films begin to agglomerate during subsequent thermal cycles (such as used for PSG, BPSG reflow), starting with minor notching at grain boundaries which later extend throughout the entire silicide layer and finally result in disconnected grains. Agglomeration of $TiSi_2$ and $CoSi_2$ is more pronounced in narrow polysilicon and single-crystal silicon lines, limiting the effectiveness of silicides in reducing the sheet resistance of narrow lines [31 − 34]. The silicide integrity can be improved by reducing the thermal budget with RTA. Also, the thermal stabilities of TiSi$_2$, CoSi$_2$, and PtSi can be significantly improved by implanting nitrogen or fluorine into the film [35, 36]. This is attributed to the presence of fluorine or nitrogen at the grain boundaries and silicide-silicon interface [14, 23].

Silicides can also be formed by co-depositing the metal with silicon, so that little or no substrate silicon is consumed during the reaction. Co-deposition can be performed from two independent sources by sputtering or CVD, or sputtered from a hot-pressed alloy target. These films are amorphous as deposited and have a higher sheet resistance than annealed films. Therefore, deposition is typically followed by furnace or rapid thermal annealing that allows the diffusion of the components to homogenize the composition and crystallize the film into the desired high-conductivity silicide phase. Annealing also permits sufficient diffusion, using the silicon substrate as an extra source or sink for silicon, to give a single-phase stoichiometric film on top of the substrate [14]. If the ratio of metal to silicon is correct, no silicon is consumed. If there is a deficiency of silicon, it will be corrected by diffusion of small amounts of silicon from the contact window.

When co-deposited, silicide is formed over the entire wafer and a masking step is required to pattern the film. In typical "polycide" processes, a refractory metal (such as tungsten) is co-deposited with silicon on top of polysilicon, and the electrode stack annealed and reactively ion etched in a multi-step process (Fig. 8.11). This allows the formation of silicide over polysilicon with a different composition and thickness than over junctions.

Co-deposition is also used to form silicides over insulators to define "local" (typically short) interconnections between circuit elements, e.g., between gate and source or drain of a MOSFET, with one additional masking step. In one variation of this process, an amorphous silicon film is deposited and patterned on top of a refractory metal layer [37, 38]. The deposited silicon film then provides the material to form silicide over insulator regions. Another method to form local interconnections is to use the titanium nitride film that is formed during annealing of Ti in a nitrogen atmosphere. Typical TiN films have resistivities ranging from 60-200 $\mu\Omega$-cm and can be patterned over insulators for short interconnections [39 − 41].

Some silicides (such as PtSi) also make good Schottky diodes when contacting high-resistivity silicon [42, 43]. Thus, if windows are opened for both the Schottky diodes and the ohmic contacts, both structures can be made at the same time.

Regardless of the method used to deposit and form the silicide, a lower-resistivity metal (typically aluminum) must be deposited and delineated to make contacts to the silicided regions and to form long-lead interconnections, as discussed in Sec. 8.3.

### Properties of Titanium and Cobalt Silicides

The silicides of titanium and cobalt exhibit faceted surface morphologies indicative of polycrystalline growth. As the anneal temperature is increased, they form intermediate crystallographic phases before the high-conductivity di-silicide is achieved. For example, at temperatures below $\simeq$ 700 °C, $TiSi_2$ forms in a predominantly metastable phase, called C49, with a resistivity of about 60 $\mu\Omega$-cm. At $\simeq$ 800 °C, the silicide is almost completely transformed into its so-called C54 stable phase having a resistivity $\simeq$ 16 $\mu\Omega$-cm [44, 45]. The terms C49 and C54 are names given to describe the crystallographic arrangement in grains. C49 has a base-centered orthorhombic arrangement with 12 atoms/cell, and a mean crystallite size of 20 nm. C54 has a face-centered orthorhombic arrangement with 24 atoms/cell, and a mean crystallite size that depends on film thickness and forming conditions (typically 50 nm for an 80-nm film [46]). The phase transformation can be observed by measuring the change in sheet resistance, but other non-destructive in-situ and ex-situ techniques have also been used. Among them are picosecond ultrasonics measurements [47], ellipsometry [48], Auger electron spectroscopy (AES) and low energy electron diffraction (LEED) [49]. These and other characterization techniques are discussed in a separate chapter.

The dependence of sheet resistance on film thickness is approximated by

$$R_S \simeq \frac{\rho}{t - t_o} ,$$
8.1

where $\rho$ is the differential resistivity obtained from a plot of sheet resistance versus thickness, $t$ is the film thickness, and $t_o$ is the thickness of an initial film which contributes to the total thickness but not to the total conductance [12]. For Ti, Co, and their silicides, the values for $\rho$ and $t_o$ are [12]

| Parameter | Ti | TiSi$_2$ | Co | CoSi$_2$ |
|---|---|---|---|---|
| $\rho$ ($\mu\Omega - cm$) | 58 | 17 | 12 | 14 |
| $t_o$ ($nm$) | 4 | 8.3 | 4.8 | 13 |

**Implantation after Silicide Formation**

In "conventional" silicides, the metal is deposited after the doped regions have been implanted and annealed. As discussed earlier, this method presents serious problems when the vertical and horizontal transistor geometries are reduced. To overcome these limitations, particularly problems caused by silicon consumption, several variations of the silicide process have been developed to define the junction after the silicide or metal is in place. Three common processes use implantation of dopants through the metal (ITM) [50 – 53], implantation through silicide (ITS) [54 – 57], and complete implantation into the silicide and its use as a solid diffusion source (SADS, Fig. 8.12) [14, 23, 54, 57]. In all cases, a thermal cycle follows implantation to diffuse and activate dopants, form silicides, remove implantation damage, or a combination of the above. The use of silicide diffusion sources has gained considerable attention because it offers several potential advantages [58]. Among them are the confinement of implantation damage to the silicide film rather than the junction and the elimination of ion channeling. This eliminates point defects in the substrate that can accelerate dopant diffusion, extended defects that contribute to junction leakage, and the uncertainties in the impurity profile due to channeling. For the silicides of most interest, at least one of the common dopants is slow within the silicides, making it difficult to use one single film as a solid source for both dopant types.

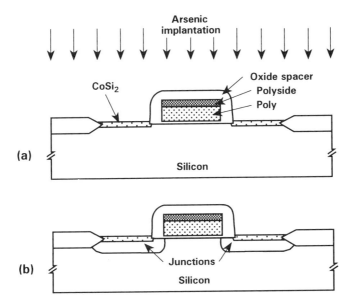

**Fig. 8.12** Junction definition after silicide formation by implanting into silicide with subsequent drive-in.

The diffusivity of boron in $CoSi_2$, for example, is considerably higher than in $TiSi_2$. It can be approximated by [59]

$$D_{Co, \text{bulk}} \simeq 0.09\, e^{-2.1(eV)/kT} \quad (cm^2/s) \, . \qquad 8.2$$

The opposite is true for arsenic which diffuses much slower in $CoSi_2$ than in $TiSi_2$. A similar problem is found when using $MoSi_2$ as a diffusion source for arsenic and boron. The high arsenic diffusivity and solubility in $MoSi_2$ and the favorable segregation of arsenic at the silicide-silicon interface make Mo silicide suitable as arsenic diffusion source and allow the formation of $n^+/p$ junctions as shallow as 100 nm with high surface carrier concentration [60]. The low boron mobility in $MoSi_2$ and $TiSi_2$, probably due to boron-molybdenum and boron-titanium compounds, make these silicides ineffective as boron diffusion sources. In practice, the rapid grain boundary diffusion of dopants is responsible for the transport of most dopant atoms from within the silicide to the surface of the silicon substrate and its redistribution.

Despite the advantages with silicides as solid diffusion sources, several problems inherent with silicides, particularly agglomeration over narrow lines, remain unsolved and have lead to the development of alternative techniques.

## 8.2 Poly-Metal Dielectric

To isolate the first metal from the substrate, a dielectric is deposited, typically planarized, and then patterned to define openings for contacts to silicon and polysilicon. This deposited insulator is commonly referred to as poly-metal dielectric (PMD). The film can consist of composite materials that must satisfy several requirements to meet the performance, density, and reliability objectives of the circuit. An ideal insulator must be contamination- and defect-free, exhibit a dielectric constant that approaches unity, a sufficiently high field strength, and infinite etch selectivity to underlaying materials (such as silicides, silicon, and polysilicon). Also, it must conformably cover steps and fill sub-0.5 $\mu$m gaps on the substrate surface, exhibit excellent adhesion to the substrate and the overlaying metal, present an infinite barrier to ionic contaminants, be easily planarized to eliminate topographies at first metal, and not react with the metal. In practice, however, dielectric properties are not ideal and a trade-off is made between there properties, depending on technology and application.

The most commonly used dielectrics for PMD are undoped silicon dioxide, doped glasses, and silicon nitride. Methods to deposit these insulators are detailed in Chap. 3. This section discusses the PMD composition and techniques to planarize its surface and define contacts to first metal.

### 8.2.1 Dielectric Composition

A typical PMD composition consists of a thin (10-50 nm) film of silicon nitride, a thick (500-1000 nm) film of phosphosilicate glass (PSG) or boro-phosphosilicate glass (BPSG) that can be densified and reflowed at 750-850 °C, and an undoped silicon dioxide film (typically a TEOS-based oxide) that covers the doped glass. The nitride film acts as an oxidation barrier. It also serves as a barrier to contaminants and can be used as an etch-stop during contact definition. PSG is typically doped with phosphine ($PH_3$) or trimethyphosphate (TMP), and BPSG with trimethylborate (TMP). Germanium doping (5-10%), combined with the more conventional boron and phosphorus doping, is being explored as one approach to reduce the planarization temperature to $\leq 750\,°C$ [61]. The reflowed doped glass conformably covers steps, fills gaps between polysilicon lines, and forms a nearly planar surface [62]. PSG films also prevent sodium ions from migrating to the substrate surface by trapping the ions in their network. PSG, however, absorbs water and, when $H_3PO_4$ forms on the glass, it attacks aluminum. Corrosion of the metal can be minimized by reducing the PSG phosphorus content to less than 6 wt%, however, at the cost of increasing the reflow temperature. When silicides are present, this temperature may not exceed 750 °C. The main purpose of the top undoped oxide is to separate the metal from the doped glass and protect it from corrosion.

## 8.2.2 Planarization Techniques

While reflowed glass reduces the insulator topography, the resulting "planarity" is not adequate for sub-0.5 $\mu$m contacts and metal pitches. For such features, lithography and etch requirements necessitate additional planarizing processes to reduce or eliminate height non-uniformities at the surface. One technique consists of depositing photoresist onto the glass and etching the resist and glass at nearly the same rate [62]. The resist fills gaps and steps and "flows" into a rather flat surface so that etching results in a flat glass surface, as illustrated in Fig. 8.13. An undoped oxide film is then deposited on top of the planar glass. Another method uses multiple deposition/etching steps of undoped glass, improving the topography in each step.

### Chemical-Mechanical Polishing

A common planarizing technique is chemical-mechanical polishing (CMP). This method is used to achieve a planar surface over the entire chip and wafer, referred to as "global planarity". The process of chemical-mechanical polishing is shown in Fig. 8.14. It consists of a rotating table that holds the wafer, an appropriate slurry, and a polishing pad that is applied to the wafer at a specified pressure. CMP is not limited to dielectrics. It is used to planarize deep and shallow trenches filled with polysilicon or oxide, and various metal films.

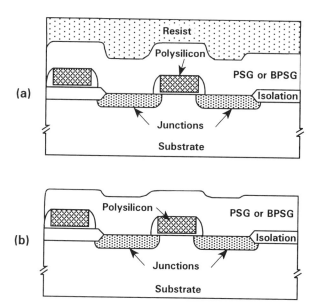

**(a)**

**(b)**

**Fig. 8.13** Planarizing technique by resist etch-back. (a) Leveled resist deposition. (b) After etch-back.

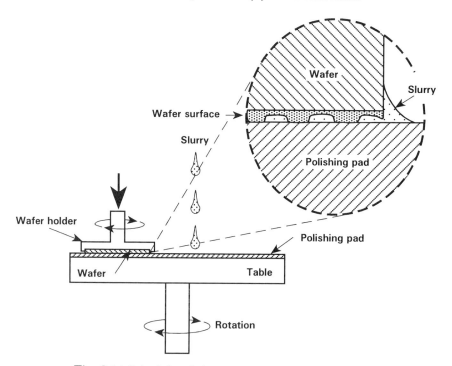

**Fig. 8.14** Principle of chemical-mechanical polishing.

Polishing results from a combination of chemical and mechanical effects. A suggested mechanism for CMP involves the formation of a chemically altered layer at the surface of the material being polished. This layer is mechanically removed from the underlying bulk material. An altered layer is then regrown on the surface, beginning the process again. For example, in $SiO_2$ polishing, the altered layer may be a hydrated oxide that can be mechanically removed or, for metal polishing, a metal oxide may be formed and removed.

The slurry composition and pad pressure determine the polishing rate. Oxide films, for example, polish twice as fast in a slurry with pH = 11 than with pH = 7. The hardness of the polishing particles should be about the same as the hardness of the film being polished to avoid damaging the film. The particle size should be uniform and the solution free of metallic contaminants. Slurry typically consists of an abrasive component and a component that chemically interacts with the surface. A typical oxide polishing slurry may consist of a colloidal suspension of oxide particles, with an average size of 30 nm, in an alkali solution (pH $\geq$ 10). A polishing rate of about 120 nm/min can be achieved with this solution. Ceria ($CeO_2$) suspensions are often used for glass polishing, particularly in the optical industry (where larger amounts of $SiO_2$ must be removed). $CeO_2$ acts as both the chemical and mechanical agent in the slurry. Very high polishing rates ($\geq$ 500 nm/min) are possible using $CeO_2$ slurry. Other abrasives, such as alumina $Al_3O_2$ may also be employed in some slurries. Figure 8.15 shows a schematic of the PMD surface before and after polishing.

A variety of polishing pads/cloths are available. They are typically grouped by their mechanical properties. Hard pads produce better planarity, while soft pads achieve better uniformity and less surface damage. The choice of pads is application dependent. For example, while soft pads are used for flat silicon substrates to avoid scratches, these pads are often not suitable for surfaces containing patterns.

Several methods to detect the polish end-point are being investigated. Some of them rely on the change in frictional forces between pad and polished surface. The most widely used method is, however, to measure the thickness of the polished film at

**550**

several intervals between polishing and determine the time needed to achieve the required polished thickness.

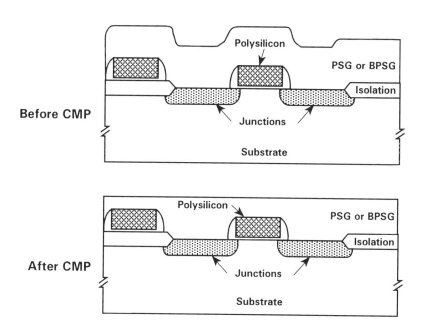

**Fig. 8.15** Poly-metal dielectric surface before and after polishing.

### 8.2.3 Contact Definition

Contacts to silicon or silicide are defined in the insulator using lithography and dry etching techniques discussed in Chaps. 4,5. Dry etching allows the definition of contact openings with high aspect ratio and vertical sidewalls. Since a planar insulator is thinner above polysilicon than above junctions (Fig. 8.15), a sufficiently large etch-rate ratio of insulator to silicon or silicide is required if contacts to both regions are to be etched simultaneously. Alternatively, the substrate surface can be coated with a thin etch-stop, such as silicon nitride, polysilicon, or aluminum oxide, prior to PMD deposition. This allows etching of contact holes to continue to single-crystal silicon without attacking silicide or polysilicon in contacts to the elevated layers. The etch-stop is then removed with an appropriate etchant (Chap. 5). When polysilicon is used as an etch stop, it is important to convert the film to silicon dioxide by, e.g., high-pressure oxidation after completing the etch process to avoid high leakage between circuits [63]. After etching, contacts must be cleaned to remove any organic material, residual oxide or debris that can cause adhesion problems or block contacts (Chap. 5).

## 8.3 Metal Interconnects

Metals used in microelectronics must satisfy several requirements to meet the objectives of circuit performance, yield and reliability. They must exhibit low series and contact resistances, negligible penetration into silicon, good adhesion to insulators and contacted surfaces, excellent uniformity, and resistance to corrosion. The metals must also form bondable films that can be delineated into uniform fine-lines of sub-halfmicron widths and spaces, withstand temperature cycling and operating current densities without failure, and be free of contaminants (such as sodium ions). The most common metals used for interconnections and contact fill are aluminum alloys and tungsten. Aluminum meets most of the above requirements but suffers from a failure mechanism at high current densities known as **electromigration,** discussed in Sec. 8.6. Adding small amounts ($\simeq$ 0.5%) of copper to aluminum (or Al-Si) increases the electromigration lifetime. Tungsten is considerably less susceptible to electromigration and can be deposited with greater conformality than physically deposited aluminum. Tungsten, however, exhibits a 2-3 times higher resistivity than aluminum and is typically used in lower-level metals. Copper has also received increased attention because of its higher conductivity and resistance to electromigration than Al. Copper is, however, difficult to delineate and must be "cladded" with a barrier metal to inhibit its diffusion into oxide and silicon where it can cause excessive leakage or shorts [64]. Table 8.4 compares important properties of the three metals.

## 8.3.1 Metal Deposition

The most widely used techniques to deposit metals are, evaporation, sputtering, and CVD (Chap. 5). Variations to these techniques are collimated sputtering and selective CVD.

**Table 8.4** Important properties of aluminum, tungsten and copper

| Property | Unit | Al | Cu | W |
|---|---|---|---|---|
| Atomic number | | 13 | 29 | 74 |
| Atomic weight | | 27.0 | 63.5 | 183.9 |
| Density | g/cm³ | 2.70 | 8.92 | 19.4 |
| Bulk resistivity | μΩ-cm | 2.6 | 1.7 | 5.5 |
| Film resistivity | μΩ-cm | 2.7-3.0[1] | 1.8-2.0 | 11-16 |
| Workfunction[2] | eV | 4.25 | 4.35 | 4.60 |
| Melting point | °C | 659 | 1083 | 3380 |
| $10^{-2}$ torr at[3] | °C | 1220 | 1260 | 3230 |
| Boiling point | °C | 2467 | 2595 | 5927 |
| Si-eutectic[4] | °C | 577 | --- | --- |

1. Value given for pure aluminum. Adding silicon to aluminum increases the resistivity from $\simeq 2.7$ μOhm-cm at 0% Si to $\simeq 4$ μOhm-cm at 3.5% Si.
2. The workfunction affects the contact barrier height. It is the energy required to free an electron from the metal surface.
3. Partial pressure at given temperature.
4. Reduces to about 525 °C with Cu added.

**Evaporation**

Evaporation of a metal source is performed by heating the metal in an evacuated chamber (Chap. 5). A vacuum of $10^{-6} - 10^{-7}$ torr is required to reduce collisions of evaporants with residual gas molecules on their way to the wafers that are placed at a certain distance from the metal source. At such pressures, the mean-free path of molecules is greater than the distance between source and substrate (Eq. 3.20). Residual gas molecules not only deflect evaporants from their straight-line trajectories but can also react with the growing metal film, contaminating it. The rate of evaporation, $N_S$, from the source is proportional to the equilibrium partial pressure of the metal at the evaporation temperature. It is given by (Eq. 3.19)

$$N_S = \frac{p_e}{\sqrt{2\pi mkT}} \quad \text{(Molecules cm}^{-2}s^{-1}), \qquad 8.3$$

where $p_e$ is the metal equilibrium partial pressure (torr), $m$ the mass of one atom or molecule, $k$ Boltzmann constant, and $T$ the metal surface temperature (in K). The wafer temperature is typically maintained below $\simeq$ 250 °C during evaporation. At such temperatures, evaporation from the substrate is negligible and the substrate deposition rate is proportional to the source evaporation rate, depending only on the geometry of the evaporation system. If the source is a small crucible placed at the center of a hemisphere (dome) on which the wafers are mounted (Fig. 8.16), the substrate deposition rate, $D$, is defined as [65, 66]

$$D = \frac{R_T}{2\pi r^2} \, , \qquad\qquad 8.4$$

where $r$ is the sphere radius and $R_T$ the rate of mass loss from the metal source, given by

$$R_T = 4.43 \times 10^{-4} \int \sqrt{\frac{M}{T}} \, p_e \, dA_S \quad (gs^{-1}) . \qquad 8.5$$

In Eq. 8.4, $M$ is the gram-molecular mass, $p_e$ is in Pa, and $A_S$ is the source area.

Source heating methods include resistive, induction, e-beam, and laser heating described in Chap. 3. The metal is placed in a non-reactive crucible, typically of boron nitride or a refractory metal.  In many applications, different materials are evaporated simultaneously to form a composite film, such as Al-Cu. Because of the different vapor pressures of the components, separate sources are used and individually controlled to yield the desired composition.

## Step Coverage

In evaporation systems, metal atoms follow essentially straight-line paths from source to wafer. Because of this directionality, shadowing of varying degrees is observed in the presence of large steps in patterned films on the wafer. This causes the metal to be thin along shadowed step edges and can seriously degrade the yield and reliability of metal interconnects. An extreme case of discontinuity is sketched in Fig. 8.17.  One method to reduce the effect of shadowing is to rotate the wafer holder (dome) in two directions and achieve more uniform coverage for a large number of wafers.

Another method is to taper edges of steps and opening, however, at the cost of increasing the metal pitch. Although there are step coverage problems with sputtered films, they tend to be less severe than with evaporated films. Due to the conformal nature of chemical vapor deposition, step coverage is usually not a problem for CVD metal films.

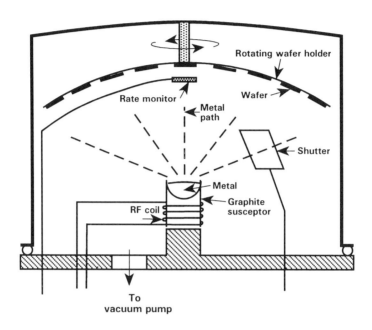

**Fig. 8.16** Schematic of a typical spherical radio-frequency evaporation system.

**Sputtering**

While evaporation is widely used to deposit aluminum and its alloys, sputtering of these materials has become more practical because of the increased deposition rate and more uniform step coverage and contact hole filling. Also, when refractory metals are used, either sputtering or CVD is required to achieve a realistic throughput in manufacturing. The principles of sputtering and collimated sputtering are discussed in Chap. 3.

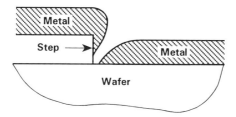

**Fig. 8.17** Metal discontinuity over step cause by shadowing in an evaporation system.

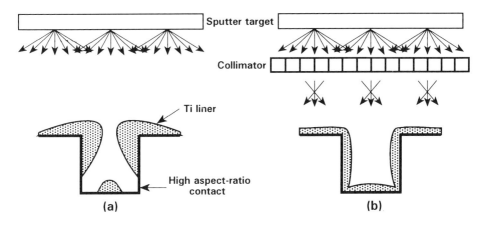

**Fig. 8.18** Effect of collimation on sputtered flux and floor coverage during liner deposition. (a) No collimator, incomplete floor coverage.(b) Collimator restricts angular divergence, resulting in more uniform floor coverage.

Collimated sputtering ensures that sufficient material covers the floor of high aspect-ratio contacts before the material coalesces at the top of the opening [67 – 69]. By interposing a collimating array between target and substrate, the divergence of the sputtered flux is restricted to an angle that depends on the aspect ratio of the collimator, the ambient pressure, and the distance between collimator and substrate (Fig. 8.18). Typically, an angular divergence of ≤ 45° is achieved by placing a collimator of 1:1 aspect ratio at close proximity to the substrate, and reducing the ambient pressure to < 4 mTorr to minimize randomization and increase the mean-free path of sputtered atoms. Sputtered atoms that are incident at an angle > 45° stick to the collimator while the flux of lower divergence proceeds to the substrate.

### Chemical-Vapor Deposition

Chemical-vapor deposition (CVD) provides more uniform step coverage and contact hole filling than evaporation or sputtering (Chap. 3). CVD is widely used to deposit tungsten, but is also applied to aluminum, copper, titanium, molybdenum, silicides and other noble and refractory metals [70 − 73].

### Tungsten Deposition

Tungsten CVD is commonly used for contact hole filling (via plugs or contact studs). It is typically done by pyrolitic decomposition of tungsten hexafluoride ($WF_6$), or by the reduction of $WF_6$ with hydrogen, silicon or silane. Pyrolitic decomposition occurs near 800 °C and results in the formation of fluorine:

$$WF_6 \rightarrow W + 3F_2 .$$

A barrier metal is deposited to inhibit the diffusion of fluorine to the silicide or silicon surface, as discussed below.

Reduction at at 250-600 °C is more typically used in microelectronics. For hydrogen, the reaction is

$$WF_6 + 3H_2 \rightarrow W + 6HF .$$

The reaction for silicon is

$$2WF_6 + 3Si_2 \rightarrow 2W + 3SiF_4 ,$$

and for silane

$$2WF_6 + 3SiH_4 \rightarrow W + 3SiF_4 .$$

When the metal is deposited over the entire wafer, a "glue" film (typically TiN) is required for its adhesion; deposition is followed by a patterning or planarizing step, described in the following section.

**Selective Tungsten CVD**

Selective metal deposition is very attractive for filling high aspect-ratio contacts and for shunting polysilicon lines. The main advantage of selectivity is the elimination of masking and etching that are otherwise required to pattern the metal. Also, step coverage and voiding is not an issue. Selective deposition is achieved by tuning the reactor conditions so that the metal nucleates only on exposed silicon, silicide, or other metal surfaces and not on insulators. The development of selective CVD is at a more advanced stage for tungsten than for aluminum.

Selective tungsten can be performed by direct deposition of W on a catalytic surface or by selective reduction of $WF_6$ on silicon, silicides or other metals [74 – 81]. The selective deposition on silicon is initiated by the highly exothermic, fast reaction of $WF_6$ with silicon to produce solid W and gaseous silicon fluoride. This is a self-limiting process because, as the thickness of tungsten on silicon increases, diffusion of reactants through W slows down considerably [82]. The final thickness formed by reduction with silicon ranges from 20-90 nm, depending on the substrate surface, deposition temperature and partial pressure of $WF_6$. Thicker films are best achieved by forming $\simeq$ 10 nm tungsten with silicon reduction, followed by hydrogen reduction whereby solid W and volatile HF are the products.

Several issues must be addressed before selective tungsten can be introduced in mass production [83]. Among them are the tungsten-silicon interface quality, the control of selectivity, and the deposition rate. Problems with the W/Si interface arise as a consequence of the silicon reduction reaction that initiates the selective deposition [84]. The reaction removes silicon from under the edge of the insulator at the base of the contact and deposits W under the insulator. This "encroachment" increases as the $WSi_6$ partial pressure and the deposition temperature increase. Also, non-uniform removal of silicon results in tunnels, referred to as "worm holes", 20-40 nm in diameter and extending up to 1 $\mu$m into silicon [81]. Higher deposition rate can be obtained with silane rather than hydrogen reduction, improving throughput and reducing silicon interface reactions with W [78]. To avoid encroachment, barrier layers, such as TiN and $WSi_x$, can be deposited before selective tungsten CVD.

Because of the different insulator heights over polysilicon and single-crystal silicon (Fig. 8.15), a polishing step may be required to remove tungsten "nail-heads" extending above polysilicon when contacts to both regions are filled simultaneously.

The basic principle of selectivity is believed to be that the higher surface energy on metals than on insulators results in the different probabilities for nucleation. By proper treatment of the insulator surface, such as removal of metal oxide residues, the density of nucleation sites on the insulator has been reduced to < 0.2/cm² [78]. The most serious problem with respect to mass production is, however, the extremely narrow process window to achieve selective deposition, and the loss of selectivity due to residual nucleation sites on insulators.

**Aluminum Deposition**

Aluminum is typically deposited from metal organic compounds. Aluminum CVD can be achieved, for example, by thermal decomposition of tri-isobutyl aluminum (TIBA) above 220 °C [85, 86]. The overall reaction is

$$2Al(C_4H_9)_3 \rightarrow 2Al + 3H_2 + 6C_4H_8 .$$

The reaction in performed in two steps. The first step at 150 °C gives

$$[(CH_3)_2CH\text{-}CH_2]_3Al \rightarrow [(CH_3)_2CH\text{-}CH_2]_2AlH + (CH_3)_2C=CH_2 .$$

It is followed by a decomposition at 250 °C giving

$$2[(CH_3)_2CH\text{-}CH_2]_2AlH \rightarrow 2Al + 3 H_2 + 4[(CH_3)_2C=CH_2] .$$

The surface is pre-treated with TiCl₄ to improve nucleation. Co-deposition of Cu with Al to reduce electromigration in aluminum is difficult to achieve with TIBA. Instead, dimethylaluminumhydride (DMAH) has been used with an appropriate Cu organometallic compound to obtain the correct Al-Cu composition. [87].

**Selective Aluminum CVD**

As mentioned above, selective aluminum deposition is not as advanced as with tungsten. Selective aluminum CVD has been deposited on Si, W, Ti, and silicides by maintaining the wafer temperature in the range 250-400 °C in a modified CVD reactor [89]. Controlling the substrate temperature during deposition is essential for selectivity to occur. Above the temperature range, selectivity is lost because of homogeneous nucleation on insulators. Deposition temperatures lower than $\simeq$250°C result in "rough" aluminum surfaces and metal islands on all surfaces [89]. For selective aluminum CVD to form in contacts to aluminum, an in-situ RF clean is required to remove native aluminum oxide prior to deposition.

**8.3.2 Contact Fill and Metal Patterning**

Contact openings are filled with an appropriate conductor, typically aluminum or tungsten, to form vertical connections to first level metal. The metal fill must form a void-free "stud" and exhibit low contact resistance to underlaying and overlaying conductors. Typical specific contact resistances that can be achieved with metal studs are in the range $10^{-8} - 5x10^{-7}$ Ohm-cm$^{-2}$, depending on the properties of the contacted surface. Heavily doped polysilicon can also be used for connecting studs, however, at the cost of increasing contact resistance. Also, polysilicon must be doped n-type when contacting n-regions and p-type for p-regions to avoid inter-diffusion and dopant compensation. When aluminum or tungsten is used, contact holes are typically "lined" with a thin film of Ti or Ti/TiN prior to stud fill. Other compositions, such as TiW and W, can also be used. The main purpose of Ti is to improve contact resistance. The TiN film is deposited to act as a diffusion barrier to elements such as silicon from and to the substrate, and fluorine generated during tungsten CVD. Other purposes of Ti or TiN are to act as a "glue" layer for tungsten adhesion, or as a wetting film to enhance aluminum reflow (described below). The liners are typically deposited by sputtering or collimated sputtering (Chap. 3).

## Aluminum Interconnect Patterning

A simple metallization technique is to deposit sequentially a barrier metal film and an aluminum alloy, and use the same film to fill contacts and define interconnects. For complete filling of contact holes, the metal thickness must be at least half the size of the contact width. Metal patterning occurs by subtractive etch or by a method known as **lift-off,** (Fig. 8.19) [90,91].

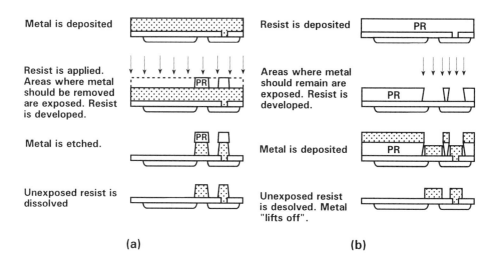

Metal is deposited

Resist is applied. Areas where metal should be removed are exposed. Resist is developed.

Metal is etched.

Unexposed resist is dissolved

Resist is deposited

Areas where metal should remain are exposed. Resist is developed.

Metal is deposited

Unexposed resist is desolved. Metal "lifts off".

(a)                                    (b)

**Fig. 8.19** Metal definition. (a) Subtractive etch. (b) Lift-off.

In subtractive etch, the metal is first deposited on the wafer and then covered with resist. The interconnect pattern is defined in the resist and the metal then etched away from regions not covered by resist. Wet or dry etching can be used for this purpose. For lift-off, the wafer is coated with resist prior to metal deposition. The pattern is defined in the resist so that regions where no metal is desired are covered by resist. The metal is then deposited and the resist dissolved, lifting off the metal covering the resist. The metal thickness and resist profile are chosen to create a discontinuity of metal at the resist pattern edges. The discontinuity allows the resist solvent to reach the lower part of the unexposed resist and lift-off the raised part of the metal film without affecting the metal pattern that adheres to the wafer. Lift-off is not only applicable to aluminum and its alloys. It can be used for titanium, platinum and other refractory metals, and composite metal stacks. The main advantage of lift-off is elimination of metal etch, increasing the patterning flexibility. Lift-off, however, has limited extendibility to sub-0.5 $\mu$m feature sizes.

### Reflowed Aluminum

Aluminum reflow is a high-temperature Al deposition process that is used to improve filling of high aspect-ratio contact or via openings [92 − 94]. Since the reflow temperature ranges from 400-550 °C, a reliable (pinhole free) barrier is required to ensure filling and avoid metal spiking. Titanium nitride is typically used as a barrier and Ti or TiN are used a wetting agent for aluminum. The reflow process can be done by depositing aluminum at low temperature and then annealing at 450-550 °C, or by depositing the metal at room temperature followed by a deposition at high temperature ("reflow-capped"). The reflow temperature can be reduced by adding germanium to aluminum [95].

### Tungsten Studs and Interconnects

A common method to fill contacts is to deposit tungsten by CVD or sputtering and then planarize the metal by etching it back to the insulator surface or by CMP, leaving tungsten only in contact openings (Fig. 8.20). CVD and etch-back processes can be performed in one tool. The deposition process must be optimized to avoid large center-line "seams" caused by the columnar morphology of the film (Fig. 8.21), and "keyhole" defects caused by premature coalescence of the film at the top of the contact (Fig. 8.22). Alternatively, selective tungsten can be used to minimize or eliminate the etch-back process. An interconnecting metal, typically an aluminum alloy, is then deposited and patterned over the studs. In some applications, tungsten is used for both the metal studs and interconnecting lines. In this case, no planarization is required before line patterning.

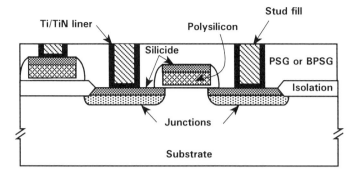

**Fig. 8.20** Schematic of a planarized metal stud.

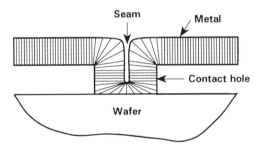

**Fig. 8.21** Columnar metal morphology causing a center-line seam.

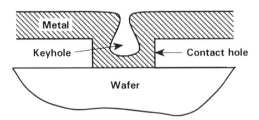

**Fig. 8.22** Keyhole defect formed by premature coalescence of metal at top of the contact.

### Damascene Process

Damascene is a jewelry fabrication term that has been adopted to refer to a microelectronics metallization process where interconnect leads are recessed in an insulator by patterning troughs in the planar dielectric and filling the troughs with metal, e.g., by collimated sputtering or CVD. The metal in the "field" is then removed by CMP, leaving troughs filled with metal (Fig. 8.23). The damascene wiring technique has been used with many different wiring materials, including W, Al alloys, Cu and Ag.

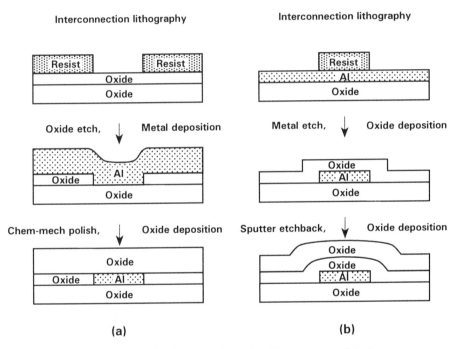

**Fig. 8.23** Metal patterning. (a) Damascene. (b) Conventional [67]

The main advantage of damascene is that it eliminates the need for etching to define the metal pattern, increasing the flexibility in the metal composition. Dry etching of Al-Cu alloys, for example, becomes more difficult as the copper content increases. When no etching is required, a larger amount of copper or other elements can be added to aluminum to improve the metal immunity to electromigration or stress migration.

## Dual-Damascene Process

Dual-damascene forms studs and interconnects with one planarization step. The process increases the density, performance, and reliability in a fully integrated wiring technology [96]. The process sequence is illustrated in Fig. 8. 24. Contacts openings and troughs for interconnects are first defined in two consecutive masking steps on a planar insulator surface (Fig. 8.24a). Contacts are selectively and partially etched in the insulator (Fig. 8.24b) (some trough masking material is removed in this step). The masking material is then etched to a depth that removes the mask from trough regions, but leaves sufficient masking material elsewhere (Fig. 8.24c). The masking material is removed, metal is deposited and polished to become level with the insulator surface (Fig. 8.24, d-f). By forming studs and interconnects with the same material, the number of interfaces between dissimilar materials is reduced, increasing the reliability of the metallization system.

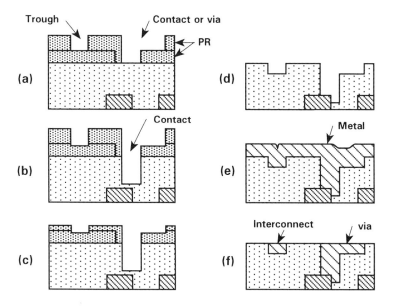

Fig. 8.24 Typical dual-damascene process. (a) Contact and troughs defined in resists. (b) Contact selectively etched in insulator. (c) Resist removed from trough regions. (d) Troughs and contacts etched in insulator. (e) Metal deposited. (f) Metal planarized by CMP.

## 8.4 Inter-Level Dielectrics

Insulators deposited between metal levels are commonly referred to as inter-level dielectrics (ILD). The desired features of ILD are similar to those of PMD. Some properties, however, are more emphasized in one than in the other, depending on applications. Table 8.5 summarizes important properties of PMD and ILD for logic applications.

**Table 8.5** Importance of some poly-metal and inter-level dielectric properties in logic designs

| Property | PMD | ILD |
|---|---|---|
| Low dielectric constant | Medium | High |
| High field strength | Medium | Medium |
| Low leakage | High | High |
| Contamination-free | High | High |
| Low defect density | High | High |
| Low water content | High | High |
| Barrier to $Na^+$ | High | Medium |
| Hot-carrier reliability | High | High |
| Gap-fill | High | High |
| Planarity | High | High |
| Thermal conductivity | High | High |
| Process at $\leq 500°C$ | Low | High |
| Etch selectivity | High | Medium |
| Adhesion | High | High |

Capacitance reduction requirements in high-performance circuits must be met by developing inter-level insulators with low dielectric constants ($\varepsilon$). The dielectric must also have good gap-fill capability in sub-0.5 $\mu$m topography and must not degrade the transistor reliability. While $\varepsilon$ must be kept low in both PMD and ILD, it is more important to reduce the ILD dielectric constant in logic designs because of the tightly packed metal patterns. Both dielectrics must exhibit low defect density, water content, and contamination levels to avoid degrading the yield and reliability of the metallization system and underlying circuit. Contamination and water can cause shifts in device characteristics or corrosion of metals. Defects can cause horizontal and vertical shorts, or "opens" (breaks) in conductor lines. A barrier to sodium ions is more important in PMD than in ILD, mainly because PMD is closer to the silicon surface. One important factor is the gap-fill between tightly packed polysilicon and metal patterns. In some logic designs, gap-fill is more important for ILD because of the small spacing between metal lines and steeper profile of their sidewalls. The deposition and planarization temperature is also an important consideration. Since PMD is deposited before first level metal, a higher temperature than $\simeq 500°C$ can be tolerated. Etch selectivity is more important for PMD than ILD because of the higher topography encountered when contact holes are etched in PMD. Finally, both PMD and ILD must exhibit good adhesion to underlaying and overlaying films.

Several materials are available for ILD. Among these are doped and undoped oxide, silicon nitride, spin-on-glass (SOG), polyimide, and various other organic materials. Two of the key features of ILD films deposited on non-planar surfaces are gap-fill and planarization. Typical surfaces upon which ILD films are deposited contain tightly packed, high aspect-ratio metal patterns. It is therefore important to completely fill the gaps between metal lines with dielectric material to avoid trapping contaminants in those regions. This task becomes more difficult as the spacing between metal lines decreases. It is also essential to smooth the insulator surface in preparation for patterning of another tightly pitched metal level on top of the insulator. There is a large variety of choices between dielectric materials, deposition, and planarizing conditions that are aimed at achieving the above two objectives with a low-cost, low thermal budget process without degrading circuit yield, reliability and performance. Some of these techniques are described in the following sections. Gap fill and planarization are not an issue with the damascene and dual-damascene processes described above, since these interconnect techniques result in a smooth surface prior to ILD deposition.

### Silicon Dioxide and Silicon Nitride

The deposition of silicon dioxide and silicon nitride is discussed in Chap. 3. In most cases, variations to these basic depositions techniques are required to achieve the desired ILD properties. They include multiple-step and multi-layer deposition combined with etch-back and/or CMP. For example, while a high deposition rate can be achieved with plasma-enhanced CVD (PECVD) oxide, conventional PECVD is not adequate to fill sub-0.5 $\mu$m gaps, because it results in voids between metal lines (Fig. 8.25a). These voids can be eliminated by depositing the film in several steps and sputter-etching between steps. Sputter-etching "facets" the oxide over vertical metal lines, improving gap-fill in a subsequent deposition step (Fig. 8.25b). This "dep-etch-dep" process results in a rather smooth surface upon which a final oxide film can be formed at a high deposition rate (Fig. 8.25c) [97].

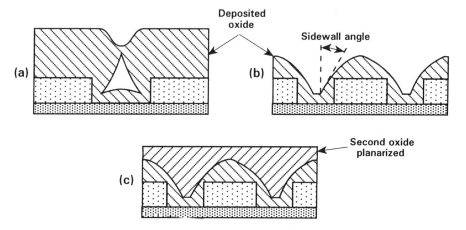

**Fig. 8.25** Elimination of voids between metal lines by a "dep-etch-dep" process. (a) Void formation with conventional CVD. (b) "Facets" formed on vertical metal lines, improving gap-fill. (c) Planar ILD.

The most common ILD insulator is a TEOS-based oxide, because of its high deposition rate at low temperature and conformal step coverage that can be achieved [98]. Several variants to TEOS deposition have been proposed. One method improves gap-fill by retarding the deposition of oxide over the metal top surface [99]. Prior to patterning, the metal is covered with Ti, W, TiN, or

**568**

TiW (Fig. 8.26a). A thin PECVD oxide film is deposited conformably over the metal contours and then etched with a $CF_4$ plasma (Fig. 8.26b). The plasma fluorine ions alter the metal surface so that the rate of of deposition of a subsequent APCVD TEOS oxide is reduced on top of the metal (Fig. 8.26c). Global planarization is achieved with a spin-on-glass, as discussed below (Fig. 8.26d,e). A final PECVD oxide is deposited at a high rate on top of the planarized surface (Fig. 8.26f).

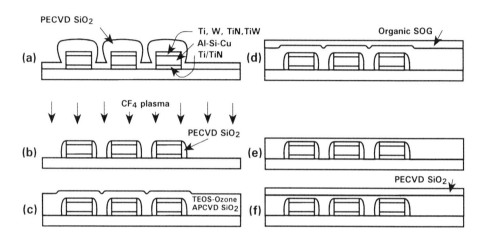

**Fig. 8.26** Gap-fill and global planarization using TEOS/ozone oxide and SOG [99]. (a) Conformal deposition of PECVD oxide. (b) Oxide etch and metal surface treatment. (c) Gap-fill with TEOS/ozone oxide. (d) SOG coating. (e) Planarization by SOG etch-back. (f) Final PECVD oxide.

Fluorine incorporated into a TEOS-base dielectric, e.g., with $NF_3$, $CF_4$, or $C_2F_6$, reduces its dielectric constant and results in more uniform gap filling [100]. The dielectric constant is found to decrease with increasing Si-F bonds, from $\simeq 4.2$ without fluorine to $\simeq 3.4$ for a Si-F/Si-O bonding ratio of 3.5% [100].

**Spin-On Glass**

Spin-on glass (SOG) is frequently used for gap fill and planarization of inter-level dielectrics in multilevel metallization [101 – 103]. The most commonly used SOG materials are $SiO_2$ based polysiloxanes. The film is typically applied to a predeposited oxide as a liquid that fills gaps and steps on the substrate. As with photoresist, the material is dispensed on a wafer and spun with a rotational speed that determines the SOG thickness. The film is then cured at $\simeq 400$ °C and typically etched back to smooth the surface in preparation for a capping oxide film on which a second interlevel metal is patterned. The capping oxide is needed to seal and protect SOG during further processing. The main purpose of the etch-back step is to leave SOG between metal lines but not on top of the metal (Fig. 8.27). Siloxane based SOG is capable of filling 0.15 $\mu$m gaps, allowing its extension to 0.25 $\mu$m designs. While a "non-etch-back" process is simpler, SOG residues left over the metal can cause degradation of contacts to metal, known as "via poisoning", as discussed below.

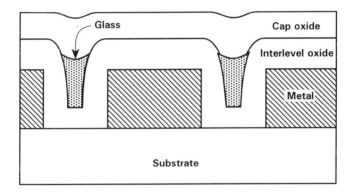

**Fig. 8.27** Gap-fill with spin-on glass. Etch-back leaves SOG between metal lines but not on top of metal.

Silicon dioxide based SOG has, however, detrimental effects on circuits, most of them related to its absorption of water and porosity [104]. In siloxane based SOG, water can reside in a free and bonded to Si-OH (silanol) state. Free water can be expelled by low-temperature ($< 450$°C) heating, but silanol-bonded requires high temperatures ($> 800$°C). Both free and bonded water will reappear in SOG after anneal if the material is again exposed to atmospheric moisture [104]. The presence of water in SOG creates several problems. Among them are, "poisoning" of vias between contacts, mobile positive charge that results in inversion of underlying p-type silicon surfaces, accelerated hot-carrier degradation in MOSFETs caused by charge trapping in the gate insulator and at its interface with silicon, and large increase in the dielectric constant causing an increase in parasitic capacitance and performance degradation.

Poisoning of vias is observed when the metal is still coated with SOG. It is attributed to moisture absorbed by SOG or left in the film due to incomplete cure. Also, resist ashing in an oxygen plasma is found to significantly increase moisture absorption [105,106]. It is believed that siloxane reacts with the oxygen plasma and the reacted SOG absorbs moisture more readily than the unreacted [106].

Water in SiO2 based materials gives rise to mobile protons whose density is in direct proportion to the amount of absorbed water [104]. The presence of hydrogen in the vicinity of the silicon surface is known to accelerate hot-carrier degradation in MOSFETs, and the mobile positive charge causes inversion on p-type silicon surfaces.

Moisture also causes a considerable increase in the dielectric constant and associated parasitic capacitance. After curing, a dielectric constant of 5-6 is measured for siloxane based SOG. This value can increase by 1-2 orders of magnitude after prolonged exposure to atmospheric moisture [104].

Several organic compounds are being developed to solve the above problems. Their discussion is, however, beyond the scope of this chapter.

**Polyimides**

Polyimides belong to a class of organic compounds derived from imidization reactions of amines and organic acids which occur in the temperature range of 130-200 °C. As with SOG these compounds are spun and cured to produce planar surfaces for multilevel metallization. The advantages of polyimide include ease of deposition and flexibility in composition; planarity as a spun film; adequate temperature tolerance; excellent weathering and mechanical wear characteristics; low pinhole density; low dielectric constant ($\simeq 3.5 - 3.9$); and low absorption of water when compared to SOG [107]. Adding fluorine to polyimide further reduces its dielectric constant. Polyimide films are used as interlevel dielectrics and protective overcoats. As with photoresist, polyimide can be used as a mask (e.g., for ion implantation) or a planarizing film by etch-back. When used as a thick overcoat (4-75 $\mu$m), the material reduces stress and also serves to block ionizing radiation (such as alpha particles) from reaching silicon. Polyimide, however, absorbs water when exposed to atmospheric moisture, increasing its dielectric constant, though the change is considerably less than with SOG.

## 8.5 Multi-Level Metals

Interconnects are now believed to be the limiting factor in "down-scaling" (shrinking) integrated circuits [84]. By using more than one metal level, the average interconnect length is reduced and with it the die size. The design of a multi-level metal (MLM) system is aimed at reducing lead resistances and capacitances without compromising yield and reliability. Such a system can be designed by repeating the techniques described earlier for via and metal patterning. For example, after patterning a second metal level, another inter-level dielectric can be deposited and planarized, followed by via and third metal patterning, and so on. Since upper level metal interconnects have longer average lengths than gate interconnects, they must have lower resistivity. Despite their higher resistivities (relative to aluminum), refractory metals, such as tungsten, are attractive for first-level metal because of their thermal and process stability. Aluminum is best suited for upper-level metals.

A schematic of a five-level metal structure is shown in Fig. 8.28. Using several metal levels poses constraints on metallization material and processing. The design of an interconnecting system is typically a trade-off between cost, yield, performance, and reliability. For example, thicker metal may be required to reduce line resistance and adequately distribute power across the chip. Thicker metal leads in turn increase line-to-line capacitance and create high aspect ratios which complicate the inter-level gap fill and planarization process. Similarly, scaled contacts and vias will have a higher aspect ratio because the contact size decreases while the dielectric thickness remains constant to maintain low parasitic capacitances, complicating the via etch and and fill processes.

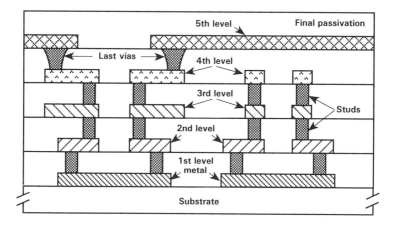

**Fig. 8.28** Illustration of a five metal level interconnect system.

## Via Designs

Conventional designs place a conductive frame ("border") much larger than the width of the conductive line on a previous level to make contact to the metal within the frame boundary. This is done to avoid etching material below the planarized level. Vias can be stacked with and without borders. Placing vias over vias without frames ("borderless"), considerably increases circuit density (Fig. 8.29). This can be achieved by controlling the etch process, or by utilizing the difference in etch rate between dissimilar materials, preventing etching below the surface of the via stud. When an etch stop is used, it is selectively removed from within the via before metallization.

## 8.6 Reliability Considerations

Failure mechanisms in metallization systems can be grouped into fours main categories: electromigration, stress migration (also called "creep"), corrosion, and surface effects on active devices. Electro- and stress-migration respectively describe the motion of metal ions under the influence of electrical current and mechanical stress. The migration of metal ions can result in an increase in line and contact resistances, "extrusion" shorts between conductor lines, breaks ("opens") in conductor lines, and leaky junctions. In most cases, the final failure arises from a combination of mechanical and electrical stress. Corrosion is typically a metal reaction induced by moisture trapped in the material or atmosphere in contact with the metal. Corrosion decreases the conductivity of metal lines and contacts and can cause breaks in conductor leads and cracks in insulators. Device degradation is a consequence of electronic traps, ionic contaminants, or stress that can be created during the metallization process. Before discussing failure mechanisms, it may be helpful to define some important terms used in reliability engineering.

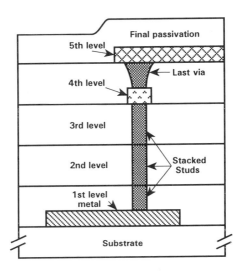

**Fig. 8.29** Area reduction by use of stacked vias.

## Definition of Reliability Terms

The reliability of a component is the probability that it will perform a specific function under specified conditions for a specific length of time. The physical or chemical process that causes a component to fail is called **failure mechanism.** The criterion for rejecting a component is referred to as **failure mode** [108]. The failure mode of a device depends on the specific function it must perform. For example, electromigration is a failure mechanism that can lead to the failure mode of a short between two adjacent metal lines. Every component will eventually fail after a certain time $t$ of operation, called **time to failure** or **lifetime.** The reciprocal of lifetime is called the **failure rate,** $\lambda$. For example, suppose that a unit consists of $10^6$ identical components and not more than one component is to fail after operating the unit for $10^5$ hours. The failure rate must then be limited to

$$\lambda < \frac{1}{10^6 \, x \, 10^5} = 10^{-11} \text{ (failure/device-hour)} . \qquad 8.6$$

It is customary to define one failure per $10^9$ device-hour as 1 FIT· (Failure unIT). With this definition, the above failure rate must be less than 0.01 FIT.

The failure rate of metallization system has been observed to exhibit three different behaviors with time (Fig. 8.30). The region on the left of the figure describes early failures due to defects in the interconnecting system. Screening of these failures is achieved by a stress test, referred to as "burn-in", that will not be further discussed here. The flat region in the middle of the curve, during which only few failures occur, is called the "useful life". The rapid degradation on the right side is referred to as the "wear-out region". Wear-out is caused by material degradation at high current densities and temperature. The design and operation rules are defined such the wear-out region is not typically reached under normal operating conditions. Accelerated stress is performed in this region to induce failures in a realistic time, as discussed below.

Several functions have been used to describe the distribution of failure modes with time. For a metallization system, the distribution is found to obey a log-normal function with

$$f(t) = \frac{1}{\sigma t \sqrt{2\pi}} \, e^{\left(-\frac{1}{2\sigma} \ln \frac{t}{t_{50}}\right)^2} , \qquad 8.9$$

where $\sigma$ is the dispersion in log-time and $t_{50}$ the median time to fail (MTF), i.e., the time required for half the samples to fail.

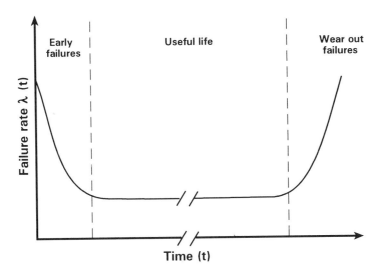

**Fig. 8.30** Failure rate versus time for typical metallization systems.

Let $F(t)$ be the probability that a device will fail, and $R(t) = 1 - F(t)$ the probability that it will survive at time $t$. The instantaneous failure rate, $\lambda(t)$, between time $t$ and $t + dt$ is

$$\lambda(t) = -\frac{1}{R(t)}\frac{dR(t)}{dt} \, , \qquad 8.7$$

A common measure of reliability is the mean time to failure (MTTF) given by

$$MTTF = \int_0^\infty t\, f(t) \, . \qquad 8.8$$

In practice, failure characteristics are studied under accelerated aging conditions for a given failure mechanism. The results are used to project the component's lifetime under operating conditions. For example, the maximum temperature and current density under normal operating conditions of a metallization system may be 85 °C and $10^5 A/cm^2$, respectively. Accelerated aging can be achieved by increasing the current density to $5x10^5 - 10^6 A/cm^2$ and the temperature to 200-300 °C. A schematic of a log-normal plot of accelerated life test data is shown in Fig. 8.31.

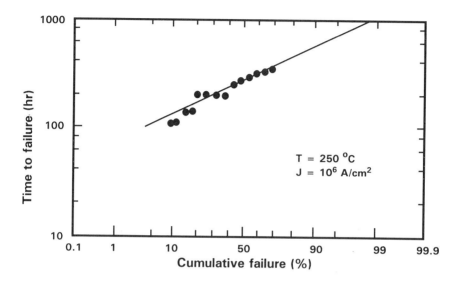

**Fig. 8.31** Log-normal distribution of accelerated life test data.

### 8.6.1 Electromigration

Electromigration (EM) is the transport of metal atoms by momentum exchange between electrons, moving under the influence of a field, and metal ions. In aluminum, EM occurs at current densities above $10^5 A/cm^2$. Electromigration has been long been identified as one major cause of three failure modes in aluminum metallization [109 − 111]. One failure mode is the pile-up of metal in some regions causing extrusion shorts between conductors. The other is condensation of voids in some regions, increasing line resistance and causing open circuits. The third mode is the formation of etch pits in contacts to silicon where electrons leave silicon and enter aluminum [112].

A simplified model to explain EM is illustrated in Fig. 8.32. When a positive metal ion is thermally activated to a saddle point in the periodic potential of the metal crystallite, two forces will affect its motion. The first force is Coulombic caused by the electric field. This force is directed toward the negative terminal and proportional to the field and ion charge. A more important force is exerted by collisions of electrons with the metal ion and transfer of momentum to the ion. At high current densities, this second force is dominant and causes the ion to move toward the positive terminal in the direction of electron flow. The vacancy left by the ion tends to move toward the negative terminal. Vacancies can condense to form voids in the metal. The concentration gradient created by the motion of aluminum causes aluminum to diffuse in the direction of decreasing concentration and plays a significant role in electromigration.

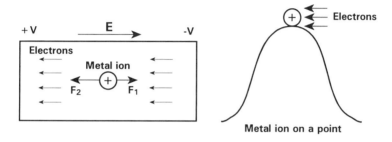

**Fig. 8.32** Model to explain electromigration.

The rate of mass transport is proportional to the electron momentum and flux, both proportional to the current density. It is found as [111]

$$m \propto J^2 e^{-E_A/kT}, \qquad 8.10$$

where $m$ is the transported mass, $J$ the current density, and $E_A$ the activation energy. The median time to failure is approximated as

$$MTF = t_{50} = AJ^{-2} e^{E_A/kT}, \qquad 8.11$$

where A is a material constant. Electromigration occurs predominantly along metal grain boundaries where the activation energy is lowest. The grain size has a strong effect on the activation energy. For aluminum based alloys, $E_A$ varies typically from 0.5 eV for small grains to 0.7 eV for large grains [113]. Copper shows about five orders of magnitude larger EM lifetime than conventional aluminum based alloys, primarily because of its large grain size [114].

The electromigration lifetime is found to increases as the metal line width decreases below $\simeq 2\mu m$ [113]. This is attributed to the reduced number of grain boundaries in the direction of current. As the line shrinks, the crystallites become aligned in a direction normal to the current, resulting in a so-called bamboo microstructure.

Electromigration is also important in metal films crossing steps where the metal thins and the current density and local resistance increase. This causes the temperature to rise, enhancing EM. When electromigration begins, it compounds the problem by further reducing the metal cross-section.

Several techniques have been used to retard electromigration in aluminum films. Among them are adding impurities, such as Cu or Ti, to the metal and using multiple layers, such as Ti under- and over-layers. One common method to improve the EM resistance is to add 0.5-4% Cu to aluminum [115]. The role of copper is attributed to its segregation at grain boundaries, suppressing vacancy formation and aluminum migration along these boundaries [116]. Other models relate the retarded diffusion of aluminum to the presence of copper within the crystallites [117], or to the modification of the grain properties by the presence of copper at the boundaries [118].

Redundant layers, such as laminates of Al-Si-Cu/TiN or Al-Si-Cu/Ti are also very efficient in increasing the metal resistance to electromigration [119 − 121]. An improvement greater than 100 times in MTF has been measured, for example, with Ti Al-Cu Ti over simple Al-Cu [120, 123]. The films are typically capped with Al-Cu to avoid oxidation of the top Ti film. Annealing at 350-400 °C results in inter-diffusion of Al and Ti to form TiAl₃ at both interfaces that is believed to increase resistance to electromigration, however, at the cost of increasing the sheet resistance of the film.

Multilevel structures, consisting of CVD W interlevel vias are found to have significantly degraded electromigration performance compared to planar samples. Composite layers of Ti-AlCu-Ti, for example, exhibit $\simeq 50X$ reduction in EM lifetime when formed over alternating studs ("stud chains") than over flat surfaces (Fig. 8.33). It is believed that the discontinuity in Cu supply at Al-Cu/W interface accounts for most of the reduction in EM resistance of W stud chains.

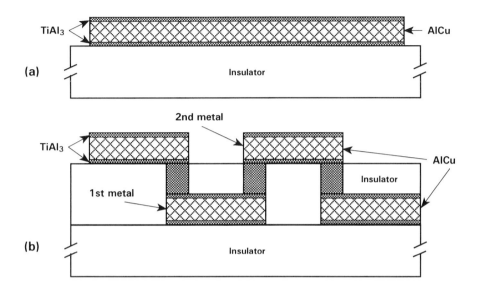

**Fig. 8.33** Cross-section of a composite metal. (a) Flat surface. (b) Stud-chain.

## 8.6.2 Stress Migration (Creep)

Interconnects typically exhibit high internal stresses during the deposition process. Also, thermal expansion coefficients in the different films of a multi level system range from $\simeq 5x10^{-7}/°C$ for $SiO_2$ to $\simeq 2.3x10^{-5}/°C$ for Al. This difference will generate additional stresses as the chip is thermally cycled. The ductility of interconnects such as AlCu will allow them to accommodate some stress [124], but atomic migration typically occurs to accommodate stress. Migration induced by mechanical stress is commonly referred to as "creep". The predominant failure mode for compressive and tensile stress is voiding in the metal that can grow large enough to sever lines [125 − 128] The magnitude of stress increases as the metal line width is reduced [128 − 132]. The time to failure can be increased by alloying [129], or by controlling the grain structure [132].

### 8.6.3 Corrosion

Corrosion of metal leads occurs in the presence of moisture. Polysilicon and silicides are rather immune to such effects. Most corrosion is caused by electrolysis induced by moisture. For example, chlorine residues react with moisture to attack aluminum following the reaction

$$2Al + 3HCl \rightarrow 2AlCl_3 + 3H_2$$

$$AlCl_3 + H2O \rightarrow Al(OH)_3 + 3HCl.$$

Corrosion continues because HCl re-appears in the second part of the reaction. The result of the reaction is a conversion of Al to aluminum hydroxide [$Al(OH)_3$], a non conductor. Also, the hydroxide increases volume, causing cracks in insulators. Continued corrosion causes breaks in interconnects. Electrolysis corrosion of aluminum is enhanced by an electric current and is typically pronounced on negative metal leads [133, 134].

Phosphorus ions leaching from PSG overcoat also react with water to form phosphoric acid that attacks aluminum [135]. As mentioned earlier, this reaction can be minimized by lowering the phosphorus content in PSG or capping the PSG film with an undoped oxide.

### Device Degradation

Several mechanisms related to multi-level metallization can degrade underlying devices. Among them are moisture, mobile charge, plasma damage, and stress. Moisture and hydrogenous species can cause device instabilities, such as threshold voltage shifts and loss of stored charge in memory structures. Serious MOSFET degradation has been observed as a consequence of mobile protons created in some dielectrics [136]. Plasma damage degrades the gate dielectric integrity and reliability in MOSFET structures and can create traps causing shifts in MOSFET characteristics and excessive junction leakage. Degradation of other structures, such as bipolar transistors, precision resistors and capacitors, is also an area of concern. Device degradation mechanisms are discussed in conjunction with device physics and characterization in a separate chapter.

# PROBLEMS

**6.1** A 0.5 $\mu$m thick aluminum film is used to make contact with silicon. The contact area is 1 x 1 $\mu m^2$ and the metal pad area is 50 x 50 $\mu m^2$. For a uniform removal of silicon over the entire contact area, calculate the thickness of silicon that will be removed after sintering at 450 ° for 20 minutes.

**6.2** A heavily-doped 0.25 $\mu$m thick polysilicon line has an average resistivity of 0.0015 $\Omega$-cm. A 50-nm thick cobalt film is deposited on top of polysilicon and silicide is formed by reacting Co and Si at 800 °C. Assuming a theoretical ratio of silicide to metal thickness and no change in the resistivity of the polysilicon film during silicidation, find the sheet resistance of the line.

**6.3** Find the sheet resistance of a metal film that is composed of 0.5 $\mu$ aluminum sandwiched between two 500-nm thick tungsten films.

**6.4** The impingement rate is the number of molecules that strike a 1 cm$^2$ surface in a unit time. Assume that all impinging molecules condense at the surface. Find the impingement rate for Cr associated with a condensation rate of 0.1 nm/s. At what oxygen partial pressure should this deposition be performed if the oxygen/chrome impingement rate ratio should be kept less than 0.1?

**6.5** The current in a 0.5 $\mu$m thick, 1 $\mu$m wide and 10 $\mu$m long aluminum line is 5 mA. The free electron density in Al is $10^{23}$ cm$^{-3}$. Assume elastic collisions between electrons and Al ions and a collision cross-sectional area of $10^{-15}$ cm$^2$ Find

a) The current density
b) The average electron drift velocity.
c) The net force exerted by electrons on the aluminum ion in the direction of current.

**6.6** The following results were obtained from two separate accelerated stress measurements. The labels were mixed, however, and it is only known that one of the two groups belongs to an electromigration test performed at a constant current density on aluminum lines. Which one is it? Calculate the median life of aluminum lines at a maximum use condition of 85 °C.

Test 1:

| Temperature (°C) | 240 | 180 |
|---|---|---|
| MTF (hr) | 150 | 2000 |

Test 2:

| Temperature (°C) | 250 | 200 |
|---|---|---|
| MTF (hr) | 20 | 86 |

**6.7** A current of 1.25 mA is passed through a tungsten stud between two aluminum lines. The tungsten stud is 0.8 $\mu$m high and has a cross-sectional area of 0.5 x 0.5 $\mu$m$^2$. The aluminum lines are 0.5 $\mu$m thick, 1 $\mu$m wide, and fully cover the stud.

a) Estimate the voltage drop across the stud.
b) Sketch the current contours and explain.

# References

**Abbreviations:**

IRPS: International Reliability Physics Symposium.
VMIC: VLSI Multilevel Interconnection Conference.
IEDM: International Electron Dvice Meeting.

[1] Semiconductor Industry Association, Technology Roadmap defined during the Semiconductor Technology Workshop, 1993.

[2] R. M. Geffken, J. G. Ryan and G. J. Slusser, *"Contact Metallurgy Development for VLSI Logic,"* IBM J. Res. Dev., 31 (6), 608 (1987).

[3] M. Hansen and A. Anderko, Constitution of Binary Alloys, McGraw Hill (1958).

[4] J. O. McCaldin and H. Sankur, *"Diffusivity and Solubility of Silicon in the Al Metallization of Integrated Circuits,"* Appl. Phys. Lett., 19, 524 (1971).

[5] B. L. Crowder and S. Zirinski, *"One Micron MOSFET VLSI Technology: Part VII-Metal Silicide Interconnection Technology - A Future Perspective,"* IEEE Trans. Electron Dev., ED-26, 369 (1979).

[6] S. P. Murarka, *"Refractory Silicides,"* IEDM Tech. Dig., 454 (1979).

[7] C. Koburger, M. Ishaq, and H. J. Geipel, *"Electrical Properties of Composite Evaporated Silicide/Polysilicon Electrodes,"* J. Electrochem. Soc., 129, 1307 (1982).

[8] M. Y. Tsai, H. H. Chao, L. M. Ephrath, B. L. Crowder, A. Cramer, R. S. Bennett, C. J. Lucchese, and M. R. Wordeman, *"One Micron Polycide (WSi₂ on Poly-Si) MOSFET Technology,"* J. Electrochem. Soc., 128, 2207 (1981).

[9] M. Y. Tsai, F. M. d'Heurle, C. S. Petersson, and R. W. Johnson, *"Properties of* WSi₂ *Film on Poly-Si,* J. Appl. Phys., 52, 5350 (1981).

[10] N. Kobayashi, S. Iwata, and N. Yamamutu, *"Refractory Metals for Silicides,"* IEDM Tech. Dig., 122 (1984).

[11] S. P. Murarka, Silicides for VLSI Applications, Academic Press, New York (1983).

[12] B. Davari, Y. Taur, D. Moy, F. M. d'Heurle, and C. Y. Ting, *"Very Shallow Junctions for Submicron CMOS Technology Using Implanted Ti for Silicidation."* in Proceeding of the 1st Ultra Large Integration Sci. Techn., S. Broydo and C. M. Osburn, Eds., 368-375, The Electrochem. Soc., Pennington (1987).

[13] R. W. Mann, L. A. Clevenger, P. D. Agnello, and F. R. White, *"Silicides and Local Interconnects for High-Performance VLSI Applications,"* To be published in IBM J. Res. Dev., June/July 1995.

[14] C. M. Osburn, *"Silicides,"* Rapid Thermal Processing Sci. and Tech., Academic Press, pp. 227-309 (1993).

[15] W. K. Chu, J. W. Mayer, H. Muller, M.-A. Nicolet, and K. N. Tu, *"Identification of the Dominant Diffusion Species in Silicide Formation,"* Appl. Phys. Lett., 25 (8), 454 (1974).

[16] F. M. D'Heurle and C. S. Peterson, *"Formation of Thin Films of* CoSi₂: *Nucleation and Diffusion Mechanisms,"* Thin Solid Films, 128, 283 (1985).

[17] S. Yanagisawa and T. Fukuyama, *"Reaction of Mo Thin Films on Si (100) Surfaces,"* J. Electrochem. Soc., 127, 1150-1156 (1980).

[18] B. El-Kareh, *"Ultrashallow Doped Film Requirements for Future Technologies,"* J. Vac. Sci. Techn. Jan./Feb. (1994).

584

[19] J. Hui, S. Wong, and J. Moll, "*Specifi Contact Resistivity of* $TiSi_2$ *to* $P^+$ *and* $N^+$ *Junctions,*" IEEE Electron Dev. Lett., EDL-6 (9), 479 (1985).

[20] J. Amano, K. Nauka, M. P. Scott, J. E. Turner, and R. Tsai, "*Junction Leakage in Titanium Self-Aligned Silicide Devices,*" Appl. Phys. Lett., 49, 737-739 (1986).

[21] M. Wittmer and T. E. Seidel, "*The Redistribution of Implanted Dopants After Metal-Silicides Formation,*" J. Appl. Phys., 49, 5826 (1978).

[22] C. L. Chu, G. Chin, K. C. Saraswat, S. S. Wong, and R. Dutton, "*Technology Limitations for* $N^+/P^+$ *Polycide Gate CMOS due to Lateral Dopant Diffusion in Silicide/Polysilicon Layers,*" IEEE Electron Dev. Lett., EDL-12 (12), 696 (1991).

[23] H. Jiang, C. M. Osburn, P. Smith, Z.-G. Xiao, D. Griffis, G. McGuire, and G. A. Rozgonyi, "*Ultra Shallow Junction Formation Using Diffusion from Silicides: I. Silicide Formation, Dopant Implantation and Depth Profiling, II. Diffusion in Silicides and Evaporation, III. Diffusion into Silicon, Thermal Stability of Silicides, and Junction Integrity,*" J. Electrochem. Soc., 139 (1), 196-218 (1992).

[24] J. W. Honeycutt and G. A. Rozgonyi, "*Enhanced Diffusion of Sb Doped Layer During Co and Ti Reactions with Silicon,*" Appl. Phys. Lett., 58 (12), 1302 (1991).

[25] M. Wittmer and K. N. Tu, "*Low-Temperature Diffusion of Dopant Atoms in Silicon During Interfacial Silicide Formation,*" Phys. Rev. B 29 (4), 2010 (1984).

[26] J. Amano, P. Merchant, T. R. Cass, J. N. Miller, and T. Koch, "*Dopant Redistribution During Titanium Silicide Formation,*" J. Appl. Phys., 59 (8), 2689 (1986).

[27] S. Batra, K. Park, S. Yoganathan, J. Lee, S. Barnejee, S. Sun, and G. Lux, "*Effects of Dopant Redistribution, Segregation, and Carrier Trapping in As-Implanted MOS Gates,*" IEEE Trans. Electron Dev., ED-37 (11), 2322 (1990).

[28] D. S. Wen, P. Smith, C. M. Osburn, and G. A. Rozgonyi, G. Lux, "*Elimination of End-of-Range Shallow Junction Implantation Damage during CMOS Titanium Silicidation,*" J. Electrochem. Soc., 136 (2), 466-471 (1989).

[29] S. C. Chen, H. Tamura, T. Hara, K. Kinoshita, K. Inoue, K. Endo, and S. Nakamura, "*Silicidation Reaction and Stress in Ti/Si,*" Jpn. J. Appl. Phys., 31 (2A), 201-205 (1992).

[30] C. M. Osburn, Q. F. Wang, M. Kellam, C. Canovai, P. L. Smith, G. E. McGuire, Z. G. Xiao, and G. A. Rozgonyi, "*Incorporation of Metal Silicides and Refractory Metals in VLSI Technology,*" Appl. Surf. Sci. 53, 291-312 (1991).

[31] Q. F. Wang, C. M. Osburn, P. L. Smith, C. A. Canovai, and G. E. McGuire, "*Thermal Stability of Thin Submicrometer Lines of* $CoSi_2$," J. Electrochem. Soc., 140 (1), 200-205 (1993).

[32] H. Norstrom, K. Meax, and P. Vandenabeele, "*Thermal Stability and Interface Bowing of Submicron* $TiSi_2$/*Polycrystalline Silicon,*" Thin Solid Films, 198, 53-66 (1991).

[33] J. B. Lasky, J. S. Nakos, O. J. Cain, and P. J. Geiss, "*Comparison of Transformation to Low-Resistivity Phase and Agglomeration of* $TiSi_2$ *and* $CoSi_2$," IEEE Trans. Electron Dev., ED-38 (2), 262-269 (1991).

[34] J. P. Gambino, E. G. Colgan, and B. Cunningham, "*The Resistance and Morphology of Submicron* $TiSi_2$ *and* $CoSi_2$ *Lines on Polysilicon,*" Abstract 216, p. 312, The Electrochem. Soc. Extended Abstracts, Phoenix, Arizona, Meeting, Oct. 13-17, 1991.

[35] Q. F. Wang, J. Y. Tsai, C. M. Osburn, R. Chapman, and G. E. McGuire, "*Improved Stability of Thin Cobalt Disilicide Films Using* $BF_2$ *Implantation,*" Appl. Phys. Lett., 61 (24), 2920 (1992).

[36] B. S. Chen and M. C. Chen, "*Thermal Stability of Cobalt Silicide Thin Films on Si(100),*" J. Appl. Phys., 74 (2), 1035-1039 (1993).

[37] D. C. Chen, S. S. Wong, P. V. Voorde, P. Merchant, T. R. Cass, J. Amano, and K.-Y. Chiu, "*A New Device Interconnect Scheme for Sub-Micron VLSI,*" IEDM Tech. Dig., p. 118 (1984).

[38] A. A. Bos, N. S. Parekh, and A. G. M. Jonkers, "*Formation of* $TiSi_2$ *From Titanium and Amorphous Silicon Layers for Local Interconnect Technology,*" Thin Solid Films, 197, 169 (1991).

[39] T. E. Tang, C.-C. Wei, R. A. Haken, T. C. Holloway, L. R. Hite, and T. G. W. Blake, "*Titanium Nitride Local Interconnect Technology for VLSI,*" IEEE Trans. Electron Dev., ED-34, 682 (1987).

[40] M. Wittmer, "*Properties and Microelectronic Applications of Thin Films of Refractory Metal Nitrides,*" J. Vac. Sci. Technol., A3, 1797 (1985).

[41] T. Okamoto, K. Tsukamoto, M. Shimizu, Y. Mishiko, and T. Matsukawa, "*Simultaneous Formation of TiN and* $TiSi_2$ *by Rapid Lamp Annealing in* $NH_3$ *Ambient for VLSI Contacts,*" Proc. IEEE Symp. VLSI Technol., 51 (1986).

[42] M. P. Lepstetter and J. M. Andrews, "*Ohmic Contacts to Silicon,*" in Ohmic Contacts to Semiconductors, B. Schwartz, Ed., The Electrochem. Soc., Princeton, New Jersey, p. 159 (1969).

[43] K. P. MacWilliams and J. D. Plummer, "*Device Physics and Technology of Complementary Silicon MESFET's for VLSI Applications,*" IEEE Trans. Electron Dev., 38 (12), 2619-2631 (1991).

[44] R. W. Mann, L. A. Clevenger, and Q. Z. Hong, "*The C49 to C54-$TiSi_2$ Transformation in Self Alaigned Silicide Applications,*" J. Appl. Phys., 73 (7), 3566-3568 (1993).

[45] J. F. Jongste, P. F. A. Alkemade, G. C. A. M. Janssen, and S. Radelaar, "*Kinetics of the Formation of C49* $TiSi_2$ *from Ti Si Multilayers as Observed by In Situ Stress Measurements,*" J. Appl. Phys., 74 (6), 3869-3879 (1993).

[46] N. I. Morimoto, J. W. Swart, and H. G. Riella, "*Analysis of the Mean Crystallite Size and Microstress in Titanium Silicide Thin Films,*" J. Vac. Sci. Technol. B, 10 (2), 586-590 (1992).

[47] N. H. Lin, R. J. Stoner, H. J. Maris, J. M. E. Harper, C. Cabral, J. M. Halbout, and G. W. Rubloff, "*Detection of Titanium Silicide Formation and Phase Transformation by Picosecond Ultrasonics,*" Mater. Res. Soc. Proc. 221-226 (1992).

[48] R. Sikora and W. Lundy, "*Phase Transformation of Titanium Silicide as Measured by Ellipsometry,*" J. Appl. Phys., 72 (3), 1160-1163 (1992).

[49] R. J. Nemanich, J. Hyeongtag, C. A. Sukow, J. W. Honeycutt, and G. A. Rozgonyi, "*Nucleation and Morphology of* $TiSi_2$ *on Si,*" Mater. Res. Soc. Proc., pp. 195-206 (1992).

[50] M. Hiriuchi and K. Yamaguchi, "*SOLID II: High-Voltage High-Gain Kilo Angstrom Channel-Length CMOSFET's Using Silicide with Self-Aligned Ultra-Shallow (US) Junction,*" IEEE Trans. Electron Dev., ED-33, 260-265 (1986).

[51] A. E. Morgan, E. K. Broadbent, M. Delfino, B. Coulman, and D. K. Sadana, "*Characterization of a Self-Aligned Cobalt Silicide Process,*" J. Electrochem. Soc., 134 (4), 925 (1987).

[52] E. Nagaswa, H. Okabayashi, and M. Morimoto, "*Mo- and Ti-Silicided Low-Resistance Shallow Junctions Formed Using the Ion Implantation Through Metal Technique,*" IEEE Trans. Electron Dev., ED-34 (3), 581-586 (1987).

[53] T. Gessner, R. Reich, W. Unger, and W. Wolke, "*The Influence of Rapid Thermal Processing on the Properties of* $MoSi_2$ *Layers Formed by Using the Ion Implantation Through Metal Technique,*" Thin Solid Films, 177, 225 (1989).

# 586

[54] D. L. Kwong, Y. H. Ku, S. K. Lee, and E. Lewis, "*Silicided Shallow Junction Formation by Ion Implantation of Impurity Ions into Silicide Layers and Subsequent Drive-in*," J. Appl. Phys., 61 (11), 5084 (1987).

[55] B.-Y. Tsui, J.-Y. Tsai, and M.-C. Chen, "*Formation of PtSi-Contacted P$^+$N Shallow Junctions by BF$_2^+$ Implantation and Low-Temperature Furnace Annealing*," J. Appl. Phys., 69 (8), 4354-4363 (1991).

[56] F. C. Shone, K. C. Saraswat, and J. P. Plummer, "*Formation of 0.1 μm N$^+$/P and P$^+$/N Junctions by Doped Silicide Technology*," IEDM Tech. Dig., 407 (1985).

[57] R. Liu, D. S. Williams, and W. T. Lynch, "*Mechanism for Process-Induced Leakage in Shallow Silicided Junctions*," IEDM Tech. Dig., p. 58 (1986).

[58] C. M. Osburn, S. Chevacharoenkul, Q. F. Wang, K. Markus, and G. E. McGuire, "*Materials and Device Issues in the Formation of Sub-100-nm Junctions*," Nucl. Inst. Meth., B74, 53-59 (1993).

[59] C. Zaring, P. Gas, B. G. Stevensson, M. Oestling, and H. J. Whitlow, "*Lattice Diffusion of Boron in Bulk Cobalt Silicide*," Thin Solid Films, 193/194, 244-247 (1990).

[60] R. Angelucci, M. Impronta, G. Pizzochero, G. Poggi, and A. Solmi, "*Shallow Junction Formation Using MoSi$_2$ as Diffusion Source*," Microelectron. Eng., 19 (1-4), 673-678 (1992).

[61] S. M. Fisher, H. Chino, K. Maeda, and Y. Nishimoto, "*Characterizing B, P, and Ge Doped Silicon Oxide Films for Interlevel Dielectrics*," Solid-State Technol., 36 (9), 55-64 (1993.

[62] A. C. Adams and C. D. Capio, "*Planarization of Phosphorus-Doped Silicon Dioxide*," J. Electrochem. Soc., 128 (2), 423-429 (1981).

[63] F. White, W. Hill, S. Esslinger, E. Payne, W. Cote, B. Chen, and K. Johnson, "*Damascene Stud Local Interconnect in CMOS Technology*," IEDM Tech. Dig., 301 (1992).

[64] J. S. H. Cho, H.-K. Kang, C. Ryu, and S. S. Wong, "*Reliability of CVD Cu Buried Interconnections*," IEDM Tech. Dig., 265 (1993)

[65] R. Glang, "*Vacuum Evaporation*," in Handbook of Thin Film technology, L. I. Maissel and R. Glang, Eds., p. 1-3, McGraw-Hill, New Yor (1983).

[66] D. B. Fraser, "*Metallization*," in VLSI Technology, S. M. Sze, Ed., McGrow-Hill, New York (1983).

[67] B. Vollmer, T. Licata, D. Resaino, and J. Ryan, *Recent Advances in the Application of Collimated Sputtering*," Thin Solid Fims, 247, 104-111 (1994).

[68] M. Sakata, H. Shimamura, S. Kobayashi, T. Kawahito, T. Kamai, and K. Abe, "*Sputtering Apparatus with Film Forming Directivity*," US Patent 4724060, 1988.

[69] S. Rossnagel, D. Mikalsen, H. Kinoshita, and J. J. Cuomo, "*Collimated Magnetron Sputter Deposition*," J. Vac. Sci. Technol. A, 9 (2) 261-265 (1991).

[70] C. F. Powell, "*Chemical Vapor Deposited Metals*," in Vapor Deposition, C. F. Powell, J. H. Oxley, and J. M. Blocher, Eds., Chap. 10, John Wiley and Sons, New York (1966).

[71] R. A. Levy and M. L. Green, "*Low Pressure Chemical Vapor Deposition of Tunsten and Aluminum for VLSI Applications*," J. Electrochem. Soc., 134, 37C-49C (1987).

[72] J. Crawford, "*Refractory Metals Pace IC Complexity*," Semiconductor International, 84-86, March 1987.

[73] S. Sachdev and R. Castellano, "*CVD Tungsten and Tungsten Silicide for VLSI Applications*," Semiconductor International, 306-310, May 1985.

[74] J. M. Shaw and J. A. Amick, "*Vapor Deposited Tungsten for Devices*," RCA Review, 31, 306 (1970).

[75] T. Ohba, S.-I. Inoue, and M. Maeda, *"Selective CVD Tungsten Silicide for VLSI Applications,"* IEDM Tech. Dig., 213 (1987)

[76] H. Kotani, T. Tsutsumi, J. Komori, and S. Nagao, *"A Highly reliable Selective CVD-W Utilizing SiH₄ Reduction for VLSI Contacts,"* IEDM Tech. Dig., 217 (1987)

[77] V. V. Lee and S. Verdonckt-Vanderbroek, *"A Selective CVD Tungsten Local Interconnect technology,"* IEDM Tech. Dig., 450 (1988)

[78] D. R. Bradbury, J. E. Turner, K. Nanka, and K. Y. Chiu, *"Selective CVD Tungsten as an Alternative to Blanket Tungsten for Submicron Plug Applications on VLSI Circuits,"* IEDM Tech. Dig., 273 (1991)

[79] R. S. Blewer, *"Progress in LPCVD Tungsten for Advanced Microelectronics Applications,"* Solid-State Technol., 117-126, Nov. 1986.

[80] E. K. Broadbent and W. T. Stacy, *"Selective Tungsten Processing by Low Process CVD Pressure,"* Solid-State Technol., 51-59, Dec. 1985.

[81] T. Mariya and H. Itoh, *"Selective CVD Tungsten and its Applications to VLSI,"* in Tungsten and other Refractory Metals for VLSI Applications, R. S. Blewer, Ed., 21-32, MRS, Pittsburgh, Pennsylvania (1986).

[82] J. A. Yarmoff and F. R. McFeely, *"Mechanism for Chemical Vapor Deposition of Tungsten on Silicon from Tungsten Hexafluoride,"* J. Appl. Phys. 63 (11), 5213-5219 (1988).

[83] R. Foster, L. Lane, and S. Tseng, *"Mass Spectroscopic Studies on Selective Tungsten Deposition - Mechanism and Reliability,"* in Tungsten and other Refractory Metals for VLSI Applications III, V. A. Wells, Ed., Material Res. Soc., Pittsburgh, Pennsylvania, 159-169 (1988).

[84] D. K. Ferry, M. N. Kozicki, and G. P. Raupp, *"Some Fundamental Issues on Metallization in VLSI,"* Proc. SPIE, Metallization: Performance and Reliability Issues for VLSI and ULSI, 1596, 2 (1991).

[85] M. J. Cooke, R. A. Heinecke, R. C. Stern, and J. W. C. Maes, *"LPCVD of Aluminum and Al-Si Alloys for Semiconductor Metallization,"* Solid-State Tech., 62-65, Dec. 1982.

[86] T. Kato, T. Ito, and M. Maeda, *"Chemical Vapor Deposition of Aluminum Enhanced by Magnetron Plasma,"* J. Electrochem. Soc., 135 (2), 455-459 (1988).

[87] T. Katagiri, E. Kondoh, N. Takeyasu, T. Nakano, H. Yamamoto, and T. Ohta, *"Metalorganic Chemical Vapor Deposition of Aluminum-Copper Alloy Film,"* Jpn. J. Appl. Phys. 32, L1078-L1080 (1993).

[88] N. Takeyasu, Y. Kawano, T. Katagiri, E. Kondoh, Y. Yamamoto, and T. Ohta, *"Characterization of Direct-Contact Via Plug Formed by Use of Selective Al-CVD,"* Extended Abstracts of the 1993 Internl. Conf. on Solid State Dev. and Mat., 180-182 (1993).

[89] T. Amazawa, H. Nakamura, and Y. Arita, *"Selective Growth of Aluminum Using a Novel CVD System,"* IEDM Tech. Dig., 442 (1988).

[90] T. Sakurai and T. Serikawa, *"Lift-off Metallization of Sputtered Al Alloy Films,"* J. Electrochem. Soc., 126, 1257 (1979).

[91] T. Batchelder, *"A Simple Metal Lift-off Process,"* Solid-State Technol., 25, 111 (1982).

[92] M. Inoue, K. Hashizume, and H. Tsuchikawa, *"The Properties of Aluminum Thin Films Sputter Deposited at Elevated Temperature,"* J. Vac. Sci. Techn., A6 (3), 1636-1641 (1988).

[93] H. Ono, Y. Ushiko, and T. Yoda, *"Development of a Planarized Al-Si Contact Filling Technology,"* Proc. IEEE VMIC, 76-82 (1990).

[94] C. S. Park, S. J. Lee, J. H. Park, J. H. Sohn, D. Chin, and J. G. Lee, *"Al PLAPH (Aluminum PLANarization by Post Heating Process for Planarized Double Level CMOS Applications,"* Proc.IEEE VMIC, 326-328 (1991).

[95] K. Kukuta, T. Kikkawa, and M. Aoki, *"Al Ge Reflow Sputtering for Submicron Contact Hole Filling,"* Proc. IEEE VMIC, 163-169 (1991).

[96] C. W. Kaanta, S. G. Bombardier, W. J. Cote, W. R. Hill, G. G. J. Kerszykowski, H. S. Landis, D. J. Poindexter, C. W. Pollard, G. H. Ross, J. G. Ryan, S. Wolff, and J. E. Cronin, *"Dual Damascene: A ULSI Wiring Technology,"* Proceeding of the 8th Intnl. Multilevel Interconnect. Conf., 144-152 (1991).

[97] G. C. Smith and A. J. Purdes, *"Sidewall-Tapered Oxide by Plasma-Enhanced Chemical Vapor Deposition,"* J. Electrochem. Soc., 132, 2721-2725 (1985).

[98] H. Kotani, M. Matsuura, A. Fujii, H. Genjou, and S. Nagao, *"Low-Temperature APCVD Oxide Using TEOS-Ozone Chemistry for Multilevel Interconnects,"* IEDM Tech. Dig., 669 (1989).

[99] M. Suzuki, T. Homma, H. Koga, T. Tanigawa, and Y. Murao, *"A Fully Planarized Multilevel Interconnection Technology Using Selective TEOS-Ozone APCVD,"* IEDM Tech. Dig., 293 (1992).

[100] M. B. Anand, T. Matsuno, M. Murota, H. Shibata, M. Kakumu, K. Mori, K. Otsuka, M. Takahashi, H. Kaji, M. Kodera, K. Itoh, R. Aoki, and M. Nagata, *"Fully Integrated Back End of the Line Interconnect Process for High Performance ULSIs,"* Proc. IEEE VMIC, 15 (1994).

[101] L. B. Vines and S. K. Gupta, *"Interlevel Dielectric Planarization with Spin-On Glass Films,"* Proc. IEEE VMIC, 506-515 (1986).

[102] C. Chiang and D. B. Fraser, *"Understanding of Spin-On Glass (SOG) Properties from their Molecular Structure,"* Proc. IEEE VMIC, 397-403 (1989).

[103] S. Lee, K. Lee, H. Oh, C. Oh, Y.-W. Kim, D. Kim, and B. Kim, *"Multilevel Metallization for ASIC Technology,"* Proc. IEEE VMIC, 59 (1994).

[104] N. Lifshitz and M. R. Pinto, *"Spin-On-Glasses in Silicon IC: Plague or Panacea?,"* Proc. SPIE, Metallization: Performance and Reliability Issues for VLSI and ULSI, 1596, 96-105 (1991).

[105] C. K. Wang, L. M. Liu, H. C. Cheng, H. C. Huang, and M. S. Lin, *"A Study of Plasma Treatments of Siloxane SOG,"* Proc. IEEE VMIC, 101 (1994).

[106] N. Rutherford, M. Camenzind, and A. Belic, *"Outgassing and Oxidative Damage in Non-Etchback Siloxane SOG Processes,* Proc. IEEE VMIC, 141 (1993).

[107] H. Eggers and K. Hieber, *"Recent Development in Multilevel Interconnect technology,"* IEDM Tech. Dig., 200 (1987).

[108] A. B. Glaser and G. E. Subak-Sharpe, *"Integrated Circuit Engineering,"* Addison-Wesley, Reading, Massachusetts (1979).

[109] J. Black, *"Electromigration Failure Modes in Aluminum Metallization for Semiconductor Devices,"* Proc. IEEE, 57 (9), 1587-1594 (1969).

[110] F. M. d'Heurle, *"Electromigration and Failure Modes in Electronics: An Introduction,"* Proc. IEEE, 59, 1409 (1971).

[111] J. Black, *"Physics of Electromigration,"* Proc. of the 12th Reliability Physics Symposium, p. 142, IEEE New York (1974).

[112] R. Holm, Electrical Contacts, Theory and Application, Springer Verlag, New York (1967).

[113] S. Vaidya, D. B. Fraser, and A. K. Sinha, *"Electromigration Resistance of Fine Line Al,"* Proc. of the 12th Reliability Physics Symposium, p. 165, IEEE New York (1980).

[114] T. Ohmi, T. Hoshi, T. Yoshie, T. Takewaki, M. Otsuki, T. Shibata, and T. Nitta *"Large-Electromigration-resistance Copper Interconnect Technology for Sub-Half Micron ULSI's,"* IEDM Tech. Dig., 285 (1991).

[115] I. Ames, F. d'Heurle, and R. Horstmann, *"Reduction of Electromigration in Aluminum Films by Copper Doping,"* IBM J. Res. Dev., 14, 461-463 (1970).

[116] F. M. d'Heurle, N. G. Ainslie, A. Ganguilee, and M. C. Shine, *"Activation Energy for Electgromigration Failure in Al Films Containing Cu,"* J. Vac. Sci. tech., 9, 289-293 (1972).

[117] R. Rosenberg, *"Inhibition of Electromigration Damage in Thin Films,"* J. Vac. Sci. tech., 9, 263-270 (1972).

[118] A Gangulee and F. M. d'Heurle, *"Effect of Alloy Additions on Electromigration Failure in Thin Al Films,"* Appl. Phys. Lett., 19, 76 (1971).

[119] T. Kikkawa, H. Aoki, E. Ikawa, and J. Dryan, *"A Quarter-Micron Interconnection technology Using Al-Si-Cu/TiN Alternate Layers,"* IEDM Tech. Dig., 281 (1991).

[120] J. J. Estabil, H. S. Rathore, and E. N. Levine, *"Electromigration Improvements with Titanium Underlay and Overlay Metallurgy,"* Proc. IEEE VMIC, 292, 1991.

[121] K. P. Rodbell, P. W. DeHaven, and J. D. Mis, *"Electromigration Behavior in Layerd Ti/AlCu/Ti Films and its Dependence on Intermetallic Structure,"* Material Reliability Issues in Microelectronics Symp., 91-97, Anaheim, California, May 1991.

[122] M Kageyama, K. Hashimoto, and H. Onoda, *"Formation of Texture Controlled Aluminum and its Migration Performance in Al Si/TiN Stacked Structure,"* Proc. IEEE IRPS, 97-101 (1991).

[123] H. S. Rathore, R. G. Filippi, R. A. Wachnik, J. J. Estabil, and T. Kwok, *"Electromigration and Current-Carrying Implications for Aluminum-Based Metallurgy with Tungsten Stud-Via Interconnections,"* Proc. SPIE, Submicron Metallization, 1805, 251-262 (1992).

[124] W. R. Runyan and K. E. Bean Semiconductor Integrated Circuit Processing Technology, Addison-Wessley, Reading, Massachusetts (1990).

[125] P. Ghate, *"Reliability of VLSI Interconnections,"* Proc. American Insti. Phys. Conf., 138, New York (1986).

[126] S. K. Groothuis and S. R. Pollack, *"Stress Related Failures Causing Open Metallization,"* Proc. IEEE IRPS, 1-8 (1987).

[127] T. D. Sullivan, J. G. Ryan, J. R. Riendeau, and D. Bouldin, *"Stress-Induced Voiding In Aluminum Alloy Metallization,"* Proc. SPIE, Metallization: Performance and Reliability Issues for VLSI and ULSI, 1596, 83-95 (1991).

[128] J. Curry, G. Fitzgibbon, Y. Guan, R. Muollo, G. Nelson, and A. Thoma, *"New Failure Mechanism in Sputtered Al-Si Films,"* Proc. IEEE IRPS, 22, 6-8 (1984).

[129] J. G. Ryan, J. B. Riendeau, S. E. Shore, G. J. Slusser, D. C. Beyar, D. P. Bouldin, and T. D. Sullivan, *"The Effect of Alloying on Stress-Induced Void Formation in Al Based Metallization,"* J. Vac. Sci. Techn., A8, 1474-1479 (1990).

[130] M. G. Fernandes, H. Kawasaki, J. L. Klein, D. Jawarani, R. Subrahmanyan, T. K. Yu, and F. Pintchovski, *"Characterization of Stress Migration in Sub-Micron Metal Interconnects,"* American Institute of Physics Conf. Proc., 305, 153-164 (1993).

[131] M. A. Korhonen, C. A. Pszkiet, and C.-Y. Li, *"Mechanics of Thermal Stress Relaxation and Stress Induced Voiding in Narrow Aluminum Based Metallizations,"* J. Appl. Phys., A8, 1474-1479 (1990).

[132] S. Shima, H. Ito, and S. Shingubara, *"Suppressing Stress-Induced and Electromigration Failures with Al/Al Stacked Structures,"* IEEE Symp. VLSI Techn., 5A1, 27-28 (1990).

[133] H. Koelmans, "*Metallization Corrosion in Si Devices by Moisture Induced Electrolysis,*" Proc. IEEE IRPS, 168-171 (1974).

[134] W. M. Paulson and R. W. Kirk, "*The Effects of Phosphorus-Doped Passivation Glasses on the Corrosion of Aluminum,*" Proc. IEEE IRPS, 172-179 (1974).

[135] N. Nagasima et al., "*Interaction Between Phosphosilicate Glass Films and Water,*" J. Electrochem Soc., 121, 434-438 (1974).

[136] S. L. Hsu, L. M. Liu, C. H. Fang, S. L. Ying, T. L. Chen, M. S. Lin, and C. Y. Chang, "*Field Inversion Created in the CMOS Double Metal Process due to PETEOS and SOG Interactions,*" IEEE Trans. Electron Dev., 40 (1) , 49-53 (1993).

# Appendix A

## Some properties of the error funtion

$$\text{erf}(x) = \frac{2}{\sqrt{\pi}} \int_0^x e^{-a^2} da$$

$$\text{erf}(x) = \frac{2}{\sqrt{\pi}} \left[ x - \frac{x^3}{3.1!} + \frac{x^5}{5.2!} \cdots \frac{(-1)^n x^{2n+1}}{(2n+1).n!} \right]$$

$$\text{erfc}(x) = 1 - \text{erf}(x) = \frac{2}{\sqrt{\pi}} \int_x^\infty x^{-u^2} du$$

$$\text{erf}(-x) = -\text{erf}(x)$$

$$\text{erf}(0) = 0$$

$$\text{erf}(\infty) = 1$$

$$\text{erf}(x) \simeq \frac{2}{\sqrt{\pi}} x \quad \text{for } x <\!< 1$$

$$\text{erfc}(x) \simeq \frac{1}{\sqrt{\pi}} \frac{e^{-x^2}}{x} \quad \text{for } x >\!> 1$$

$$\frac{d\,\text{erf}(x)}{dx} = \frac{2}{\sqrt{\pi}} e^{-x^2}$$

$$\int_0^x \text{erfc}(x')\,dx' = x\,\text{erf}\,x + \frac{1}{\sqrt{\pi}}(1 - e^{-x^2})$$

$$\int_0^\infty \text{erfc}(x)\,dx = \frac{1}{\sqrt{\pi}}$$

Approximation for $x > 0$ ( with error $< \pm 1.5 \times 10^{-7}$):

$$\text{erf}(x) \simeq 1 - (a_1 T + a_2 T^2 + a_3 T^3 + a_4 T^4 + a_5 T^5)\, e^{-x^2}$$

where

$$T = \frac{1}{1 + 0.3275911x}$$

$a_1 = 0.254829592$

$a_2 = -0.284496736$

$a_3 = 1.421413741$

$a_4 = -1.453152027$

$a_5 = 1.061405429$

# Appendix B

## Multiples and submultiples of units

| Prefix | Symbol | Numerical value | Power |
|--------|--------|-----------------|-------|
| atto | a | 0.000,000,000,000,000,001 | $10^{-18}$ |
| femto | f | 0.000,000,000,000,001 | $10^{-15}$ |
| pico | p | 0.000,000,000,001 | $10^{-12}$ |
| nano | n | 0.000,000,001 | $10^{-9}$ |
| micro | $\mu$ | 0.000,001 | $10^{-6}$ |
| milli | m | 0.001 | $10^{-3}$ |
| centi | c | 0.1 | $10^{-1}$ |
| deka | dk | 10 | $10^{1}$ |
| hecto | dk | 100 | $10^{2}$ |
| kilo | k | 1000 | $10^{3}$ |
| mega | M | 1,000,000 | $10^{6}$ |
| giga | G | 1,000,000,000 | $10^{9}$ |
| tera | T | 1,000,000,000,000 | $10^{12}$ |
| peta | P | 1,000,000,000,000,000 | $10^{15}$ |
| exa | E | 1,000,000,000,000,000,000 | $10^{18}$ |

## The Greek alphabet

| Alpha | A | $\alpha$ | Nu | N | $\nu$ |
|-------|---|----------|----|----|-------|
| Beta | B | $\beta$ | Xi | $\Xi$ | $\xi$ |
| Gamma | $\Gamma$ | $\gamma$ | Omicron | O | $o$ |
| Delta | $\Delta$ | $\delta$ | Pi | $\Pi$ | $\pi$ |
| Epsilon | E | $\varepsilon$ | Rho | P | $\rho$ |
| Zeta | Z | $\zeta$ | Sigma | $\Sigma$ | $\sigma$ |
| Eta | H | $\eta$ | Tau | T | $\tau$ |
| Theta | $\Theta$ | $\theta$ | Upsilon | $\Upsilon$ | $\upsilon$ |
| Iota | I | $\iota$ | Phi | $\Phi$ | $\phi$ |
| Kappa | K | $\kappa$ | Chi | X | $\chi$ |
| Lambda | $\Lambda$ | $\lambda$ | Psi | $\Psi$ | $\psi$ |
| Mu | M | $\mu$ | Omega | $\Omega$ | $\omega$ |

## Conversion factors

### Change of base

$\log_e N = \log_e 10 \; \log_{10} N = 2.3026 \log_{10} N$

### Length

|            | METER      | cm         | $\mu$m     | Å          | nm         |
|------------|------------|------------|------------|------------|------------|
| 1 METER    | 1          | $10^2$     | $10^6$     | $10^{10}$  | $10^9$     |
| 1 centimeter | $10^{-2}$ | 1         | $10^4$     | $10^8$     | $10^7$     |
| 1 micrometer | $10^{-6}$ | $10^{-4}$ | 1         | $10^4$     | $10^3$     |
| 1 Angstrom | $10^{-10}$ | $10^{-8}$  | $10^{-4}$  | 1          | $10^{-1}$  |
| 1 nanometer | $10^{-9}$ | $10^{-7}$  | $10^{-3}$  | 10         | 1          |

1 m = 100 cm = 39.3701 in = 3.20839 Ft = 1.093613 Yd

1 in = 1000 mils = 2.54 cm = 25,400 $\mu$m

1 km = 0.62137 MI

1 mil = 25.4 $\mu$m = 0.00254 cm

1 yd = 0.9144 m

### Area

|          | $m^2$          | $cm^2$    | $\mu m^2$       | $in.^2$         |
|----------|----------------|-----------|-----------------|-----------------|
| 1 $m^2$  | 1              | $10^4$    | $10^{12}$       | 1550            |
| 1 $cm^2$ | $10^{-4}$      | 1         | $10^8$          | 0.155           |
| 1 $\mu m^2$ | $10^{-12}$  | $10^{-8}$ | 1               | $1.55x10^{-9}$  |
| 1 $in.^2$ | $6.45x10^{-4}$ | 6.45     | $6.45x10^8$     | 1               |

### Volume

|          | $m^3$          | $cm^3$    | $\mu m^3$       | $in.^3$         |
|----------|----------------|-----------|-----------------|-----------------|
| 1 $m^3$  | 1              | $10^6$    | $10^{18}$       | $6.1x10^4$      |
| 1 $cm^3$ | $10^{-6}$      | 1         | $10^{12}$       | $6.1x10^{-2}$   |
| 1 $\mu m^3$ | $10^{-18}$  | $10^{-12}$ | 1              | $6.1x10^{-14}$  |
| 1 $in.^3$ | $1.64x10^{-5}$ | 16.4     | $1.64x10^{13}$  | 1               |

## Energy

|      | J | eV | kg | cal |
|------|------|------|------|------|
| 1 J  | 1 | $6.24x10^{18}$ | $1.17x10^{-17}$ | 0.24 |
| 1 eV | $1.6x10^{-19}$ | 1 | $1.78x10^{-36}$ | $3.83x10^{-20}$ |
| 1 kg | $8.99x10^{16}$ | $5.61x10^{35}$ | 1 | $2.15x10^{16}$ |
| 1 cal | 4.19 | $2.61x10^{19}$ | $4.66x10^{-17}$ | 1 |

# Index

# Fundamental constants

| Quantity | Symbol | Value | Unit |
|---|---|---|---|
| Atomic mass unit | u | $1.661x10^{-27}$ | kg |
| Avogadro's constant | $N_A$ | $6.022x10^{23}$ | $mol^{-1}$ |
| Boltzmann constant | k | $1.381x10^{-23}$ | $J\ K^{-1}$ |
|  | k | $8.615x10^{-5}$ | eV/K |
|  | k | $3.298x10^{-24}$ | cal/K |
| Electron rest mass | $m_0$ | $9.1095x10^{-31}$ | kg |
|  | $m_0$ | $0.511x10^6$ | eV |
| Elementary charge | q | $1.602x10^{-19}$ | C |
|  | q | $4.802x10^{-10}$ | $cm^{3/2}g^{1/2}s^{-1}$ |
| Gas constant | R | 1.9872 | $cal\ mol^{-1}K^{-1}$ |
|  | R | $5.191x10^{19}$ | eV/K mol |
|  | R | 8.3144 | J/K mol |
| Permeability of vaccum | $\mu_0$ | $4\pi x10^{-7}$ | H/m |
|  | $\mu_0$ | $12.566x10^{-7}$ | H/m |
| Permittivity in vacuum | $\varepsilon_0$ | $8.85418x10^{-12}$ | F/m |
| Planck constant | h | $6.626x10^{-34}$ | J/Hz |
|  | h | $4.135x10^{-15}$ | eV s |
|  | $\hbar$ | $1.0546x10^{-34}$ | Js |
|  | $\hbar$ | $6.580x10^{-16}$ | eV s |
| Proton rest mass | $m_p$ | $1.6725x10^{-27}$ | kg |
|  | $m_9$ | $9.3826x10^8$ | eV |
| Speed of light in vacuum | c | $2.9979x10^8$ | m/s |
| Standard atmosphere | At | 1.01325 | $N/m^2$ |
| Thermal voltage (300 K) | kT/q | 0.0259 | V |
| Wavel. of 1-eV quantum | $\lambda$ | 1.23977 | $\mu m$ |

$u = (10^{-3}kg\ mol^{-1})/N_A$

$k = R/N_A$

$\varepsilon_0 = (\mu_0 c^2)^{-1}$

$\hbar = h/2\pi$

# Periodic Table of the Elements

| Group IA | IIA | IIIA | IVA | VA | VIA | VIIA | VIIIA | | | IB | IIB | IIIB | IVB | VB | VIB | VIIB | VIII |
|---|---|---|---|---|---|---|---|---|---|---|---|---|---|---|---|---|---|
| 1 $1.00794$ **H** Hydrogen | | | | | | | | | | | | | | | | | 2 $4.00260$ **He** Helium |
| 3 $6.941$ **Li** Lithium | 4 $9.01218$ **Be** Beryllium | | | | | | | | | | | 5 $10.811$ **B** Boron | 6 $12.011$ **C** Carbon | 7 $14.0067$ **N** Nitrogen | 8 $15.9994$ **O** Oxygen | 9 $18.99840$ **F** Fluorine | 10 $20.1797$ **Ne** Neon |
| 11 $22.98977$ **Na** Sodium | 12 $24.305$ **Mg** Magnesium | | | | | | | | | | | 13 $26.98154$ **Al** Aluminum | 14 $28.0855$ **Si** Silicon | 15 $30.97376$ **P** Phosphorus | 16 $32.066$ **S** Sulfur | 17 $35.4527$ **Cl** Chlorine | 18 $39.948$ **Ar** Argon |
| 19 $39.0983$ **K** Potassium | 20 $40.078$ **Ca** Calcium | 21 $44.9559$ **Sc** Scandium | 22 $47.88$ **Ti** Titanium | 23 $50.9415$ **V** Vanadium | 24 $51.996$ **Cr** Chromium | 25 $54.9380$ **Mn** Manganese | 26 $55.847$ **Fe** Iron | 27 $58.9332$ **Co** Cobalt | 28 $58.6934$ **Ni** Nickel | 29 $63.546$ **Cu** Copper | 30 $65.39$ **Zn** Zinc | 31 $69.723$ **Ga** Gallium | 32 $72.61$ **Ge** Germanium | 33 $74.9216$ **As** Arsenic | 34 $78.96$ **Se** Selenium | 35 $79.904$ **Br** Bromine | 36 $83.80$ **Kr** Krypton |
| 37 $85.4678$ **Rb** Rubidium | 38 $87.62$ **Sr** Strontium | 39 $88.9059$ **Y** Yttrium | 40 $91.224$ **Zr** Zirconium | 41 $92.9064$ **Nb** Niobium | 42 $95.94$ **Mo** Molybdenum | 43 $(98)$ **Tc** Technetium | 44 $101.07$ **Ru** Ruthenium | 45 $102.9055$ **Rh** Rhodium | 46 $106.42$ **Pd** Palladium | 47 $107.868$ **Ag** Silver | 48 $112.41$ **Cd** Cadmium | 49 $114.82$ **In** Indium | 50 $118.710$ **Sn** Tin | 51 $121.757$ **Sb** Antimony | 52 $127.60$ **Te** Tellurium | 53 $126.9045$ **I** Iodine | 54 $131.29$ **Xe** Xenon |
| 55 $132.9054$ **Cs** Cesium | 56 $137.33$ **Ba** Barium | 57 $138.9055$ **La** Lanthanum | 72 $178.49$ **Hf** Hafnium | 73 $180.9479$ **Ta** Tantalum | 74 $183.85$ **W** Tungsten | 75 $186.207$ **Re** Rhenium | 76 $190.2$ **Os** Osmium | 77 $192.22$ **Ir** Iridium | 78 $195.08$ **Pt** Platinum | 79 $196.9665$ **Au** Gold | 80 $200.59$ **Hg** Mercury | 81 $204.383$ **Ti** Thallium | 82 $207.2$ **Pb** Lead | 83 $208.9804$ **Bi** Bismuth | 84 $(209)$ **Po** Polonium | 85 $(210)$ **At** Astatine | 86 $(210)$ **Rn** Radon |
| 87 $(223)$ **Fr** Francium | 88 $226.0254$ **Ra** Radium | 89 $(227)$ **Ac** Actinium | 104 $(261)$ **Unq** (Unnilquadium) | 105 $(262)$ **Unp** (Unnilpentium) | 106 $(263)$ **Unh** (Unnilhexium) | 107 $(262)$ **Uns** (nnilseptium) | 108 $(265)$ **Uno** (Unniloctium) | 109 $(266)$ **Une** (Unnilennium) | | | | | | | | | |

Atomic
Weight

Atomic
Number — 14 $28.0855$ — Symbol

**Si**
Silicon — Name

Gases

Solids — Liquids

Synthetically prepared

| 58 $140.12$ **Ce** Cerium | 59 $140.9077$ **Pr** Praseodymium | 60 $144.24$ **Nd** Neodymium | 61 $(145)$ **Pm** Promethium | 62 $150.36$ **Sm** Samarium | 63 $151.965$ **Eu** Europium | 64 $157.25$ **Gd** Gadolinium | 65 $158.9253$ **Tb** Terbium | 66 $162.50$ **Dy** Dysprosium | 67 $164.9303$ **Ho** Holmium | 68 $167.26$ **Er** Erbium | 69 $168.9342$ **Tm** Thulium | 70 $173.04$ **Yb** Ytterbium | 71 $174.967$ **Lu** Lutetium |
|---|---|---|---|---|---|---|---|---|---|---|---|---|---|
| 90 $232.0381$ **Th** Thorium | 91 $231.0359$ **Pa** Protactinium | 92 $238.029$ **U** Uranium | 93 $237.0482$ **Np** Neptunium | 94 $(244)$ **Pu** Plutonium | 95 $(244)$ **Am** Americium | 96 $(243)$ **Cm** Curium | 97 $(247)$ **Bk** Berkelium | 98 $(247)$ **Cf** Californium | 99 $(251)$ **Es** Einsteinium | 100 $(257)$ **Fm** Fermium | 101 $(257)$ **Md** Mendelevium | 102 $(259)$ **No** Nobelium | 103 $(260)$ **Lr** Lawrencium |